ANALYTICAL AND PHYSICAL ELECTROCHEMISTRY

FUNDAMENTAL SCIENCES Chemistry

ANALYTICAL AND PHYSICAL ELECTROCHEMISTRY

Hubert H. Girault

Translated by Magnus Parsons

EPFL Press
A Swiss publishing company distributed by Marcel Dekker, Inc.

MARCEL DEKKER

Headquarters
Marcel Dekker, Inc., 270 Madison Avenue, New York, NY 10016, U.S.A.
tel: 212-696-9000; fax: 212-685-4540

Distribution and Customer Service
Marcel Dekker, Inc., Cimarron Road, Monticello, New York 12701, U.S.A.
tel: 800-228-1160; fax: 845-796-1772

Eastern Hemisphere Distribution
Marcel Dekker AG, Hutgasse 4, Postfach 812, CH-4001 Basel, Switzerland
tel: 41-61-260-6300; fax: 41-61-260-6333

www.dekker.com

Library of Congress Cataloging-in-Publication Data
A catalog record for this book is available from the Library of Congress.

EPFL Press

is a label owned by Presses polytechniques et universitaires romandes, a Swiss academic publishing company whose main purpose is to publish the teaching and research works of the Ecole polytechnique fédérale de Lausanne.

Presses polytechniques et universitaires romandes,
EPFL – Centre Midi, CH-1015 Lausanne, Switzerland
E-Mail: ppur@epfl.ch
Phone: 021 / 693 21 30
Fax: 021 / 693 40 27

www.epflpress.org

ISBN 2-940222-03-7 (EPFL Press)
ISBN 0-8247-5357-7 (Marcel Dekker, Inc.)

Printed in Italy

To Freya-Merret, Jan-Torben and Jördis

PREFACE

This book is a translation of a textbook entitled 'Electrochimie Physique et Analytique' (Presses Polytechniques Universitaires Romandes, 2001). The original goal was to gather in a single book the physical bases of electroanalytical techniques, including electrophoretic methods. Indeed, most of the textbooks dedicated to electrochemistry cover either the physical or the analytical aspects.

As science becomes more and more interdisciplinary, a thorough comprehension of the fundamental aspects becomes more important. The book is therefore intended to provide in a rigorous manner an introduction to the concepts underlying the electrochemical methods of separation (capillary electrophoresis, gel electrophoresis, ion chromatography, etc.) and of analysis (potentiometry, conductometry and amperometry).

My first thanks go to Magnus Parsons (Isle of Sky, Scotland) who did the translation. The present text has been thoroughly reviewed again by Prof. Roger Parsons (FRS), and I wish to thank most sincerely Roger for his support over all these years.

My thanks also go to the reading committee composed of Drs. Henrik Jensen, Jean-Pierre Abid, Maurizio Caragno, Debi Pant and Jördis Tietje-Girault.

I also thank all the team at Fontis Media (Lausanne, Switzerland) for producing this book, and in particular Thierry Lenzin for his patience and meticulous editing.

PREFACE TO THE ORIGINAL FRENCH VERSION

"ELECTROCHIMIE PHYSIQUE ET ANALYTIQUE"

For historical reasons, physical electrochemistry and analytical electrochemistry are often taught separately. The purpose of this course book is to bring these two subjects together in a single volume, so as to bridge the fundamental physical aspects to the analytical applications of electrochemistry.

The philosophy of this book has been to publish *in extenso* all the mathematical derivations in a rigorous and detailed manner, in such a way that the readers can understand rather then accept the physical origins of the main electroanalytical principles.

By publishing this book, I express my thanks to all those who have taught me the way through electrochemistry:

- From my early years in France, I wish to thank all the teachers from the Ecole Nationale d'Electrochimie et d'Electrométallurgie de Grenoble (ENSEEG) for developing my interest in electrochemistry, and of course I thank my parents for their financial and moral support.
- From my years in England, my most profound gratitude goes to Sir Graham Hills for both his scientific and political approach to Science, as well as to Lady Mary Hills for her friendship from the very beginning of my thesis. My admiration goes to Professor Martin Fleischmann (FRS), whose creative force has always been a source of inspiration, and to Professor Roger Parsons (FRS) whose intellectual rigor and mastery of thermodynamics can be found, I hope, in these pages. I would not forget Professor David Schiffrin who has taught me so much and with whom I spent several fruitful years. Thanks to them, I acquired during these years in Southampton a certain comprehension of classical physical electrochemistry.
- From my years in Scotland begins the period of my interest in analytical electrochemistry. I owe much to Drs Graham Heath and Lesley Yellowlees who helped me discover another type of electrochemistry, and I insist on expressing my sincere admiration to Professor John Knox (FRS) for his very scientific approach to chromatography and capillary electrophoresis.
- From my years in Switzerland, I thank Professor Michael Grätzel for his support when I arrived in Lausanne.

As a textbook, this work has been tried and tested on a series of undergraduate classes, and I thank all those students and teaching assistants who helped me with their comments to smooth out the difficulties. In particular I would like to thank Dr. Rosaria Ferrigno for her constructive criticisms; Dr. Pierre-François Brevet, Dr. Frédéric Reymond, Dr. David Fermin and Dr. Joël Rossier for their advice; and Dr. Olivier Bagel for having carried out the experiments whose results have served to illustrate several of the methods described here.

A detailed review of the work was carried out by Professors Jean-Paul Diard (ENSEEG, France) and Roger Parsons (Southampton, UK), and I thank them for their work. For the preparation of the original French version of this text, I thank the PPUR for their work in a collaboration that was both cordial and fruitful.

Finally, more than thanks must go to Dr. Jördis Tietje-Girault for her infallible support over the course of the years ever since our first meeting in the laboratory of Professor Graham Hills.

TABLE OF CONTENTS

	PREFACE		vii
	PREFACE TO THE ORIGINAL FRENCH VERSION		viii
CHAPTER 1	ELECTROCHEMICAL POTENTIAL		1
	1.1	Electrochemical potential of ions	1
	1.2	Electrochemical potential of electrons	23
CHAPTER 2	ELECTROCHEMICAL EQUILIBRIA		33
	2.1	Redox reactions at metallic electrodes	33
	2.2	Cells and accumulators	50
	2.3	Pourbaix diagrams	54
	2.4	Electrochemical equilibria at the interface between two electrolytes	57
	2.5	Analytical applications of potentiometry	63
	2.6	Ion exchange membranes	78
		Appendix : The respiratory chain	82
CHAPTER 3	ELECTROLYTE SOLUTIONS		83
	3.1	Liquids	83
	3.2	Thermodynamic aspects of solvation	85
	3.3	Structural aspects of ionic solvation	100
	3.4	Ion-ion interactions	105
	3.5	Ion pairs	123
	3.6	Computational methods	129
		Appendix	131
CHAPTER 4	TRANSPORT IN SOLUTION		133
	4.1	Transport in electrolyte solutions	133
	4.2	Conductivity of electrolyte solutions	138
	4.3	Influence of concentration on conductivity	145
	4.4	Dielectric friction	152
	4.5	Thermodynamics of irreversible systems	160
	4.6	Statistical aspects of diffusion	163
		Appendix: Elements of fluid mechanics	171

CHAPTER 5 ELECTRIFIED INTERFACES 177
5.1 Interfacial tension 177
5.2 Interfacial thermodynamics 180
5.3 Thermodynamics of electrified interfaces 184
5.4 Spatial distribution of polarisation charges 194
5.5 Structure of electrochemical interfaces 215

CHAPTER 6 ELECTROKINETIC PHENOMENA AND ELECTROCHEMICAL SEPARATION METHODS 221
6.1 Electrokinetic phenomena 221
6.2 Capillary electrophoresis 229
6.3 Electrophoretic methods of analytical separation 239
6.4 Electrophoretic separation of biopolymers 244
6.5 Ion chromatography 256
6.6 Industrial methods of electrochemical separation 261

CHAPTER 7 STEADY STATE AMPEROMETRY 265
7.1 Electrochemical kinetics 265
7.2 Current controlled by the kinetics of the redox reactions 268
7.3 Reversible systems: Current limited by diffusion 275
7.4 Electrodes with a diffusion layer of controlled thickness 280
7.5 Quasi-reversible systems: Current limited by kinetics and diffusion 288
7.6 Irreversible systems: Current limited by kinetics and diffusion 294
7.7 Quasi-reversible systems: Current limited by diffusion, migration and kinetics 295
7.8 Experimental aspects of amperometry 296

CHAPTER 8 PULSE VOLTAMMETRY 301
8.1 Chronoamperometry following a potential step 301
8.2 Polarography 311
8.3 Square wave voltammetry 319
8.4 Stripping voltammetry
8.5 Thin layer voltammetry 333
8.6 Amperometric detectors for chromatograpy 337

CHAPTER 9 ELECTROCHEMICAL IMPEDANCE 339
9.1 Transfer function 339
9.2 Elementary circuits 343
9.3 Impedance of an electrochemical system 351
9.4 AC voltammetry 368

CHAPTER 10 CYCLIC VOLTAMMETRY 375
10.1 Electrochemically reversible reactions with semi-infinite linear diffusion 375
10.2 Influence of the kinetics 382
10.3 EC reactions 385

10.4 Electron transfer at liquid | liquid interfaces 396
10.5 Assisted ion transfer at liquid | liquid interfaces 399
10.6 Surface reactions 402
10.7 Hemi-spherical diffusion 404
10.8 Voltabsorptometry 407
10.9 Semi-integration 408

ANNEX A VECTOR ANALYSIS 411
1. Coordinate systems 411
2. Circulation of the field vector 412
3. The vector gradient 413
4 Flux of the field vector 414
5 The Green-Ostrogradski theorem 414

ANNEX B Work functions and standard redox potentials 417

SYMBOLS 425

INDEX 429

CHAPTER 1

ELECTROCHEMICAL POTENTIAL

1.1 ELECTROCHEMICAL POTENTIAL OF IONS

The chemical potential is the main thermodynamic tool used to treat chemical equilibria. It allows us to predict whether a reaction can happen spontaneously, or to predict the composition of reactants and products at equilibrium. In this book, we shall consider electrochemical reactions that involve charged species, such as electrons and ions. In order to be able to call on the thermochemical methodology, it is convenient to define first of all the notion of electrochemical potential, which will be the essential tool used for characterising the reactions at electrodes as well as the partition equilibria between phases. To do this, let us recall first of all, what a chemical potential is, and in particular the chemical potential of a species in solution.

1.1.1 Chemical potential

Thermodynamic definition

Let us consider a phase composed of chemical species j. By adding to this phase one mole of a chemical species i whilst keeping the extensive properties of the phase constant, i.e. the properties linked to its dimensions (V, S, n_j), we increase the internal energy U of the phase. In effect, we are adding the kinetic energy E_{trans}, the rotational energy E_{rot} and the vibrational energy E_{vib} if i is a molecule, the interaction energy between the species E_{int}, perhaps the electronic energy E_{el} if we have excited electronic states and the energy linked to the atomic mass of the atoms E_{mass} if we consider radiochemical aspects, such that:

$$U = E_{trans} + E_{rot} + E_{vib} + E_{el} + E_{int} + E_{mass} \tag{1.1}$$

Thus, we define the ***chemical potential*** of the species i as being the increase in internal energy due to the addition of this species

$$\mu_i = \left(\frac{\partial U}{\partial n_i}\right)_{V,S,n_{j\neq i}} \tag{1.2}$$

In general, the variation in internal energy can be written in the form of a differential:

$$\mathrm{d}U = -p\mathrm{d}V + T\mathrm{d}S + \sum_i \mu_i \,\mathrm{d}n_i \tag{1.3}$$

Having defined the Gibbs energy G as a function of the internal energy

$$G = U + pV - TS \tag{1.4}$$

we can see, by taking the differential of each term of this equation and by replacing $\mathrm{d}U$ by the equation (1.3), that

$$\mathrm{d}G = V\mathrm{d}p - S\mathrm{d}T + \sum_i \mu_i \,\mathrm{d}n_i \tag{1.5}$$

This expression gives a definition of the chemical potential, which is in fact easier to use experimentally

$$\mu_i = \left(\frac{\partial G}{\partial n_i}\right)_{T,p,n_{j\neq i}} = \overline{G}_i \tag{1.6}$$

In other words, the chemical potential μ_i is equal to the work which must be supplied keeping T & p constant in order to transfer one mole of the species i from a vacuum to a phase, except for the volume work. By definition, it represents the partial molar Gibbs energy $\overline{G}_i$. In the case of a pure gas, the chemical potential is in fact the molar Gibbs energy

$$\mu = \left(\frac{\partial G}{\partial n}\right)_{T,p} = G_\mathrm{m} = \frac{G}{n} \tag{1.7}$$

Before treating the chemical potential of a species in the gas phase, let's look, by way of an example, at the influence of pressure on the molar Gibbs energy.

EXAMPLE

Let us calculate the variation in Gibbs energy associated with the isothermal compression from 1 to 2 bars ($T = 298$ K) of (1) water treated as an incompressible liquid and (2) vapour treated as an ideal gas.
Considering one mole, we have the molar quantities

$$\Delta G_\mathrm{m} = \int V_\mathrm{m}\,\mathrm{d}p$$

For water as a liquid, the molar volume ($V_\mathrm{m} = 18$ cm^3·mol^{-1}) is constant if we use the hypothesis that liquid water is incompressible. Thus we have:

$$\Delta G_\mathrm{m} = V_\mathrm{m}\Delta p = 18\cdot 10^{-6}(\mathrm{m^3{\cdot}mol^{-1}})\times 10^5(\mathrm{Pa}) = 1.8\ \mathrm{J\cdot mol^{-1}}$$

For water as vapour, considered as an ideal gas, the molar volume depends on the pressure,

$V_\mathrm{m} = RT/p$

from which we get

$$\Delta G_m = \int_{p_1}^{p_2} V_m \, dp = RT \int_{p_1}^{p_2} \frac{dp}{p} = RT \ln\left(\frac{p_2}{p_1}\right) = 1.7 \text{ kJ} \cdot \text{mol}^{-1}$$

which is a thousand times greater.

Chemical potential in the gas phase

For an ideal gas ($pV = nRT$), we express the chemical potential μ for a given temperature with respect to a ***standard pressure*** value defined when the pressure has the standard value $p^\ominus$ of 1 bar (=100kPa). Thus by integration, the chemical potential for a given pressure p is linked to the standard chemical potential by:

$$\mu(T) - \mu^\ominus(T) = \int_{p^\ominus}^{p} V dp = \int_{p^\ominus}^{p} \frac{RT}{p} dp \tag{1.8}$$

that is

$$\mu(T) = \mu^\ominus(T) + RT \ln\left(\frac{p}{p^\ominus}\right) \tag{1.9}$$

Remember that an ideal gas is one in which the molecules do not have any interaction energy, and consequently a real gas can only be considered in this manner at low pressures. The chemical potential tends towards negative infinity when the pressure tends to zero because the entropy tends to infinity and because $\mu = G_m = H_m - TS_m$.

When the pressure is sufficiently high, the interactions between the gas molecules can no longer be ignored. These are attractive at medium pressures and the chemical potential of the real gas is therefore below what it would be if the gas behaved as an ideal one. On the other hand, at high pressures, the interactions are mostly repulsive, and in this case the chemical potential of a real gas is higher than it would be if it behaved as an ideal one.

These deviations of the behaviour of a real gas with respect to an ideal gas are taken into account by adding a correcting factor to the expression (1.9) for the chemical potential:

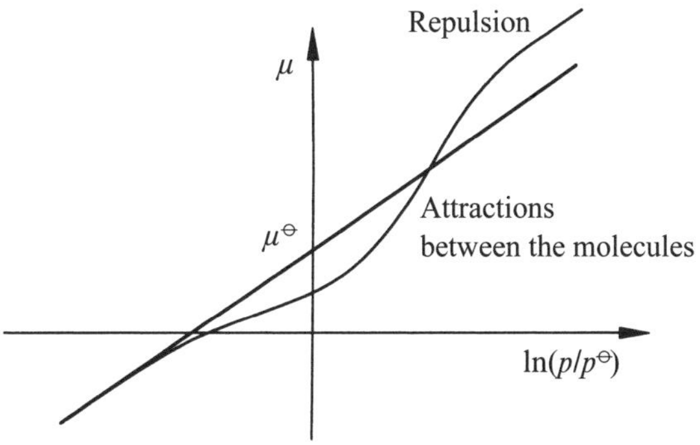

Fig. 1.1 Variation of chemical potential with pressure.

$$\mu(T) = \mu^{\ominus}(T) + RT\ln\left(\frac{p}{p^{\ominus}}\right) + RT\ln\varphi = \mu^{\ominus}(T) + RT\ln\left(\frac{f}{p^{\ominus}}\right) \quad (1.10)$$

where φ is called the ***fugacity coefficient*** (dimensionless) and $f = \varphi p$ is the ***fugacity***.

The term $RT\ln\varphi$ represents the energy of interaction between the molecules. Given that gases tend towards behaving ideally at low pressures, we can see that $\varphi \to 1$ when $p \to 0$.

The reasoning developed above for a pure gas can be applied equally to ideal mixtures of ideal gases. The chemical potential of the constituent *i* of an ideal mixture of gases is therefore given by

$$\mu_i(T) = \mu_i^{\ominus}(T) + RT\ln\left(\frac{p_i}{p^{\ominus}}\right) = \mu_i^{\ominus}(T) + RT\ln\frac{p}{p^{\ominus}} + RT\ln y_i \quad (1.11)$$

with p_i being the partial pressure of the constituent *i* and y_i the mole fraction. The standard state of a constituent *i* corresponds to the pure gas *i* considered as ideal and at the standard pressure of 1 bar.

Chemical potential in the liquid phase

In a liquid phase, the molecules are too close to one another to allow the hypothesis used in the case of ideal gases, i.e. that the intermolecular forces can be neglected. We define an ideal solution as a solution in which the molecules of the various constituents are so similar that a molecule of one constituent may be replaced by a molecule of another without altering the spatial structure of the solution (e.g. the volume) or the average interaction energy. In the case of a binary mixture A and B, this means that A and B have approximately the same size, and that the energy of the interactions A-A, A-B and B-B are almost equal (for example a benzene-toluene mixture).

When there is an equilibrium between a liquid phase and its vapour, the chemical potential of all the constituents is the same in both phases. If the solution is ideal, its constituents obey Raoult's Law $p_i = x_i\, p_i^*$ with p_i being the partial pressure of the constituent *i* and p_i^* the saturation vapour pressure of the pure liquid. By analogy with ideal gases, we define a solution as ideal if the chemical potential of its constituents can be written as a function of the mole fraction x_i in the liquid

$$\mu_i(T) = \mu_i^{\ominus,ideal}(T) + RT\ln x_i \quad (1.12)$$

The equality of the chemical potentials between the vapour phase and the liquid phase leads to

$$\mu_i^{\ominus,ideal}(T) = \mu_i^{\ominus}(T) + RT\ln\left(\frac{p_i^*}{p^{\ominus}}\right) \quad (1.13)$$

In the case of the benzene-toluene mixture, Raoult's law is obeyed for all values of the mole fractions (ideal solution).

Other types of ideal solutions are the binary mixtures A-B in which the molecules are not identical but, where one constituent is present in a much greater quantity than the other. If A is in the majority, it becomes the solvent and B the solute. Such a solution is ideal in as much as the replacement of a molecule of A by one of B or vice-versa has little effect on the properties of the solution, given the dilution of B in A. We call this particular type of ideal solution an ideally dilute solution.

Mole fraction scale

At the molecular level, we can say that in an ideally dilute solution, the solute molecules do not interact with each other, but only interact with the molecules of the solvent that surrounds them. Here again, we have the analogy with the ideal gases. In an ideally dilute solution, that is to say that when $x_A \rightarrow 1$, the solvent obeys Raoult's law. The chemical potential of the solvent A is then written as

$$\mu_A(T) = \mu_A^{\ominus,ideal}(T) + RT \ln x_A \tag{1.14}$$

The deviation from the ideal behaviour (for example when the concentration of B is no longer negligible in relation to that of A) can be taken into account by adding a correction term to the expression for the chemical potential

$$\mu_A(T) = \mu_A^{\ominus,ideal}(T) + RT \ln x_A + RT \ln \gamma_A \tag{1.15}$$

Since solutions become ideal when $x_A \rightarrow 1$, we can see that at this limit $\gamma_A \rightarrow 1$.

As far as the solute is concerned, it obeys Henry's law $p_B = x_B K_B$, where p_B is the partial pressure of the solute B, x_B the mole fraction of B in the liquid and K_B the Henry constant which has the dimension of a pressure.

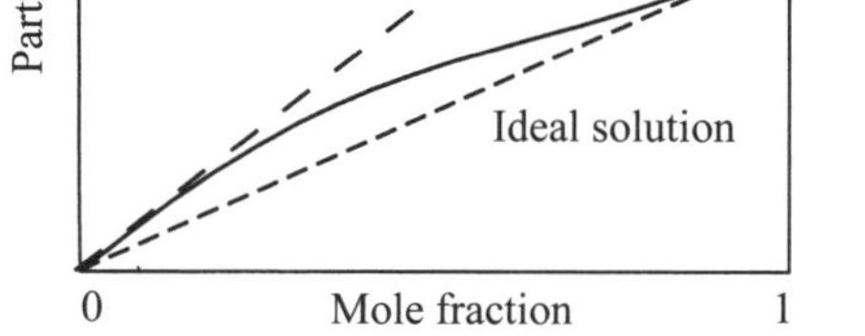

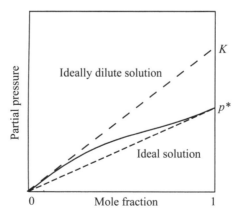

Fig. 1.2 Diagram of the partial pressure for a binary system. For small mole fractions (solute), the partial pressure is proportional to the mole fraction (Henry's law). For mole fractions approaching unity (solvent), the partial pressure is proportional to the mole fraction (Raoult's law).

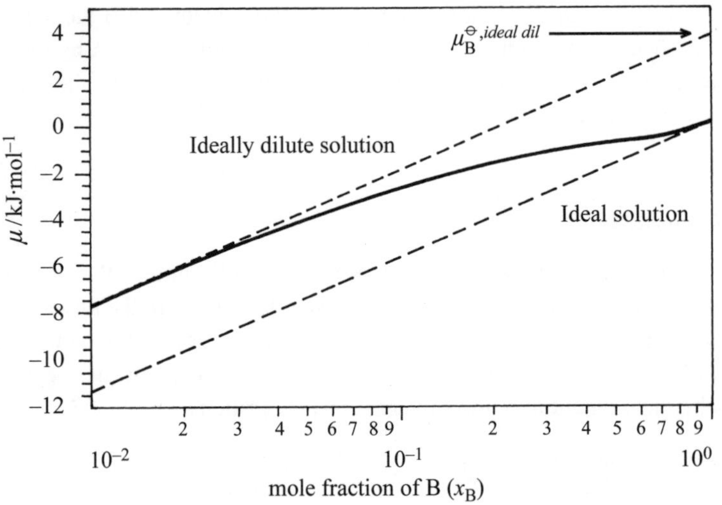

Fig. 1.3 Variation of the chemical potential and the definition of the standard chemical potential at 25°C on the scale of mole fractions for an ideally dilute solution.

The chemical potential of the solute is then written as

$$\mu_B(T) = \mu_B^{\ominus,ideal\ dil}(T) + RT \ln x_B \tag{1.16}$$

with $\mu_B^{\ominus,ideal\ dil}$ being the standard chemical potential on the mole fraction scale for ideally dilute solutions. It is important to note that the standard state is an imaginary state which we would have at the limit $x_B \to 1$, that is to say an extrapolation of the chemical potential of the solute from the infinitely dilute case to its pure state. In other words, the standard state is a pure solution of B that would behave like an ideally dilute solution, i.e. a pure solution in which the molecules do not interact. The equality of the chemical potentials of B between the vapour phase and the liquid phase gives

$$\mu_B^{\ominus,ideal\ dil}(T) = \mu_B^{\ominus}(T) + RT \ln\left(\frac{K_B}{p^{\ominus}}\right) \tag{1.17}$$

The deviations from the ideal behaviour can also be taken into account by adding a correction term to the chemical potential.

$$\mu_B(T) = \mu_B^{\ominus,ideal\ dil}(T) + RT \ln x_B + RT \ln \gamma_B \tag{1.18}$$

The term $RT \ln \gamma_B$ then represents the work of interaction of the solute molecules among themselves. If the solute is a salt, the predominant energy of interaction will be the electrostatic one. We will show later on in this book that it is possible to model this interaction energy using statistical mechanics (see §3.4.2, the Debye-Hückel theory).

Given that solutions become ideally dilute when $x_B \to 0$, we can see then that $\gamma_B \to 1$. The product $\gamma_B x_B$ is called the ***activity*** a_B of B and γ_B the ***activity coefficient***. The activity a_B is a sort of effective mole fraction.

Molality scale

For dilute solutions, we often use scales of molality (number of moles per kg of solvent) or of molarity (number of moles per litre of solution). In the case of molalities (scale of composition independent of the temperature) defined by

$$m_B = \frac{n_B}{n_A M_A} \tag{1.19}$$

where M_A is the molar mass of the solvent (kg·mol^{-1}), we have

$$x_B = \frac{n_B}{n_A + n_B} = \frac{n_A M_A m_B}{n_A + n_B} = x_A M_A m_B \tag{1.20}$$

which, substituted into the expression for chemical potential (1.18), leads to

$$\mu_B(T) = \left[\mu_B^{\ominus,ideal\ dil}(T) + RT\ln\left(m^{\ominus}\ M_A\right)\right] + RT\ln\left(\frac{\gamma_B\ x_A\ m_B}{m^{\ominus}}\right) \tag{1.21}$$

where $m^{\ominus}$ is the ***standard molality*** whose value is 1 mol·kg^{-1}.

In fact, we can re-write this equation in the form

$$\mu_B(T) = \mu_B^{\ominus,m}(T) + RT\ln\left(\gamma_B^m \frac{m_B}{m^{\ominus}}\right) \tag{1.22}$$

where $\mu_B^{\ominus,m}$ is the standard chemical potential in the molality scale and γ_B^m is the activity coefficient also in the molality scale.

Molarity scale

To express the mole fraction of a constituent as a function of the molar concentration defined by

$$c_B = \frac{n_B}{V} \tag{1.23}$$

where V is the volume of the phase, we can first write

$$m_B = \frac{n_B}{n_A M_A} = \frac{n_B}{(n_A M_A + n_B M_B) - n_B M_B} = \frac{c_B}{d - c_B M_B} \tag{1.24}$$

where d is the density of the solution in kg·l^{-1}. Combining equations (1.18), (1.20) and (1.24), we obtain

$$\mu_B(T) = \mu_B^{\ominus,c}(T) + RT\ln\left(\gamma_B^c \frac{c_B}{c^{\ominus}}\right) \tag{1.25}$$

where $c^{\ominus}$ is the ***standard molarity*** of 1 mol·l^{-1}. The standard chemical potential in the molarity scale is defined as a function of the molar volume V_{mA} of solvent by

$$\mu_{\mathrm{B}}^{\ominus,c}(T) = \mu_{\mathrm{B}}^{\ominus,ideal\ dil}(T) + RT\ln\left(V_{\mathrm{mA}}c^{\ominus}\right) \tag{1.26}$$

and the activity coefficient by

$$\gamma_{\mathrm{B}}^{c} = \frac{x_{\mathrm{A}}d_0\gamma_{\mathrm{B}}}{d - c_{\mathrm{B}}M_{\mathrm{B}}} = \gamma_{\mathrm{B}}^{m}\frac{d_0 m_{\mathrm{B}}}{c_{\mathrm{B}}} \tag{1.27}$$

where d_0 is the density of the pure solvent ($= M_{\mathrm{A}}/V_{\mathrm{mA}}$).

In the case of dilute solutions, the activity coefficients in the molality and molarity scales are equal.

IMPORTANT NOTE

In the rest of this book, we shall mainly treat chemical potentials in the molarity scale, and we shall write equation (1.25) ignoring the 'mute' term $c^{\ominus}$ and will then have it in the following simplified form

$$\mu_{\mathrm{B}}(T) = \mu_{\mathrm{B}}^{\ominus}(T) + RT\ln\left(\gamma_{\mathrm{B}}c_{\mathrm{B}}\right) = \mu_{\mathrm{B}}^{\ominus}(T) + RT\ln a_{\mathrm{B}} \tag{1.28}$$

it being understood that the logarithmic term is dimensionless and that the standard chemical potential and the activity coefficient are relative to the molarity scale.

Application of chemical potentials

Chemical potentials are important tools for studying the behaviour of chemical reactions. Let us consider the following reaction

$$a\mathrm{A} + b\mathrm{B} + c\mathrm{C} + \ldots \rightleftarrows x\mathrm{X} + y\mathrm{Y} + z\mathrm{Z}$$

The Gibbs energy of this reaction is defined as the work to add the products minus the work to add the reactants. So the Gibbs energy of a reaction can be written as a linear combination of the chemical potentials :

$$\Delta G_{\mathrm{reaction}} = \sum_{\mathrm{products}} \nu_i\mu_i - \sum_{\mathrm{reactants}} \nu_i\mu_i \tag{1.29}$$

where ν_i represent the stœchiometric coefficients. This definition shows that at equilibrium, the Gibbs energy of a reaction is zero because the work to add the products cancels out the work to remove the reactants. Concerning chemical equilibria, it is perhaps a good idea to remember at this point the difference between the thermodynamic reversibility and the chemical reversibility of a reaction. The former corresponds to an infinitely slow transformation with a quasi-equilibrium existing at each infinitesimal stage of the reaction. The latter relates to the feasibility of the reverse reaction.

In the case of charged species, the chemical potential also represents the work necessary to bring this species from vacuum into a phase, but this displacement implies an electrostatic work if the phase is at a potential different from that of vacuum. In order to be able to quantify this work, we need to recall some basic electrostatics.

1.1.2 External potential

Basic electrostatics

Considering two charges q_1 and q_2, in a vacuum, the force exerted by q_1 on q_2, is given by Coulomb's law, written as

$$\boldsymbol{f} = q_2 \frac{q_1}{4\pi\varepsilon_0 r^2}\hat{\boldsymbol{r}} = q_2\boldsymbol{E} \tag{1.30}$$

where $\hat{\boldsymbol{r}}$ represents the unit vector. The proportionality constant $1/4\pi\varepsilon_0$ is due to the SI units system, and has units of V·m·C^{-1} or m·F^{-1} ($4\pi\varepsilon_0 = 1.111265 \cdot 10^{-10}/\text{F·m}^{-1}$). ε_0 is called the ***permittivity*** of vacuum.

By definition, the electric field is expressed as the gradient of electrical potential (see the Annex A on vectorial analysis)

$$\boldsymbol{E} = -\,\mathbf{grad}V \tag{1.31}$$

Using spherical coordinates centred on the charge q_1, a simple integration of equation (1.31) using equation (1.30) shows that the potential at the distance r from this charge is

$$V(r) = \frac{q_1}{4\pi\varepsilon_0 r} \tag{1.32}$$

if the potential is taken equal to zero when $r \to \infty$. For a discontinuous distribution of point charges, the electric fields are additive, and consequently, the total potential is the sum of the potentials generated by each charge q_i.

$$V_{\text{total}} = \frac{1}{4\pi\varepsilon_0}\sum_i \frac{q_i}{r_i} \tag{1.33}$$

The Gauss theorem

To calculate the potential due to a charged object, a relatively simple method is to apply Gauss's theorem that shows that the flux of an electric field coming out of a closed surface is equal to the charge contained inside the surface divided by the permittivity of the medium.

In the case of a spherical conducting object of radius R and total charge Q, the flux leaving a concentric sphere of radius r is

$$4\pi r^2 \boldsymbol{E}\cdot\hat{\boldsymbol{r}} = \frac{Q}{\varepsilon_0} \tag{1.34}$$

where $\hat{\boldsymbol{r}}$ represents the unit radial vector. By integration between infinity ($V_\infty = 0$) and r, we deduce the potential at the distance r from the centre of the object

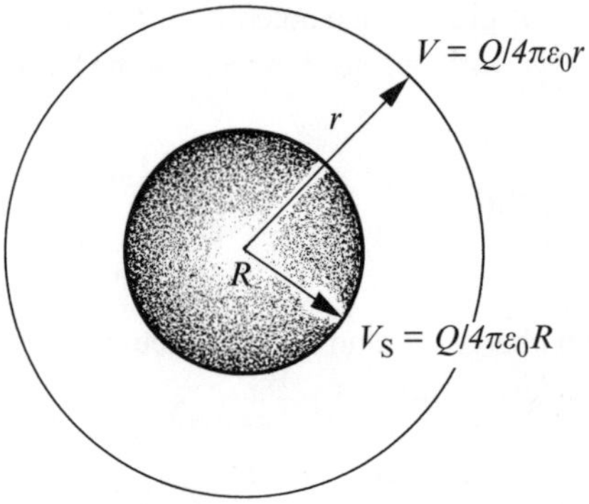

Fig. 1.4 Electrical potential around a spherical object of radius R having a charge Q. The Gauss surface is here defined as the outer concentric sphere of radius r.

$$V = \frac{Q}{4\pi\varepsilon_0 r} \quad \text{for } r > R \tag{1.35}$$

The potential at the surface of the object is then V_S as shown in Figure 1.4. The potential and the charge of a conductor are proportional

$$Q = CV \tag{1.36}$$

The constant of proportionality C is called the ***capacitance*** of the object.

Another way of writing Gauss's theorem is to apply the Green-Ostrogradski theorem in order to find what is often called ***Maxwell's first equation***, which links the divergence of the electric field leaving a surface to the volumic charge density ρ contained in the volume defined by this surface

$$\operatorname{div} \boldsymbol{E} \quad (= \nabla \cdot \boldsymbol{E}) = \frac{\rho}{\varepsilon_0} \tag{1.37}$$

Coulomb's theorem

In this book, we shall consider conductors and metallic electrodes. To evaluate an electric field near a conductor, it is useful to use Coulomb's theorem to show that near to a conductor at equilibrium, close to a point where the surface charge density is σ, the electric field is normal to the surface and is expressed by

$$\boldsymbol{E} = \frac{\sigma}{\varepsilon_0}\hat{\boldsymbol{n}} \tag{1.38}$$

where $\hat{\boldsymbol{n}}$ represents the unit normal vector to the surface. This equation is demonstrated by taking a Gauss surface that surrounds a surface element of area dS as shown in Figure 1.5, knowing that there is no electric field inside a conductor at equilibrium. In fact, if there was a field inside this conductor, currents would be circulating in it. Applying Gauss's theorem to the inside of the conductor shows that it is electrically neutral at equilibrium.

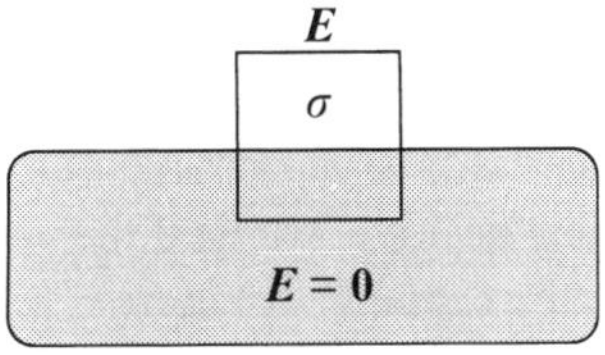

Fig. 1.5 Gauss's surface surrounding a surface element of a charged conductor whose surface charge density is σ. By symmetry, the field is normal to the surface.

Note that the electric field near to a uniformly charged plane where the charge is spread over the two faces can be written as

$$\boldsymbol{E} = \frac{\sigma}{2\varepsilon_0}\hat{\boldsymbol{n}} \tag{1.39}$$

due to the symmetry of the two sides of the plane.

EXAMPLE

Let us calculate the capacity of a planar capacitor consisting of two conducting plates with a surface area S whose surface charge densities on the internal faces are respectively σ and $-\sigma$, and separated by a distance d.

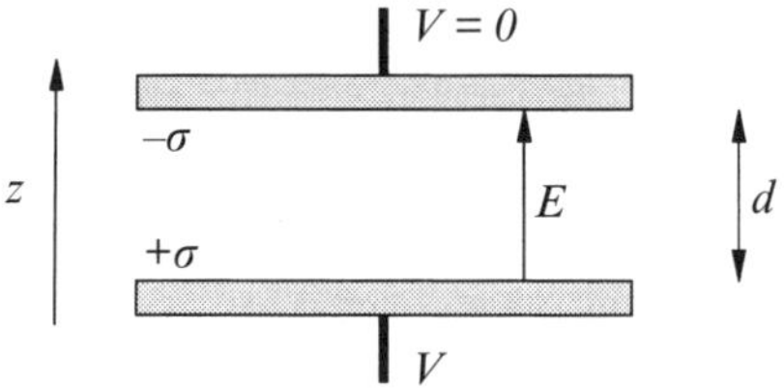

The projection of the electric field on the z axis is given by Coulomb's theorem. The field is constant between the two plates and is written as

$$E = \frac{\sigma}{\varepsilon_0} = \frac{V}{d}$$

where V is the potential difference at the terminals of the capacitor. The charge of the capacitor is written thus

$$Q = \sigma S = \frac{\varepsilon_0 S}{d}V = CV$$

Outer potential and the Volta potential difference

By definition, a ***potential difference*** (p.d.) is the difference in potential between two points. However, by an abuse of language, there is a tendency to use the term potential to designate a potential difference.

The p.d. between the exterior of the surface of a charged object and a vacuum is called the ***outer potential*** and is designated by the Greek letter ψ in the Lange convention. The outer potential is a measurable quantity. If the object is positively charged, the outer potential is positive, whilst if the object is negatively charged, then so is the outer potential. In the example of Figure 1.4, the outer potential corresponds to the potential just at the surface of the sphere, being V_S.

The difference of two outer potentials between two charged objects A and B is called the ***Volta potential difference***.

1.1.3 Surface potential

For every condensed phase, the structure of the surface is different from the internal structure. In particular, the coordination number of the surface molecules is lower and this translates into the fact that these molecules have a higher potential energy than those found within the phase. To compensate for this difference in potential energy between the surface and the bulk, the surface region organises itself in such a way as to minimise this difference of potential energy.

In order to treat the electrostatic consequences of this surface reorganisation, it is useful first of all to review briefly the polarisation of matter and introduce the notion of relative permittivity.

Polarisation of matter

Given that an atom possesses a positive charge at the nucleus, surrounded by a cloud of electrons, the application of an electric field causes a shift δ between the centre of the positive charges q and negative charges $-q$. The resultant dipole moment is written as

$$\boldsymbol{p} = q\boldsymbol{\delta} \tag{1.40}$$

If there are N atoms per unit volume, the dipole moment per unit volume will therefore be $Nq\delta$. By definition, we will call this volumic dipole moment the polarisation vector $\boldsymbol{P}$

$$\boldsymbol{P} = N\boldsymbol{p} \tag{1.41}$$

As a first approximation, we will make the hypothesis that the polarisation vector is proportional to the electric field that induces it

$$\boldsymbol{P} = \chi\, \varepsilon_0\, \boldsymbol{E} \tag{1.42}$$

where χ is called the ***electric susceptibility*** and expresses the ease with which the electrons can move, that of course depends on the atoms contained in the dielectric material.

Relative permittivity

Take a flat capacitor made of two conducting plates separated by a vacuum. Its capacity that links the charge to the potential of the capacitor is given by

$$C = \frac{\varepsilon_0 S}{d} \tag{1.43}$$

If the space between the plates is now filled with a dielectric material, we observe that the capacity is greater, which means that the potential is smaller for the same charge, or again that the electric field between the plates is weaker. If we consider the atoms of the dielectric near to the plates, we can define a surface density of the polarisation charge σ_{pol}.

To obtain this quantity, we can calculate the resultant dipole moment per unit surface area, which is on one hand the product of the surface density of polarisation charge and the thickness of the capacitor, and on the other hand is equal to the product of the volumic dipole moment and the volume per unit area

$$\sigma_{\text{pol}} d = \frac{PSd}{S} = Pd \tag{1.44}$$

where Sd is the volume between the plates of the flat capacitor. Thus, we can see that the surface density of polarisation charge is equal to the magnitude of the volumic dipole moment vector. σ_{pol} is of the opposite sign to the free charge σ accumulated on the internal faces of the conducting plates of the flat capacitor.

The polarisation charge is induced by the free charge. If we discharge the capacitor, the free charge will disappear by conduction in the contact wires, while the polarisation charge will disappear by relaxation. Applying Gauss's theorem to the surface indicated by the dotted part of Figure 1.6, the electric field in the dielectric is then given by

$$E = \frac{\sigma_{\text{free}} - \sigma_{\text{pol}}}{\varepsilon_0} \tag{1.45}$$

or again

$$E = \frac{\sigma_{\text{free}} - P}{\varepsilon_0} = \frac{\sigma_{\text{free}}}{\varepsilon_0} \frac{1}{1+\chi} \tag{1.46}$$

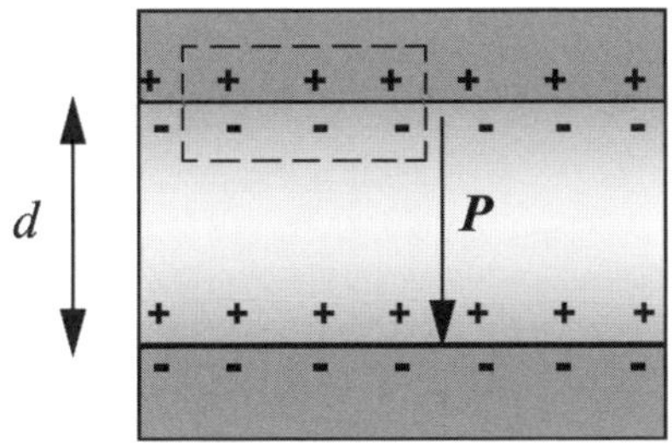

Fig. 1.6 Polarisation charges inside a flat capacitor made of two metal plates and filled by a dielectric.

Given that the electric charge is uniform in the dielectric, the potential between the terminals of the capacitor is simply

$$V = Ed = \frac{\sigma_{\text{free}}\, d}{\varepsilon_0 (1+\chi)} \tag{1.47}$$

and the capacitance

$$C = \frac{\varepsilon_0\, S\, (1+\chi)}{d} = \frac{\varepsilon_0 \varepsilon_r\, S}{d} \tag{1.48}$$

where ε_r is a proportionality factor which links the capacity of a capacitor filled with a dielectric material to that of the same capacitor in a vacuum. ε_r is called the ***relative permittivity*** or ***dielectric constant*** and is defined as a dimensionless number. The terminology dielectric constant is not very appropriate as ε_r is not constant, as it depends on the frequency of the potential applied on the terminals of the capacitor and on the temperature.

In the case of liquids, the relative permittivity varies from about 2 for non-polar solvents such as alkanes, to more than 100 for formamide ($HCONH_2$). Water has a static or low frequency relative permittivity of about 78.4 at 20°C. This high value is due to the coercive effects of the dipolar molecules.

We also define the ***permittivity*** of a medium ε as

$$\varepsilon = \varepsilon_0 \varepsilon_r \tag{1.49}$$

If we apply a sinusoidal field to the terminals of the capacitor, the relative permittivity remains equal to the static value as the frequency increases as shown in Figure 1.7.

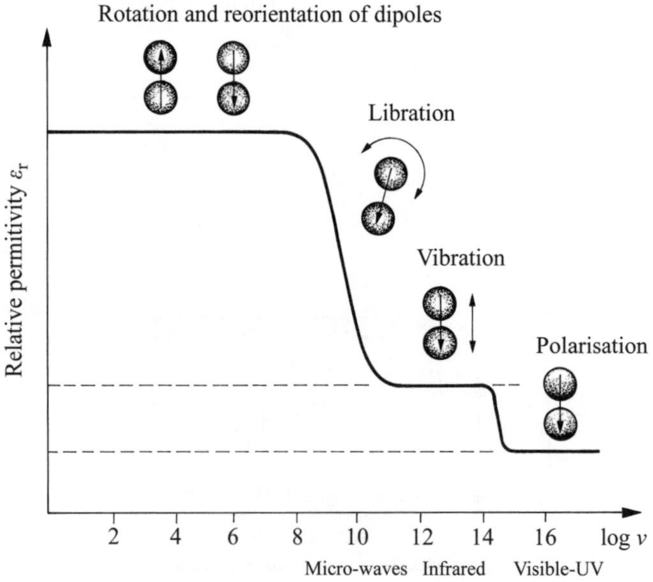

Fig. 1.7 Variation of the relative permittivity of water as a function of the frequency of the applied electric field.

This means that even at radio frequencies, the water molecules can reorient themselves at the frequency of the field imposed. At a frequency of about $\nu = 10^8$ Hz, the relative permittivity decreases because the molecules have too much inertia to be able to follow these repetitive 'flip-flops'. The dipoles oscillate in a movement that is called ***libration***.

At higher frequencies, the dipolar molecules are 'frozen', and only the atoms of each molecule try to follow the electric field. In this band of infrared frequency where the vibrational movements dominate, the relative permitivity has a value of about 5.9. At even higher frequencies in the UV-VIS part of the spectrum ($\nu > 10^{14}$ Hz), the nuclei of the atoms 'give up' and in their turn remain 'frozen'. Only the electrons continue to oscillate with the field. The relative permittivity, sometimes called the ***optical relative permittvity*** has then a value of 1.8.

Electric displacement vector

Applying Maxwell's first equation, we can see that it is possible to distinguish the charges inside a Gauss surface, between the free charge and a charge due to a non-uniform polarisation

$$\mathrm{div}\,\boldsymbol{E} = \frac{\rho_{\mathrm{free}} + \rho_{\mathrm{pol}}}{\varepsilon_0} = \frac{\rho_{\mathrm{free}}}{\varepsilon_0} - \frac{\mathrm{div}\,\boldsymbol{P}}{\varepsilon_0} \tag{1.50}$$

In the case of linear systems where the polarisation vector is proportional to the electric field, we can write

$$\mathrm{div}\left(\boldsymbol{E} + \frac{\boldsymbol{P}}{\varepsilon_0}\right) = \frac{\rho_{\mathrm{free}}}{\varepsilon_0} \tag{1.51}$$

or again, defining an electric displacement vector $\boldsymbol{D}$,

$$\boldsymbol{D} = \varepsilon_0\,\boldsymbol{E} + \boldsymbol{P} = \varepsilon\,\boldsymbol{E} \tag{1.52}$$

we thus have

$$\mathrm{div}\boldsymbol{D} = \rho_{\mathrm{free}} \tag{1.53}$$

The definition of the electric displacement vector is useful to express what happens at the contact surface between two dielectric materials. In effect, the application of Gauss's theorem allows us to show easily that the normal component at the surface of the vector $\boldsymbol{D}$ presents a discontinuity at the surface of separation of the two media if it carries a free surface charge density σ

$$\boldsymbol{D}_1 \cdot \hat{\boldsymbol{n}} - \boldsymbol{D}_2 \cdot \hat{\boldsymbol{n}} = \sigma \tag{1.54}$$

It is also possible to show that the tangential component of the electrical field is continuous at the separation surface of the two media.

Thus, at a metal | dielectric junction, the electric field is zero in the metal and consequently, the tangential components are also zero on both sides of the surface. The electric field in the dielectric is therefore normal at the surface. This is important when we consider the distribution of the electrical potential at the surface of an electrode in solution.

Surface of polar liquids

In polar solvents, the intermolecular forces are, on average isotropic because of thermal agitation, and no particular orientation of the dipoles takes place.

When a new surface is created, the molecules inside the solution with a strong negative potential energy have to be transferred towards the interface where their potential energy is higher (fewer intermolecular interactions).

In order to minimise this increase in potential energy, the dipole molecules at the surface have a tendency to align themselves. In effect, the dipole-dipole interaction energies are more negative when they are aligned in the same direction. The difference in potential across this layer of oriented dipoles is called the ***surface potential***. This is designated by the Greek letter χ (Note that this is the same symbol used for electric susceptibility), which gives it the name 'chi potential' (Lange's convention) used sometimes.

This surface potential is considered as positive if the potential increases from the exterior towards the interior of the phase (dipoles oriented towards the interior). This quantity is not measurable, but can be estimated by various approximations.

The potential drop across a layer of aligned dipoles is equivalent to that of a capacitor

$$\chi = Q/C \tag{1.55}$$

where C is the capacitance for a flat capacitor given by

$$C = \frac{\varepsilon_0 \varepsilon_r S}{d} \tag{1.56}$$

where ε_0 is the permittivity of the vacuum, ε_r the relative permittivity of the layer of dipoles, S the surface area and d the separation distance between the charged plates, which is the length of the dipoles.

In the case of n perfectly aligned dipoles the surface potential is then

$$\chi = \frac{n}{S}\frac{qd}{\varepsilon_0 \varepsilon_r} = \frac{N_s p}{\varepsilon_0 \varepsilon_r} \tag{1.57}$$

where N_s is the density of dipoles per unit surface area and p the dipole moment.

In the case of solvents that have a large number of hydrogen bonds, the surface molecules also try to minimise their excess potential energy by optimising the number

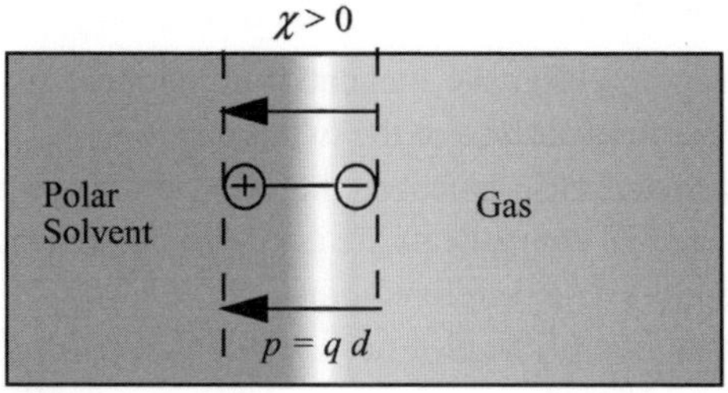

Fig. 1.8 Surface potential of water represented by aligned dipoles.

of hydrogen bonds. In the case of water, non-linear surface spectroscopy studies have shown that molecules at the air | water interface have a highly ordered structure.

EXAMPLE

Let us calculate the surface potential χ of water, knowing that the dipole moment of a molecule of water in a vacuum is 1.85 Debye. Compare this value to the one deduced experimentally at about 0.13 V.

We have seen that water has a relative permittivity of about 78.4 at 20°C, and that this high value is due to the coercive effect of the dipolar molecules. At a frequency of about $\nu = 10^8$ Hz, this value falls to 5.9 because the molecules can no longer follow the change of polarity of the capacitor, and keep the same orientation. For this example, we shall consider that the relative permittivity of a monolayer of oriented water is 5.9.

The number of water molecules per m^2 is the number of water molecules per m^3 to the power 2/3

$$N_S = \left[\frac{N_A}{V_{mH_2O}}\right]^{2/3} = \left[\frac{6.02 \cdot 10^{23}}{18 \cdot 10^{-6}}\right]^{2/3} = 1.03 \cdot 10^{19} \, m^{-2}$$

where N_A is the ***Avogadro constant*** (Often by an abuse of language, N_A is called Avogadro's number, which is erroneous, because by definition, and contrary to a constant, a number has no units.) The molecular area per molecule is therefore 10^{-19} m^2, which is reasonable, given that the average diameter of a molecule of water is about 300 pm.

The dipole moment of water in a vacuum is

$$p = 1.85 \text{ Debye} = 1.85 \cdot 10^{-21} \ / \ (3 \cdot 10^8) = 6.17 \cdot 10^{-30} \text{ C} \cdot \text{m}$$

We can note in passing that the dipole moment of water in the liquid phase is greater, approximately equal to 2.6 Debye, because of interactions between the molecules. For the surface molecules, this value diminishes towards the value in the gas phase.

Using the hypothesis that we can use the value in the gas phase as a first approximation at the interface, we can calculate the surface potential of a perfectly oriented monolayer of water

$$\chi = \frac{N_S p}{\varepsilon_0 \varepsilon_r} = \frac{(1.03 \cdot 10^{19}) \cdot (6.17 \cdot 10^{-30})}{(8.8542 \cdot 10^{-12}) \cdot 5.9} = 1.22 \text{ V}$$

It is about ten times the measured value, which illustrates that the electrostatic considerations are not the only ones to take into account; the thermal agitation, and the optimisation of the number of hydrogen bonds also plays a very important role at the surface of water.

Adsorbed ionised surfactants

In the case of electrolyte solutions, the difference of the penetration of cations and anions into the surface layer may result in the formation of 'ionic dipoles', which generate a surface potential. This difference of position with respect to the interface is large when one of the ions is an amphiphile (half-lipophilic, half-hydrophilic) as with cationic or anionic surfactants. In this case, the amphiphilic ion is adsorbed in a monolayer at the surface. This adsorption then generates a large surface potential.

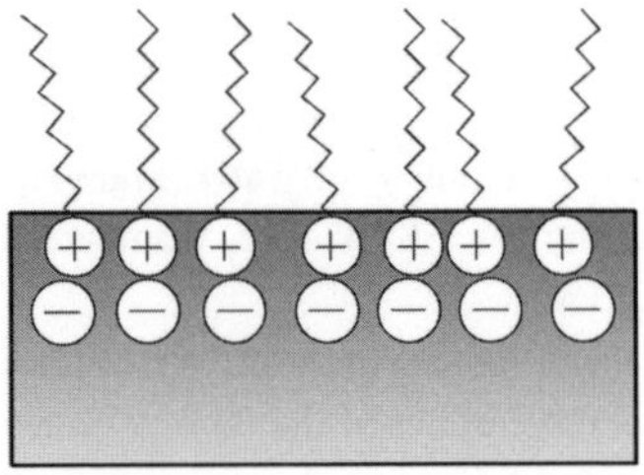

Fig. 1.9 Specific adsorption of cationic surfactants at the air|electrolyte solution interface leading to the formation of a negative surface potential.

Metal surfaces

The free or conduction electrons inside a metal are submitted to a force field coming from ions fixed within a symmetrical network. When a new surface is created, the electrons and the ions at the surface are subject to asymmetric forces. The electrons are attracted to the vacuum, since an expansion of the electron cloud towards the exterior induces a fall in their kinetic energy, but they are then subject to a coulombic force pulling them back. In effect, it is possible to model a metal using a uniform distribution of localised positive charges and mobile electrons (the Jellium model). This electronic spill-over translates into the formation of a dipolar surface region as shown in Figure 1.10. The resulting surface potential is positive.

In conclusion, whatever the nature of the phase under consideration, we can say that the surface potential in electrified phases results, in general, from the anisotropy of the forces exerted on the species situated at the surface.

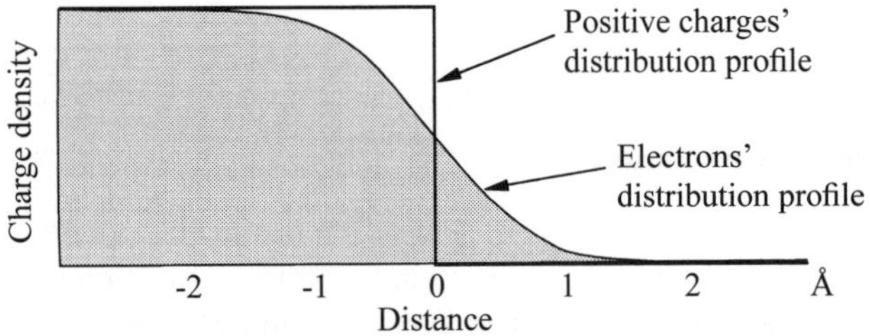

Fig. 1.10 Electron spill-over at the surface of a metal according to the Jellium model.

1.1.4 Inner potential or the Galvani potential

The surface potential and the outer potential have been defined from physical models. And so the work to bring one mole of ions from a vacuum towards a charged phase comprises two terms:

- an electrostatic term associated with the crossing of the layer of interfacial orientated dipoles $z_i F\chi$;
- an electrostatic term associated with the charge on the phase $z_i F\psi$.

Effectively, the charge of one mole of ions is $z_i F$, F being the ***Faraday constant*** which is defined by $F = N_A e = 96\ 485\ \text{C·mol}^{-1}$, N_A being Avogadro's constant and e the elementary charge, $e = 1.60218 \cdot 10^{-19}$ C.

Thus, we can define the ***inner potential*** of a phase as the sum of the surface and outer potentials.

$$\phi = \chi + \psi \tag{1.58}$$

It is worth emphasizing that these potentials are really potential differences. The inner potential of a phase is in fact the potential difference between the bulk of the phase and the vacuum.

For any conducting condensed phase, it is possible to define an inner potential that is constant across the whole of the interior of the phase. Excess charges can only exist at the surface of a phase.

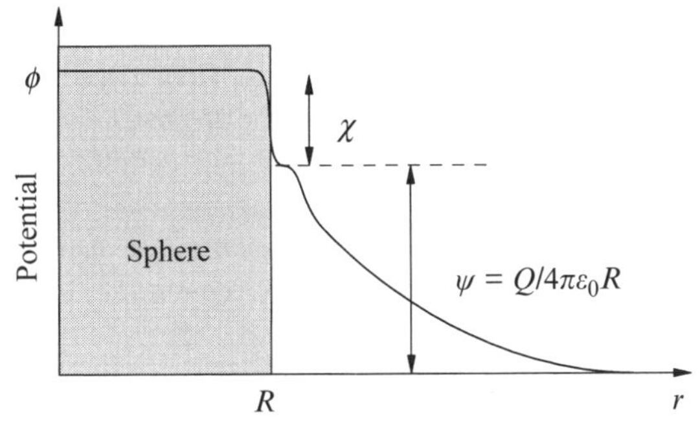

Fig. 1.11 Radial distribution of potential for a metal sphere of radius R carrying a positive charge Q, illustrating the contributions of the outer potential and the surface potential. The inner potential is constant inside the sphere.

1.1.5 The Galvani potential difference

If the Galvani potential is constant inside each phase, this supposes that all the differences in Galvani potential between two phases in contact happen in the interfacial region. This distribution causes accumulations of charges (positive or negative) generating a potential difference $\Delta_\alpha^\beta g(\text{charge})$ and some eventual dipolar

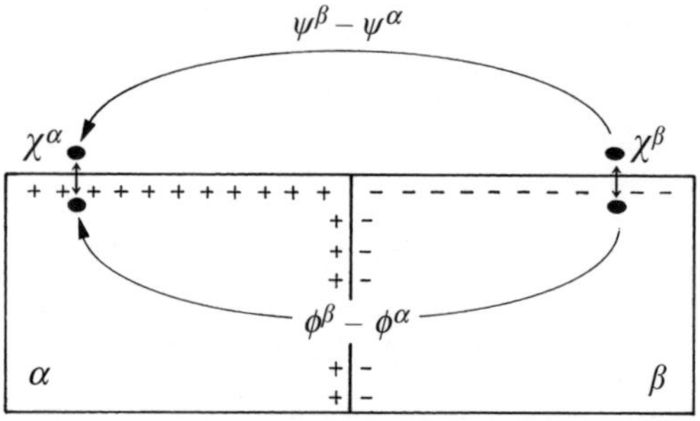

Fig. 1.12 Surface potential, the Volta potential and the Galvani potential differences for two phases in contact.

contributions generating a potential difference $\Delta_\alpha^\beta g$(dip).The latter p.d. should not to be confused with the surface potentials which are only defined for interfaces between a phase and an inert gas whose pressure may be very low (e.g. metal | vacuum)

$$\Delta_\alpha^\beta \phi \quad = \quad \phi^\beta - \phi^\alpha \quad = \quad \Delta_\alpha^\beta g(\text{dip}) \;+\; \Delta_\alpha^\beta g(\text{charge}) \tag{1.59}$$

This question of potential distribution between two phases in contact will be taken up in more detail in chapter 5.

1.1.6 Electrochemical potential

We shall now link these electrical potentials to a thermodynamic quantity. For this, let us consider the transfer of one mole of ions i from a vacuum into a phase. We will call the work required to effect this transfer the ***electrochemical potential*** which we shall designate by $\tilde{\mu}_i$. In fact, strictly speaking, this definition of electrochemical potential is none other than the one for the chemical potential given by equation (1.6), the prefix electro- reminding us only that we are dealing with charged species and that the inner potential of the phase intervenes with the increase in internal energy when ions are added to a phase.

In order to understand better the notion of electrochemical potential, it is more useful to come back to the initial definition of the work required to bring ions from a vacuum into a phase. This work comprises 3 separate terms:

- a chemical term which includes all the short-distance interactions between the ion and its environment (ion-dipole interactions, ion-dipole induced reactions, dispersion forces etc.);
- an electrostatic term linked to the crossing of the layer of oriented interfacial dipoles, $z_iF\chi$;
- an electrostatic term linked to the charge of the phase, $z_iF\psi$.

And so, in the case of a mole of ions, the electrochemical potential is written as

$$\tilde{\mu}_i \quad = \quad \mu_i + z_iF\chi + z_iF\psi \quad = \quad \mu_i + z_iF\phi \quad = \quad \mu_i^\ominus + RT\ln a_i + z_iF\phi \tag{1.60}$$

Given that strictly speaking the electrochemical potential of the ion is in fact a chemical potential, we ought to call μ_i the chemical term or the chemical contribution of the electrochemical potential. Even though often done by abuse of language, it is incorrect to call μ_i the chemical potential of the ion. However, given the absence of official vocabulary to designate the chemical term, we shall in this book commit the abuse of language, and call μ_i the chemical potential of the ion. The chemical potential can be decomposed as before, into a standard term and an activity term. Again, by abuse of language, $\mu_i^{\ominus}$ represents the standard chemical potential of the ion.

If we use equation (1.3) relative to the variation of internal energy, we can write it

$$\begin{aligned} \mathrm{d}U &= -p\mathrm{d}V + T\mathrm{d}S + \sum_i \mu_i \,\mathrm{d}n_i + \sum_i \phi \,\mathrm{d}q_i \\ &= -p\mathrm{d}V + T\mathrm{d}S + \sum_i (\mu_i + z_i F\phi)\,\mathrm{d}n_i = -p\mathrm{d}V + T\mathrm{d}S + \sum_i \tilde{\mu}_i \,\mathrm{d}n_i \end{aligned} \tag{1.61}$$

where $\mathrm{d}q_i$ represents the variations in charge associated with the addition of the species i. The term $\phi\ \mathrm{d}q_i$ represents the electrical work with respect to the addition of charges to a phase having an inner potential ϕ. Implicitly, we are making the hypothesis that the phase is large enough so that adding charges does not modify the inner potential significantly.

Thus, the general definition of electrochemical potential could be given by

$$\tilde{\mu}_i = \left(\frac{\partial G}{\partial n_i}\right)_{p,T,n_{j\neq i}} \tag{1.62}$$

The electrochemical potential is a measure of the work needed to move one mole of ions from a vacuum into a phase, at constant pressure and temperature. Nevertheless, absolutely speaking, it is impossible to add ions to a phase without changing its charge and therefore its outer potential. Consequently, the notion of electrochemical potential defined by equation (1.62) is a virtual one.

In effect, the goal of thermodynamics is to establish a relation between measurable experimental quantities. We have just seen that this is not the case for the electrochemical potential, which is a value that cannot be determined experimentally. The notion of the electrochemical potential of a charged species is therefore an abstract notion which is a very useful mathematical tool for treating electrochemical phenomena, but which cannot, strictly speaking, be considered as a thermodynamic value.

To conclude, we define the electrochemical potential of an ion as a virtual quantity defined by equations (1.60) & (1.62). Note the abuse of language in calling the chemical contribution of equation (1.60) the chemical potential. Like the chemical potential of neutral species, the electrochemical potential of charged species has the essential property of being a quantity that is independent of the phases when these are in contact equilibrium.

It is also possible to define the ***standard electrochemical potential***, i.e. the electrochemical potential of a charged species in a standard state, for example on the molarity scale for different values of inner potential.

$$\tilde{\mu}_i^{\ominus} = \mu_i^{\ominus} + z_i F \phi \tag{1.63}$$

The standard electrochemical potential to a certain extent represents the part of the electrochemical potential that is independent of the concentration. This definition will be useful in chapter 7.

1.1.7 Chemical potential of a salt

In the case of a salt $C_{\nu^+}^{z+} A_{\nu^-}^{z-}$ in solution, the chemical potential of the salt is a measurable quantity corresponding to the solvation energy of that salt. If the dissolved salt is totally dissociated, this chemical potential of the salt can be expressed as a linear combination of the electrochemical potentials of its constituent ions

$$\mu_{\text{salt}} = \mu_{C_{\nu^+}^{z+} A_{\nu^-}^{z-}} = \nu^+ \tilde{\mu}_{C^{z+}} + \nu^- \tilde{\mu}_{A^{z-}} \tag{1.64}$$

By developing, we get

$$\mu_{\text{salt}} = \nu^+ \mu^{\ominus}_{C^{z+}} + \nu^- \mu^{\ominus}_{A^{z-}} + RT \ln\left[a^{\nu^+}_{C^{z+}} \, a^{\nu^-}_{A^{z-}} \right] \tag{1.65}$$

the terms linked to the inner potential of the phase can be eliminated because of the electroneutrality of the solution $[\nu^+ z^+ + \nu^- z^- = 0]$.

For the activity coefficients on the molarity scale, we can define the activity coefficient of the dissociated salt as

$$\gamma_{\text{salt}} = \gamma^{\nu^+}_{C^{z+}} \, \gamma^{\nu^-}_{A^{z-}} \tag{1.66}$$

We then define the mean ionic coefficient as the arithmetic mean of the ionic activity coefficients

$$\gamma_{\pm} = \gamma_{\text{salt}}^{1/\nu} \tag{1.67}$$

with $\nu = \nu^+ + \nu^-$. The term relating to the concentration is often a source of errors. By defining the stœchiometric concentration c_{salt} as the number of moles of salt dissolved per litre of solvent,

$$c_{\text{salt}} = n_{\text{salt}} / V_{\text{solvent}} \tag{1.68}$$

the term concerning the product of the ionic concentrations in equation (1.65) can be expressed in the form

$$c^{\nu^+}_{C^{z+}} \, c^{\nu^-}_{A^{z-}} = \left(\nu^+ c_{\text{salt}}\right)^{\nu^+} \left(\nu^- c_{\text{salt}}\right)^{\nu^-} = \left(\nu_{\pm} c_{\text{salt}}\right)^{\nu} \tag{1.69}$$

with $\nu_{\pm}{}^{\nu} = \left(\nu^+\right)^{\nu^+} \left(\nu^-\right)^{\nu^-}$.

In summary, we can express the chemical potential of a dissociated salt in the molarity scale in terms of the standard chemical potentials of the respective ions, and the mean activity coefficient

$$\mu_{\text{salt}} = \left[\nu^+ \mu^{\ominus}_{C^{z+}} + \nu^- \mu^{\ominus}_{A^{z-}} \right] + \nu RT \ln\left(\gamma_{\pm} \nu_{\pm} c_{\text{salt}}\right) \tag{1.70}$$

EXAMPLE

The chemical potential on the scale of concentrations of $MgCl_2$ is defined by

$$\mu_{MgCl_2} = \tilde{\mu}_{Mg^{2+}} + 2\tilde{\mu}_{Cl^-}$$

and consequently

$$\mu^{\ominus}_{MgCl_2} = \tilde{\mu}^{\ominus}_{Mg^{2+}} + 2\tilde{\mu}^{\ominus}_{Cl^-}$$

The activity coefficient of the salt is therefore

$$\gamma_{MgCl_2} = \gamma_{Mg^{2+}}\gamma^2_{Cl^-} = \gamma^3_{\pm}$$

The average stœchiometric coefficient is then

$$\nu_{\pm}^{\ 3} = 1 \cdot 2^2 = 4$$

Thus, the chemical potential of the salt is written as

$$\mu_{MgCl_2} = \left[\tilde{\mu}^{\ominus}_{Mg^{2+}} + 2\tilde{\mu}^{\ominus}_{Cl^-}\right] + RT\ln\left(4\gamma_{MgCl_2}c^3_{MgCl_2}\right)$$

1.1.8 Real chemical potential

Given that ψ can be measured, it has been proposed to define the ***real chemical potential*** of the ion i in the phase as the electrochemical potential in the non-charged phase, that is to say

$$\alpha_i = \mu_i + z_i F\chi \tag{1.71}$$

Thus, the electrochemical potential can be defined as the sum of the real chemical potential and an electrostatic term related to the charge of the phase.

1.2 ELECTROCHEMICAL POTENTIAL OF ELECTRONS

1.2.1 Band structure

A solid crystal can be considered as a giant molecule and an approximation of the electronic wave function can be obtained by using the theory of molecular orbitals (MO). As with a classical molecule, the MOs are obtained using the linear combination of atomic orbitals (AO). Take for example the case of sodium, whose electronic structure is $1s^2$, $2s^2$, $2p^6$, $3s^1$ and suppose that the crystal contains N atoms. When two atoms of sodium are close to one another to form the dimer Na_2, the two AOs, $1s$ combine to form two MOs $\sigma 1s$ and σ^*1s. For Na_2, the overlap of the atomic orbitals

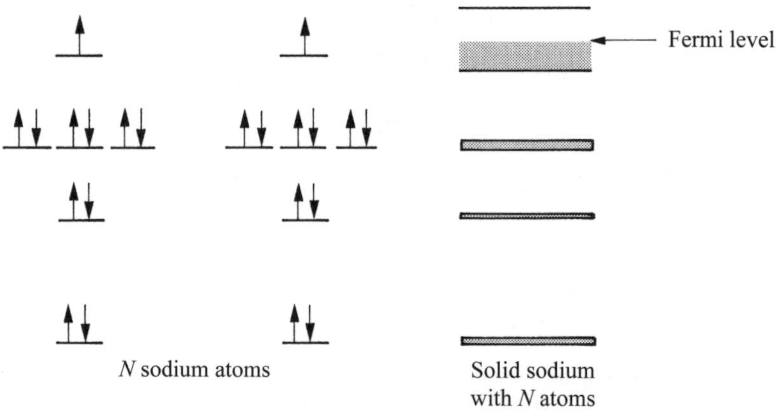

Fig. 1.13 Distribution of the electronic energy levels for sodium.

close to the nucleus is very weak, and consequently the energy level of the MOs is almost identical to that of the AO 1*s* of the isolated atoms. For the *N* atoms of the crystal, the *N* AOs 1*s* form *N* MOs whose energies are more or less identical to those of the AO 1*s* of the sodium atom. In the same way, the electrons of levels 2*s* and 2*p* interact and form *N* MOs whose energies are more or less identical to those of the AOs 2*s* and 2*p*. On the other hand, the electrons 3*s*, interact and form delocalised MOs. The energy difference between the upper and lower MOs is roughly the same as that separating the MOs $\sigma 3s$ and $\sigma^* 3s$ (a few eV) in Na_2. If the number of atoms in the crystal *N* is large, the energy levels of the MOs will be so close that they will form a continuous band of energy levels.

To summarize, we can say that an atomic energy level gives rise to a large band of energy levels if there is overlap of the orbitals. In the case of copper, whose isolated atom has the structure $1s^2$, $2s^2$, $2p^6$, $3s^2$, $3p^6$, $3d^{10}$ and $4s^1$, only the levels 3*d* and 4*s* give rise to band a structure. The band structure in solids is seen in the X-ray emission spectra as rather diffuse lines, whilst the lines caused by transitions of inner orbitals are very sharp.

For an electron to be mobile in a solid, there have to be vacant levels inside a band. The highest filled band is called the ***valence band*** and the lowest empty band is called the ***conduction band***.
Three cases are then possible according to the relative positions of the bands:

- The conduction and valence bands overlap. The number of mobile charge carriers is then large and the solid has a conductivity known as metallic, e.g. Na, Cu, ...
- The conduction band is empty at absolute zero. The conduction and valence bands are separated by a band gap greater than 5 eV. The solid is known as an insulator, because a simple thermal excitation ($\approx kT = 0.025$ eV at 300 K) is not sufficient to move an electron from the valence band to the conduction band.
- If the band gap is of the order of a few eV, thermal excitation is then possible. The

conduction will be weak, but not negligible, and will increase with temperature. The solid is then called a semiconductor.

Note that the electrical conductivity of a metal decreases with an increase in temperature, whereas that of a semiconductor increases.

1.2.2 The Fermi-Dirac distribution

Because of Pauli's exclusion principle, the distribution of electrons between the different energy levels does not obey Boltzmann's statistical law but the Fermi-Dirac law. The probability $P(E)$ that an energy level between E and $E + \mathrm{d}E$ will be occupied is given by

$$P(E) = f(E)\mathrm{d}E = \frac{\mathrm{d}E}{1+\exp\left[(E-\tilde{\mu}^{\mathrm{M}}_{e^-/\mathrm{pze}})/kT\right]} \tag{1.72}$$

as shown in Figure 1.14. $\tilde{\mu}^{\mathrm{M}}_{e^-/\mathrm{pze}}$ is the chemical potential of the electron expressed on the scale where the lowest electronic state (point of zero energy) is taken as the origin (see Figure 1.15).

At absolute zero ($T = 0$ K), all the electronic states are filled in sequential order, and the energy which corresponds to the Highest Occupied Molecular Orbital (HOMO) at 0 K is called the ***Fermi energy*** E_f. Thus, the probability $P(E)$ of the filling of an electronic state has a value of 1 if $E < E_f$ and 0 if $E > E_f$. At this temperature, the chemical potential is then equal to the Fermi energy ($\tilde{\mu}^{\mathrm{M}}_{e^-/\mathrm{pze}}$). Given that, for a metal, the electrochemical potential of the electron is hardly affected by the temperature, it is often called by an abuse of language the ***Fermi level***, at all temperatures ($\tilde{\mu}^{\mathrm{M}}_{e^-/\mathrm{pze}}(T) \cong E_f$).

At higher temperatures, the probability curve shows an inflexion point for $E = \tilde{\mu}^{\mathrm{M}}_{e^-/\mathrm{pze}}(T)$, but, taking into account the low value of kT relative to $\tilde{\mu}^{\mathrm{M}}_{e^-/\mathrm{pze}}(T)$ at the ambient temperature, the Fermi-Dirac function reaches levels very different from

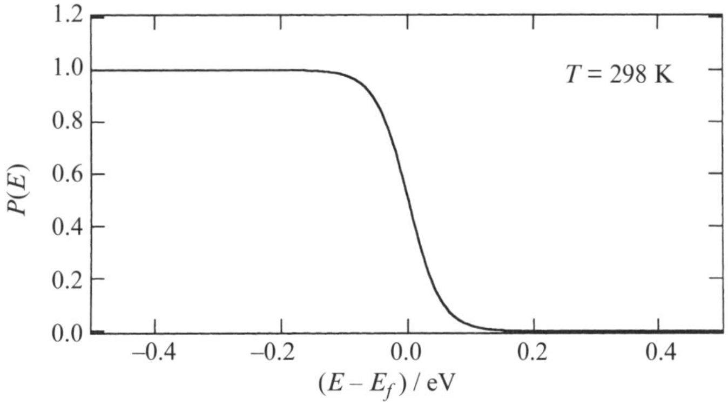

Fig. 1.14 The Fermi-Dirac distribution function.

0 and 1 only in a very restricted energy region around the Fermi level.

The Fermi level is thus defined as the energy level having an occupation probability of 1/2.

The Fermi-Dirac equation gives the probability that a given energy state will be occupied by an electron. To know the number of electrons per unit volume $n(E)$ possessing an energy E, this probability is multiplied by the density of states – occupied or not – having that energy.

1.2.3 Work function

The work needed to add one mole of electrons to a metal is by definition the electrochemical potential of the electron

$$\tilde{\mu}_{e^-}^{\mathrm{M}} = \mu_{e^-}^{\mathrm{M}} - F\phi^{\mathrm{M}} = (\mu_{e^-}^{\mathrm{M}} - F\chi^{\mathrm{M}}) - F\psi^{\mathrm{M}} = \alpha_{e^-}^{\mathrm{M}} - F\psi^{\mathrm{M}} \tag{1.73}$$

where the electrochemical potential, the surface potential, the outer and inner potentials are defined with respect to a vacuum. Note that the chemical potential of the electron does not involve the definition of a standard state and we do not need to define an activity. Effectively, the number of electrons in a metal is linked to its nature, and the chemical term depends only on the nature of the metal.

In solid state physics, the work necessary to extract an electron from an uncharged metal can, for example, be measured by photoelectric emission. This work of extraction of an electron called the ***work function*** is designated by the Greek capital letter Φ (not to be confused with the inner potential ϕ). By definition, we can see that the work function is the opposite of the real chemical potential defined by eqn.(1.71).

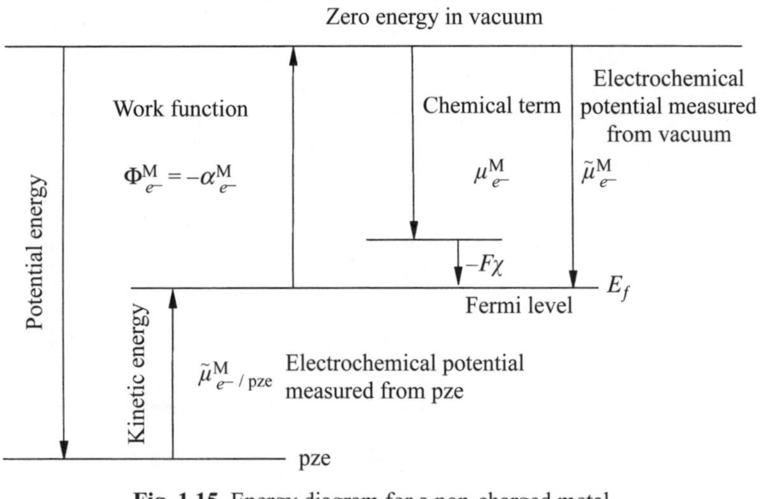

Fig. 1.15 Energy diagram for a non-charged metal.

$$\Phi^{\mathrm{M}}_{e^-} = -\alpha^{\mathrm{M}}_{e^-} \tag{1.74}$$

The work function is defined as the energy necessary to extract an electron from the Fermi level of an uncharged metal. As a first approximation, we can say that the Fermi level is invariable with temperature. It is important to note that the work function depends on the structure of the surface at the atomic level and must consequently be defined for each crystal face. The work function of metals varies from 2.3 eV for potassium to 5.3 eV for gold (see Annex B).

1.2.4 Contact between two metals

When two uncharged metals are put in contact with each other, the electrons flow from the metal M^1 with the smaller work function towards the metal M^2 with the larger work function, until the Fermi levels are equal in both metals, this being the thermodynamic equilibrium (see Figure 1.16). Thus, at equilibrium, M^1 becomes positively charged and M^2 becomes negatively charged.

The electrochemical equilibrium is described by the equality of the electrochemical potential of the electrons.

$$\tilde{\mu}^{\mathrm{M}^1}_{e^-} = \tilde{\mu}^{\mathrm{M}^2}_{e^-} \tag{1.75}$$

By applying the definition of the electrochemical potential (1.73), we can calculate the Volta potential difference.

$$\psi^{\mathrm{M}^2} - \psi^{\mathrm{M}^1} = -(\Phi^{\mathrm{M}^2} - \Phi^{\mathrm{M}^1})/F \tag{1.76}$$

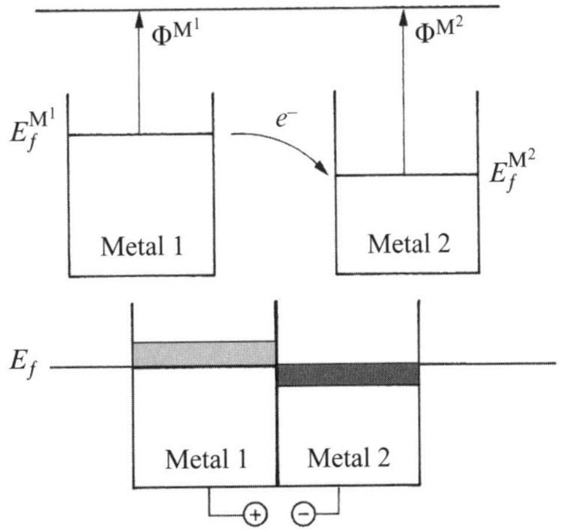

Fig. 1.16 Contact between two metals having different work functions.

EXAMPLE

Let us calculate the Volta potential difference resulting from putting in contact a piece of gold and a piece of silver, knowing that the work functions for these two metals are 5.32 and 4.30 eV respectively

$$\psi^{Ag} - \psi^{Au} = -(\Phi^{Ag} - \Phi^{Au})/F$$

It is interesting to dwell for a moment on the units. The eV is a unit of energy which corresponds to the displacement work of an elementary charge $1.6 \cdot 10^{-19}$ C under the action of a potential difference of 1 V. Thus 1 eV corresponds to $1.6 \cdot 10^{-19}$ J. In equation (1.76), we are dealing with one mole of electrons and the work function must be multiplied by Avogadro's constant. Faraday's constant being the charge corresponding to a mole of elementary charges, we simply have:

$$\psi^{Ag} - \psi^{Au} = -\frac{N_A e(\Phi^{Ag} - \Phi^{Au})}{N_A e} = 1.02 \text{ V}$$

1.2.5 The Kelvin probe

The Kelvin probe is used to measure differences in work function and indirectly differences in surface potential.

Let us consider two metal pieces α and β separated by an air gap and linked together by an electrical circuit comprising a voltage source (E) and a current measuring device (A), as shown in Figure 1.17. To operate the set-up, the voltage source E is adjusted in such a way that no current flows through the circuit, thus ensuring that the pieces of metal α and β are not charged, and thus ensuring the equality of the external potentials of the two phases α and β

$$\psi^{\alpha} = \psi^{\beta} \tag{1.77}$$

Two methods exist to experimentally create this condition. In the first approach, a γ-radiation source can be used to ionise the gas between the two metals α and β. If the two metals α and β have different outer potentials ψ, thus creating an electric field within the gap, the cations in the gas phase move towards the negatively charged metal, and conversely the anions move towards the positively charged metal. This movement of ions produces a current. The voltage source is then varied until the current becomes zero, where the equality (1.77) is verified.

In the second approach, given that AC currents are easier to measure than DC ones, the phase β is linked to a vibrator such that the distance d between the two pieces of metals α and β forming a capacitor varies at a frequency ω. If the two metals α and β have different outer potentials ψ, a capacitive current of frequency ω associated to the harmonic variation of the capacitor (see eqn. (1.48)) can be measured, for example using an oscilloscope. Similarly as above, the voltage source is adjusted so that the alternating current having the frequency of the vibrator becomes zero.

The voltage E provided by the source is a compensation potential that is by definition equal to the inner potential difference between the copper wires that

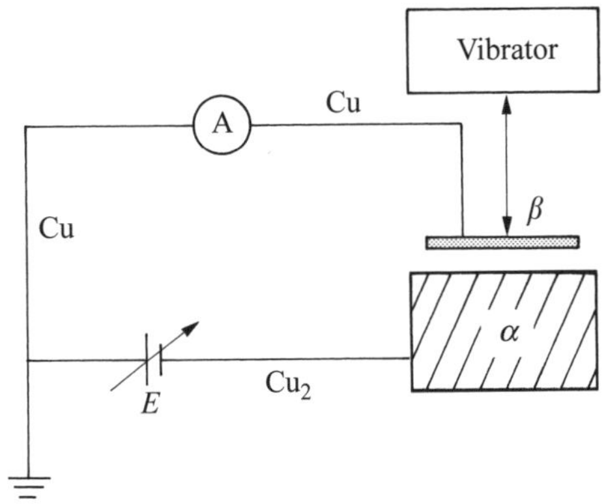

Fig. 1.17 Vibrating electrode for the measurement of Volta potential differences.

connect the source to the metal plates β and α, copper wires which we shall call Cu and Cu_2 respectively. Thus

$$E = \phi^{Cu_2} - \phi^{Cu} = \left(\tilde{\mu}_{e^-}^{Cu} - \tilde{\mu}_{e^-}^{Cu_2}\right) / F \tag{1.78}$$

since the chemical potential of the electron in the two copper wires is the same. For metals in contact, the electrochemical potential of electrons in each metal is equal (equation (1.75)), so that (1.78) becomes

$$E = \left(\tilde{\mu}_{e^-}^{\beta} - \tilde{\mu}_{e^-}^{\alpha}\right) / F \tag{1.79}$$

and using the condition (1.77) obtained experimentally, we conclude

$$E = \left(\Phi^{\alpha} - \Phi^{\beta}\right) / F \tag{1.80}$$

Then it follows from (1.76) that E is equal to the Volta potential difference between α and β as if they were in contact.

This apparatus can also be used to measure differences in surface potential. In the above example, the metal sample α has a compensation potential E_{I}, and let us suppose that we pass only above the metal α a gas whose molecules can adsorb onto the metal (e.g. adsorbtion of carbon monoxide on platinum). The surface potential of α is then modified and a new value of the compensation potential E_{II} is measured.

$$E_{\mathrm{II}} = \left(\Phi^{\alpha_{\mathrm{II}}} - \Phi^{\beta}\right) / F \tag{1.81}$$

If it can be assumed that the work function of β remains unchanged then

$$E_{\mathrm{II}} - E_{\mathrm{I}} = \left(\Phi^{\alpha_{\mathrm{II}}} - \Phi^{\alpha_{\mathrm{I}}}\right) / F = \chi^{\alpha_{\mathrm{II}}} - \chi^{\alpha_{\mathrm{I}}} \tag{1.82}$$

since the chemical potential of the electron is a constant property of the bulk metal α.

If the vibrating electrode is a very fine point, we can use a Kelvin probe to image the surfaces to show the arrangement of molecules.

1.2.6 Electrochemical potential of the electron in solution

Up until now we have looked at the electrochemical potential of the electron in a solid. Given its extreme reactivity, the electron can exist in aqueous solution as a dissolved species only for relatively short periods and consequently we shall not look at the electrochemical potential of the dissolved electron. Nevertheless, it can sometimes be useful to use the rather abstract notion of electrochemical potential or even the notion of the Fermi level for the electron in solution knowing that it resides on a reduced species.

The energy levels of the electron will then depend on the reduced and the oxidised species. The major problem of energy levels in solution is that contrary to the case of a metal, these fluctuate with the polarisation fluctuations of the solvent.

As a first approximation, we can make the hypothesis that the distribution of the energy levels is Gaussian and centred on the most probable value. We then have for the oxidised species

$$W(E) \quad = \quad \frac{1}{\sqrt{4\pi\lambda kT}} \exp^{-\frac{(E-E_{\mathrm{ox}})^2}{4\pi\lambda kT}} \tag{1.83}$$

and similarly for the reduced species

$$W(E) \quad = \quad \frac{1}{\sqrt{4\pi\lambda kT}} \exp^{-\frac{(E-E_{\mathrm{red}})^2}{4\pi\lambda kT}} \tag{1.84}$$

where λ represents the reorganisation energy of the solvent.

And so we call the half-sum of the most probable energies of the oxidised and reduced species the ***Fermi level in solution*** or the ***redox Fermi level***.

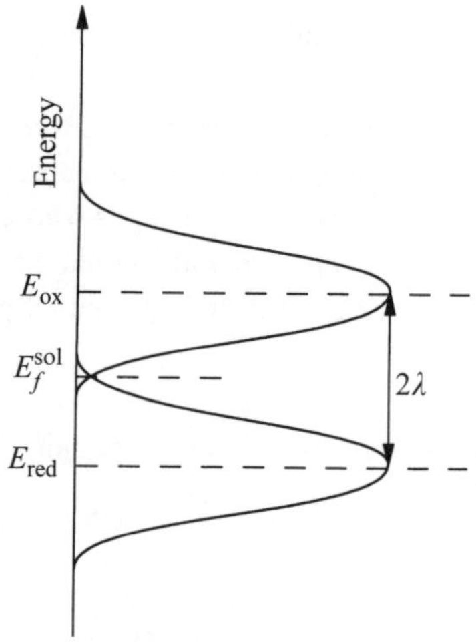

Fig. 1.18 Energy diagram for a redox couple in solution

We must be careful of the fact that these energy distributions around an average value are linked to temporal fluctuations and do not represent bands of levels like those in a solid. The most probable energy values are electronic states of the solvated species and the energy difference between the two levels is of the order of eV.

By definition the Fermi level of the electron in solution measured from the vacuum level corresponds to the electrochemical potential of the electron in solution

$$E_f^{\text{sol}} = \tilde{\mu}_{e^-}^{\text{S}} \tag{1.85}$$

CHAPTER 2

ELECTROCHEMICAL EQUILIBRIA

In the first chapter, we have established the basic thermodynamic principles to deal with electrochemical systems. Now, we are going to use the concept of electrochemical potential to study different types of electrochemical equilibria such as heterogeneous redox reactions, ion distribution, etc.

2.1 REDOX REACTIONS AT METALLIC ELECTRODES

2.1.1 Galvani potential difference between an electrode and a solution

In the same way that the Gibbs energy, ΔG, of a chemical equilibrium

$$a\mathrm{A} + b\mathrm{B} \rightleftarrows c\mathrm{C}$$

is a linear combination of chemical potentials as described by equation (1.29):

$$\Delta G = c\mu_{\mathrm{C}} - (b\mu_{\mathrm{B}} + a\mu_{\mathrm{A}}) = 0 \tag{2.1}$$

an electrochemical equilibrium of a redox reaction in solution at a metal electrode

$$\mathrm{Ox}^{\mathrm{S}} + ne^{-\mathrm{M}} \rightleftarrows \mathrm{Red}^{\mathrm{S}}$$

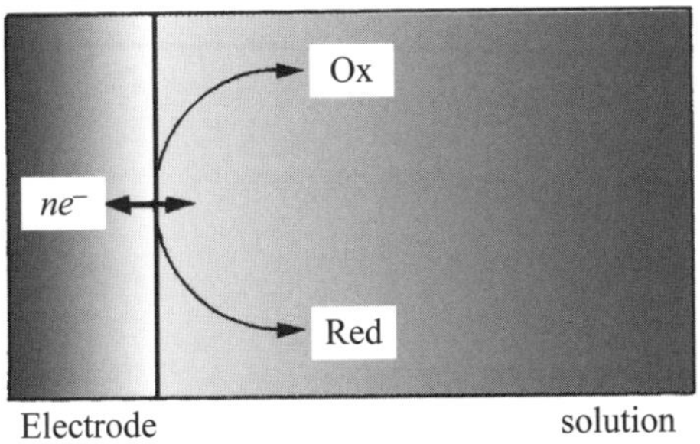

Fig. 2.1 Redox reaction at a metal electrode in a solution containing a redox couple.

e.g. the Fe^{II}/Fe^{III} redox couple at a gold electrode ($Fe^{III} + e^{-Au} \rightleftarrows Fe^{II}$), can be described as a linear combination of the electrochemical potentials of the reactant species, the electron being considered as a species in its own right.

$$\Delta\tilde{G} \;=\; \tilde{\mu}_{\rm red}^{\rm S} - (n\tilde{\mu}_{e^-}^{\rm M} + \tilde{\mu}_{\rm ox}^{\rm S}) \;=\; 0 \tag{2.2}$$

where $\Delta\tilde{G}$ represents ***the electrochemical Gibbs energy***, that is to say the sum of the work necessary to add the products of the reaction and to extract the reactants. The adjective 'electrochemical' qualifying the Gibbs energy is to remind us that we are dealing with charged species.

By using equation (1.60) that defines the electrochemical potential in a chemical term and an electrical term, we can develop equation (2.2) to obtain

$$\left[\mu_{\rm red}^{\ominus,\rm S} + RT\ln a_{\rm red}^{\rm S} + z_{\rm red}F\phi^{\rm S}\right] - n\left[\mu_{e^-}^{\rm M} - F\phi^{\rm M}\right] - \left[\mu_{\rm ox}^{\ominus,\rm S} + RT\ln a_{\rm ox}^{\rm S} + z_{\rm ox}F\phi^{\rm S}\right] \;=\; 0 \tag{2.3}$$

where $\mu_i^{\ominus,\rm S}$ represents the standard chemical potential of the species i in solution in the molarity scale (or molality) and $a_i^{\rm S}$ the activity of i in the molarity scale (or molality). Furthermore, noting that the number of electrons exchanged during the reaction is $n = z_{\rm ox} - z_{\rm red}$, we then have

$$\begin{aligned} nF(\phi^{\rm M} - \phi^{\rm S}) \;&=\; (\mu_{\rm ox}^{\ominus,\rm S} - \mu_{\rm red}^{\ominus,\rm S} + n\mu_{e^-}^{\rm M}) \;+\; RT\ln\left(\frac{a_{\rm ox}^{\rm S}}{a_{\rm red}^{\rm S}}\right) \\ &=\; -\Delta G^{\ominus} \;+\; RT\ln\left(\frac{a_{\rm ox}^{\rm S}}{a_{\rm red}^{\rm S}}\right) \;=\; -\Delta G \end{aligned} \tag{2.4}$$

where ΔG is the chemical contribution of the Gibbs energy of the reduction of O into R at an electrode, and $\Delta G^{\ominus}$ is the standard Gibbs energy. The difference $\phi^{\rm M} - \phi^{\rm S}$ is the difference in Galvani potential between the metal and the solution. The term $F(\phi^{\rm M} - \phi^{\rm S})$ is the electrical work required to transfer one mole of elementary charges from inside the metal to the bulk of the solution.

Equation (2.4) illustrates the major characteristic of redox reactions, i.e. the direct relation between the Galvani potential difference, between the electrode and the solution, and the concentrations of the species in solution. In the case of a chemical equilibrium in solution,

$$\rm A + B \rightleftarrows C$$

the equilibrium constant $K_{\rm e}$ is defined uniquely by

$$K_{\rm e} \;=\; \frac{a_{\rm C}}{a_{\rm A}a_{\rm B}} \;=\; \exp^{-\Delta G^{\ominus}/RT} \tag{2.5}$$

whilst for a redox equilibrium, we have

$$K_{\rm e} \;=\; \frac{a_{\rm red}^{\rm S}}{a_{\rm ox}^{\rm S}} \;=\; \exp^{-\Delta G^{\ominus}/RT}\exp^{-nF(\phi^{\rm M}-\phi^{\rm S})/RT} \tag{2.6}$$

and consequently the equilibrium constant for the redox equilibrium is not a proper constant, but depends on the difference in inner potential between the electrode and the solution. Thus for a redox reaction at an electrode, we can fix the concentrations of the oxidised and reduced species in solution and impose *de facto* the Galvani potential difference, or else, using a potentiostat (see §7.1), fix experimentally the Galvani potential difference, thus imposing the ratio of the activities of the species in solution. On the other hand, it is not possible to fix independently both the Galvani potential difference and the concentrations in solutions.

2.1.2 The Nernst equation

Empirical approach

By the end of the 19th century, thirty years before the concept of electrochemical potential had been introduced, it had been observed experimentally that the difference in potential between a ***working electrode,*** at which a redox reaction occurs, and a ***reference electrode*** (see §2.5.1) obeys a law of the type

$$E = V_{\text{working}} - V_{\text{reference}} = E^{\ominus} + \frac{RT}{nF}\ln\left(\frac{a_{\text{ox}}^{\text{S}}}{a_{\text{red}}^{\text{S}}}\right) \tag{2.7}$$

In fact, to measure a potential difference between the working and the reference electrodes, we use a high impedance voltmeter that ensures that practically no current passes through the circuit, and therefore no net electrode reactions can take place to disturb the equilibrium. If the reference electrode is a standard hydrogen electrode (SHE, *vide infra*) as illustrated in Figure 2.2, this relation is more commonly known in the form

$$E_{\text{SHE}} = \left[E_{\text{ox/red}}^{\ominus}\right]_{\text{SHE}} + \frac{RT}{nF}\ln\left(\frac{a_{\text{ox}}^{\text{S}}}{a_{\text{red}}^{\text{S}}}\right) \tag{2.8}$$

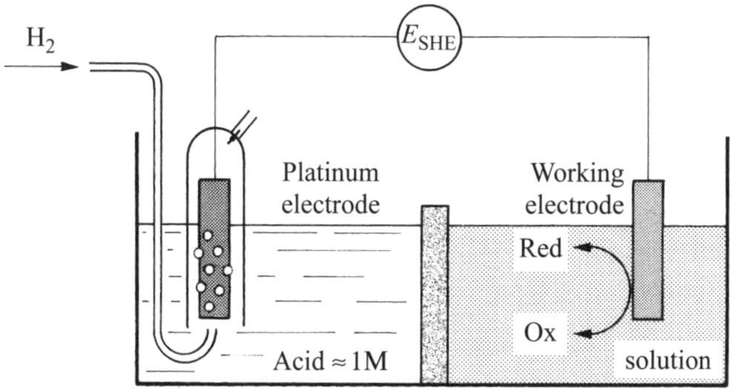

Fig. 2.2 Electrochemical cell for measuring the electrode potential on the SHE scale. A salt bridge is an ionic conductor introduced to physically separate the two solutions, but keeping at the same time their inner potentials equal or almost equal.

In this case, the ***electrode potential E*** is the potential difference between the terminals of the two electrodes, and not the difference in Galvani potential between the working electrode phase and the solution phase containing the dissolved ox and red species. So the zero of this scale of electrode potential is a convention linked to the choice of the reference electrode. History has made the reference electrode of choice the ***standard hydrogen electrode*** (SHE) defined such that the proton activity in solution should be unity and the fugacity of the hydrogen should be the standard pressure of 1 bar (=100 kPa). In a practical and not terribly rigorous manner, we can make a hydrogen electrode by bubbling hydrogen through an acid solution of pH zero. (Working at an atmospheric pressure of 101 325 Pa rather than at standard pressure introduces an error of 0.17 mV).

A posteriori derivation of the Nernst equation

Nowadays, we can derive the Nernst equation from the inner potentials of the phases used. In order to do so, we have to define a complete system, including the copper wires going to the terminals of the voltmeter.

Let us consider the following electrochemical cell

$$\mathrm{Cu^{I}} \mid \mathrm{Pt} \mid \mathrm{H^{+}}, \tfrac{1}{2}\mathrm{H_2} .... \| .... \mathrm{ox}, \mathrm{red} \mid \mathrm{M} \mid \mathrm{Cu^{II}}$$

Such an electrochemical cell is called a ***galvanic cell*** or ***battery***. By convention, the p.d. of a galvanic cell is defined as the potential of the right terminal less that of the left one. Thus, if we connect a voltmeter to the terminals of the cell above, the p.d. is by definition the difference in Galvani potentials between the terminals, here represented by the copper wires

$$E \quad = \quad V_{\mathrm{right}} - V_{\mathrm{left}} \quad = \quad \phi^{\mathrm{Cu^{II}}} - \phi^{\mathrm{Cu^{I}}} \tag{2.9}$$

It is important to note that the two terminals of a galvanic cell must always be made of identical materials. The symbol | stands for an electrochemical interface and the symbol || stands for a ***liquid junction*** across which the potential difference is negligible, which means that the inner potentials of the solutions on either side of this junction are equal (see §4.5.2 for more detail). In fact, the official IUPAC (International Union of Pure & Applied Chemistry) symbol is a vertical double bar of dotted lines.

We have thus four electrochemical equilibria:

- at the contact $\mathrm{M} \mid \mathrm{Cu^{II}}$ between the copper wire and the working electrode, there is an electronic equilibrium which implies that the chemical potentials of the electron in the two metal phases are equal

$$\tilde{\mu}_{e^-}^{\mathrm{Cu^{II}}} \quad = \quad \tilde{\mu}_{e^-}^{\mathrm{M}} \tag{2.10}$$

- at the working electrode, there is the equilibrium of the redox reaction

$$\tilde{\mu}_{\mathrm{ox}}^{\mathrm{S}} \quad + \quad n\,\tilde{\mu}_{e^-}^{\mathrm{M}} \quad = \quad \tilde{\mu}_{\mathrm{red}}^{\mathrm{S}} \tag{2.11}$$

• at the platinum electrode, there is the reduction of the proton and the oxidation of hydrogen,

$$\tilde{\mu}_{H^+}^{S} + \tilde{\mu}_{e^-}^{Pt} = \tfrac{1}{2}\mu_{H_2} \tag{2.12}$$

• the contact between the platinum electrode and the copper wire gives us

$$\tilde{\mu}_{e^-}^{Pt} = \tilde{\mu}_{e^-}^{Cu^I} \tag{2.13}$$

We can develop each one of these equations using equation (1.60), giving us

$$F(\phi^{Cu^{II}} - \phi^{M}) = \mu_{e^-}^{Cu^{II}} - \mu_{e^-}^{M} \tag{2.14}$$

$$nF(\phi^{M} - \phi^{S}) = (\mu_{ox}^{\ominus,S} - \mu_{red}^{\ominus,S} + n\,\mu_{e^-}^{M}) + RT\ln\left(\frac{a_{ox}^{S}}{a_{red}^{S}}\right) \tag{2.15}$$

$$F(\phi^{Pt} - \phi^{S}) = \left(\mu_{H^+}^{\ominus,S} - \tfrac{1}{2}\mu_{H_2}^{\ominus} + \mu_{e^-}^{Pt}\right) + RT\ln\left[a_{H^+}^{S}\left(\frac{p^{\ominus}}{f_{H_2}}\right)^{1/2}\right] \tag{2.16}$$

$$F(\phi^{Pt} - \phi^{Cu^I}) = \mu_{e^-}^{Pt} - \mu_{e^-}^{Cu^I} \tag{2.17}$$

If we make the hypothesis that the inner potential of the solution is the same on both sides of the liquid junction, the potential difference between the copper wires can be obtained by combining equations (2.14) to (2.17)

$$nF\left(\phi^{Cu^{II}} - \phi^{Cu^{I}}\right) = (\mu_{ox}^{\ominus,S} - \mu_{red}^{\ominus,S} - n\,\mu_{H^+}^{\ominus,S} + \frac{n}{2}\mu_{H_2}^{\ominus}) + RT\ln\left[\frac{a_{ox}^{S}}{a_{red}^{S}a_{H^+}^{S}}\left(\frac{f_{H_2}}{p^{\ominus}}\right)^{1/2}\right] \tag{2.18}$$

In the case of a SHE reference electrode (the activity of the proton in solution being unity and the hydrogen fugacity being the standard pressure of 1 bar), we then arrive at the Nernst equation (2.8).

$$\begin{aligned} E_{SHE} &= nF\left(\phi^{Cu^{II}} - \phi^{Cu^{I}}\right) = (\mu_{ox}^{\ominus,S} - \mu_{red}^{\ominus,S} - n\,\mu_{H^+}^{\ominus,S} + \frac{n}{2}\mu_{H_2}^{\ominus}) + RT\ln\left(\frac{a_{ox}^{S}}{a_{red}^{S}}\right) \\ &= \left[E_{ox/red}^{\ominus}\right]_{SHE} + \frac{RT}{nF}\ln\left(\frac{a_{ox}^{S}}{a_{red}^{S}}\right) \end{aligned} \tag{2.19}$$

Thermodynamic approach

We can also rationalise the Nernst equation (2.8) using a thermodynamic approach, by considering the redox equilibria taking place at the two electrodes of the cell shown in Figure 2.2:

(I) $\text{ox} + n\,e^- \rightleftarrows \text{red}$ at the working electrode

(II) $\frac{n}{2}H_2 \rightleftarrows n\,H^+ + n\,e^-$ at the platinum reference electrode

Even though we have a redox equilibrium at each electrode, the sum of these two electrochemical reactions gives a reaction which is not in equilibrium.

$$\text{(III)}\quad \text{ox} + \tfrac{n}{2}\,\text{H}_2 \rightleftarrows n\,\text{H}^+ + \text{red}$$

In effect, equilibrium could only be obtained if the electrons were able to circulate freely in the external circuit, but this reactive path is blocked by the high impedance of the voltmeter.

The Gibbs energy of this virtual reaction in solution is then

$$\Delta G_\text{r} = \Delta G_\text{r}^\ominus + RT\ln\left[\frac{a_\text{red}^\text{S}\left(a_{\text{H}^+}^\text{S}\right)^n\left(p^\ominus\right)^{n/2}}{a_\text{ox}^\text{S}\, f_{\text{H}_2}^{n/2}}\right] \tag{2.20}$$

Comparison of equation (2.20) with equations (2.18) and (2.19) allows us to link the electrode potential E_SHE (which is an experimental value) to the Gibbs energy of the virtual reaction (III)

$$\begin{aligned} E_\text{SHE} &= -\frac{\Delta G_\text{r}}{nF} = -\frac{\Delta G_\text{r}^\ominus}{nF} + \frac{RT}{nF}\ln\left[\frac{a_\text{ox}^\text{S}\, f_{\text{H}_2}^{n/2}}{a_\text{red}^\text{S}\left(a_{\text{H}^+}^\text{S}\right)^n\left(p^\ominus\right)^{n/2}}\right] \\ &= \left[E_\text{ox/red}^\ominus\right]_\text{SHE} + \frac{RT}{nF}\ln\left(\frac{a_\text{ox}^\text{S}}{a_\text{red}^\text{S}}\right) \end{aligned} \tag{2.21}$$

The ***standard redox potential*** of reduction $[E^\ominus{}_\text{ox/red}]_\text{SHE}$ for the redox couple ox/red can then be defined from the standard Gibbs energy of the virtual equilibrium

$$\left[E_\text{ox/red}^\ominus\right]_\text{SHE} = -\Delta G^\ominus / nF = \left[\mu_\text{ox}^{\ominus,\text{S}} - \mu_\text{red}^{\ominus,\text{S}} - n\,\mu_{\text{H}^+}^{\ominus,\text{S}} + \frac{n}{2}\mu_{\text{H}_2}^\ominus\right] / nF \tag{2.22}$$

$[E^\ominus{}_\text{ox/red}]_\text{SHE}$ is therefore a measure of the work necessary to transfer an electron between a metal electrode and a redox couple in solution, on a scale relative to the proton/hydrogen couple. Expression (2.21) has the advantage of highlighting the Gibbs energy of reaction. If the virtual reaction (III) is exergonic ($\Delta G_\text{r} < 0$), this means that the hydrogen can reduce the oxidised species. The reaction can then produce work, which in the case of the apparatus in Figure 2.2 means that, if the voltmeter is replaced by a resistor, the work of the reaction can then be recuperated in the electric circuit linking the two electrodes. This electrical work corresponds to the passage of the electrons from the working electrode to the platinum electrode and is equal to $-nFE_\text{SHE}$.

2.1.3 Standard redox potential

The standard redox potential scale shows us whether a reaction can take place. Consider, for example, the dissolution of zinc in an acid:

$$\text{Zn} + 2\text{H}^+ \longrightarrow \text{Zn}^{2+} + \text{H}_2$$

Since the standard redox potential for the reduction of Zn^{2+} is negative, the oxidation reaction of zinc dissolution takes place spontaneously. Nevertheless, the two redox half-reactions

$$Zn \rightleftarrows Zn^{2+} + 2e^-$$

$$2H^+ + 2e^- \rightleftarrows H_2$$

can take place in two separate places if an electrical circuit is used.

Let's take two beakers, one containing sodium nitrate, with a zinc electrode in it, and the other containing nitric acid with a platinum electrode in it. The two beakers are linked by a salt bridge containing sodium nitrate, and the two electrodes are connected together by a conducting wire. The Gibbs energy of reaction for this experimental setup is identical to that of the reaction dissolving zinc directly in acid. Note that in the system illustrated in Figure 2.3, the zinc electrode will dissolve, even though it is not directly in the acid. Thus, we have made an electrochemical battery capable of supplying energy. This setup differs from the one used to demonstrate the Nernst equation in that there is free passage of current in the circuit.

By definition, the standard redox potential for the reduction of a proton in water is zero, at all temperatures.

$$\left[E^{\ominus}_{H^+/\frac{1}{2}H_2}\right]_{SHE} = 0\,V$$

The more positive the standard redox potential of a redox couple, the more difficult the oxidation is; the reduced species is then the stable species. For example, if we consider the Au^+/Au pair, metallic gold is stable in the environment.

$$\left[E^{\ominus}_{Au^+/Au}\right]_{SHE} = 1.83\,V$$

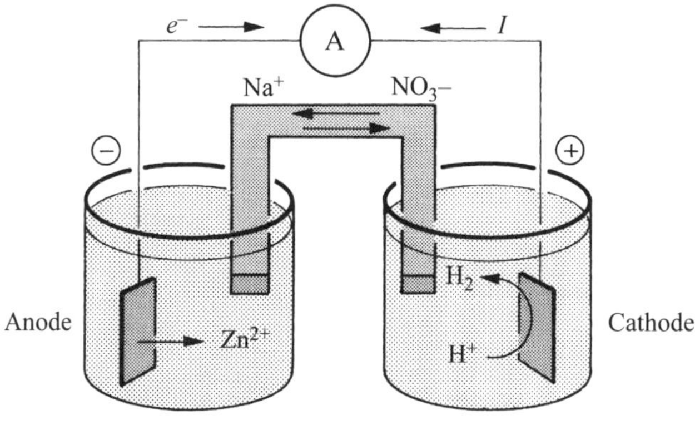

Fig. 2.3 Electrochemical dissolution of zinc. The beaker on the left and the salt bridge are filled with a solution of sodium nitrate, and the beaker on the right is filled with nitric acid.

Inversely, the more negative the standard redox potential of a redox couple, the more difficult the reduction is; the oxidised species is then generally stable in the natural state, such as the Li^+ cation in the Li^+/Li pair whose standard redox potential is

$$\left[E^{\ominus}_{Li^+/Li}\right]_{SHE} = -3.045\ V$$

The oxidised species of a redox pair with a high standard redox potential can be used as an oxidant. For example, permanganate the standard potential of which is

$$\left[E^{\ominus}_{MnO_4^-,Mn^{2+}}\right]_{SHE} = 1.51\ V$$

is used in organic chemistry or analytical chemistry as an oxidising species. Inversely, the reduced species of a redox pair whose standard redox potential is negative can be used as a reducing agent. To compare accurately the oxidising/reducing abilities of a redox couple, the corresponding Gibbs energies should be calculated as illustrated in the example below.

Standard redox potentials play a major role in bioenergetics, and an example of the different redox pairs involved in the respiratory chain is given in the appendix (see page 82).

EXAMPLE

Starting from standard redox potentials, let's explain why $FeCl_3$ is used for dissolving copper in the manufacturing process of printed circuits, knowing that

$$\left[E^{\ominus}_{Fe^{3+}/Fe^{2+}}\right]_{SHE} = 0.77\ V \qquad \left[E^{\ominus}_{Cu^{2+}/Cu}\right]_{SHE} = 0.34\ V$$

In the same way, let's explain why $FeCl_3$ does not attack the gold contacts, knowing that

$$\left[E^{\ominus}_{Au^+/Au}\right]_{SHE} = 1.83\ V$$

Look at the equilibria

$$\begin{array}{llll} 2\ x & Fe^{3+} + e^- & \rightleftarrows\ Fe^{2+} & [I] \\ & Cu & \rightleftarrows\ Cu^{2+} + 2\,e^- & [II] \\ \hline & \multicolumn{2}{l}{2\,Fe^{3+} + Cu \rightleftarrows 2\,Fe^{2+} + Cu^{2+}} & \end{array}$$

Using equation (2.22), the standard Gibbs energy of this resulting equilibrium is then written as

$$\begin{aligned} \Delta G^{\ominus} &= 2\Delta\tilde{G}^{\ominus}_{I} + \Delta\tilde{G}^{\ominus}_{II} = 2\left[\mu^{\ominus}_{Fe^{2+}} - \mu^{\ominus}_{Fe^{3+}}\right] + \left[\mu^{\ominus}_{Cu^{2+}} - \mu^{\ominus}_{Cu}\right] \\ &= -2F\left[E^{\ominus}_{Fe^{3+}/Fe^{2+}}\right]_{SHE} + 2F\left[E^{\ominus}_{Cu^{2+}/Cu}\right]_{SHE} = -83\ kJ\cdot mol^{-1} \end{aligned}$$

The negative value obtained shows that the reaction is ***exergonic***.

In the case of gold, we have

$$\begin{array}{lcll} Fe^{3+} + e^- & \rightleftarrows & Fe^{2+} & [I] \\ Au & \rightleftarrows & Au^+ + e^- & [III] \\ \hline Fe^{3+} + Au & \rightleftarrows & Fe^{2+} + Au^+ & \end{array}$$

The standard Gibbs energy of this resulting equilibrium is written in the same manner

$$\Delta G^\ominus = \Delta\tilde{G}_{I}^\ominus + \Delta\tilde{G}_{III}^\ominus = -F\left[E^\ominus_{Fe^{3+}/Fe^{2+}}\right]_{SHE} + F\left[E^\ominus_{Au^+/Au}\right]_{SHE} = 102\ \mathrm{kJ\cdot mol^{-1}} > 0$$

Conversely, this positive value shows that the reaction is ***endergonic*** and therefore not favourable from a thermodynamic point of view.

Standard cell potential

The standard redox potential $E^\ominus$ is related to a Gibbs energy that is a function of state, and has consequently additive properties. Thus, from tabulated values of standard redox potentials, we can calculate the standard potential of galvanic cells involving different redox pairs. For example, the standard potential $E^\ominus_{cell}$ of the following cell

$$Cu^{I} \mid Ag \mid Ag^+ ....Zn^{2+} \mid Zn \mid Cu^{II}$$

can be calculated, knowing the standard redox potentials for the Zn^{2+}/Zn and Ag^+/Ag couples

$$\left[E^\ominus_{Zn^{2+}/Zn}\right]_{SHE} = -0.763\ \mathrm{V} \quad \text{and} \quad \left[E^\ominus_{Ag^+/Ag}\right]_{SHE} = 0.799\ \mathrm{V}$$

In fact, the potential of the above cell can be considered as a series of two cells similar to the one illustrated in Figure 2.2

$$Cu^{I} \mid Ag \mid Ag^+..\|..H^+,\tfrac{1}{2}H_2 \mid Pt \mid Cu \mid Pt \mid H^+,\tfrac{1}{2}H_2..\|..Zn^{2+} \mid Zn \mid Cu^{II}$$

The potential measured is then the difference of the respective Nernst equations

$$E = \left(\left[E^\ominus_{Zn^{2+}/Zn}\right]_{SHE} + \frac{RT}{2F}\ln\left[\frac{a_{Zn^{2+}}}{a_{Zn}}\right]\right) - \left(\left[E^\ominus_{Ag^+/Ag}\right]_{SHE} + \frac{RT}{F}\ln\left[\frac{a_{Ag^+}}{a_{Ag}}\right]\right) \qquad (2.23)$$

If the metal electrodes are made of pure zinc and pure silver, the activities of these pure metals are unity. The classic error is to take the activity of a solid electrode as unity; this is incorrect if the electrode is an alloy. Equation (2.23) then reduces to:

$$E = \left[E^\ominus_{Zn^{2+}/Zn}\right]_{SHE} - \left[E^\ominus_{Ag^+/Ag}\right]_{SHE} + \frac{RT}{F}\ln\left[\frac{\sqrt{a_{Zn^{2+}}}}{a_{Ag^+}}\right] \qquad (2.24)$$

The standard potential of the cell is simply

$$E^{\ominus}_{\text{cell}} = \left[E^{\ominus}_{\text{Zn}^{2+}/\text{Zn}}\right]_{\text{SHE}} - \left[E^{\ominus}_{\text{Ag}^{+}/\text{Ag}}\right]_{\text{SHE}} = -1.562\text{V}$$

Formal redox potential

The thermodynamic approach of electrochemistry in solutions relies on the thermodynamic notion of activities, rather than on experimental quantities such as concentrations that can be directly measured. To alleviate this problem, it is customary to define the ***formal redox potential*** or ***apparent standard redox potential*** by

$$\left[E^{\ominus\prime}_{\text{ox/red}}\right]_{\text{SHE}} = \left[E^{\ominus}_{\text{ox/red}}\right]_{\text{SHE}} + \frac{RT}{nF}\ln\left(\frac{\gamma_{\text{ox}}}{\gamma_{\text{red}}}\right) \tag{2.25}$$

This notation has the advantage of being able to show the Nernst equation in a form that is more directly related to experimental conditions

$$E_{\text{SHE}} = \left[E^{\ominus\prime}_{\text{ox/red}}\right]_{\text{SHE}} + \frac{RT}{nF}\ln\left(\frac{c_{\text{ox}}}{c_{\text{red}}}\right) \tag{2.26}$$

It is important to note that the formal redox potential depends on experimental conditions such as the ionic strength of the solution. Of course, in dilute solutions the activity coefficients tend to unity and activities can be replaced by concentrations. In this case, the formal redox potential tends to the standard redox potential value.

2.1.4 Measuring standard redox potentials

Measuring directly with a standard hydrogen electrode

In certain cases, it is possible to measure directly the standard redox potential by assembling galvanic cells. The cell

$$\text{Cu}^{\text{I}} \mid \text{Pt} \mid \tfrac{1}{2}\text{H}_2,\ \text{HCl} \mid \text{AgCl} \mid \text{Ag} \mid \text{Cu}^{\text{II}}$$

is a classic example of this. The potential at the terminals of this cell is simply given by the difference of the respective Nernst equations:

$$E = \left(\left[E^{\ominus}_{\text{Ag}^{+}/\text{Ag}}\right]_{\text{SHE}} + \frac{RT}{F}\ln\left[\frac{a_{\text{Ag}^{+}}}{a_{\text{Ag}}}\right]\right) - \left(\left[E^{\ominus}_{\text{H}^{+}/\frac{1}{2}\text{H}_2}\right]_{\text{SHE}} + \frac{RT}{F}\ln\left[a_{\text{H}^{+}}\left(\frac{f_{\text{H}_2}}{p^{\ominus}}\right)^{-1/2}\right]\right) \tag{2.27}$$

If the silver electrode is made of pure silver, the silver activity is unity. Also, by convention, the standard redox potential of the $H^{+}/\frac{1}{2}\,H_2$ pair is zero and equation (2.27) reduces to :

$$E = \left[E^{\ominus}_{\text{Ag}^{+}/\text{Ag}}\right]_{\text{SHE}} + \frac{RT}{F}\ln\left[\frac{a_{\text{Ag}^{+}}}{a_{\text{H}^{+}}}\left(\frac{f_{\text{H}_2}}{p^{\ominus}}\right)^{1/2}\right] \tag{2.28}$$

In as much as the silver chloride is not very soluble, we can introduce the solubility product K_S of the silver chloride (K_S=1.77·10^{-10}) in equation (2.28) in such a way that the activity of chloride ions is taken into account. The solubility product is defined as the equilibrium constant of the dissolution equilibrium

$$\mathrm{AgCl} \rightleftarrows \mathrm{Ag^+} + \mathrm{Cl^-}$$

Then, as long as AgCl is a pure substance, we have

$$K_S = \frac{a_{Ag^+}a_{Cl^-}}{a_{AgCl}} = a_{Ag^+}a_{Cl^-} \cong \frac{c_{Ag^+}c_{Cl^-}}{c^{\ominus 2}} \qquad (2.29)$$

By substituting in equation (2.28), we obtain

$$\begin{aligned} E &= \left[E^{\ominus}_{Ag^+/Ag}\right]_{SHE} + \frac{RT}{F}\ln\left[\frac{K_S}{a_{H^+}a_{Cl^-}}\left(\frac{f_{H_2}}{p^{\ominus}}\right)^{1/2}\right] \\ &= \left\{\left[E^{\ominus}_{Ag^+/Ag}\right]_{SHE} + \frac{RT}{F}\ln K_S\right\} + \frac{RT}{F}\ln\left(\frac{f_{H_2}}{p^{\ominus}}\right)^{1/2} - \frac{RT}{F}\ln\left(a_{H^+}a_{Cl^-}\right) \end{aligned} \qquad (2.30)$$

By re-grouping the constant terms in equation (2.30), we can define the standard redox potential for the silver|silver chloride couple

$$\left[E^{\ominus}_{AgCl/Ag}\right]_{SHE} = \left[E^{\ominus}_{Ag^+/Ag}\right]_{SHE} + \frac{RT}{F}\ln K_S = 0.799 - 0.577 = 0.222\ \mathrm{V} \qquad (2.31)$$

If the fugacity of hydrogen is kept constant (for example at the standard pressure value so that the second term of equation (2.30) is equal to zero), we see that the cell voltage given by equation (2.30) is a direct function of the concentration of HCl. By extrapolation to dilute concentrations for which the activity coefficients tend to unity, we can obtain experimentally the standard redox potential value from

$$\left[E^{\ominus}_{AgCl/Ag}\right]_{SHE} = \lim_{c_{HCl}\to 0}\left(E + \frac{RT}{F}\ln\left(c_{H^+}c_{Cl^-}\right)\right) \qquad (2.32)$$

The example below illustrates this approach of measuring a standard redox potential.

EXAMPLE

G.J. Hills and D.J.G Ives (*J. Chem. Soc.*, 311(1951)) made precise measurements of the potentials of cells comprising both a hydrogen electrode and a calomel electrode.

$$\mathrm{Cu^I} \mid \mathrm{Pt} \mid \tfrac{1}{2}\mathrm{H_2}(f=p^{\ominus}),\ \ \mathrm{HCl} \mid \tfrac{1}{2}\mathrm{Hg_2Cl_2} \mid \mathrm{Hg} \mid \mathrm{Cu^{II}}$$

The values obtained are:

m/mmol·kg^{-1}	1.6077	3.0769	5.0403	7.6938	10.9474
E / V	0.60080	0.56825	0.54366	0.52267	0.50532
m/mmol·kg^{-1}	13.968	18.872	25.067	37.690	51.645
E / V	0.49339	0.47870	0.46490	0.44516	0.42994
m/mmol·kg^{-1}	64.718	75.081	94.276	119.304	
E / V	0.41906	0.41187	0.40088	0.38948	

Let's calculate, from these data, the standard redox potential of the calomel (Hg_2Cl_2) electrode described in more detail in §2.51. As for the Ag|AgCl electrode described above (see equation (2.30)), the potential of the cell is written as

$$E = \left[E^{\ominus}_{Hg^+/Hg}\right]_{SHE} + \frac{RT}{F}\ln\left[\frac{a_{Hg^+}}{a_{H^+}}\left(\frac{f_{H_2}}{p^{\ominus}}\right)^{1/2}\right]$$

The fugacity of the hydrogen being the standard pressure of 1 bar, we have, expressing the activities on the molality scale:

$$E = \left[E^{\ominus}_{\frac{1}{2}Hg_2Cl_2/Hg}\right]_{SHE} - \frac{RT}{F}\ln\left[a_{H^+}a_{Cl^-}\right]$$

$$= \left[E^{\ominus}_{\frac{1}{2}Hg_2Cl_2/Hg}\right]_{SHE} - \frac{RT}{F}\ln\left[\gamma^m_{H^+}\gamma^m_{Cl^-}\right] - \frac{RT}{F}\ln\left[m_{H^+}m_{Cl^-}\right]$$

Thus in order to obtain the standard redox potential for the calomel/mercury couple, we can plot $E+2RT/\mathrm{F}\ln[m_{HCl}]$ as a function of the molality of hydrochloric acid, and the standard value will be found by extrapolation to zero concentration.

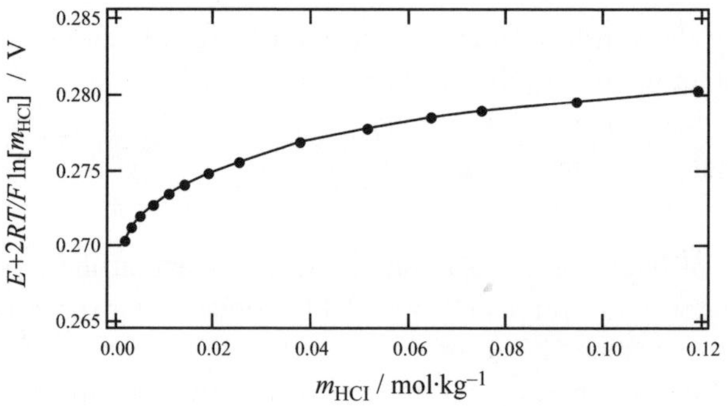

Extrapolating from such a curve is difficult, but we can say that the standard redox potential is found between 265 and 270 mV. We shall see in chapter 3 how it is possible to refine this estimate using the Debye-Hückel theory.

Measuring directly with a reference electrode

In fact, the hydrogen electrode is not the most practical reference electrode to use. From an experimental point of view, the current preference is a silver|silver chloride electrode which is simple to construct and does not involve toxic components such as mercury as in the calomel electrode. The use of reference electrodes often involves liquid junctions across which a potential difference E_{LJ} is established. (The method for calculating this p.d. will be covered in detail in Chapter 4). In a first approximation, this term is often considered as negligible.

The galvanic cell of a redox system compared to a silver/silver chloride reference electrode, as illustrated in Figure 2.4, is written as

$$Cu^{I} \mid Ag \mid AgCl \mid KCl_{sat} .... \| .... ox, red \mid M \mid Cu^{II}$$

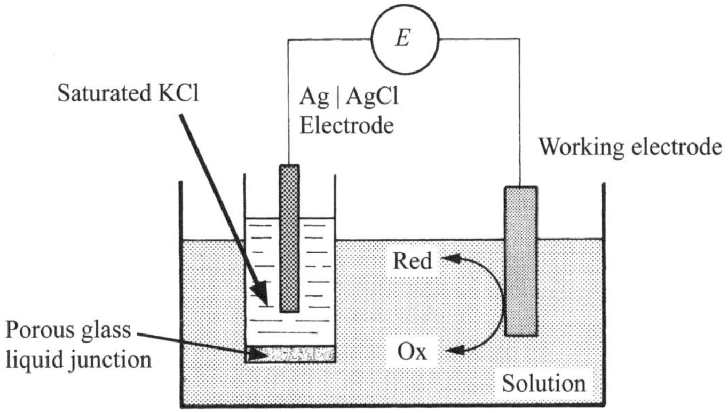

Fig. 2.4 Electrochemical cell for measuring the standard redox reduction potential with an Ag | AgCl reference electrode. The salt bridge is made of porous glass.

Using equation (2.31), the equilibrium potential for such a cell is

$$E = \left(\left[E^{\ominus}_{ox/red}\right]_{SHE} + \frac{RT}{nF}\ln\left[\frac{a_{ox}}{a_{red}}\right]\right) - \left(\left[E^{\ominus}_{AgCl/Ag}\right]_{SHE} - \frac{RT}{F}\ln a_{Cl^-}\right) \tag{2.33}$$

When the silver electrode, covered with a fine layer of silver chloride is in contact with a saturated KCl solution, we can regroup the constant terms, including those relative to the activity of the chloride ions, and write

$$E = \left[E^{\ominus}_{ox/red}\right]_{Ag|AgCl|KCl_{sat}} + \frac{RT}{nF}\ln\left[\frac{a_{ox}}{a_{red}}\right] \tag{2.34}$$

$\left[E^{\ominus}_{ox/red}\right]_{Ag|AgCl|KCl_{sat}}$ represents the standard redox potential of reduction for the ox/red couple on the scale of the 'silver | silver chloride | saturated KCl' reference electrode.

As before, to measure the standard redox potential on the silver chloride/silver scale, we measure the formal redox potential at different concentrations and then extrapolate to infinite dilution.

Spectroelectrochemical measurements

Another method for measuring a formal redox potential value consists of coupling electrochemistry with absorption spectroscopy. Instead of fixing the concentrations of the reduced and oxidised species and measuring the cell potential at equilibrium, with the aid of an instrument called a potentiostat (see §7.1), we fix the potential of the cell and measure by UV-VIS or infra-red absorption, the concentrations of the species in solution. The acronym for this technique is ***OTTLE*** standing for Optically Transparent Thin Layer Electrode.

Let's look at a cell of the type illustrated in Figure 2.5. The volume of solution in the UV-VIS cuvette is very small so that a significant change in the ratio of concentration c_{ox} / c_{red} is caused by a very small amount of electric charge passed

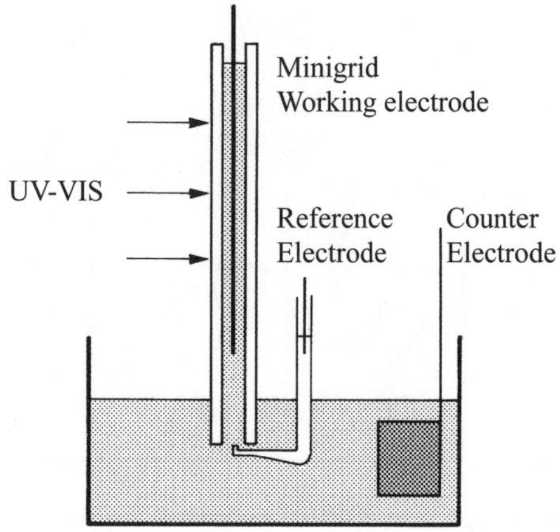

Fig. 2.5 Diagram of an OTTLE electrode. The working electrode is a fine metallic grid placed in a UV-VIS cell with a short optical path.

at the semi-transparent electrode. The latter can either be made of a metal deposited as a thin layer by evaporation, or a very fine minigrid made of a wire that is a few micrometres in diameter.

By applying an electrode potential different from the equilibrium potential, a new ratio c_{ox} / c_{red} is quickly established at the working electrode, then in the whole of the thin layer of solution crossed by the light. Thus, at each potential, we can record the UV-VIS absorption spectrum, e.g. the one shown in Figure 2.6, for the reaction

$$OsCl_6^- + e^- \rightleftarrows OsCl_6^{2-}$$

and therefore calculate the ratio c_{ox} / c_{red}. In this way, we can determine the standard redox potential of new molecules such as organo-metallic or organic compounds. From a practical point of view, it is interesting to vary the value of the potential applied, first in one direction, then in the other. If the results of the return sweep do not correspond with those of the forward sweep, then the reaction is not chemically reversible. Also, the presence of isosbestic points (absorbance points which do not depend on the ratio c_{ox} / c_{red}) guarantee that we are talking about an elementary electron transfer reaction concerning only one oxidised species and one reduced species.

If the redox pair cannot react on the electrode itself, as in the case of proteins where the redox centre is wrapped up inside the protein, we can use mediators whose role is to provide a shuttle for the electrons.

EXAMPLE

From the absorbance values in Figure 2.6 measured at 410 nm, let's calculate the standard redox reduction potential for this reaction. To do this, we shall use the Beer-Lambert law which expresses the ratio between the concentration of the absorbing species and the absorbance of the solution. Also we can consider that at 1.1 V the species are completely oxidised, and are completely reduced at 0.75 V. So at 1.1 V, we have maximum

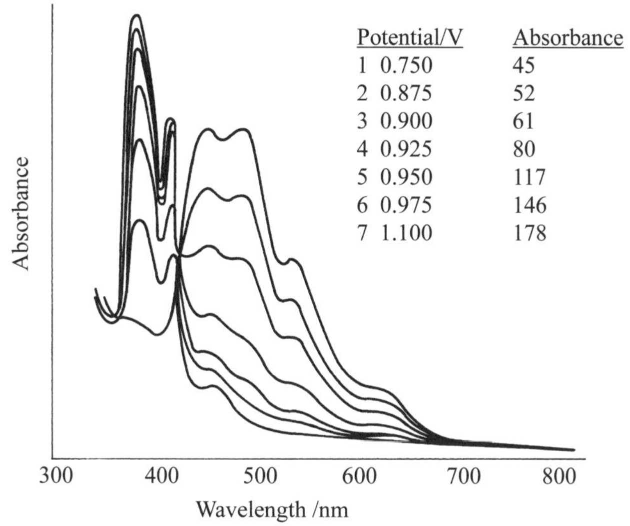

Fig. 2.6 Absorbance as a function of applied potential for the redox couple $OsCl_6^-/OsCl_6^{2-}$ (Personal communication from Dr. L. Yellowlees, University of Edinburgh, Scotland).

absorbance of 178, whilst the absorbance at 0.75 V represents the base line. Starting from intermediate values, we can calculate the Nernst potential.

For each value of the potential, we can estimate the ratio c_{ox} / c_{red} knowing that the sum $c_{ox} + c_{red}$ is a constant.

Potential/V	c_{ox} / c_{red}					$\log(c_{ox} / c_{red})$
0.875	(52 – 45)	/	(178 – 52)	=	0.055	–1.255
0.900	(61 – 45)	/	(178 – 61)	=	0.1367	–0.864
0.925	(80 – 45)	/	(178 – 80)	=	0.357	–0.447
0.950	(117 – 45)	/	(178 – 117)	=	1.80	0.072
0.975	(146 – 45)	/	(178 – 146)	=	3.15	0.499

Plotting these on a graph, we have a straight line whose Y-axis at the origin gives us the formal reduction potential. This method is relatively precise and simple to use

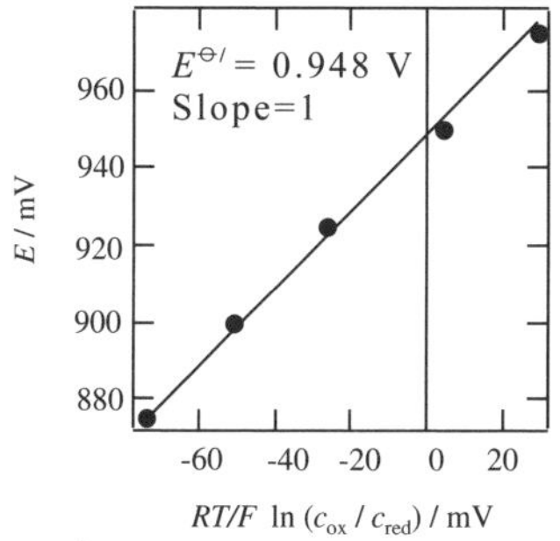

Amperometric measurements

The standard redox potential can be determined using amperometric methods such as those described in chapters 7, 8 and 10. The most commonly used method is cyclic voltammetry via the measurement of the half-wave potential, which is equal to the half-sum of the peak potentials (see §10.1).

Calculation of the standard redox potentials

When the direct measurement of the standard redox potential is difficult, the recommended method is to use standard thermodynamic data tables of the formation enthalpy and entropy of the reduced and oxidised species to calculate directly the standard redox potential from equation (2.22). There is an example of this in chapter 3 (see page 91).

2.1.5 Absolute redox potential

The standard redox potential defined by the Nernst equation is a relative value linked to the choice of the reference electrode. Nevertheless, there are situations in which it would be preferable to have an absolute redox potential, having for its origin the electron at rest in a vacuum, in order to create an absolute scale of redox potentials.

Considering the Galvani potential difference between the metal and the solution as given by equation (2.4), it would be logical to define the ***absolute standard redox potential*** as the difference between the standard chemical potentials of the oxidised and reduced species

$$\left[E^{\ominus}_{\text{ox/red}}\right]_{\text{abs}} = \phi^{\text{M}} - \phi^{\text{S}} - (\mu^{\text{M}}_{e^-}/F) = \left(\mu^{\ominus,\text{S}}_{\text{ox}} - \mu^{\ominus,\text{S}}_{\text{red}}\right)/nF \qquad (2.35)$$

$F\left[E^{\ominus}_{\text{ox/red}}\right]_{\text{abs}}$ then represents the sum of the work necessary to extract an electron from the Fermi level of the charged metal ($=-\tilde{\mu}_{e^-}$) and the electrostatic contribution of the electrochemical potential of the electron in solution ($-F\phi^{\text{S}}$). $[E^{\ominus}_{\text{ox/red}}]_{\text{abs}}$ then depends only on the standard chemical potentials of the redox pair.

Unfortunately, this approach is experimentally unfeasible because it is impossible to measure an absolute value of a Galvani potential. Another definition consists of not taking into account the electrostatic contribution due to the surface potential of the electrolyte. In this case, using the definition of the real standard chemical potential given by equation (1.71), we have

$$\left[E^{\ominus}_{\text{ox/red}}\right]_{\text{abs}} = \phi^{\text{M}} - \phi^{\text{S}} - (\mu^{\text{M}}_{e^-}/F) + \chi^{\text{S}} = \left(\alpha^{\ominus,\text{S}}_{\text{ox}} - \alpha^{\ominus,\text{S}}_{\text{red}}\right)/nF \qquad (2.36)$$

This expression has the advantage to express the absolute standard redox potential as the difference between the real standard chemical potentials, which can be measured. Equation (2.36) can be reorganised as a combination of experimentally accessible quantities

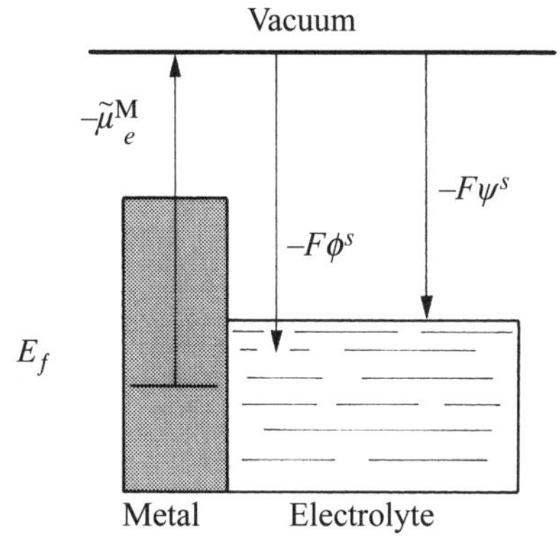

Fig. 2.7 Energy diagram of the metal|electrolyte interface.

$$\begin{aligned}\left[E^{\ominus}_{ox/red}\right]_{abs} &= \phi^{M} - \phi^{S} - (\mu^{M}_{e^-}/F) + \chi^{S} = -\tilde{\mu}^{M}_{e^-}/F - \psi^{S} \\ &= -\alpha^{M}_{e^-}/F + \left(\psi^{M} - \psi^{S}\right)\end{aligned} \tag{2.37}$$

$F\left[E^{\ominus}_{ox/red}\right]_{abs}$ then stands for the work necessary to extract an electron from the Fermi level of the metal and transfer it to the surface of the electrolyte. $-\alpha^{M}_{e^-}$ stands for the work of extraction of the electron from the non-charged metal, and $\psi^{M} - \psi^{S}$ stands for the Volta potential difference between the metal and the electrolyte containing the redox pair.

The definition (2.37) of the absolute standard redox potential allows us to define the Fermi level of the electron in solution as it is presented in §1.2.6. In fact, by considering the equality of the electrochemical potentials of the electron in the metal and in solution, we can write

$$E_f^{sol} = \tilde{\mu}^{S}_{e^-} = \tilde{\mu}^{M}_{e^-} = -F\left[E^{\ominus}_{ox/red}\right]_{abs,\ q_s=0} \tag{2.38}$$

Thus, when the charge on the solution is zero, i.e. when the outer potential ψ^{S} of the solution is zero, the concepts of the Fermi level of the electron in solution and the absolute standard potential of the electrode meet.

In the case of the standard hydrogen electrode that serves as the origin of the SHE scale of standard redox potentials, we can calculate the absolute standard redox potential

$$\begin{aligned}\left[E^{\ominus}_{ox/red}\right]_{abs} &= \phi^{M} - \phi^{S} - (\mu^{M}_{e^-}/F) + \chi^{S} = \left(\mu^{\ominus,S}_{H^+} - \tfrac{1}{2}\mu^{\ominus,G}_{H_2}\right)/F + \chi^{S} \\ &= \left(\alpha^{\ominus,S}_{H^+} - \tfrac{1}{2}\mu^{\ominus,G}_{H_2}\right)/F\end{aligned} \tag{2.39}$$

by determining the standard chemical potential of the gaseous hydrogen molecule and the standard real chemical potential of the aqueous proton. In order to calculate this difference, it is important to take the same reference state, e.g. the proton and the electron without interaction in a vacuum. In this case, the standard chemical potential of gaseous molecular hydrogen can be estimated from the ionisation energy of the hydrogen atom and the covalent bond energy of the hydrogen molecule

$$H^+_g + e^-_g \rightleftarrows H_g \qquad \Delta G^\ominus = -13.613 \text{ eV}$$

$$H_g \rightleftarrows \tfrac{1}{2}H_{2g} \qquad \Delta G^\ominus = -2.107 \text{ eV}$$

whilst for the standard real potential of the aqueous proton we have

$$H^+_g \rightleftarrows H^+_{aq} \qquad \alpha^{\ominus,S}_{H^+} \cong -11.276 \text{ eV} = -1090 \text{ kJ}\cdot\text{mol}^{-1}$$

And so we have

$$\left[E^\ominus_{H^+/\frac{1}{2}H_2}\right]_{abs} \cong [-11.276-(-13.613-2.107)] = 4.44 \pm 0.05 \text{ V} \qquad (2.40)$$

(Numerical values : R. Parsons in *Standard potentials in aqueous solutions* edited by A.J. Bard, R. Parsons & J. Jordan, Marcel Dekker, New York).

The two scales of potentials are such that

$$\left[E^\ominus_{ox/red}\right]_{abs} \cong \left[E^\ominus_{ox/red}\right]_{SHE} + 4.44 \text{ V} \qquad (2.41)$$

2.2 CELLS AND ACCUMULATORS

2.2.1 Voltaic cells

Voltaic cells are redox assemblies similar to that shown in Figure 2.3. Historically, we can cite the ***Daniell cell*** invented in 1836 that comprises a zinc anode and a copper cathode immersed respectively in solutions of zinc salts and copper salts. The cell reaction can be written as

$$Zn + Cu^{2+} \leftrightarrows Zn^{2+} + Cu$$

and the standard potential of the cell is then given by

$$E^\ominus_{Cell} = E^\ominus_{Cathode} - E^\ominus_{Anode} = \left[E^\ominus_{Cu^{2+}/Cu}\right]_{SHE} - \left[E^\ominus_{Zn^{2+}/Zn}\right]_{SHE} = 1.1 \text{ V}$$

In a general manner, the ***cathode*** is defined as the electrode where the reduction reaction takes place and the ***anode*** as the electrode where the oxidation reaction takes place. In the case of a cell generating energy, the cathode will be the positive terminal and the anode the negative terminal. On the contrary, for an electrolytic cell consuming energy, the cathode will be the negative terminal and the anode the positive terminal.

The best-known commercialised cell is the ***Leclanché cell*** that can be described as

$$Zn_s \mid Zn^{2+}, NH_4Cl \mid MnO_{2\,s}, Mn_2O_{3\,s} \mid C$$

The anode is made of zinc and the cathode of carbon. During the discharge, the degree of oxidation of the manganese goes from IV ($MnO_{2\,s}$) to III ($Mn_2O_{3\,s}$). The ammonium chloride serves at the same time as an electrolyte and as a buffer, the cathodic reduction generating hydroxide ions. The p.d. of this battery is 1.55 V at the outset. This reaction is not chemically reversible because the dissolved zinc can exist in several forms such as $Zn(OH)_2$, $Zn(OH)Cl$, $Zn(NH_3)_2$ or even in the precipitated form $ZnO{\cdot}Mn_2O_3$.

The electrode reactions in this cell are far from being simple. The degree of oxidation of the MnO_2 and its purity are two of the reasons for the difficulty in predicting accurately the nature of the reactions. In a schematic way, we can write:

$$Zn + 2\,MnO_2 + H_2O \longrightarrow Zn^{2+} + Mn_2O_3 + 2\,OH^-$$

The Leclanché cell has the advantage of being cheap to make, and millions are manufactured every year. Six Leclanché cells can be mounted in series to make a 9 V battery.

By replacing the electrolyte with KOH, it becomes a Leclanché cell known as 'alkaline'. The electrode reactions are then

$$Zn + 2\,MnO_2 + H_2O \longrightarrow Mn_2O_{3\,s} + Zn(OH)_2$$

or

$$Zn + MnO_2 + OH \longrightarrow MnO(OH) + ZnO$$

These batteries are better adapted for continuous regular discharges and have a longer lifespan than ordinary Leclanché cells. However, there are greater technological problems with the casing, because of the use of the KOH.

Among other types of commercial batteries, we can mention mercury or silver button batteries that are mostly used in watches and other small devices.

$$Zn + HgO + H_2O \longrightarrow Zn^{2+} + Hg + 2\,OH^-$$

$$Zn + Ag_2O + H_2O \longrightarrow Zn^{2+} + 2\,Ag + 2\,OH^-$$

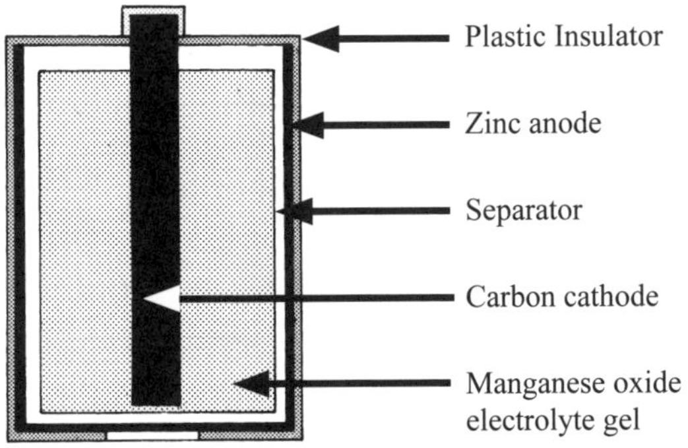

Fig. 2.8 Construction of a Leclanché cell.

2.2.2 Accumulators

Of all the electrochemical accumulators, the lead acid accumulator is without any doubt the most widespread, with several million units being sold every year for the automobile market alone. The chain can be schematically represented by:

$Pb_s \mid PbSO_{4\,s} \mid H_2SO_4 \mid PbSO_{4\,s}\,,\, PbO_{2\,s} \mid Pb_s$

When the molarity of the acid is 2M, the p.d. is about 2 V. Six cells in series are used to make 12 V batteries. These have a capacity of up to 100 A·h and are also able to deliver for a few seconds very heavy currents, e.g. 400-450 A, necessary to start a car. The electrode reactions of such an accumulator are in fact very complex, and a great deal of technology is involved in the production of modern accumulators.

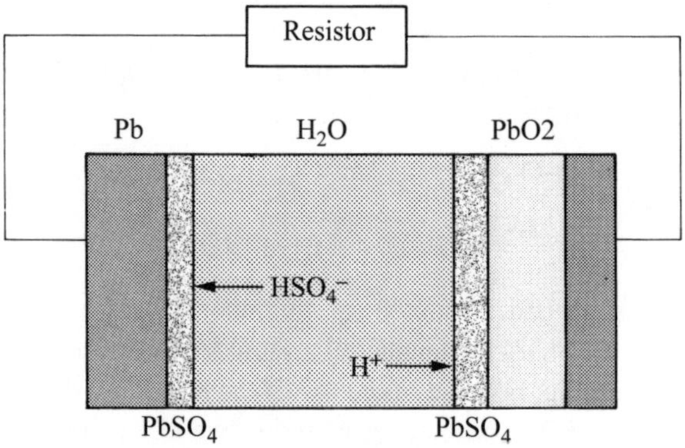

Fig. 2.9 Schematic representation of the working principle of a lead acid accumulator.

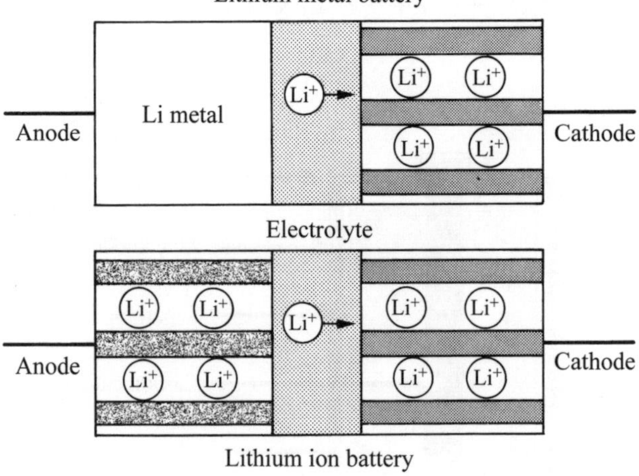

Fig. 2.10 Schematic diagram of the principle of a lithium battery (top) and a lithium ion battery (bottom).

Another accumulator that has been widely commercialised and above all used for high power applications is the Ni/Cd (nickel/cadmium) accumulator whose electrode reactions can be summarised in the form:

$$2\,NiO(OH) + Cd + 2\,H_2O \longrightarrow 2\,Ni(OH)_2 + Cd(OH)_2$$

The theoretical p.d. on open circuit is about 1.3 V.

More recently, Ni/MH (nickel/metal hydride) and lithium accumulators have been developed for the needs of portable electronic devices and electric cars. Lithium batteries offer a great amount of energy for a given mass and volume. The material of the cathode is often an insertion-composite such as $LiMnO_2$. The electrolyte can either be a liquid or a polymer and the anode is either directly lithium metal or an alloy, or possibly an insertion-composite based on graphite for example. The potential of the cells vary between 3 V and 4 V according to the materials used for the electrodes.

2.2.3 Fuel cells

Another type of electrochemical power generating devices is the fuel cell, where the reactants are supplied from outside the cell. Since the first work of Grove, who showed in 1839 that a battery with hydrogen and oxygen as reactants could produce electricity, there are now three main categories of fuel cells:

- Low temperature cells (<100°C) such as the one in Figure 2.11 comprising two porous electrodes separated by a cation exchange membrane (see §2.6). These cells are intended particularly to be used for powering electric vehicles.

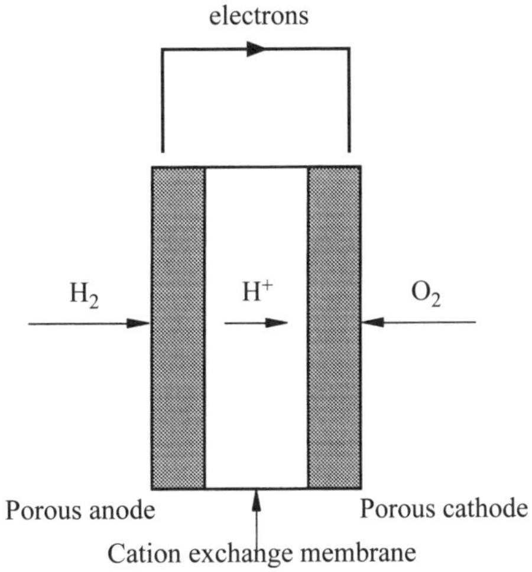

Fig. 2.11 Schematic diagram of a low-temperature fuel cell. Hydrogen and oxygen are introduced via the porous electrodes. The electrochemical reactions happen at the electrode|membrane interface (oxidation of hydrogen into protons at the anode, reduction of oxygen and the production of water at the cathode).

The power needed varies from a few dozen kW for a car, up to about 250 kW for a bus.

- Medium temperature cells (<200°C) function in a similar way to low-temperature cells, but the membrane is replaced by a layer of phosphoric acid in a porous support. These cells, which are mainly intended for large-scale electricity distribution, have a power of several MW. Their thermal and electrical yield is of the order of 40%.
- High temperature cells, using a ceramic ionic conductor as electrolyte when the temperature is above several hundred degrees. The principal fuel is hydrogen, and in this case the maximum voltage is 1.23 V.

2.3 POURBAIX DIAGRAMS

We have seen that the standard redox potential can be used to evaluate whether a substance oxidises or reduces more or less easily. In the case of metals, the oxidation is often accompanied by acid-base reactions, and, in order to determine the conditions for the stability of a metal at any given state of oxidation, it is useful to draw a ***potential-pH diagram*** also known as a ***Pourbaix diagram***.

For a given element, the limits between the different zones of stability are linked to equations of electrochemical equilibria, or of acid-base equilibria. To establish these zone diagrams, we have to fix arbitrarily a concentration for the dissolved species such as the ions. In the case of an electrochemical equilibrium between two dissolved species, the separation limit will correspond to the condition $c_{ox} = c_{red}$.

EXAMPLE

Let's take for example the iron | water system at 25°C, which introduces the thermodynamic bases of corrosion.

In order to simplify, we will only consider the following species:

Fe, Fe^{2+}, Fe^{3+}, $Fe(OH)_{2\ solid}$, $Fe(OH)_{3\ solid}$

We shall also ignore the oxidation state +VI that only occurs in the case of very alkaline pH.

The electrochemical equilibria to consider are :

$$Fe^{2+} + 2e^- \rightleftarrows Fe \qquad E^{\ominus}_{Fe^{2+}/Fe} = -0.44\ \text{V} \qquad \text{(I)}$$

$$Fe^{3+} + e^- \rightleftarrows Fe^{2+} \qquad E^{\ominus}_{Fe^{3+}/Fe^{2+}} = 0.77\ \text{V} \qquad \text{(II)}$$

$$Fe^{2+} + 2\,OH^- \rightleftarrows Fe(OH)_2 \qquad pK_{s3} = 38 \qquad \text{(III)}$$

$$Fe^{3+} + 3\,OH^- \rightleftarrows Fe(OH)_3 \qquad pK_{s2} = 15.1 \qquad \text{(IV)}$$

$$Fe(OH)_2 + 2\,H^+ + 2\,e^- \rightleftarrows Fe + 2\,H_2O \qquad \text{(V)}$$

$$Fe(OH)_3 + 3\,H^+ + e^- \rightleftarrows Fe^{2+} + 3\,H_2O \qquad \text{(VI)}$$

$$Fe(OH)_3 + H^+ + e^- \rightleftarrows Fe(OH)_2 + H_2O \qquad \text{(VII)}$$

Ignoring the activity coefficients and taking an arbitrary concentration of dissolved species equal to 0.01 M, the Nernst equations related to the first two redox equilibria are written respectively

Fe | Fe^{2+} :

$$E = E^{\ominus}_{Fe^{2+}/Fe} + \frac{RT}{2F}\ln c_{Fe^{2+}} = -0.499 \text{ V} \tag{I}$$

Fe^{2+} | Fe^{3+} :

$$E = E^{\ominus}_{Fe^{3+}/Fe^{2+}} + \frac{RT}{F}\ln\left(\frac{c_{Fe^{3+}}}{c_{Fe^{2+}}}\right) = 0.77 \text{ V} \tag{II}$$

These two equations translate into straight horizontal lines as illustrated in the Figure below. The position of straight line I is a function of the arbitrary choice of concentration (e.g. $c = 10^{-2}$ M), whilst the position of straight line II is constant, being fixed by the condition of equality of the two concentrations.
The acid/base equilibria of the iron oxides are controlled by the solubility products K_S.

Fe^{2+} | Fe $(OH)_2$:

$$K_{S2} = c_{Fe^{2+}}c^2_{OH^-} = 10^{-28}\frac{c_{Fe^{2+}}}{c^2_{H^+}} = 10^{-15.1}$$

which yields the following equation for a vertical line

$$pH = 6.45 - 0.5\log c_{Fe^{2+}} \tag{III}$$

Fe^{3+} | Fe $(OH)_3$:

$$K_{S3} = c_{Fe^{3+}}c^3_{OH^-} = 10^{-42}\frac{c_{Fe^{3+}}}{c^3_{H^+}} = 10^{-38}$$

which yields

$$pH = \left(4 - \log c_{Fe^{3+}}\right)/3 \tag{IV}$$

For the arbitrarily chosen concentration of 0.01 M, the limits between the zones of stability of the ions and their respective hydroxides give vertical straight lines at pH=7.45 and pH=2.0.
The limits due to the three redox equilibria (V-VII), where the oxidised form is in the form of a hydroxide, we obtain from equations (I-IV):

Fe | Fe $(OH)_2$:

$$\begin{aligned} E &= E^{\ominus}_{Fe^{2+}/Fe} + \frac{RT}{2F}\ln c_{Fe^{2+}} = E^{\ominus}_{Fe^{2+}/Fe} + \frac{RT}{2F}\ln\left(K_{S2}\cdot 10^{28}\cdot c^2_{H^+}\right) \\ &= -0.44 + 0.38 - 0.059\text{pH} = -0.06 - 0.059\text{pH} \end{aligned} \tag{V}$$

Line (V) has a negative slope of 59 mV/pH unit.

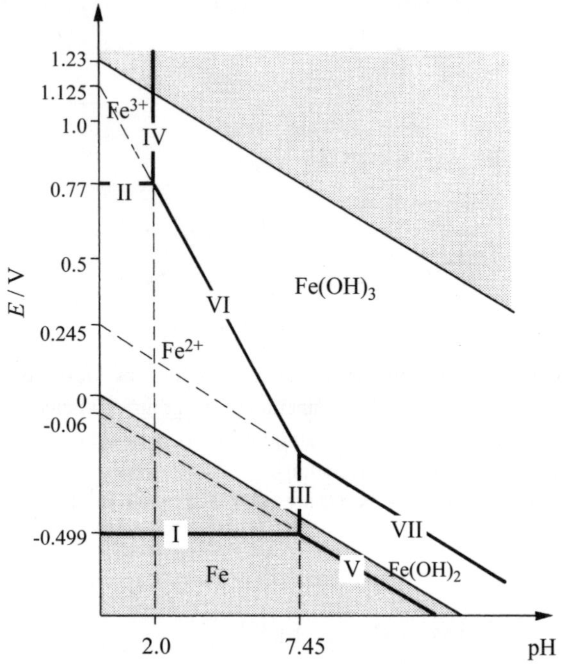

Pourbaix diagram of iron in an aqueous medium

Fe^{2+} | Fe $(OH)_3$:

$$E = E^{\ominus}_{Fe^{3+}/Fe^{2+}} + \frac{RT}{F}\ln\left(\frac{c_{Fe^{3+}}}{c_{Fe^{2+}}}\right) = E^{\ominus}_{Fe^{3+}/Fe^{2+}} + \frac{RT}{F}\ln\left(\frac{K_{S3}\cdot 10^{42}\cdot c^3_{H^+}}{c_{Fe^{2+}}}\right) \qquad \text{(VI)}$$

$$= 0.77 + 0.355 - 0.177\text{pH} = 1.125 - 0.177\text{pH}$$

Line (VI) has a negative slope of 177 mV/pH unit.

$Fe(OH)_2$ | Fe $(OH)_3$:

$$E = E^{\ominus}_{Fe^{3+}/Fe^{2+}} + \frac{RT}{F}\ln\left(\frac{c_{Fe^{3+}}}{c_{Fe^{2+}}}\right) = E^{\ominus}_{Fe^{3+}/Fe^{2+}} + \frac{RT}{F}\ln\left(\frac{K_{S3}\cdot 10^{42}\cdot c^3_{H^+}}{K_{S2}\cdot 10^{28}\cdot c^2_{H^+}}\right) \qquad \text{(VII)}$$

$$= 0.77 - 0.525 - 0.059\text{pH} = 0.245 - 0.059\text{pH}$$

Line (VII) has a negative slope of 59 mV/pH unit.

By drawing the seven lines obtained by this method, we can define the different zones of stability of the different species as illustrated in the diagram above.
In order to understand the stability zones of iron in an aqueous environment, we need to superimpose on this diagram the redox equilibria of water at atmospheric pressure

$$H^+ + e^- \rightleftarrows \tfrac{1}{2}H_2$$

$$E = \left[E^{\ominus}_{H^+/\frac{1}{2}H_2}\right]_{SHE} + \frac{RT}{F}\ln a_{H^+} = -0.059\text{pH}$$

$$O_2 + 4\,H^+ + 4\,e^- \rightleftarrows 2\,H_2O$$

$$E = \left[E^{\ominus}_{O_2/H_2O}\right]_{SHE} + \frac{RT}{4F}\ln a^4_{H^+} = 1.23 - 0.059\text{pH}$$

Thermodynamics shows us that above the line $E = 1.23 - 0.059\text{pH}$, the species are reduced by the water with oxygen given off, and that below the line $E = -0.059\text{pH}$, the species are oxidised with hydrogen given off. Even if this is correct in theory, the speeds of oxidation and reduction of the water can be very slow, and certain elements can thus appear to be stable outside of their stability zone.

2.4 ELECTROCHEMICAL EQUILIBRIA AT THE INTERFACE BETWEEN TWO ELECTROLYTES

2.4.1 The Nernst equation for the distribution of ions between two phases

We have seen how the notion of electrochemical potential allows the derivation of the Nernst equation for redox reactions at an electrode, and we shall now see how to treat the ionic distribution between two phases, e.g. two immiscible electrolyte solutions.

We can define the ***standard Gibbs transfer energy*** of a species i as the difference in standard chemical potentials between an aqueous electrolyte (w) and a non-aqueous electrolyte in an organic phase (o)

$$\Delta G^{\ominus,w\to o}_{tr,i} = \mu^{\ominus,o}_i - \mu^{\ominus,w}_i \tag{2.42}$$

At equilibrium, we have equality of the electrochemical potentials of the species i in the adjacent phases and by developing these for each phase using equation (1.60), we obtain

$$\mu^{\ominus,w}_i + RT\ln a^w_i + z_iF\phi^w = \mu^{\ominus,o}_i + RT\ln a^o_i + z_iF\phi^o \tag{2.43}$$

The Galvani potential difference between the two phases is then written as

$$\Delta^w_o\phi = \phi^w - \phi^o = \Delta^w_o\phi^{\ominus}_i + \frac{RT}{z_iF}\ln\left(\frac{a^o_i}{a^w_i}\right) \tag{2.44}$$

with $\Delta^w_o\phi^{\ominus}_i$ the ***standard transfer potential*** equal to

$$\Delta^w_o\phi^{\ominus}_i = \frac{\Delta G^{\ominus,w\to o}_{tr,i}}{z_iF} \tag{2.45}$$

which expresses in a voltage scale the standard Gibbs energy of transfer.

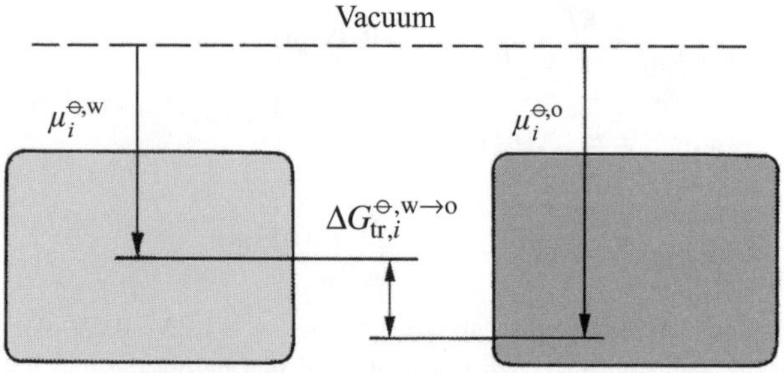

Fig. 2.12 The ion i has a more negative standard chemical potential in the organic phase than in the aqueous phase. When the two phases are put in contact, the ion has a tendency to transfer from the aqueous to the organic phase.

The Nernst equation for the distribution of an ion between two immiscible electrolyte solutions is expressed directly in terms of a difference of Galvani potentials. This Nernst equation has important applications in the domains of ion-selective electrodes, hydrometallurgy, and phase transfer catalysis.

It is customary in pharmacology to define the ***partition coefficient*** also called the ***distribution coefficient*** of an ion by

$$\ln P_i = \ln\left(\frac{a_i^{\mathrm{o}}}{a_i^{\mathrm{w}}}\right) = \ln P_i^{\ominus} + \frac{z_i F}{RT}\Delta_{\mathrm{o}}^{\mathrm{w}}\phi \tag{2.46}$$

We can see that the distribution coefficient of an ion depends on the Galvani potential difference between the two phases. This is why, in the case of the distribution of an ion between two phases, it is more judicious to define the ***standard distribution coefficient***

$$\ln P_i^{\ominus} = -\frac{\mu_i^{\ominus,\mathrm{o}} - \mu_i^{\ominus,\mathrm{w}}}{RT} = -\frac{\Delta G_{\mathrm{tr},i}^{\ominus,\mathrm{w}\to\mathrm{o}}}{RT} = -\frac{z_i F}{RT}\Delta_{\mathrm{o}}^{\mathrm{w}}\phi_i^{\ominus} \tag{2.47}$$

2.4.2 Distribution potential

Let's consider the distribution of a salt C^+A^- between two immiscible liquids. At equilibrium, we have an equality of the electrochemical potentials of both the cation and the anion:

$$\mu_{\mathrm{C}^+}^{\ominus,\mathrm{w}} + RT\ln a_{\mathrm{C}^+}^{\mathrm{w}} + F\phi^{\mathrm{w}} = \mu_{\mathrm{C}^+}^{\ominus,\mathrm{o}} + RT\ln a_{\mathrm{C}^+}^{\mathrm{o}} + F\phi^{\mathrm{o}} \tag{2.48}$$

$$\mu_{\mathrm{A}^-}^{\ominus,\mathrm{w}} + RT\ln a_{\mathrm{A}^-}^{\mathrm{w}} - F\phi^{\mathrm{w}} = \mu_{\mathrm{A}^-}^{\ominus,\mathrm{o}} + RT\ln a_{\mathrm{A}^-}^{\mathrm{o}} - F\phi^{\mathrm{o}} \tag{2.49}$$

By substitution, we can see that the distribution of the salt between the two phases polarises the interface, and that the Galvani potential difference is given by

$$2F\Delta_{\rm o}^{\rm w}\phi \;=\; \left(\mu_{\rm C^+}^{\ominus,\rm o}-\mu_{\rm C^+}^{\ominus,\rm w}\right)-\left(\mu_{\rm A^-}^{\ominus,\rm o}-\mu_{\rm A^-}^{\ominus,\rm w}\right)+RT\ln\left(\frac{a_{\rm C^+}^{\rm o}a_{\rm A^-}^{\rm w}}{a_{\rm C^+}^{\rm w}a_{\rm A^-}^{\rm o}}\right) \tag{2.50}$$

Taking into account the electroneutrality in each phase ($c_{\rm C^+}=c_{\rm A^-}$), this equation reduces to

$$\Delta_{\rm o}^{\rm w}\phi \;=\; \frac{\Delta_{\rm o}^{\rm w}\phi_{\rm C^+}^{\ominus}+\Delta_{\rm o}^{\rm w}\phi_{\rm A^-}^{\ominus}}{2}+\frac{RT}{2F}\ln\left(\frac{\gamma_{\rm C^+}^{\rm o}\gamma_{\rm A^-}^{\rm w}}{\gamma_{\rm C^+}^{\rm w}\gamma_{\rm A^-}^{\rm o}}\right) \tag{2.51}$$

In the case of dilute solutions, the second term in this equation is negligible.

The Galvani potential difference imposed by the distribution of a salt is independent of the volume of the phases in contact, and is called the ***distribution potential***.

EXAMPLE

To illustrate equation (2.51), let's calculate the distribution potential at the interface water | 1,2-dichloroethane (DCE) for the distribution of different salts having the following standard Gibbs transfer energies

$$\Delta G_{\rm tr,Na^+}^{\ominus,\rm w\to DCE}=56\ \rm kJ\cdot mol^{-1}\,,\quad \Delta G_{\rm tr,Cl^-}^{\ominus,\rm w\to DCE}=50\ \rm kJ\cdot mol^{-1}$$

$$\Delta G_{\rm tr,TBA^+}^{\ominus,\rm w\to DCE}=-22\ \rm kJ\cdot mol^{-1}\,,\quad \Delta G_{\rm tr,TPB^-}^{\ominus,\rm w\to DCE}=-33\ \rm kJ\cdot mol^{-1}$$

where TBA^+=tetrabutylammonium and TPB^-=tetraphenylborate.

Thus, the distribution of NaCl, TBACl and NaTPB gives respectively the following distribution potentials ($F\cong 10^5\ \rm C{\cdot}mol^{-1}$):

$$\Delta_{\rm DCE}^{\rm w}\phi_{\rm NaCl}=30\ \rm mV,\quad \Delta_{\rm DCE}^{\rm w}\phi_{\rm TBACl}=-360\ \rm mV$$

$$\Delta_{\rm DCE}^{\rm w}\phi_{\rm NaTPB}=445\ \rm mV,\quad \Delta_{\rm DCE}^{\rm w}\phi_{\rm TBATPB}=55\ \rm mV$$

These results show clearly that the distribution of a hydrophilic salt such as NaCl which is not very soluble in DCE or a lipophilic salt such as TBATPB which is not very soluble in water, gives rise to weak polarisations, whilst the distribution of a salt with a lipophilic cation and a hydrophilic anion gives rise to a strong negative polarisation. Inversely, the distribution of a salt with a hydrophilic cation and a lipophilic anion gives rise to a strong positive polarisation.

2.4.3 Distribution of an acid

A problem that is often encountered in preparative or pharmacological chemistry is linked to the distribution of acid or base molecules. To treat this problem, let's look at the simple case of the distribution of an acid between two phases.

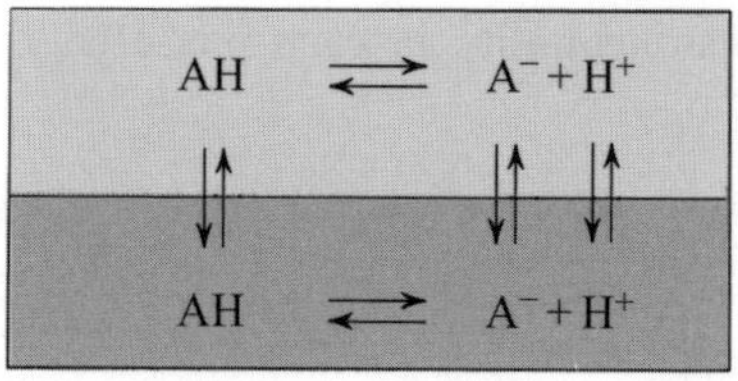

Fig. 2.13 Distribution of an acid in neutral and ionised form.

At the equilibrium, we have equality of the electrochemical potentials of the two ions A^- and H^+ in the two phases (ignoring the presence of OH^- ions)

$$\mu_{H^+}^{\ominus,w} + RT\ln a_{H^+}^{w} + F\phi^{w} = \mu_{H^+}^{\ominus,o} + RT\ln a_{H^+}^{o} + F\phi^{o} \tag{2.52}$$

$$\mu_{A^-}^{\ominus,w} + RT\ln a_{A^-}^{w} - F\phi^{w} = \mu_{A^-}^{\ominus,o} + RT\ln a_{A^-}^{o} - F\phi^{o} \tag{2.53}$$

and we have the equality of the chemical potentials of the acid AH in the two phases,

$$\mu_{AH}^{\ominus,w} + RT\ln a_{AH}^{w} = \mu_{AH}^{\ominus,o} + RT\ln a_{AH}^{o} \tag{2.54}$$

We should also consider the acid-base equilibria in both phases

$$\mu_{H^+}^{\ominus,w} + RT\ln a_{H^+}^{w} + \mu_{A^-}^{\ominus,w} + RT\ln a_{A^-}^{w} = \mu_{AH}^{\ominus,w} + RT\ln a_{AH}^{w} \tag{2.55}$$

$$\mu_{H^+}^{\ominus,o} + RT\ln a_{H^+}^{o} + \mu_{A^-}^{\ominus,o} + RT\ln a_{A^-}^{o} = \mu_{AH}^{\ominus,o} + RT\ln a_{AH}^{o} \tag{2.56}$$

The Galvani potential difference is in this case given by the distribution potential of the A^- and H^+ ions.

The acidity constant in the organic phase is linked to that in the aqueous phase by

$$K_a^{o} = \frac{a_{A^-}^{o} a_{H^+}^{o}}{a_{AH}^{o}} = K_a^{w}\frac{P_{A^-}P_{H^+}}{P_{AH}^{\ominus}} = K_a^{w}\frac{P_{A^-}^{\ominus}P_{H^+}^{\ominus}}{P_{AH}^{\ominus}} \tag{2.57}$$

This equation shows that to calculate the pK_a of an acid in the organic phase, knowing its pK_a value in water, we need to know the standard distribution coefficients of the various species involved.

2.4.4 Distribution diagrams

On the basis of the concept of Pourbaix diagrams, it is possible to make zone diagrams for the distribution of ionisable species such as acids or bases. To illustrate this, let's first consider the distribution diagram for a hydrophilic AH acid in a biphasic water/organic solvent system.

At a high aqueous pH, the acid is in the anionic form and can exist in both phases according to the Galvani potential difference. The Nernst equation for the distribution of the anion, ignoring the activity coefficients is written as

$$\Delta_o^w\phi = \Delta_o^w\phi_{A^-}^{\ominus} - \frac{RT}{F}\ln\left(\frac{c_{A^-}^o}{c_{A^-}^w}\right) \tag{2.58}$$

Thus, the separation limit between the anionic form in water and the organic solvent ($c_{A^-}^w = c_{A^-}^o$) is a horizontal straight line. As in the Pourbaix diagrams, the separation limit between the acid and basic forms in water is a vertical line given by

$$\mathrm{pH} = \mathrm{p}K_a^w \tag{2.59}$$

Finally, the line separating the neutral acid in water and the anion A_o^- in the organic phase is given by including the acidity constant in equation (2.58) to give

$$\Delta_o^w\phi = \left[\Delta_o^w\phi_{A^-}^{\ominus} + \frac{RT}{F}\ln K_a^w\right] - \frac{RT}{F}\ln\left(\frac{c_{A^-}^o a_{H^+}^w}{c_{AH}^w}\right) \tag{2.60}$$

As in the Pourbaix diagrams, we obtain a delimiting line that depends on the pH. The distribution diagram of a hydrophilic acid is shown in Figure 2.14.

If the AH acid is lipophilic, we have to take into account the distribution of the acid in the organic phase

$$K_a^w = \frac{a_{A^-}^w a_{H^+}^w}{a_{AH}^w} = \frac{a_{A^-}^w a_{H^+}^w}{a_{AH}^o} P_{AH}^{\ominus} \tag{2.61}$$

and, neglecting the activity coefficients, the separation limit between the aqueous anion A_w^- and the neutral form in the organic solvent is described by

$$\mathrm{pH} = \mathrm{p}K_a^w + \log P_{AH}^{\ominus} \tag{2.62}$$

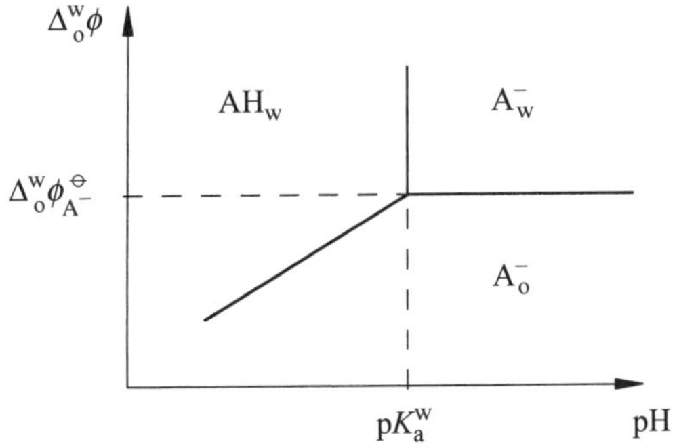

Fig. 2.14 Distribution diagram for a hydrophilic acid.

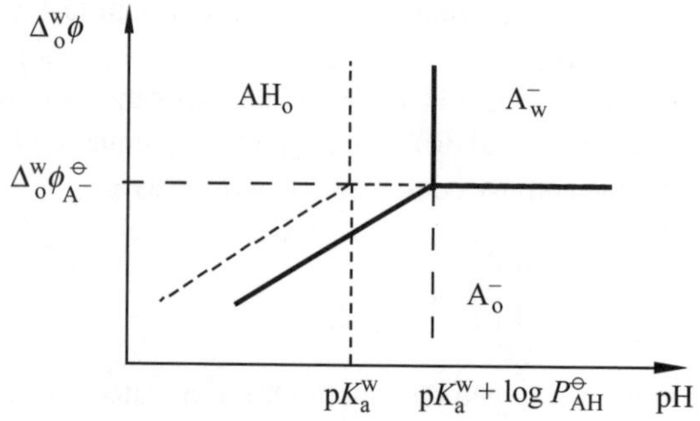

Fig. 2.15 Distribution diagram for a lipophilic acid.

The separation limit between the two ionic forms is still the one given by the Nernst equation for the distribution of the anion. The separation limit between the anion and the acid in the solvent is given by

$$\Delta_o^w\phi = \left[\Delta_o^w\phi_{A^-}^{\ominus} + \frac{RT}{F}\ln\frac{K_a^w}{P_{AH}^{\ominus}}\right] - \frac{RT}{F}\ln\left(\frac{c_{A^-}^o a_{H^+}^w}{c_{AH}^o}\right) \tag{2.63}$$

Again, this limit depends on the pH. The diagram in Figure 2.15 shows that the more lipophilic the AH acid is, the smaller is the stability zone of the anion A_w^-.

2.4.5 Redox equilibria at the liquid | liquid interface

Let's consider the transfer of electrons between an oxidised species O_1 in a phase α and a reduced species R_2 in a phase β as illustrated in Figure 2.16.

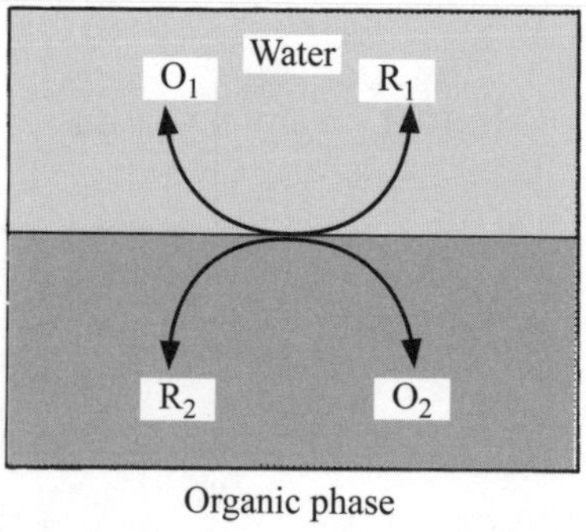

Fig. 2.16 Heterogeneous redox reaction at a liquid | liquid interface

At equilibrium

$$O_1^w + R_2^o \rightleftarrows R_1^w + O_2^o$$

we have the following equality of the electrochemical potentials

$$\tilde{\mu}_{R_1}^w + \tilde{\mu}_{O_2}^o = \tilde{\mu}_{O_1}^w + \tilde{\mu}_{R_2}^o \tag{2.64}$$

Developing this, we obtain the equivalent of the Nernst equation for this reaction of electron transfer at the interface, i.e.

$$\Delta_o^w \phi = \Delta_o^w \phi_{ET}^{\ominus} + \frac{RT}{nF} \ln\left(\frac{a_{R_1}^w a_{O_2}^o}{a_{O_1}^w a_{R_2}^o}\right) \tag{2.65}$$

with $\Delta_o^w \phi^{\ominus}$ the standard redox potential for the interfacial transfer of electrons

$$\Delta_o^w \phi_{ET}^{\ominus} = \left[\tilde{\mu}_{R_1}^w + \tilde{\mu}_{O_2}^o - \tilde{\mu}_{O_1}^w - \tilde{\mu}_{R_2}^o\right] / nF \tag{2.66}$$

It is interesting to bring into this equation the standard redox potentials defined with respect to the standard hydrogen electrode in water. To do this, we shall use equations (2.19) & (2.42) to get

$$\Delta_o^w \phi_{ET}^{\ominus} = \left[E_{O_2/R_2}^{\ominus}\right]_{SHE} - \left[E_{O_1/R_1}^{\ominus}\right]_{SHE} + \left[\Delta G_{tr,O_2}^{\ominus,w\to o} - \Delta G_{tr,R_2}^{\ominus,w\to o}\right] / nF \tag{2.67}$$

2.5 ANALYTICAL APPLICATIONS OF POTENTIOMETRY

2.5.1 Reference electrodes

By definition a reference electrode is an electrode for which the Galvani potential difference $\phi^M - \phi^S$ between the metal and the solution is constant. For it to be stable, only a negligible current can be allowed to pass through the reference electrode, not to disturb the conditions for equilibrium.

Calomel electrode

To illustrate the principle of a reference electrode, we are going to study the calomel electrode based on the redox couple $\frac{1}{2}Hg_2Cl_2/Hg$. This electrode was for a long time the most commonly used, before giving way to the one more often used nowadays, which is the silver/silver chloride electrode based on the redox couple AgCl/Ag. Hg_2Cl_2 is a solid salt called calomel, and a calomel electrode is fabricated by placing calomel and mercury in contact. An electrical contact is made to the mercury using a metal that does not form alloys with the mercury itself, such as iridium or even platinum. The mercury covered by calomel is kept in a small glass tube using for example some glass wool. This assembly as illustrated in Figure 2.17, is placed in a solution of KCl contained in a tube forming the body of the reference electrode. A glass frit is used to form a liquid junction between the internal solution of KCl and the

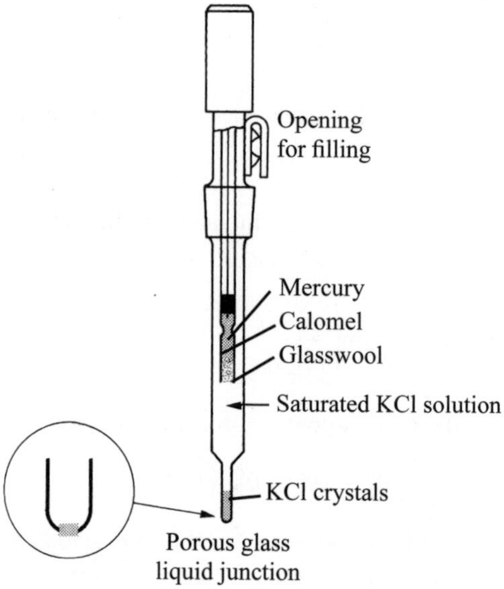

Fig. 2.17 Calomel electrode with a liquid junction (Copyright Metrohm, CH).

solution for which we want to measure the inner potential. The electrochemical cell for a calomel electrode is

$$\text{Solution} \,||\, \text{KCl}_{\text{sat}} \text{ (Internal Solution)} \,|\, \tfrac{1}{2}\text{Hg}_2\text{Cl}_2 \,|\, \text{Hg} \,|\, \text{Pt} \,|\, \text{Cu}$$

The electrochemical equilibria to consider are:

- the electronic equilibria between the copper wire, the platinum contact and the mercury,

$$\tilde{\mu}_{e^-}^{\text{Cu}} = \tilde{\mu}_{e^-}^{\text{Pt}} = \tilde{\mu}_{e^-}^{\text{Hg}} \tag{2.68}$$

- the redox equilibrium related to oxidation of mercury

$$\tilde{\mu}_{\text{Hg}^+} + \tilde{\mu}_{e^-}^{\text{Hg}} = \mu_{\text{Hg}} \tag{2.69}$$

- the solubility of calomel.

$$\tilde{\mu}_{\text{Hg}^+} + \tilde{\mu}_{\text{Cl}^-} = \tfrac{1}{2}\mu_{\text{Hg}_2\text{Cl}_2} \tag{2.70}$$

These last two equations can be combined to give

$$\tilde{\mu}_{\text{Cl}^-} + \mu_{\text{Hg}} = \tilde{\mu}_{e-}^{\text{Cu}} + \tfrac{1}{2}\mu_{\text{Hg}_2\text{Cl}_2} \tag{2.71}$$

which can be expressed as

$$\mu_{\text{Cl}^-}^{\ominus,\text{IS}} + RT\ln a_{\text{Cl}^-} - F\phi^{\text{IS}} + \mu_{\text{Hg}} = \mu_{e-}^{\text{Cu}} - F\phi^{\text{Cu}} + \tfrac{1}{2}\mu_{\text{Hg}_2\text{Cl}_2} \tag{2.72}$$

The difference of Galvani potentials between the copper wire and the internal solution (IS) then reads

$$F\left(\phi^{\mathrm{Cu}}-\phi^{\mathrm{IS}}\right) = \left[\mu_{e^-}^{\mathrm{Cu}}+\tfrac{1}{2}\mu_{\mathrm{Hg_2Cl_2}}-\mu_{\mathrm{Cl^-}}^{\ominus,\mathrm{IS}}-\mu_{\mathrm{Hg}}\right] - RT\ln a_{\mathrm{Cl^-}} \qquad (2.73)$$

Thus, when the internal solution is saturated in KCl, $a_{\mathrm{Cl^-}}$ is constant and therefore $\phi^{\mathrm{Cu}}-\phi^{\mathrm{IS}}$ is also constant.

When a calomel reference electrode is used to measure an electrode potential with a setup as the one illustrated in Figure 2.4, the electrochemical cell can be written as

$$\mathrm{Cu^I} \mid \mathrm{Hg} \mid \tfrac{1}{2}\mathrm{Hg_2Cl_2} \mid \mathrm{KCl_{sat}}....\|....\mathrm{ox,red} \mid \mathrm{M} \mid \mathrm{Cu^{II}}$$

The equilibrium potential E measured relative to a calomel electrode in a saturated solution of KCl is then

$$\begin{aligned} E &= \phi^{\mathrm{Cu^{II}}}-\phi^{\mathrm{Cu^I}} = \left(\phi^{\mathrm{Cu^{II}}}-\phi^{\mathrm{IS}}\right)+\left(\phi^{\mathrm{IS}}-\phi^{\mathrm{S}}\right)+\left(\phi^{\mathrm{S}}-\phi^{\mathrm{Cu^I}}\right) \\ &= \left[E^{\ominus}_{\mathrm{ox/red}}\right]_{\mathrm{SHE}}+\frac{RT}{nF}\ln\left(\frac{a_{\mathrm{ox}}}{a_{\mathrm{red}}}\right)-\left[E^{\ominus}_{\frac{1}{2}\mathrm{Hg_2Cl_2/Hg}}\right]_{\mathrm{SHE}}+\frac{RT}{F}\ln a_{\mathrm{Cl^-}} \qquad (2.74) \\ &= \left[E^{\ominus}_{\mathrm{ox/red}}\right]_{\mathrm{Hg|\frac{1}{2}Hg_2Cl_2|KCl_{sat}}}+\frac{RT}{nF}\ln\left(\frac{a_{\mathrm{ox}}}{a_{\mathrm{red}}}\right) \end{aligned}$$

Figure 2.18 shows a schematic potential distribution across a cell used to measure an electrode potential E with respect to a calomel reference electrode. The voltmeter measures the difference of the inner potentials between the two copper wires ($E = \phi^{\mathrm{Cu^{II}}}-\phi^{\mathrm{Cu^I}}$). The Galvani potential difference between the copper wire (I) and the internal solution $\phi^{\mathrm{Cu^I}}-\phi^{\mathrm{IS}}$ is constant as described by equation (2.73) and the Galvani potential difference between the copper wire (II) and the solution $\phi^{\mathrm{Cu^{II}}}-\phi^{\mathrm{S}}$ is determined by the ratio of activities of ox and red in the solution as given by equation (2.4). The liquid junction potential $\phi^{\mathrm{IS}}-\phi^{\mathrm{S}}$ is assumed to be negligible (see below).

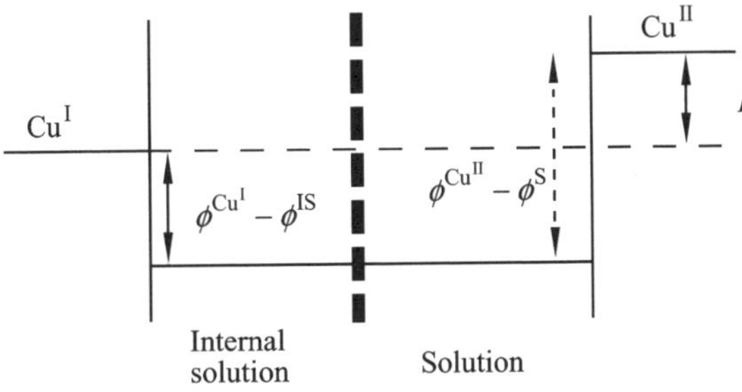

Fig. 2.18 Schematic representation of the potential distribution in a cell. For clarity, the potential in the metal M and in the reference electrode is not shown.

Silver | silver chloride electrode

For a silver | silver chloride electrode for which the electrochemical cell reads

Solution || KCl_{sat} (internal solution) | AgCl | Ag | Cu

we can show by the same reasoning as above that

$$F\left(\phi^{Cu}-\phi^{IS}\right) = \left[\mu_{e^-}^{Cu}+\mu_{AgCl}-\mu_{Cl^-}^{\ominus,S}-\mu_{Ag}\right] - RT\ln a_{Cl^-} \tag{2.75}$$

In the same way, when the KCl solution is saturated, a_{Cl^-} is constant and therefore $\phi^{Cu}-\phi^{IS}$ is also constant.

For the cell

$$Cu^{I} \mid Ag \mid AgCl \mid KCl_{sat}....\|....ox,red \mid M \mid Cu^{II}$$

we have already seen (equation (2.33)) that the equilibrium potential is

$$\begin{aligned} E &= \phi^{Cu^{II}}-\phi^{Cu^{I}} = \left(\phi^{Cu^{II}}-\phi^{IS}\right)+\left(\phi^{IS}-\phi^{S}\right)+\left(\phi^{S}-\phi^{Cu^{I}}\right) \\ &= \left[E^{\ominus}_{ox/red}\right]_{SHE}+\frac{RT}{nF}\ln\left(\frac{a_{ox}}{a_{red}}\right)-\left[E^{\ominus}_{AgCl/Ag}\right]_{SHE}+\frac{RT}{F}\ln a_{Cl^-} \end{aligned} \tag{2.76}$$

or else

$$E = \left[E^{\ominus}_{ox/red}\right]_{Ag|AgCl|KCl_{sat}} + \frac{RT}{nF}\ln\left(\frac{a_{ox}}{a_{red}}\right) \tag{2.77}$$

Table 2.1 shows the standard redox potentials at different temperatures of the main commercialised reference electrodes.

Table 2.1 Potential (Volt) of reference electrodes on the SHE scale. Compilation by Skoog, Holler & Nieman, *Principles of instrumental analysis*, Saunders College Publishing, 1998.

T/°C	Calomel 3.5 M KCl	Calomel Saturated KCl	Ag | AgCl 3.5 M KCl	Ag | AgCl Saturated KCl
15	0.254	0.2511	0.212	0.209
20	0.252	0.2479	0.208	0.204
25	0.250	0.2444	0.205	0.199
30	0.248	0.2411	0.201	0.194
35	0.246	0.2376	0.197	0.189

Liquid junction potential

The liquid junction between the electrolyte containing the redox species being studied and the internal solution of saturated KCl is in general supported by a glass frit. We can calculate the values of the potential differences across a liquid junction between two aqueous solutions S_1 and S_2, by calling on the thermodynamics of irreversible processes

$$\phi^{S_2} - \phi^{S_1} = \frac{-RT}{F} \int_{S_1}^{S_2} \sum_i \frac{t_i}{z_i} \,\mathrm{d}(\ln a_i) \tag{2.78}$$

where t_i is the transport number of the species i across the junction (see §4.5). In the case of a junction between a concentrated KCl solution and any other not-too-concentrated electrolyte solution (<<1M), the dominant diffusion salt transfer is that of KCl out of the internal solution. Given that these two ions have approximately the same mobility, their transport numbers are nearly equal and the potential difference across a liquid junction between a not-too-concentrated electrolyte solution and a saturated solution of KCl is almost zero, a few mV at most. A liquid junction whose potential difference is not negligible is indicated by a dotted vertical bar ⁝.

From a practical point of view, it is always preferable that the level of the internal solution inside the reference electrode body should be above that of the solution in which it is operating, in order to minimise the risk of contamination of the internal solution. Also, in the case of saturated KCl electrodes, it is important to check that KCl crystals are always present in the electrode.

One of the main causes of errors when using reference electrodes comes from the liquid junction that can be 'blocked' in the glass frit following precipitations or contaminations.

2.5.2 Potentiometric Titration

Nernst's law for redox reactions is the basis for the ***potentiometric titration*** technique. The principle of this method is to titrate a redox species, e.g. hydroquinone, by adding an oxidising or reducing agent, according to the analyte, and by measuring the potential of an inert working electrode (platinum, vitreous carbon, etc.) compared to a reference electrode.

For example, we can titrate Fe^{II} with a standard solution of Ce^{IV} since

$$Fe^{III} + e^- \rightleftarrows Fe^{II} \qquad E^{\ominus}_{SHE} = 0.771\ \mathrm{V}$$

$$Ce^{4+} + e^- \rightleftarrows Ce^{3+} \qquad E^{\ominus}_{SHE} = 1.44\ \mathrm{V}$$

The potential of a platinum working electrode measured with respect to an SHE reference electrode will then be

$$E_{Pt} = 0.771 + 0.059\log\left(\frac{c_{Fe^{3+}}}{c_{Fe^{2+}}}\right) = 1.44 + 0.059\log\left(\frac{c_{Ce^{4+}}}{c_{Ce^{3+}}}\right) \tag{2.79}$$

By drawing a curve of the electrode potential as a function of the concentration of Fe^{3+}, we have for the titration of an 1M Fe^{2+} solution

$$E_{Pt} = 0.771 + 0.059\log\left(\frac{x}{1-x}\right) \tag{2.80}$$

We can see that when x is more or less equal to 0.5, the potential is approximately equal to the standard redox potential of the redox couple titrated. This is the ***redox buffer effect*** illustrated in Figure 2.19.

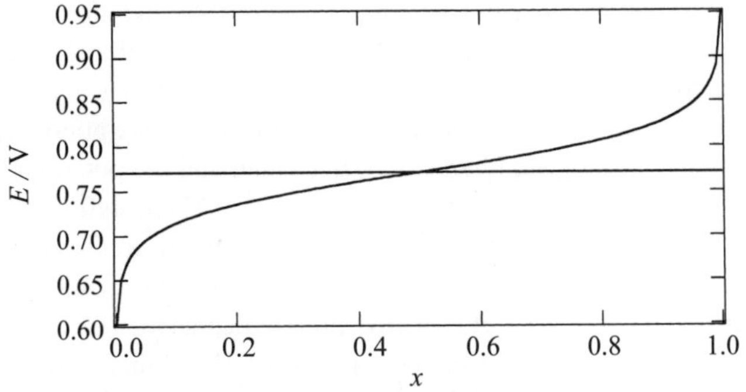

Fig. 2.19 The redox buffer effect.

On the contrary, if we plot the electrode potential curve as a function of added volumc of titrant, wc obtain a titration curve similar to those for acid-base titrations. The equivalence point, which is the inflection point of the curve, is then determined by tracing the derivative as shown in Figure 2.20.

Titration analysis methods, although very old, are still widely used today in industry, because they are very precise, reliable and can often be automated.

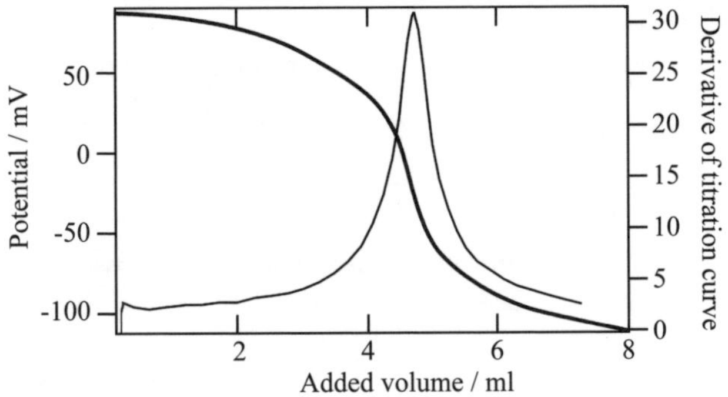

Fig. 2.20 Titration of an iodine solution with thiosulfate on a platinum electrode. The potential scale is arbitrary.

EXAMPLE

Let's calculate the titration curve for a solution 1 containing a reduced species R_1 in a volume V_1 by a solution 2 containing an oxidant O_2.
The equations to consider are (1) the Nernst equations for the two redox couples O_1/R_1 and O_2/R_2,(2) the equations of mass conservation for these redox couples during the addition of the titrant solution 2, and (3) the mass balance of the redox reaction. Thus, by neglecting the activity coefficients, we have the following system of equations:

$$\frac{c_{O_1}}{c_{R_1}} = \exp\left[\frac{F}{RT}(E - E^{\ominus}_{O_1/R_1})\right]$$

$$\frac{c_{O_2}}{c_{R_2}} = \exp\left[\frac{F}{RT}(E - E^{\ominus}_{O_2/R_2})\right]$$

$$c_{O_1} + c_{R_1} = c_1 V_1 / [V_1 + V_a]$$

Conservation of species 1 and 2

$$c_{O_1} + c_{R_1} = c_1 V_1 / [V_1 + V_a]$$

$$c_{O_2} + c_{R_2} = c_2 V_a / [V_1 + V_a]$$

Mass balance for the reaction

$$V_1 c_1 - (V_1 + V_a)\cdot c_{R_1} = V_a c_2 - (V_1 + V_a)\cdot c_{O_2}$$

where c_1 and c_2 are the initial concentrations of the reduced species 1 and the oxidised species 2, V_1 is the initial volume of solution 1 and V_a the added volume of solution 2.
As with pH titrations, it is much easier to calculate the added volume as a function of the potential than to calculate the electrode potential as a function of the added volume. We can easily show, e.g. with the help of a calculation software, that the solution of this system is

$$V_a = \frac{1+\exp\left[\frac{F}{RT}\left(E - E^{\ominus}_{O_2/R_2}\right)\right]}{1+\exp\left[\frac{F}{RT}\left(E - E^{\ominus}_{O_1/R_1}\right)\right]} \frac{c_1 V_1}{c_2} \exp\left[\frac{F}{RT}\left(E - E^{\ominus}_{O_1/R_1}\right)\right]$$

Thus we obtain the following curve

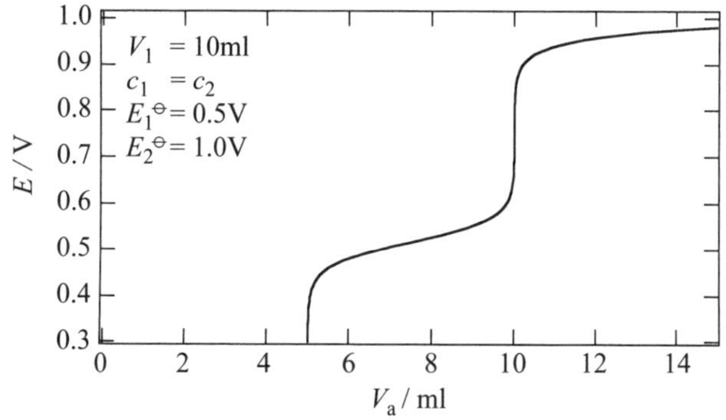

Of course, in this calculation the volumes calculated for potential values very much below $E^{\ominus}_{O_1/R_1}$ have no physical significance. In order to find the exact curve, the system must be solved to calculate E as a function of V_a. Nevertheless, this simplified approach can be used to illustrate the redox buffer effect and to calculate the end-point titration volume.

2.5.3 Ion selective electrodes

The Nernst equation for the distribution of ions is the basic concept for ***ion selective electrodes*** (ISE). An ISE is made up of three parts: the analyte, the membrane and the reference electrolyte. With the help of two reference electrodes, we can measure the relative Galvani potential difference between the analyte and the reference electrolyte as shown in Figure 2.21.

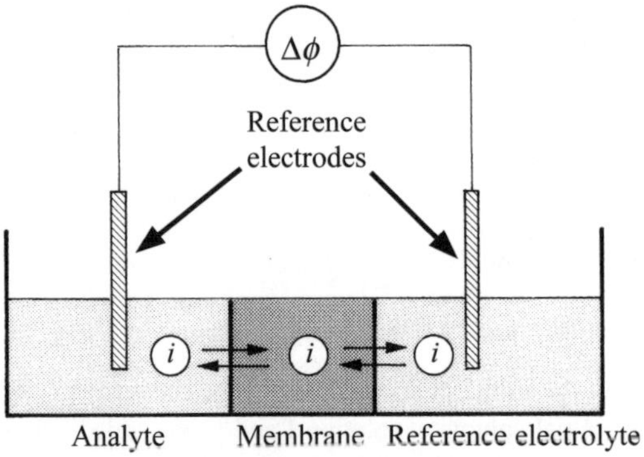

Fig. 2.21 Diagram of an ion selective electrode.

The membrane can be either

- Glass, such as a borosilicate conducting glass for the detection of Na^+ and H^+ ions, the main application being of course for the pH electrode.
- Hydrophobic polymer gel such as plasticized PVC or a supported organic liquid, for the detection of alkali metals by using natural selective ionophores such as the heterocyclic antibiotic valinomycin or synthetic selective ionophores such as dibenzo-18-crown-6 (see Figure 2.22) for the potassium ion.
- Crystal, such as a mono-crystalline crystal e.g. lanthane fluoride LaF_3 for the detection of fluoride, or a polycrystalline crystal such as silver sulfite Ag_2S for the detection of sulfite ions.

Thermodynamic approach

The most important criterion for the correct working of an ISE is the establishment of an interfacial thermodynamic equilibrium between the three phases for the ionic species to be determined. This is written as

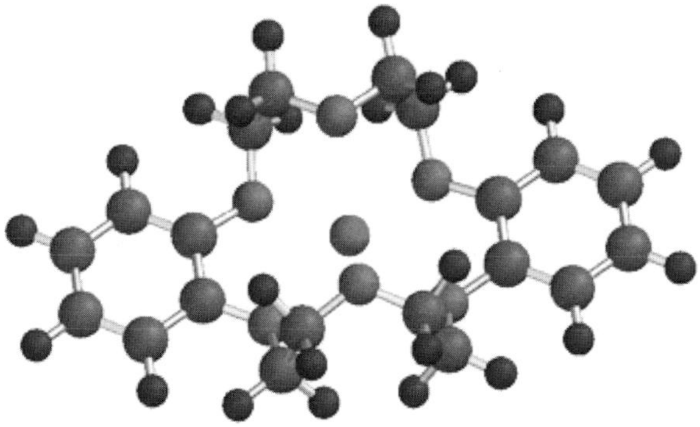

Fig. 2.22 Complexation of the potassium ion by the dibenzo-18-crown-6 molecule.

$$\tilde{\mu}_i^{\text{Analyte}} = \tilde{\mu}_i^{\text{Membrane}} = \tilde{\mu}_i^{\text{Reference}} \tag{2.81}$$

The difference in Galvani potential $\phi^{\text{R}} - \phi^{\text{A}}$ between the reference compartment and the analyte is expressed by

$$\phi^{\text{R}} - \phi^{\text{A}} = \left(\phi^{\text{R}} - \phi^{\text{M}}\right) + \left(\phi^{\text{M}} - \phi^{\text{A}}\right) \tag{2.82}$$

By using equation (2.44), we obtain

$$\phi^{\text{R}} - \phi^{\text{A}} = \Delta_{\text{M}}^{\text{R}}\phi_i^{\ominus} + \frac{RT}{z_iF}\ln\left(\frac{a_i^{\text{M}}}{a_i^{\text{R}}}\right) + \Delta_{\text{A}}^{\text{M}}\phi_i^{\ominus} + \frac{RT}{z_iF}\ln\left(\frac{a_i^{\text{A}}}{a_i^{\text{M}}}\right) = \frac{RT}{z_iF}\ln\left(\frac{a_i^{\text{A}}}{a_i^{\text{R}}}\right) \tag{2.83}$$

Table 2.2 Applications of glass and membrane ion selective electrodes (Copyright, Metrohm CH)

Analyte	Measuring Range / M	Interfering species	Examples
H^+	10^{-14} ... 1	Li^+	Measurement of pH and titrations
Na^+	10^{-5} ... 1	pH>pNa +4, Ag^+, Li^+, K^+	Water, clinical analysis, ...
K^+	10^{-6} ... 1	Cs^+, NH_4^+, H^+, Na^+	Soil, fertilizer, wine, clinical analysis , ...
Ca^{2+}	5 10^{-7} ... 1	Na^+, Pb^{2+}, Fe^{2+}, Zn^{2+} Mg^{2+}	Soil, food, milk, ..
NO_3^-	7 10^{-6} ... 1	Br^-, NO_2^-, Cl^-, OAc^-	Soil, vegetables, water, food, ...
BF_4^-	7 10^{-6} ... 1	NO_3^-, SO_4^{2-}, ClO_4^-, F^-, OAc^-	Surface active agents, plating baths....

Table 2.3 Applications of crystalline ion selective electrodes (Copyright, Metrohm CH).

Analyte	Measuring Range / M	Interfering species	Examples
F^-	10^{-6} ... sat	OH^-	Cosmetic products , pharmaceuticals, manure, ...
Cl^-	5 10^{-5} ... 1	Br^-, I^-, S^{2-}, CN^-, NH_3, $S_2O_3^-$	Food, paper,...
Br^-	5 10^{-6} ... 1	Cl^-, I^-, S^{2-}, CN^-, NH_3, $S_2O_3^-$	Clinical analysis, gasoline, Chemical industry
I^-	5 10^{-8} ... 1	Cl^-, Br^-, S^{2-}, CN^-, NH_3, $S_2O_3^-$	Clinical analysis, determination of Hg
CN^-	8 10^{-6} ... 10^{-2}	Br^-, I^-, Cl^-	Water, plating baths
SCN^-	5 10^{-6} ... 1	Cl^-, Br^-, S^{2-}, CN^-, $S_2O_3^-$	Plating baths
S^{2-}, Ag^+	10^{-7} ... 1	Proteins	S^{2-}: Food, paper Ag^+: plating baths
Cu^{2+}	10^{-8}...10^{-1}	Ag^+, Hg^{2+}, S^{2-}	Water, plating baths
Cd^{2+}	10^{-7}...10^{-1}	Ag^+, Hg^{2+}, Cu^{2+}	Soil, water, plating baths
Pb^{2+}	10^{-6}...10^{-1}	Ag^+, Hg^{2+}, Cu^{2+}	Soil, water, plating baths

because the sum of the standard transfer potentials $\Delta_M^R \phi_i^\ominus$ and $\Delta_A^M \phi_i^\ominus$ is zero. Thus, for the setup illustrated in Figure 2.21, the potential measured is written as

$$E = \phi^{Cu^{II}} - \phi^R + \phi^R - \phi^A + \phi^A - \phi^{Cu^I} \tag{2.84}$$

If the two reference electrodes are identical, e.g. two $Ag|AgCl|KCl_{sat}||...$ electrodes, then the terms $\phi^{Cu^{II}} - \phi^R$ and $\phi^A - \phi^{Cu^I}$ given by equation (2.75) cancel out, and so the potential measured is directly $\phi^R - \phi^A$. Figure 2.23 illustrates the distribution of potential for an ion selective electrode comprising one membrane and two identical reference electrodes.

Given that the concentration of the species i to be determined in the analyte is fixed in the reference compartment, the response of an ISE is $60/z$ mV per decade of concentration of the species i in the analyte.

Notice that the concentration of the ion i in the membrane is not involved. Nevertheless, this should be high enough so that an equilibrium between the three phases can be established. It is also important to note that the equilibrium condition (2.81) applies above all to the distribution of the ion i, assuming that the distribution

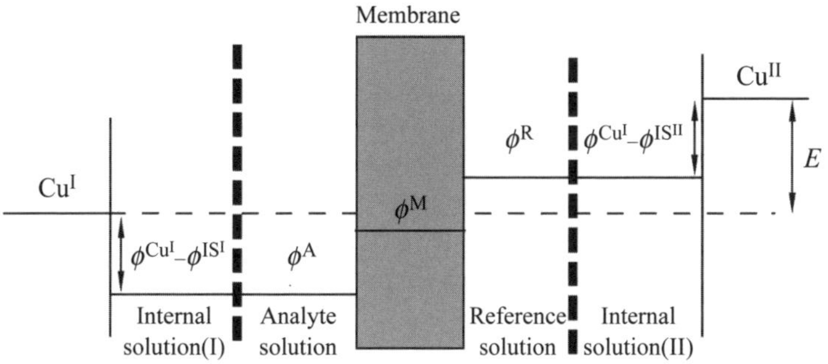

Fig. 2.23 Potential distribution for an ion selective electrodes comprising two identical reference electrodes.

of the other ions can be neglected, i.e. that the concentrations of the other hydrophilic ions are negligible in the membrane and that the concentrations of the lipophilic ions are negligible in the two aqueous compartments. If this is not the case, diffusion of the different salts through the membrane will take place. When salt transport cannot be neglected, the equilibrium condition (2.81) applies to the activities of the species i at the boundaries and equation (2.83) applies to the surface concentrations at the membrane. However, the diffusion of a salt from one phase to the other is a slow process in relation to the time of the experiment, and a potentiometric measurement can be done with frequent calibration to take into account the departure from the ideal behaviour.

pH electrode

The pH is defined by the equation

$$\mathrm{pH} \quad = \quad -\log a_{\mathrm{H}^+} \tag{2.85}$$

The most direct method for measuring the pH of a solution is to measure the potential of a hydrogen electrode working with a hydrogen fugacity equal to the standard pressure of 1 bar, versus a standard hydrogen electrode (SHE) as shown in the cell below.

$$\mathrm{Cu}^{\mathrm{I}} \mid \mathrm{Pt} \mid \mathrm{H}^+(a_{\mathrm{H}^+}=1), \tfrac{1}{2}\mathrm{H}_2(p^{\ominus}) .... \| .... \tfrac{1}{2}\mathrm{H}_2(p^{\ominus}), \mathrm{H}^+ \mid \mathrm{Pt} \mid \mathrm{Cu}^{\mathrm{II}}$$

Neglecting the potential of the liquid junction between the two solutions, the equality of the hydrogen fugacities in the two compartments gives us directly

$$E_{\mathrm{SHE}} \quad = \quad \frac{RT}{F}\ln a_{\mathrm{H}^+} \quad = \quad -\frac{RT\ln 10}{F}\mathrm{pH} \tag{2.86}$$

If, in place of a standard hydrogen reference electrode, we use a hydrogen reference electrode operating with a buffer solution of a given pH ($\mathrm{pH}_{\mathrm{buffer}}$) then, by neglecting the potential of the liquid junction, and by keeping the hydrogen fugacity equal in the two compartments, we have

$$E = \frac{RT\ln 10}{F}\left[\mathrm{pH} - \mathrm{pH}_{\mathrm{buffer}}\right] \tag{2.87}$$

Nevertheless, as we have already seen, hydrogen electrodes are not practical to use, and glass electrodes are most commonly used. They are often found in the form of combined electrodes as illustrated in Figure 2.24. The internal compartment of the electrode contains a solution of constant pH with a reference electrode (e.g. Ag|AgCl). The external compartment contains a reference electrode in contact with the analyte via a liquid junction.

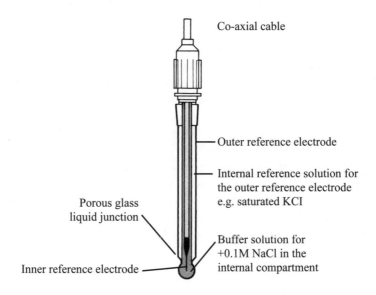

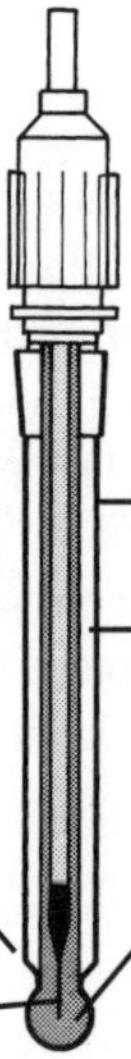

Fig. 2.24 Glass pH electrode (Copyright, Metrohm, CH).

The composition of glass pH electrodes varies with the manufacturer. The major components are SiO_2, Na_2O, CaO with a certain proportion of Li and Ba. The mobility of the cations Na^+ and Li^+ ensures the passage of a very small current in the glass and allows a thin layer to form on the surfaces of the glass in contact with the solutions where the alkali cations are replaced by protons. This thin layer, sometimes called the hydrated glass layer, is the basis of the functioning of the pH electrode.

The pH electrode requires frequent calibrations in buffer solutions. Institutions such as IUPAC (the International Union of Pure & Applied Chemistry) or the NIST (National Institute of Standards and Technology, USA) have established lists of reference buffers, and certified standard buffer solutions are available on the market. The measuring cell of a pH electrode is written as

$$\mathrm{Ag}\,|\,\mathrm{AgCl}\,|\,\mathrm{KCl} \geq 3.5\ \mathrm{M}\,||\,\mathrm{Solution}\,|\,\mathrm{Glass}\,|\,\mathrm{Buffer\ Solution} + 0.1\ \mathrm{M\ NaCl}\,|\,\mathrm{AgCl}\,|\,\mathrm{Ag}$$

Modelling the response of a glass electrode remains difficult, and the simple model of the response of an ion selective electrode developed above (equation (2.83)) is not really applicable in that it is not certain that the protons can be in equilibrium throughout the glass. Other models are based on the concept of hydrated glass layers where protons and sodium ions coexist.

The difference in Galvani potential $\phi^{R} - \phi^{A}$ between the reference buffer solution and the analyte is then given by

$$\phi^{R} - \phi^{A} = \left(\phi^{R} - \phi^{HGR}\right) + \left(\phi^{HGR} - \phi^{G}\right) + \left(\phi^{G} - \phi^{HGA}\right) + \left(\phi^{HGA} - \phi^{A}\right) \quad (2.88)$$

where the exponents HGR and HGA indicate the layers of hydrated glass on the reference solution side and the analyte side.

Equation (2.44) for the equilibrium of the proton between the aqueous solution and the adjacent layer of hydrated glass allows us to write

$$\left(\phi^{R} - \phi^{HGR}\right) + \left(\phi^{HGA} - \phi^{A}\right) = \Delta_{HGR}^{R}\phi_{H^+}^{\ominus} + \frac{RT}{F}\ln\left(\frac{a_{H^+}^{HGR}}{a_{H^+}^{R}}\right) + \Delta_{A}^{HGA}\phi_{H^+}^{\ominus} + \frac{RT}{F}\ln\left(\frac{a_{H^+}^{A}}{a_{H^+}^{HGA}}\right)$$

$$= \frac{RT}{F}\ln\left(\frac{a_{H^+}^{HGR}a_{H^+}^{A}}{a_{H^+}^{R}a_{H^+}^{HGA}}\right) \quad (2.89)$$

and the one for the equilibrium of sodium between the layer of hydrated glass and the glass

$$\left(\phi^{HGR} - \phi^{G}\right) + \left(\phi^{G} - \phi^{HGA}\right) = \Delta_{G}^{HGR}\phi_{Na^+}^{\ominus} + \frac{RT}{F}\ln\left(\frac{a_{Na^+}^{G}}{a_{Na^+}^{HGR}}\right) + \Delta_{HGA}^{G}\phi_{Na^+}^{\ominus} + \frac{RT}{F}\ln\left(\frac{a_{Na^+}^{HGA}}{a_{Na^+}^{G}}\right)$$

$$= \frac{RT}{F}\ln\left(\frac{a_{Na^+}^{HGA}}{a_{Na^+}^{HGR}}\right) \quad (2.90)$$

from which we obtain:

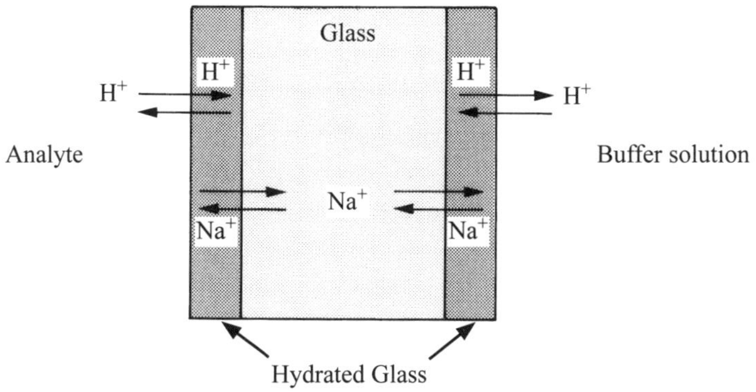

Fig. 2.25 Ionic equilibria in a glass electrode.

$$\phi^{R} - \phi^{A} = \frac{RT}{F}\ln\left(\frac{a_{H^+}^{A}}{a_{H^+}^{R}}\right) + \frac{RT}{F}\ln\left(\frac{a_{H^+}^{HGR}\, a_{Na^+}^{HGA}}{a_{H^+}^{HGA}\, a_{Na^+}^{HGR}}\right) \tag{2.91}$$

The second term in equation (2.91) is sometimes called the asymmetry potential that can be measured by calibration. This corresponds to the exchange of ions.

$$H^{+HGA} + Na^{+HGR} \rightleftarrows Na^{+HGA} + H^{+HGR}$$

This term varies little, but its fluctuations are not negligible when the electrode is being used. Fortunately, these variations, also called 'drifts', in the response of the electrode are slow and may be compensated for by regular calibration.

The Nikolsky equation

In the presence of an interfering ion j that can also be partially dissolved in the membrane, the Galvani potential difference is not imposed solely by the distribution of the principal ion i but also by the distribution of the interfering ion j, and equation (2.83) becomes

$$\phi^{R} - \phi^{A} = \frac{RT}{z_i F}\ln\left(\frac{a_i^{A} + K_{ij}^{pot}\left(a_j^{A}\right)^{z_i/z_j}}{a_i^{R}}\right) \tag{2.92}$$

where K_{ij}^{pot} is the selectivity coefficient of the principal ion i with respect to the interfering ion j. This is an empirical relation known as the ***Nikolsky equation*** and the selectivity coefficients are measured directly from the response of the ISE.

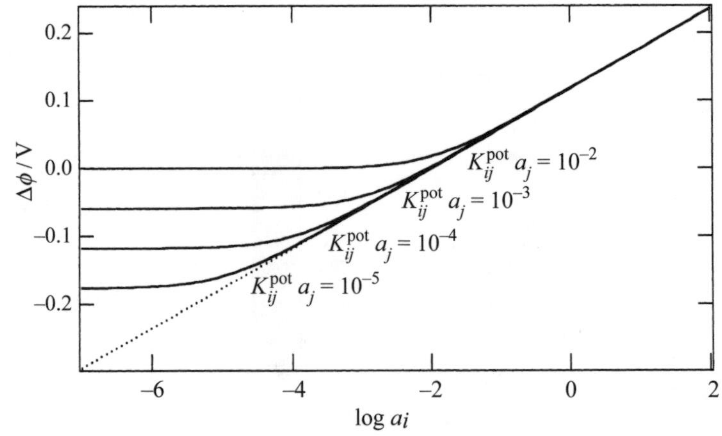

Fig. 2.26 Response of an ion selective electrode in the presence of interfering ions.

Let's demonstrate the Nikolsky equation in the case where the principal ion and the interfering ion are monovalent. To do this, let's consider the equilibrium relative to the penetration of the interfering ion in the membrane by an exchange reaction that can be written as

$$i^{\mathrm{M}} + j^{\mathrm{A}} \rightleftarrows i^{\mathrm{A}} + j^{\mathrm{M}}$$

The equilibrium constant is independent of the Galvani potential difference between the membrane and the aqueous analyte solution and is given by

$$K_{ij} = \frac{a_i^{\mathrm{A}} a_j^{\mathrm{M}}}{a_i^{\mathrm{M}} a_j^{\mathrm{A}}} = \exp\left(-\frac{\Delta G_{\mathrm{tr},i}^{\ominus,\mathrm{M}\to\mathrm{A}} - \Delta G_{\mathrm{tr},j}^{\ominus,\mathrm{M}\to\mathrm{A}}}{RT}\right) \tag{2.93}$$

The Galvani potential difference across the membrane | analyte interface is then

$$\Delta_{\mathrm{A}}^{\mathrm{M}}\phi = \frac{\Delta G_{\mathrm{tr},i}^{\ominus,\mathrm{M}\to\mathrm{A}}}{F} + \frac{RT}{F}\ln\left(\frac{a_i^{\mathrm{A}}}{a_i^{\mathrm{M}}}\right) = \frac{\Delta G_{\mathrm{tr},j}^{\ominus,\mathrm{M}\to\mathrm{A}}}{F} + \frac{RT}{F}\ln\left(\frac{a_j^{\mathrm{A}}}{a_j^{\mathrm{M}}}\right) \tag{2.94}$$

The electroneutrality condition in the membrane reads

$$c_i^{\mathrm{M}} + c_j^{\mathrm{M}} = c_{\mathrm{X}^-}^{\mathrm{M}} \tag{2.95}$$

Thus, the activity ratio between the analyte and the membrane can be expressed as

$$\frac{a_i^{\mathrm{A}}}{a_i^{\mathrm{M}}} = \frac{a_i^{\mathrm{A}}\left[c_i^{\mathrm{M}} + c_j^{\mathrm{M}}\right]}{\gamma_i^{\mathrm{M}} c_i^{\mathrm{M}} c_{\mathrm{X}^-}^{\mathrm{M}}} = \frac{a_i^{\mathrm{A}}\left[1 + \dfrac{c_j^{\mathrm{M}}}{c_i^{\mathrm{M}}}\right]}{\gamma_i^{\mathrm{M}} c_{\mathrm{X}^-}^{\mathrm{M}}} = \frac{a_i^{\mathrm{A}}\left[1 + K_{ij}\dfrac{\gamma_i^{\mathrm{M}}}{\gamma_j^{\mathrm{M}}}\dfrac{a_j^{\mathrm{A}}}{a_i^{\mathrm{A}}}\right]}{\gamma_i^{\mathrm{M}} c_{\mathrm{X}^-}^{\mathrm{M}}} \tag{2.96}$$

The Galvani potential difference across the membrane | analyte interface is then

$$\Delta_{\mathrm{A}}^{\mathrm{M}}\phi = \frac{\Delta G_{\mathrm{tr},i}^{\ominus,\mathrm{M}\to\mathrm{A}}}{F} + \frac{RT}{F}\ln\left(\frac{a_i^{\mathrm{A}} + K_{ij}\dfrac{\gamma_i^{\mathrm{M}}}{\gamma_j^{\mathrm{M}}} a_j^{\mathrm{A}}}{\gamma_i^{\mathrm{M}} c_{\mathrm{X}^-}^{\mathrm{M}}}\right) \tag{2.97}$$

In the same way, the Galvani potential difference across the membrane | reference electrolyte interface is also written as

$$\begin{aligned}\Delta_{\mathrm{M}}^{\mathrm{R}}\phi &= -\frac{\Delta G_{\mathrm{tr},i}^{\ominus,\mathrm{M}\to\mathrm{R}}}{F} + \frac{RT}{F}\ln\left(\frac{\gamma_i^{\mathrm{M}} c_{\mathrm{X}^-}^{\mathrm{M}}}{a_i^{\mathrm{R}} + K_{ij}\dfrac{\gamma_i^{\mathrm{M}}}{\gamma_j^{\mathrm{M}}} a_j^{\mathrm{R}}}\right) \\ &\cong -\frac{\Delta G_{\mathrm{tr},i}^{\ominus,\mathrm{M}\to\mathrm{R}}}{F} + \frac{RT}{F}\ln\left(\frac{\gamma_i^{\mathrm{M}} c_{\mathrm{X}^-}^{\mathrm{M}}}{a_i^{\mathrm{R}}}\right)\end{aligned} \tag{2.98}$$

with the hypothesis that the concentration of the interfering ion in the reference electrolyte is negligible. Finally, the potential at the terminals of the ISE is obtained by adding equations (2.97) and (2.98) where the elimination of the standard Gibbs energies gives

$$\Delta_{\mathrm{A}}^{\mathrm{R}}\phi = \frac{RT}{F}\ln\left(\frac{a_i^{\mathrm{A}} + K_{ij}\dfrac{\gamma_i^{\mathrm{M}}}{\gamma_j^{\mathrm{M}}}a_j^{\mathrm{A}}}{\gamma_i^{\mathrm{M}}c_{\mathrm{X}^-}^{\mathrm{M}}}\right) + \frac{RT}{F}\ln\left(\frac{\gamma_i^{\mathrm{M}}c_{\mathrm{X}^-}^{\mathrm{M}}}{a_i^{\mathrm{R}}}\right) = \frac{RT}{F}\ln\left(\frac{a_i^{\mathrm{A}} + K_{ij}\dfrac{\gamma_i^{\mathrm{M}}}{\gamma_j^{\mathrm{M}}}a_j^{\mathrm{A}}}{a_i^{\mathrm{R}}}\right) \tag{2.99}$$

If the concentration of the principal ion in the membrane is quasi-constant, then the activity coefficient is itself quasi-constant, and if, furthermore, we consider the activity coefficient of the interfering ion as almost equal to unity, then the selectivity coefficient K_{ij}^{pot} is given by

$$K_{ij}^{\mathrm{pot}} = K_{ij}\gamma_i^{\mathrm{M}} \tag{2.100}$$

2.5.4 Ion Selective Field Effect Transistor (ISFET)

Let us consider a *p*-type field effect transistor with *n*-type source and drain. The principle of the field effect is to polarise the metal gate of the transistor with a positive potential with respect to the *p*-type semiconductor. In this way, just under the insulating layer (often made of silicon dioxide SiO_2 or silicon nitride Si_3N_4), the semiconductor becomes *n*-type by the accumulation of electrons caused by the positive polarisation of the gate. As soon as this inversion from *p*-type to *n*-type has taken place under the gate, a current flows between the source and the drain.

In a membrane ISFET the metal gate is replaced by a membrane that is similar to the ones used in ISEs. In this way, the Nernst's law for the distribution of an ion *i* across the analyte|membrane interface (see equation (2.44)) is used to control the field effect in the transistor.

2.6 ION EXCHANGE MEMBRANES

2.6.1 Structure

Membranes play an ever increasing role in separation processes. The driving force for any separation process is always a gradient of electrochemical potential. These gradients can be generated either by pressure gradients (microfiltration, ultrafiltration, inverse osmosis), by concentration gradients (dialysis), or by electrical potential gradients.

Ion exchange membranes are principally polymer membranes treated by grafting with anionic or cationic groups. The anionic groups are often sulfonates $(-SO_3^-)$

or carboxylates $(-COO^-)$, whilst the cationic groups are principally tertiary amines $(-N^+(CH_3)_3)$.

These polymers are generally strongly reticulated in order to avoid excessive swelling on contact with solutions. They come in a granular form, for example for ionic chromatography, or in the form of sheets with a polymer support incorporated. The characteristics of ion exchange membranes used in industry are the following:

- high charge selectivity
- good ionic conductivity
- good mechanical stability
- a charge density of 1 to 2 molal (1 mole of ions per kilogram of polymer).

2.6.2 Donnan potential

A cation exchange membrane is impermeable to anions, while an anion exchange membrane is impermeable to cations (with the exception of protons). So, if a cation exchange membrane separates two electrolyte solutions (e.g. NaCl) and if a potential is applied to the terminals of the membrane, the electric current across the membrane will be carried only by the mobile cations. The exclusion of the mobile anion Cl^- from the membrane is due to the ***Donnan exclusion principle***. In order to understand this principle, let's consider a cation exchange membrane with anionic fixed charges R^- and sodium counter-ions in equilibrium with adjacent solutions of NaCl and calculate the concentration of chloride ions in the membrane at equilibrium.

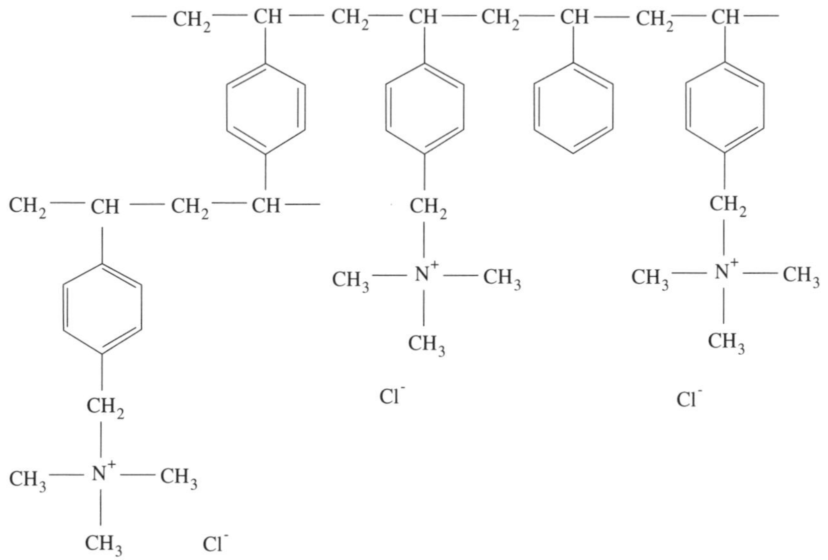

Fig. 2.27 Anion exchange polymer.

The equality of electrochemical potentials of the mobile ions between the membrane phase and the liquid phase is written as

$$\mu_i^{\ominus,\mathrm{m}} + RT\ln c_i^{\mathrm{m}} + RT\ln \gamma_i^{\mathrm{m}} + z_i F\,\phi^{\mathrm{m}} \;=\; \mu_i^{\ominus,\mathrm{s}} + RT\ln c_i^{\mathrm{s}} + RT\ln \gamma_i^{\mathrm{s}} + z_i F\,\phi^{\mathrm{s}} \tag{2.101}$$

Considering the standard states to be equal in the membrane and in the aqueous solution ($\mu_i^{\ominus,\mathrm{m}} = \mu_i^{\ominus,\mathrm{s}}$) because the solvent is common to the two phases, we can see that the Galvani potential difference between the two phases is then

$$\phi^{\mathrm{m}} - \phi^{\mathrm{s}} \;=\; \frac{RT}{z_i F}\ln\left(\frac{\gamma_i^{\mathrm{s}}\,c_i^{\mathrm{s}}}{\gamma_i^{\mathrm{m}}\,c_i^{\mathrm{m}}}\right) \;=\; E_D \tag{2.102}$$

In general, the cation molality inside the membrane (e.g. Na^+) is larger than in solution. This means that we have a difference in inner potential between the solution and the membrane. This is called the ***Donnan potential***.

To estimate the chloride concentration in the membrane, we therefore write that

$$E_D \;=\; \frac{RT}{F}\ln\left(\frac{a_{\mathrm{Na}^+}^{\mathrm{s}}}{a_{\mathrm{Na}^+}^{\mathrm{m}}}\right) \;=\; -\frac{RT}{F}\ln\left(\frac{a_{\mathrm{Cl}^-}^{\mathrm{s}}}{a_{\mathrm{Cl}^-}^{\mathrm{m}}}\right) \tag{2.103}$$

or again

$$a_{\mathrm{Na}^+}^{\mathrm{s}}\,a_{\mathrm{Cl}^-}^{\mathrm{s}} \;=\; a_{\mathrm{Na}^+}^{\mathrm{m}}\,a_{\mathrm{Cl}^-}^{\mathrm{m}} \tag{2.104}$$

Ignoring in a first approximation the activity coefficients, this equation simplifies to

$$c_{\mathrm{Na}^+}^{\mathrm{s}}\,c_{\mathrm{Cl}^-}^{\mathrm{s}} \;=\; c_{\mathrm{Na}^+}^{\mathrm{m}}\,c_{\mathrm{Cl}^-}^{\mathrm{m}} \tag{2.105}$$

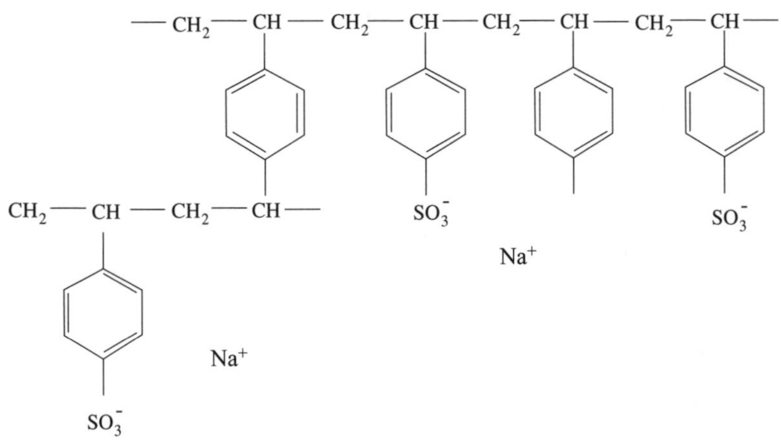

Fig. 2.28 Cation exchange polymer.

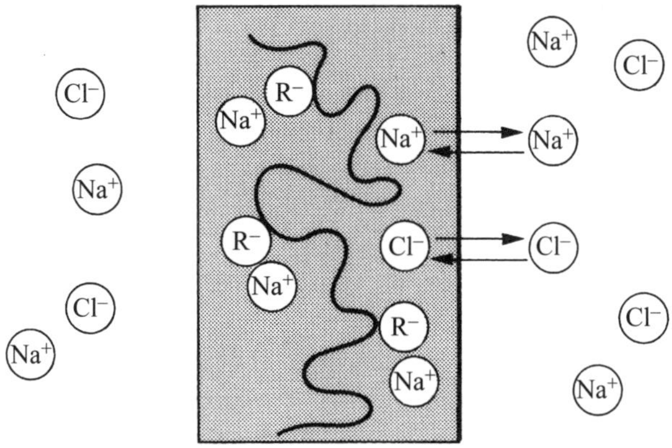

Fig. 2.29 Cation exchange membrane.

The electro-neutrality of the two phases is written as

$$c^{m}_{Na^+} = c^{m}_{Cl^-} + c^{m}_{R^-} \tag{2.106}$$

where R^- is the anion fixed in the membrane

$$c^{s}_{Na^+} = c^{s}_{Cl^-} \tag{2.107}$$

By combining (2.104), (2.105) and (2.106), we obtain a quadratic equation for $c^{m}_{Cl^-}$

$$\left(c^{m}_{Cl^-}\right)^2 + c^{m}_{Cl^-} c^{m}_{R^-} = \left(c^{s}_{Cl^-}\right)^2 \tag{2.108}$$

Resolving this equation gives

$$c^{m}_{Cl^-} = \frac{-c^{m}_{R^-} + \sqrt{\left(c^{m}_{R^-}\right)^2 + 4\left(c^{s}_{Cl^-}\right)^2}}{2} = \frac{c^{m}_{R^-}}{2}\left(\sqrt{1 + 4\left(\frac{c^{s}_{Cl^-}}{c^{m}_{R^-}}\right)^2} - 1\right) \tag{2.109}$$

If $c^{s}_{Cl^-} << c^{m}_{R^-}$, then :

$$\sqrt{1 + 4\left(\frac{c^{s}_{Cl^-}}{c^{m}_{R^-}}\right)^2} = 1 + 2\left(\frac{c^{s}_{Cl^-}}{c^{m}_{R^-}}\right)^2 \tag{2.110}$$

and

$$c^{m}_{Cl^-} = \left(c^{s}_{Cl^-}\right)^2 \ / \ c^{m}_{R^-} \tag{2.111}$$

The condition $c^{m}_{R^-} > c^{s}_{Cl^-}$ applies above all to membranes in which the space accessible to the electrolyte is restricted by narrow pores.

APPENDIX : THE RESPIRATORY CHAIN

Standard redox potentials play an important role in redox metabolisms such as the respiratory chain, photosynthesis and in a more general way, bioenergetic systems.

It is important to note that the standard redox potentials in biology are not always expressed in relation to the standard hydrogen electrode ($a_{H^+} = 1$), but in relation to a hydrogen electrode working at the physiological pH of 7. Thus the two scales differ by 420 mV; the standard redox potential of oxygen on the biological scale being 0.82 V instead of 1.24 V on the SHE scale.

Also, the standard redox potentials of biological systems have mostly been measured in aqueous media even if the redox species operates *in vivo* in a membrane environment.

Nonetheless, Figure 2.30 shows that in this chain, the two electrons generated by the oxidation of NADH tumble down this cascade of redox reactions eventually to reduce a molecule of oxygen. This transfer of electrons therefore happens in an exergonic fashion. The first stage of the chain is an enzyme reaction involving the enzyme NADH dehydrogenase whose prosthetic group is the flavine mononucleotide FMN. In fact, this first stage is itself a chain of redox reactions.

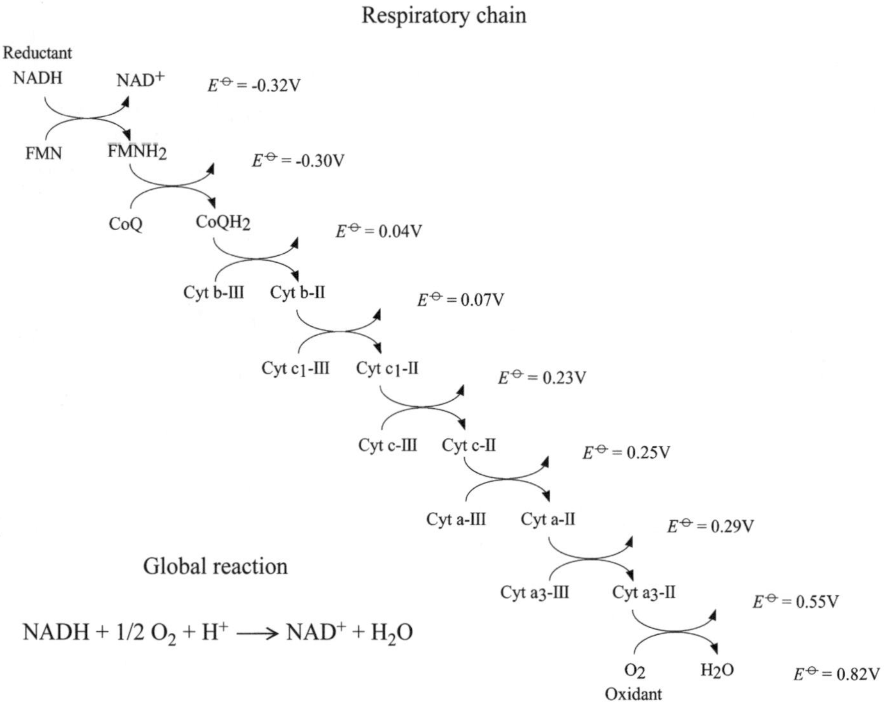

Fig. 2.30 Successive reactions taking place in the respiratory chain.

CHAPTER 3

ELECTROLYTE SOLUTIONS

3.1 LIQUIDS

3.1.1 Radial distribution function

To consider the structure of a liquid, it is useful to define the radial distribution function $g(r)$ which is the representation of the probability of finding a given atom or molecule in a shell of thickness dr from a central atom or molecule. The quantity $n(r)$dr is then the number of atoms or molecules whose distance from the origin is between r and r+dr and is given by

$$n(r)\mathrm{d}r \;= \rho\, g(r)\, 4\pi r^2 \mathrm{d}r \tag{3.1}$$

where ρ is the density of the liquid.

In the case of a crystalline solid, the distribution function is regular, and represents the order of the lattice. The distribution function of solid mercury clearly illustrates the order existing at short and long distances around the central atom. In the case of a liquid, the probability of finding an atom far away from the central atom becomes uniform, and this value is usually taken to normalise the radial distribution function of the liquid as illustrated in Figure 3.1 for liquid mercury. This figure shows that, for a liquid, there exists an order at short distances around the central atom which disappears very quickly after a few atomic radii.

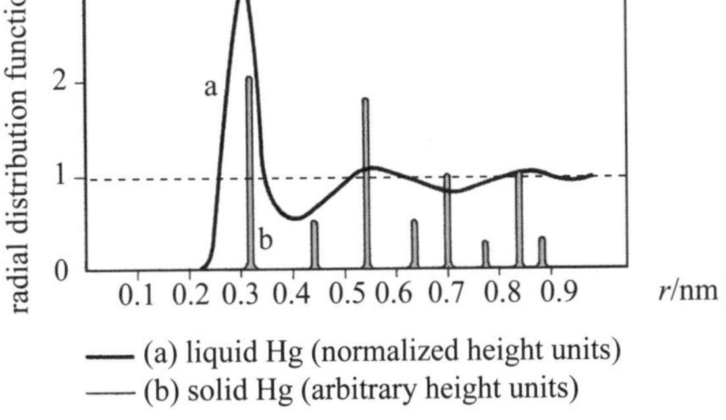

Fig. 3.1 Distribution function for liquid and solid mercury (Adapted from D.Tabor: Gases, liquids and solids, Cambridge University Press, 1969, Cambridge).

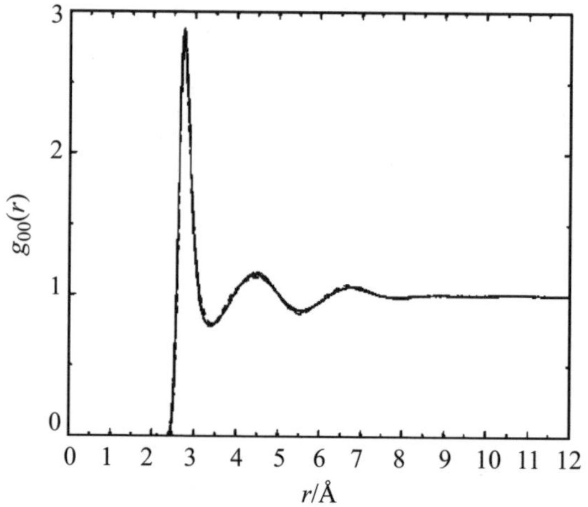

Fig. 3.2 Distribution function for liquid water at room temperature, r represents the distance between two oxygen atoms (G. Hura et al, *Phys. Chem. Chem. Phys.*, 5 (2003) 1981-1991, Reproduced by permission of the PCCP Owner Societies).

This type of distribution function is characteristic of liquids. Even for a liquid as complex as water, we can see in Figure 3.2 that the order does not extend above three molecular layers.

We can therefore conclude that a liquid differs from a solid by the value of the structural order. In a solid, this order extends over all the molecules with a few faults spread through the network. In a liquid, the structural order extends only to sub-ensembles, whose size varies constantly. Thus the coordination number of a molecule in a liquid varies perpetually (between 4 and 5 in the case of water), whilst it is fixed in a solid.

3.1.2 Water

Water is the commonest solvent surrounding us, and above all the major constituent of our body (55 % of their mass, 70-80 % of the mass of our brain). Furthermore, water is a solvent with physical properties that are rather special. From a geometric point of view, water contains 8 electrons distributed in hybrid orbitals sp^3, like methane and ammonia. The presence of two orbitals and two OH bonds facilitates a tetrahedral coordination of H_2O, which is the reason for its specific properties. In the solid state, there are a large number of crystalline structures according to temperature and pressure. The form existing at 0°C and atmospheric pressure is ***Ice I***, which has a hexagonal structure comprising cavities formed by six water molecules as illustrated in Figure 3.3. Each molecule is then tetracoordinated, and this results in a density of only 0.924 g/ml, rather amazing on first sight.

At room temperature, water is a liquid, unlike similar molecules such as CH_4, NH_3 and H_2S that are gaseous. This liquid state indicates the presence of very strong interactions that maintain cohesion and induce a certain organisation of the liquid. The

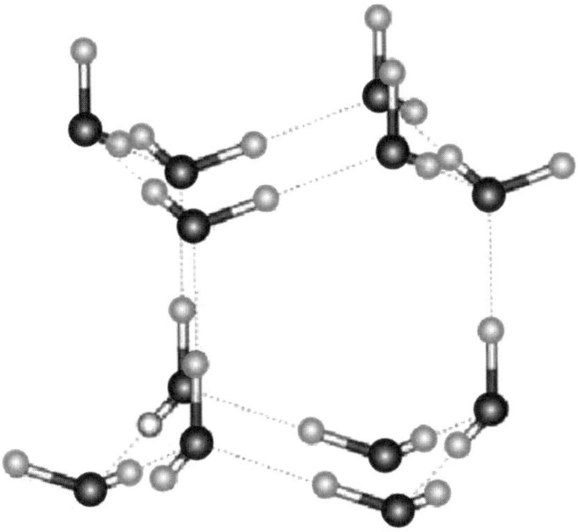

Fig. 3.3 Structure of ice I. Each atom of oxygen is surrounded by four atoms of hydrogen, two for the molecule at 101pm and two for two neighbouring molecules at 175pm (from http://www.sbu.ac.uk/water).

hydrogen bonds play an essential role in the structure of water, whether it is liquid or solid.

In the case of liquid water, each molecule is surrounded by four or five molecules, which is very few compared to a model of rigid spheres where each sphere is surrounded by twelve neighbours, and is also few compared to simple liquids like argon where each atom is surrounded by at least eight neighbours.

There are many models of liquid water. We can cite the Frank model, called 'the mixture model', where water is considered as a mixture of aggregates of molecules organised on the basis of the structure of ice and of free molecules. This model is based on the idea that the hydrogen bonds are highly directional. The other family of models considers water as a mixture of two or several types of water molecules, e.g. a mixture of tetra- and penta-coordinated molecules.

3.2 THERMODYNAMIC ASPECTS OF SOLVATION

We have seen that the work done in transferring an ion from a vacuum to a phase is by definition the electrochemical potential, and now we shall see how this value is linked to thermodynamic data of formation and solvation. The numerical values given in this paragraph come from various sources (which explains why sometimes there is a difference in the values) and are only given generally as an indication of the orders of magnitude involved.

3.2.1 Standard enthalpies of ion formation

The thermochemical data of formation such as the ***standard enthalpy of formation*** are expressed from gaseous, liquid or solid reference states (the most stable state of the element considered at 298 K). For example, the standard formation enthalpy of liquid water corresponds to the reaction

$$H_{2(g)} + \tfrac{1}{2}O_{2(g)} \longrightarrow H_2O_{(l)} \qquad \Delta H_f^{\ominus}(H_2O_{(l)}) = -285.8\ \mathrm{kJ \cdot mol^{-1}}$$

IMPORTANT NOTE

The tables of thermodynamic data give the formation enthalpies and the Gibbs energies of formation, but give the absolute standard molar entropies for which the reference state is a perfect crystal at 0 K, except for ions in solution for which the reference state is the standard molar entropy of the proton in solution, taken to be zero.

The Born-Haber cycle for $NaCl_{(s)}$

The standard formation enthalpy of a solid may be found by several methods. Let's take the classic case of sodium chloride NaCl. The general reaction can be written as

$$Na_{(s)} + \tfrac{1}{2}Cl_{2(g)} \longrightarrow NaCl_{(s)}$$

In fact, this equation may be considered as the sum of five elementary reactions: the sublimation of solid sodium, the ionisation of atomic sodium, the atomisation of molecular chlorine, the electron affinity of atomic chlorine and finally the formation of $NaCl_{(s)}$ from the ions in the gas phase.

$$Na_{(s)} \longrightarrow Na_{(g)} \qquad \Delta H_{sub}^{\ominus}(Na) = 107.3\ \mathrm{kJ \cdot mol^{-1}}$$

$$Na_{(g)} \longrightarrow Na^+_{(g)} + e^-_{(g)} \qquad \Delta H_i^{\ominus}(Na) = 495.8\ \mathrm{kJ \cdot mol^{-1}}$$

$$\tfrac{1}{2}Cl_{2(g)} \longrightarrow Cl_{(g)} \qquad \tfrac{1}{2}\Delta H^{\ominus}(Cl-Cl) = 121\ \mathrm{kJ \cdot mol^{-1}}$$

$$Cl_{(g)} + e^-_{(g)} \longrightarrow Cl^-_{(g)} \qquad \Delta H_{ea}^{\ominus}(Cl) = -348.7\ \mathrm{kJ \cdot mol^{-1}}$$

$$Na^+_{(g)} + Cl^-_{(g)} \longrightarrow NaCl_{(s)} \qquad \Delta H_R^{\ominus}(NaCl) = -786.5\ \mathrm{kJ \cdot mol^{-1}}$$

The change in standard enthalpy of the fifth reaction is the ***standard reticulation enthalpy*** (or lattice formation enthalpy) which corresponds to the formation of a crystalline solid from ions in the gas phase.

$$M^+_{(g)} + X^-_{(g)} \longrightarrow MX_{(s)} \qquad \Delta H_R^{\ominus} \leq 0$$

All standard reticulation enthalpies are negative. The energy of reticulation is defined as the reticulation enthalpy at $T = 0$ K. The ***lattice energy*** (or lattice dissociation energy) or network energy is positive and equal to the opposite of the reticulation energy.

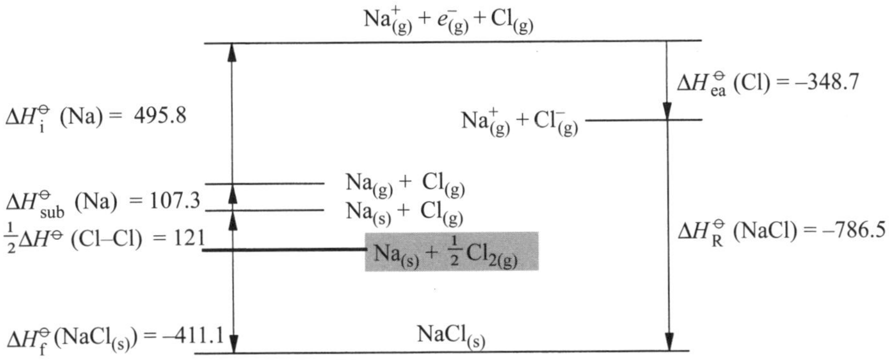

Fig. 3.4 The Born-Haber cycle for solid NaCl. Numerical values in kJ·mol^{-1}.

Figure 3.4 illustrates how to calculate the standard enthalpy of formation of $NaCl_{(s)}$ using experimentally accessible values.

The Born-Haber cycle for $HCl_{(aq)}$

Born-Haber cycles can also be used for determining the standard formation enthalpies of species in solution, which in the case of acids and salts is the standard formation enthalpy of the ions.

$$\tfrac{1}{2}\,Cl_{2(g)} + e^-_{(g)} \longrightarrow Cl^-_{(aq)} \qquad \Delta H^\ominus_f\,(Cl^-_{(aq)})$$

where the standard state in solution is in general either 1 mol·l^{-1} or 1 mol·kg^{-1}.

Because it is impossible, respecting the neutrality of the phases, to measure directly the formation enthalpies of the isolated ions in solution, it is necessary to make an arbitrary choice of scale, and a generally used convention is to take ***the standard formation enthalpy of the aqueous proton as equal to zero at all temperatures***.

$$\tfrac{1}{2}\,H_{2(g)} \longrightarrow H^+_{(aq)} + e^-_{(g)} \qquad \Delta H^\ominus_f\,(H^+_{(aq)}) = 0$$

In the same way, we make the hypothesis that the Gibbs energy of formation is also zero at all temperatures

$$\Delta G^\ominus_f\,(H^+_{(aq)}) = 0$$

and consequently that the absolute standard molar entropy is also zero. Be careful, this last hypothesis is often a source of confusion.

$$S^\ominus\,(H^+_{(aq)}) = 0$$

Thus, using these hypotheses, the standard formation enthalpy of the chloride anion in solution is equal to the standard formation enthalpy of $HCl_{(aq)}$ and can be determined by measuring experimentally the standard hydration enthalpy of $HCl_{(g)}$,

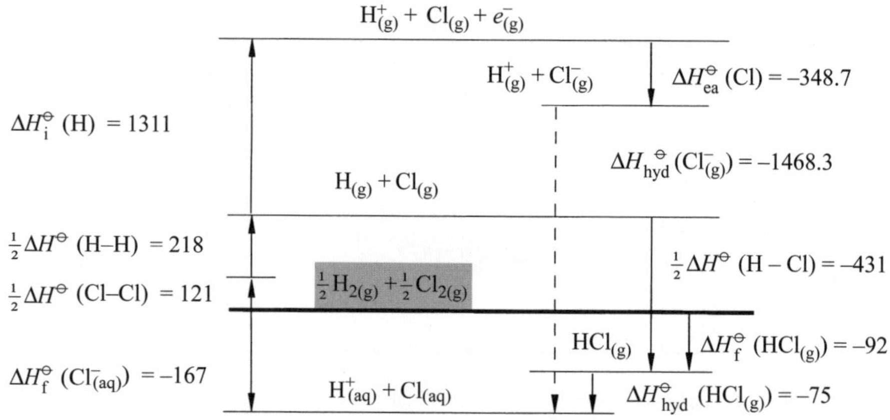

Fig. 3.5 The Born-Haber cycle for aqueous HCl. Numerical values in kJ·mol^{-1}. The dotted arrow relates to the standard hydration enthalpies.

the standard formation energy of $HCl_{(g)}$ being determined from the bond energies of H_2, Cl_2 and HCl as illustrated in Figure 3.5.

Actually, it is possible to measure the hydration enthalpies of salts (e.g.$NaCl_{(s)} \longrightarrow Na^+_{(aq)} + Cl^-_{(aq)}$) and acids (e.g. $HCl_{(g)} \longrightarrow H^+_{(aq)} + Cl^-_{(aq)}$). From an experimental point of view, the method consists of placing a precise quantity of the salt in a glass bulb, immersing this in the solvent and measuring by calorimetry the heat generated by the dissolution of the salt when the bulb is broken. Table 3.1 shows some values obtained by this method. Note in passing that these values can be either positive or negative.

Table 3.1 Standard hydration enthalpies of salts in kJ·mol^{-1}
(NBS Tables of chemical thermodynamic properties,
J. Phys. & Chem. Reference Data, 11 (1982)).

	F^-	Cl^-	Br^-	I^-	OH^-	CO_3^{2-}	NO_3^-	SO_4^{2-}
Li^+	4.9	–37	–48.8	–63.3	–23.6	–18.2	–2.7	–29.8
Na^+	1.9	3.89	–0.6	–7.5	–44.5	–26.7	20.4	–2.4
K^+	–17.74	17.22	19.9	20.3	–57.1	–30.9	34.9	23.8
NH_4^+	–1.2	14.8	16	13.7			25.69	6.6
Mg^{2+}	–17.7	–160	–185.6	–213.2	2.3	–25.3	–90.9	–91.2
Ca^{2+}	11.5	–81.3	–103.1	–119.7	–16.7	–13.1	–19.2	–18.0

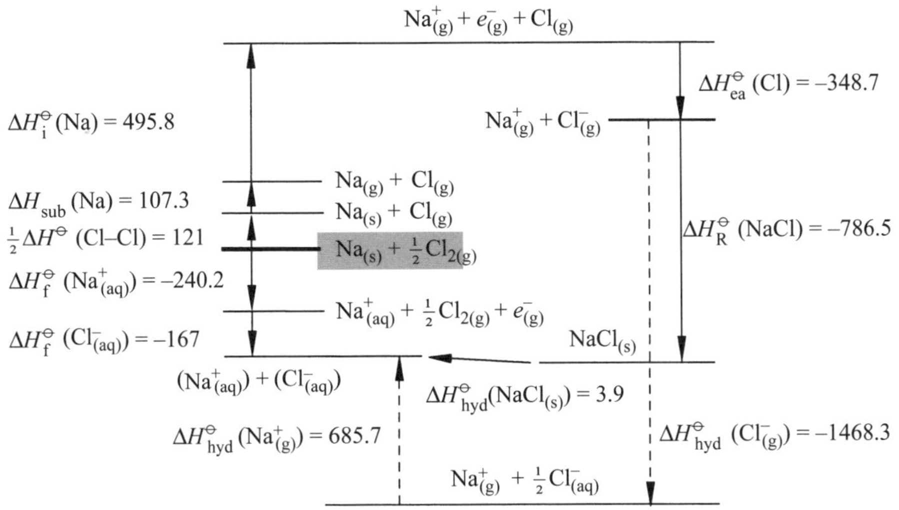

Fig. 3.6 The Born-Haber cycle for aqueous NaCl. Numerical values in kJ·mol^{-1}. The dotted arrows relate to the standard hydration enthalpies of the ions.

The Born-Haber cycle of $NaCl_{(aq)}$

Knowing the standard formation enthalpy of the ion $Cl^-_{(aq)}$, a Born-Haber cycle for NaCl dissolved in water allows the determination of the standard formation enthalpy of $Na^+_{(aq)}$ from the hydration enthalpy of the salt NaCl. as shown in Figure 3.6.

Electrochemical measurement of the standard formation enthalpies

The Gibbs energy of formation of an ion in the aqueous phase is, for certain species, a quantity that can be measured by electrochemical methods. Still looking at sodium, let's consider the reaction

$$\tfrac{1}{2}H_{2(g)} + Na^+_{(aq)} \rightleftarrows H^+_{(aq)} + Na_{(s)}$$

The standard Gibbs energy of this reaction is equal to the opposite of the standard Gibbs energy of formation of the sodium ion in aqueous solution. Thus, using equation (2.22), it follows that

$$\Delta G^\ominus_f (Na^+_{(aq)}) = F\left[E^\ominus_{Na^+/Na}\right]_{SHE} \tag{3.2}$$

It is important to note that the choice of the arbitrary scale of standard formation enthalpies of ions in solution $(\Delta H^\ominus_f(H^+_{(aq)}) = 0 \text{ kJ·mol}^{-1})$ is completely compatible with the arbitrary choice of the scale of standard redox potentials, i.e. $\left[E^\ominus_{H^+/\frac{1}{2}H_2}\right]_{SHE} = 0$ V.

The standard formation entropy is then given by the temperature coefficient of the standard redox potential

$$\Delta S_f^\ominus = -\left(\frac{\partial \Delta G_f^\ominus}{\partial T}\right)_p = -nF\left(\frac{\partial \left[E^\ominus\right]_{SHE}}{\partial T}\right)_p \tag{3.3}$$

We deduce from this that the standard formation enthalpy of the sodium ion in solution is given by

$$\begin{aligned}\Delta H_f^\ominus(Na^+_{(aq)}) &= \Delta G_f^\ominus(Na^+_{(aq)}) + T\Delta S_f^\ominus(Na^+_{(aq)}) \\ &= F\left(\left[E^\ominus_{Na^+/Na}\right]_{SHE} - T\left(\frac{\partial \left[E^\ominus_{Na^+/Na}\right]_{SHE}}{\partial T}\right)_p\right)\end{aligned} \tag{3.4}$$

EXAMPLE

Let's calculate the standard formation enthalpy of aqueous sodium, knowing that at 25°C $[E^\ominus_{Na^+/Na}]_{SHE} = -2.714$ V and that

$$\left(\frac{\partial \left[E^\ominus_{Na^+/Na}\right]_{SHE}}{\partial T}\right)_p = -0.000772 \ \mathrm{V \cdot K^{-1}}$$

By substituting in equation (3.2), we find the standard Gibbs energy of formation

$$\Delta G_f^\ominus(Na^+_{(aq)}) = -261.86 \ \mathrm{kJ \cdot mol^{-1}}$$

and by substituting in equation (3.4), we then deduce the standard formation enthalpy

$$\Delta H_f^\ominus(Na^+_{(aq)}) = -239.7 \ \mathrm{kJ \cdot mol^{-1}}$$

Note that it is also possible to calculate the formation entropy from the standard molar entropies, which are

$$S^\ominus(Na^+_{(aq)}) = 60.25 \ \mathrm{J \cdot mol^{-1} \cdot K^{-1}}$$

$$S^\ominus(Na_{(s)}) = 51.45 \ \mathrm{J \cdot mol^{-1} \cdot K^{-1}}$$

$$S^\ominus(H_{2(g)}) = 130.684 \ \mathrm{J \cdot mol^{-1} \cdot K^{-1}}$$

to obtain

$$\Delta S_f^\ominus(Na^+_{(aq)}) = -\left[51.45 - 60.25 - \tfrac{1}{2} \cdot 130.684\right] = 74.14 \ \mathrm{J \cdot mol^{-1} \cdot K^{-1}}$$

which can be compared to the value obtained from the temperature variation of the standard redox potential given by equation (3.3)

$$\Delta S_f^\ominus(Na^+_{(aq)}) = 96485 \cdot 0.000772 = 74.48 \ \mathrm{J \cdot mol^{-1} \cdot K^{-1}}$$

Calculation of the standard redox potential from thermodynamic data

To calculate a standard redox potential, we can calculate the Gibbs energy of the equilibrium

$$\text{ox} + \tfrac{n}{2}\,H_2 \rightleftarrows n\,H^+ + \text{red}$$

and obtain the standard redox potential from equation (2.22).

EXAMPLE:

Take, for example, the reduction of ferric ions to ferrous ions

$$Fe^{3+}_{(aq)} + e^- \rightleftarrows Fe^{2+}_{(aq)}$$

and calculate the Gibbs energy of the equilibrium

$$Fe^{3+}_{(aq)} + \tfrac{1}{2}H_{2(g)} \rightleftarrows Fe^{2+}_{(aq)} + H^+_{(aq)}$$

We can break this equilibrium down into elementary equilibria by making formation reactions appear

$$Fe_s \rightleftarrows Fe^{3+}_{(aq)} + 3e^- \qquad \Delta G_f^\ominus(Fe^{3+}_{(aq)}) = -4.6\ \text{kJ}\cdot\text{mol}^{-1} \qquad \text{(I)}$$

$$Fe_s \rightleftarrows Fe^{2+}_{(aq)} + 2e^- \qquad \Delta G_f^\ominus(Fe^{2+}_{(aq)}) = -78.9\ \text{kJ}\cdot\text{mol}^{-1} \qquad \text{(II)}$$

$$\tfrac{1}{2}H_{2(g)} \rightleftarrows H^+_{(aq)} + e^- \qquad \Delta G_f^\ominus(H^+_{(aq)}) = 0\ \text{kJ}\cdot\text{mol}^{-1} \qquad \text{(III)}$$

The numerical application of equation (2.22) gives us

$$\left[E^\ominus_{Fe^{III}/Fe^{II}}\right]_{SHE} = -\Delta G^\ominus / F = -\left[\Delta G_f^\ominus(Fe^{2+}_{(aq)}) + \Delta G_f^\ominus(H^+_{(aq)}) - \Delta G_f^\ominus(Fe^{3+}_{(aq)})\right] / F$$

$$= -[-78\,900 + 4\,600]/96\,485 = 0.77\ \text{V}$$

3.2.2 Standard enthalpies of ionic solvation

The ***standard solvation enthalpy***, e.g. of the sodium ion, corresponds to the reaction

$$Na^+_{(g)} \longrightarrow Na^+_{(aq)} \qquad \Delta H^\ominus_{sol}$$

The standard solvation enthalpy corresponds to the transfer of the ion from the gas in the standard state ($p^\ominus = 1$ bar) to the solution in the generally considered standard state in the scale of molalities ($\text{mol}\cdot\text{kg}^{-1}$) or of molarities ($\text{mol}\cdot\text{l}^{-1}$). When water is the solvent, the solvation enthalpy is called the ***hydration enthalpy***.

As for standard formation enthalpies, the standard solvation enthalpies are in general given on a scale where the ***standard solvation enthalpy of the proton is taken to be zero***.

$$H^+_{(g)} \longrightarrow H^+_{(aq)} \qquad \Delta H^\ominus_{hyd}(H^+) = 0\ \text{kJ}\cdot\text{mol}^{-1}$$

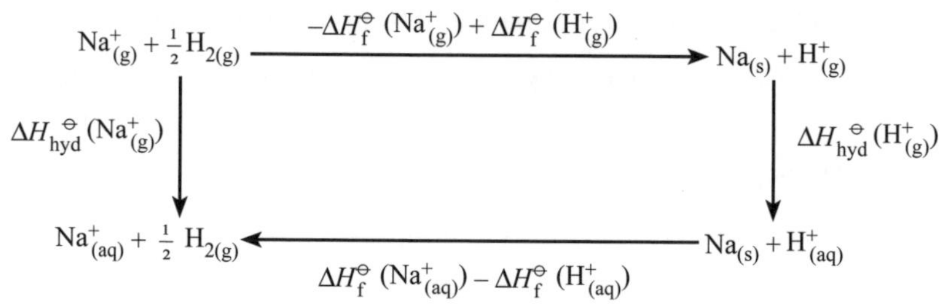

Fig. 3.7 Thermodynamic cycle for the hydration of sodium.

This is equivalent to saying that the hydration enthalpy corresponds to the transfer of an ion from the gas phase to the aqueous phase, associated to the transfer of a proton from the aqueous phase to the gas phase, as illustrated in Figure 3.7.

Using the cycle of Figure 3.7, the standard hydration enthalpy of Na^+ reads

$$\Delta H_{hyd}^\ominus(Na^+) \quad = \quad \Delta H_f^\ominus(Na^+_{(aq)}) - \Delta H_f^\ominus(Na^+_{(g)}) + \Delta H_f^\ominus(H^+_{(g)})$$

Knowing that the standard formation enthalpy of the gaseous proton is 1529 $kJ\cdot mol^{-1}$, and that the standard formation enthalpies of gaseous and aqueous sodium ions are equal to 603.1 and – 240.2 $kJ\cdot mol^{-1}$ respectively, we deduce that the standard hydration enthalpy of sodium is 685.7 $kJ\cdot mol^{-1}$ on the proton scale. The ionic standard hydration enthalpy can also be calculated from thermodynamic cycles such as the one illustrated in Figure 3.6 for aqueous NaCl.

In order to have an absolute scale of standard hydration enthalpy, we need to know the standard enthalpy of the reaction having the proton at rest in the gas phase as the origin and the solvated proton as the final state. From an experimental point of view, we can for example make the hypothesis that the hydration enthalpy of a proton is about the same as the attachment energy of a proton to a group of molecules of water in the gas phase. Via this approximation, which neglects the fact that the aqueous phase has a surface potential, we get

$$\Delta H_{hyd}^\ominus(H^+) = -1090 \text{ kJ} \cdot \text{mol}^{-1}$$

Other experimental procedures based on other approximations have been suggested for evaluating the standard hydration enthalpy of the proton, and the values obtained vary ± 30 $kJ\cdot mol^{-1}$ around the above cited value which is often taken as a reference. It is clear that if, in the cycle of Figure 3.7, we take $\Delta H_{hyd}^\ominus(H^+) = -1090$ $kJ\cdot mol^{-1}$, we then obtain the standard hydration enthalpy of sodium on the absolute scale i.e. 686 – 1090 = – 404 $kJ\cdot mol^{-1}$ (see Table 3.3).

The fact that Gibbs energies and enthalpies of solvation of ionic species are not measurable is a major problem. In effect, it is not possible to add or extract an ion without disturbing the electroneutrality of the liquid phase. In the case of organic solvents, it is cumbersome to use the proton scale since the weak dissociation of acids in these solvents make the experiments difficult. Another commonly used approach consists either of considering that the solvation enthalpy of an ion is equal to that of a similar neutral atom or molecule (e.g. ferrocene-ferricinium) or of considering that the Gibbs energies of solvation of a similar cation and anion are equal (e.g. tetraphenylarsonium tetraphenylborate). In the latter case, we have:

$$\Delta G^{\ominus}_{\text{solv,TPA}^+} = \Delta G^{\ominus}_{\text{solv,TPB}^-} = \tfrac{1}{2}\Delta G^{\ominus}_{\text{solv,TPA}^+\text{TPB}^-} \tag{3.5}$$

Note that these last two methods provide also relative solvation enthalpy scales and not absolute ones.

3.2.3 The Born model

We saw in chapter 1 that the electrochemical potential $\tilde{\mu}_i$ of a charged species i included a chemical contribution μ_i (called by an abuse of language the chemical potential) which takes into account all the short distance interactions between the ion and the solvent, and two electrostatic contributions linked to the surface potential ($z_iF\chi$) on one hand, and to the external potential of the phase ($z_iF\psi$) on the other.

In 1920, Born proposed a model of ionic solvation based on a cycle as shown in Figure 3.8, which avoided the problems linked to the surface potential and the external potential. In this cycle, the work of transfer w_t of an ion i from a vacuum to a solvent phase (or from a gas in the standard state) can be considered as the sum of the work w_d of discharging the ion in vacuum to form a neutral sphere of the same size as that of the ion, the work w_n of transferring this neutral sphere from the vacuum to the phase, and the work w_c of charging this sphere in the solvent phase considered as a continuum dielectric. Thus, the overall work to transfer an ion is written as

$$w_t = w_d + w_n + w_c \tag{3.6}$$

The work w_n to transfer an uncharged sphere can be taken a priori as the chemical potential of a rare gas of the same size in the given solvent. The solvation energies of

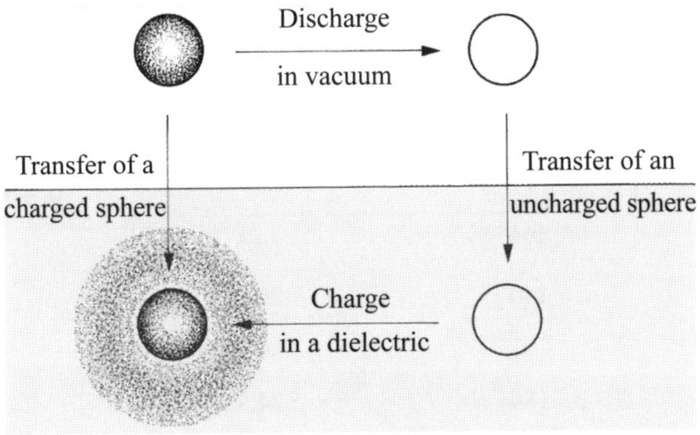

Fig. 3.8 The Born solvation model.

rare gases are always positive since they correspond to a work of cavity formation in the solvent. Nevertheless, it is noticeable (Table 3.2) that the Gibbs energy of solvation of the rare gases decreases as their size increases. This is due to the fact that the polarisability of the atoms increases with their size, resulting in a greater atom-solvent interaction energy, translating into a negative contribution to the solvation energy.

Another approach for evaluating w_n consists of considering that the work of formation of a cavity is equivalent to the work of formation of a spherical surface with a radius corresponding to that of the ion. From the definition of the surface tension given in §5.1, this work comprises a term relative to the creation of a new spherical surface $4\pi r^2\gamma$ and a term relative to the volume of the cavity and the pressure $4/3\pi r^3 p$.

The work of charging a neutral sphere of radius r in vacuum from the charge $q = 0$ to the charge of the ion $q = ze$ where z is the charge number of the ion and e the elementary charge, is defined as

$$w = \int_0^{ze} V(q)\,\mathrm{d}q \tag{3.7}$$

where $V(q)$ is the electrostatic potential generated by the charge on the sphere, which is itself a function of the charge q and the radius r_{ion}

$$V(q) = \frac{q}{4\pi\varepsilon_0 r_{\text{ion}}} \tag{3.8}$$

Thus, the work of discharging the ion in vacuum is

$$w_{\text{d}} = -\int_0^{ze} \frac{q}{4\pi\varepsilon_0 r_{\text{ion}}}\,\mathrm{d}q = -\left[\frac{q^2}{8\pi\varepsilon_0 r_{\text{ion}}}\right]_0^{ze} = -\frac{z^2e^2}{8\pi\varepsilon_0 r_{\text{ion}}} \tag{3.9}$$

In a similar way, the work w_{d} of charging the sphere in a dielectric medium with a relative permittivity ε_{r} is

$$w_{\text{c}} = \frac{z^2e^2}{8\pi\varepsilon_0\varepsilon_{\text{r}} r_{\text{ion}}} \tag{3.10}$$

Table 3.2 Values of standard hydration and solvation energies in benzene for the rare gases (M.H. Abraham & J. Liszi, *J. Chem. Soc. Faraday Trans*. I, **74** (1978) 1604).

Gas	Radius / pm	$\Delta G^{\ominus}_{\text{hyd}}$ / kJ·mol^{-1}	$\Delta G^{\ominus}_{\text{sol}}$ / kJ·mol^{-1} Benzene [ε_{r}=2.3]
He	129	29.5	23.5
Ne	140	29.0	22.5
Ar	171	26.2	17.4
Kr	180	24.9	14.6
Xe	203	23.4	11.1

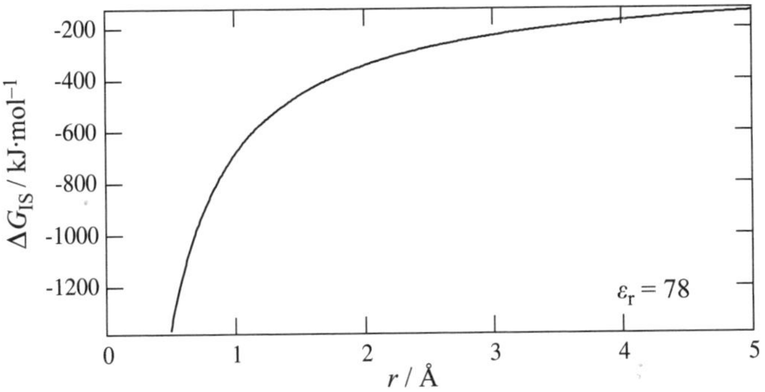

Fig. 3.9 Variation in ion-solvent interaction energy as a function of the ionic radius for a univalent ion.

In conclusion, we can write that

$$w_t - w_n = w_d + w_c = -\frac{z^2 e^2}{8\pi\varepsilon_0 r_{ion}}\left(1 - \frac{1}{\varepsilon_r}\right) \qquad (3.11)$$

This difference $w_t - w_n$ represents the ion-solvent interaction Gibbs energy, ΔG_{IS}, in as much as it represents the contribution of the ionic charge on the short range interactions. Since the electrostatic contribution is the dominant factor compared to w_n, ΔG_{IS} can be considered in a first approximation equal to the chemical potential of the ion, the Gibbs hydration energy corresponding to the real chemical potential.

$$\Delta G_{IS} = -\frac{z^2 e^2 N_A}{8\pi\varepsilon_0 r_{ion}}\left(1 - \frac{1}{\varepsilon_r}\right) \qquad (3.12)$$

Given that the relative permittivity of solvents varies from 2 to more than 100, we can see that this Gibbs energy is always negative and consequently the solvent stabilises the ion, as shown in the graph in Figure 3.9 for the case of aqueous solutions.

These values of ion-solvent interaction energy can be compared to the energies of covalent bonds such as the energies of the carbon-carbon bonds which are 348, 612 and 962 kJ·mol^{-1} for single, double and triple bonds respectively. As can be seen on the graph of Figure 3.10, for an ionic radius of 2 Å, the ion-solvent interaction Gibbs energy is about 300 kJ·mol^{-1} for a large range of solvents, that is almost as much energy as for a single covalent bond.

An important hypothesis to note in this theory is that it neglects the dielectric saturation and assumes that the relative permittivity around the ion is the same as that in the bulk of the solution. In fact, near an ion, there is first of all a solvation layer of oriented dipoles ($\varepsilon_r \approx 2$), then a less structured layer which ensures the junction between the order imposed in the solvation layer and the order present in the solvent (see Figure 3.12).

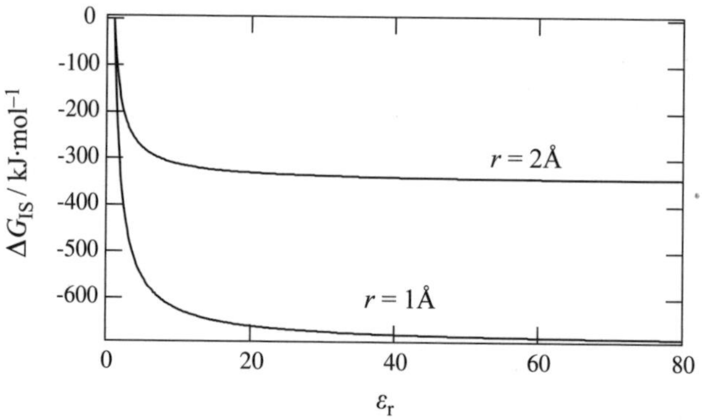

Fig. 3.10 Variation of ion-solvent interaction energy as a function of the relative permittivity of the medium for a univalent ion.

Tests on the Born model corroborate it quite well. The Gibbs energy of ionic solvation is usually found to be inversely proportional to the relative permittivity and, for a given solvent, the solvation Gibbs energy is also found to be inversely proportional to the radius. It should be noticed, however, that the values of ionic radii do not correspond to the crystallographic ones but to a 'corrected' value. For water, the correction is +0.85 Å for the cations and +0.10 Å for the anions, as shown in Figure 3.11. This correction of radii values means that the Born model predicts ion-solvent interaction values larger than those observed experimentally.

The numerical values presented in Table 3.3 are given only as an example. Values published in the literature can vary considerably, the main reason for the differences coming from the absolute hydration values used for the proton.

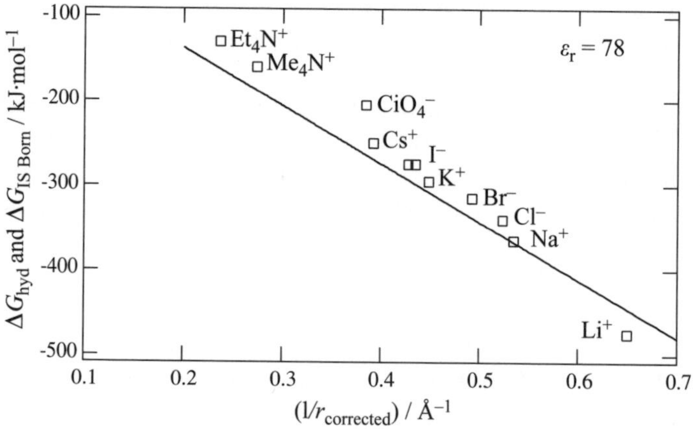

Fig. 3.11 Comparison of the ion-solvent interaction energy calculated and the experimental Gibbs energy of hydration values with modified ionic radii (□).

Table 3.3 Standard absolute Gibbs energies, standard absolute enthalpies, standard hydration entropies of ions and standard partial molar volumes, Y. Marcus in *Liquid/liquid interfaces Theory and methods*, p 39-61, Volkov & Deamer Eds, CRC Press, 1996, Boca Raton (USA). The standard states are 1 bar for the gas phase and 1 mol·l^{-1} for the aqueous phase at 298.5 K.

Cations	Radius / pm	$\Delta G^{\ominus}_{hyd}$/ kJ·mol^{-1}	$\Delta H^{\ominus}_{hyd}$/ kJ·mol^{-1}	$\Delta S^{\ominus}_{hyd}$/ J·mol^{-1}·K^{-1}	$\bar{V}^{\ominus}_{m}$/ cm^3·mol^{-1}
H^+		–1055	–1090	–131	–5.5
Li^+	69	–475	–530	–161	–6.4
Na^+	102	–365	–415	–130	–6.7
K^+	138	–295	–330	–93	3.5
Rb^+	149	–275	–305	–84	8.6
Cs^+	170	–250	–280	–78	15.8
NH_4^+	148	–285	–325	–131	12.4
Me_4N^+	280	–160	–215	–163	84.1
Et_4N^+	337	–130	–205	–241	143.6
Mg^{2+}	72	–1830	–1945	–350	–32.2
Ca^{2+}	100	–1505	–1600	–271	–28.9
Fe^{2+}	78	–1840	–1970	–381	–30.2
Ni^{2+}	69	–1980	–2115	–370	–35
Fe^{3+}	65	–4265	–4460	–576	–53
Anions					
F^-	133	–465	–510	–156	4.3
Cl^-	181	–340	–365	–94	23.3
Br^-	196	–315	–335	–78	30.2
I^-	220	–275	–290	–55	41.7
OH^-	133	–430	–520	–180	–0.2
NO_3^-	179	–300	–310	–95	34.5
ClO_4^-	250	–205	–245	–76	49.6

3.2.4 Enthalpy and entropy of the ion-solvent interaction

The ion solvent interaction entropy can be found from the Gibbs energy ΔG_{IS} by

$$\left(\frac{\partial \Delta G_{\text{IS}}}{\partial T}\right)_{p,\mu_i} = -\Delta S_{\text{IS}} \tag{3.13}$$

Bringing the Born equation (3.12) into equation (3.13), we have

$$\Delta S_{\text{IS}} = \frac{N_{\text{A}} z^2 e^2}{8\pi\varepsilon_0 r} \frac{1}{\varepsilon_{\text{r}}^2}\left(\frac{\partial \varepsilon_{\text{r}}}{\partial T}\right) \tag{3.14}$$

from which we can deduce the enthalpy of the ion-solvent interaction

$$\Delta H_{\text{IS}} = -\frac{N_{\text{A}} z^2 e^2}{8\pi\varepsilon_0 r}\left[1 - \frac{1}{\varepsilon_{\text{r}}} - \frac{T}{\varepsilon_{\text{r}}^2}\frac{\partial \varepsilon_{\text{r}}}{\partial T}\right] \tag{3.15}$$

In the case of water $\partial\varepsilon_{\text{r}} / \partial T = -0.3595$, and the third term of equation (3.15) is approximately -0.0176 at 298 K.

3.2.5 Electrostatic Gibbs energy

Another way to calculate the work of charging an ion in a dielectric medium is to consider the electrostatic Gibbs energy

$$w_{\text{c}} = \frac{1}{2}\iiint_V \boldsymbol{D}\cdot\boldsymbol{E}\,\mathrm{d}v \tag{3.16}$$

This energy corresponds to the energy stored in the dielectric when the ion is charged. In fact, if we consider a dipole as made of two charges of opposite sign attached by a spring, the electric field generated by the ion increases the distance between the two charges and consequently the dipolar electric field. Thus, we can say that the presence of an ion in a dielectric increases its internal energy. This energy corresponds to the work of charging the ion in a dielectric. The equation (3.16) is almost similar to the one used in the Born model for calculating the charging work in the dielectric, the only difference being the integration variable (the charge in the first one, the volume in the second).

This second approach is more appropriate when we wish to calculate the solvation Gibbs energy of an ion close to a wall. In fact, when the ion approaches the wall, it interacts with its image. However, this type of problem is beyond the scope of this book. Let's look at the simple case of an ion in a large volume of dielectric, and develop equation (3.16)

$$w_{\text{c}} = \frac{1}{2}\iiint_V \boldsymbol{D}\cdot\boldsymbol{E}\,\mathrm{d}v = \frac{\varepsilon_0\varepsilon_{\text{r}}}{2}\iiint_V \boldsymbol{E}^2\,\mathrm{d}v = -\frac{\varepsilon_0\varepsilon_{\text{r}}}{2}\iiint_V \boldsymbol{E}\cdot\textbf{grad}\phi\,\mathrm{d}v \tag{3.17}$$

Using the identity

$$\mathrm{div}(\phi\boldsymbol{E}) = \textbf{grad}\phi\cdot\boldsymbol{E} + \phi\,\mathrm{div}\,\boldsymbol{E} \tag{3.18}$$

we obtain

$$w_c = \frac{\varepsilon_0\varepsilon_r}{2}\iiint_V \phi \operatorname{div} \boldsymbol{E} \, \mathrm{d}v - \frac{\varepsilon_0\varepsilon_r}{2}\iiint_V \operatorname{div}(\phi\boldsymbol{E}) \, \mathrm{d}v \tag{3.19}$$

By applying the Poisson equation to the volume external to the ion

$$\operatorname{div} \boldsymbol{E} = 0 \tag{3.20}$$

we obtain, using the Green-Ostrogradski theorem

$$\begin{aligned} w_c &= -\frac{\varepsilon_0\varepsilon_r}{2}\iiint_V \operatorname{div}(\phi\boldsymbol{E}) \, \mathrm{d}v = -\frac{\varepsilon_0\varepsilon_r}{2}\iint_S \phi \, \boldsymbol{E}\cdot\hat{\boldsymbol{n}} \; \mathrm{d}s \\ &= \frac{\varepsilon_0\varepsilon_r}{2}\iint_S \phi \, \mathbf{grad}\phi\cdot\hat{\boldsymbol{n}} \; \mathrm{d}s \end{aligned} \tag{3.21}$$

where the integration surface is the surface of the ion ($\hat{\boldsymbol{r}} = \hat{\boldsymbol{n}}$).

In the homogeneous dielectric medium where the ion of a charge ze is located, the potential generated by the ion is

$$\phi = \frac{ze}{4\pi\varepsilon_0\varepsilon_r r} \tag{3.22}$$

and the integral in equation (3.21) is then written as

$$I = \left(\frac{ze}{4\pi\varepsilon_0\varepsilon_r}\right)^2 \iint_S \frac{1}{r}\,\mathbf{grad}\frac{1}{r}\cdot\hat{\boldsymbol{n}} \; \mathrm{d}s \tag{3.23}$$

The integration surface is the one from which the field lines come, i.e. the surface of the ion, considered as spherical. Knowing, as shown in §4.4.1, that

$$\mathbf{grad}\, r = \hat{\boldsymbol{r}} \tag{3.24}$$

and therefore that

$$\mathbf{grad}\left(\frac{1}{r}\right) = -\frac{\hat{\boldsymbol{r}}}{r^2} \tag{3.25}$$

we can calculate the integral in spherical coordinates

$$\int_0^{\pi}\int_0^{2\pi} \frac{1}{r^3}\, r\sin\theta \, \mathrm{d}\theta \; r\mathrm{d}\phi = \frac{4\pi}{r} \tag{3.26}$$

In conclusion, the volumic electrostatic Gibbs energy due to the presence of an ion is written as:

$$w_c = \frac{(ze)^2}{8\pi\varepsilon_0\varepsilon_r r_{ion}} \tag{3.27}$$

Thus we find again equation (3.10).

3.3 STRUCTURAL ASPECTS OF IONIC SOLVATION

The presence of an ion in a polar solvent causes an orientation of the dipolar molecules around the ion. This zone where the ion induces an order by reducing the rotation freedom of the molecules is in general called the ***solvation layer***, and for water the ***hydration layer*** as shown schematically in Figure 3.12.

This induced order differs from the structural order of the solvent, and between these two different orders, there is a disorganised region. If the number of molecules of water is greater in the hydration layer than in the disorganised layer, the ion is called a ***structure-maker*** (Li^+, F^-, Mg^{2+}) whereas in the opposite case, it is called a ***structure-breaker*** (Cs^+, Rb^+, ClO_4^-). A structure-maker ion is characterised by a very negative hydration energy and a negative standard partial molar volume (see Table 3.3).

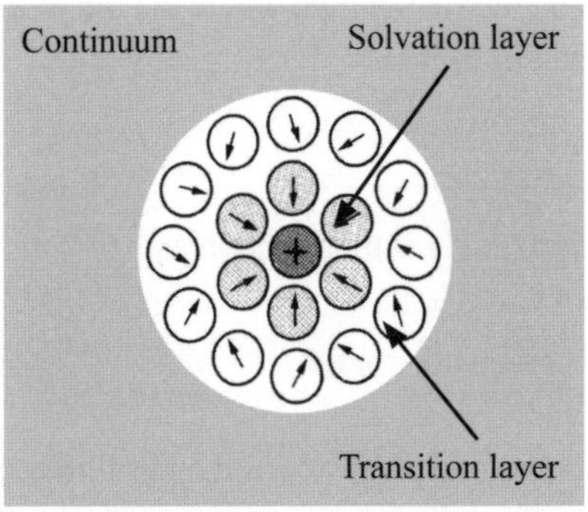

Fig. 3.12 Two-layer solvation model in a continuum.

3.3.1 Solvation time

By various methods, such as nuclear magnetic resonance relaxation, it is possible to measure the times of residence of the solvent molecules in the first solvation layer. The residence time of a molecule of water around an ion varies enormously as shown in Figure 3.13. The shortest times, of the order of picoseconds are around the Cl^-, I^-, Br^- anions or cations such as Me_4N^+; for the alkaline ions the residence time is of the order of nanoseconds, whilst for doubly-charged ions the times vary from 10 ns for Ca^{2+}, 100 ns for Fe^{2+}, 10 μs for Ni^{2+} and Mg^{2+}, up to several days for Cr^{3+}.

3.3.2 Solvation number

Numerous methods have been used to estimate the number of solvent molecules forming the first solvation layer. Among others, we can mention (UV-VIS, Raman and NMR spectroscopies), methods based on transport in solution, and radiotracer measurements. The values obtained are highly dependent on the method used and vary mostly between 4 and 9, the value 6 being the most frequent.

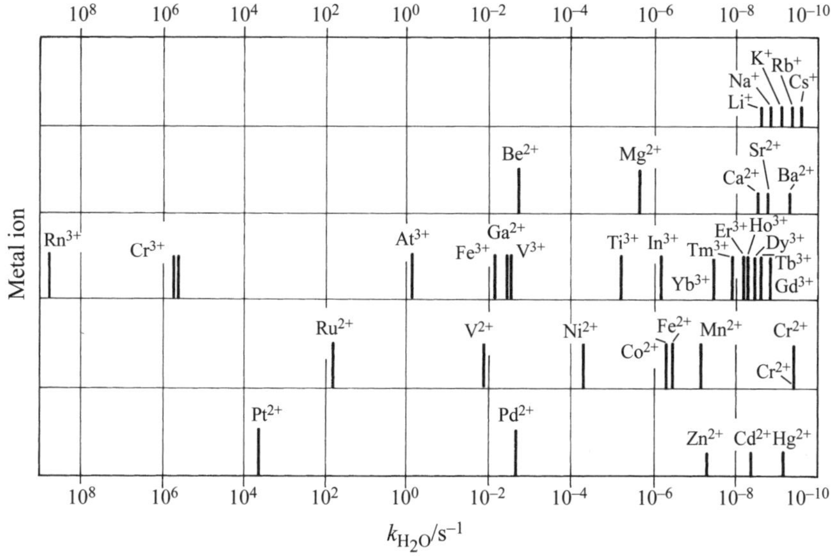

Fig. 3.13 Exchange rate constant for a molecule of water around a metal ion (S.F. Lincoln and A.E. Merbach, *Adv. Inorg. Chem.*, Vol. 42 (1995) 1-87, Academic Press, with permission from Elsevier Science).

Table 3.4 Ion-n water molecules interaction enthalpies in $kJ \cdot mol^{-1}$ (A. Gerschel, *Liaisons intermoléculaires*, Interéditions, 1995, Paris).

Ion	$n = 1$	$n = 2$	$n = 3$	$n = 4$	$n = 5$	$n = 6$	Liquid
Li^+	–142	–250	–338	–405	–463	–513	–515
Na^+	–100	–181	–248	–305	–357	–401	–405
F^-	–97	–166	–223	–280	–335		–505
Cl^-	–55	–108	–157	–203			–370

The interaction energy of an ion with the solvent molecules surrounding it is an essential parameter in estimating the solvation enthalpy. In this respect, it is interesting to compare the interaction enthalpies of an ion in the gas phase with clusters of solvent molecules.

From the values in Table 3.4, we can see that for the cations, the hydration energy corresponds by and large to the interaction energy with the nearby hydration molecules, whereas for the anions, this interaction energy only represents about half of the hydration energy. In order to understand these results better, we need to study the ion-dipole interactions in more depth.

3.3.3 Potential of a dipole

Let's look at a dipole made up of charges $-e$ and $+e$ separated by a distance $2l$, the polarisation vector $\boldsymbol{p}$ pointing from $-e$ towards $+e$ as in Figure 3.14. At a point M in space, the distances to the charges $-e$ and $+e$ are respectively r_- and r_+.

In vacuum, the potential created by this dipole is therefore

$$V(\mathrm{M}) = \frac{e}{4\pi\,\varepsilon_0\,r_+} - \frac{e}{4\pi\,\varepsilon_0\,r_-} \tag{3.28}$$

To calculate the potential created by an ideal dipole ($l \to 0$ when $\boldsymbol{p}$ remains constant), we write the potential $V(\mathrm{M})$ in the form

$$V(\mathrm{M}) = \frac{1}{4\pi\,\varepsilon_0}\left[\frac{e}{\|\boldsymbol{r}-\boldsymbol{l}\|} - \frac{e}{\|\boldsymbol{r}+\boldsymbol{l}\|}\right] \tag{3.29}$$

since

$$\boldsymbol{r}_+ = \boldsymbol{r}-\boldsymbol{l} \qquad \text{and} \qquad \boldsymbol{r}_- = \boldsymbol{r}+\boldsymbol{l} \tag{3.30}$$

Given that l is small compared to r, we can do a second order series expansion

$$\frac{1}{\|\boldsymbol{r}-\boldsymbol{l}\|} = \frac{1}{\sqrt{(\boldsymbol{r}-\boldsymbol{l})^2}} = \frac{1}{\sqrt{\boldsymbol{r}^2 - 2\,\boldsymbol{r}\cdot\boldsymbol{l} + \boldsymbol{l}^2}} = \frac{1}{r}\left[1+\frac{\boldsymbol{r}\cdot\boldsymbol{l}}{r^2} - \frac{\boldsymbol{l}^2}{2r^2} + \ldots\right] \tag{3.31}$$

and

$$\frac{1}{\|\boldsymbol{r}+\boldsymbol{l}\|} = \frac{1}{r}\left[1-\frac{\boldsymbol{r}\cdot\boldsymbol{l}}{r^2} - \frac{\boldsymbol{l}^2}{2r^2} + \ldots\right] \tag{3.32}$$

from which wc obtain

$$V(\mathrm{M}) = \frac{e}{4\pi\,\varepsilon_0}\,\frac{2\,\boldsymbol{r}\cdot\boldsymbol{l}}{r^3} = \frac{\boldsymbol{p}\cdot\boldsymbol{r}}{4\pi\,\varepsilon_0\,r^3} \tag{3.33}$$

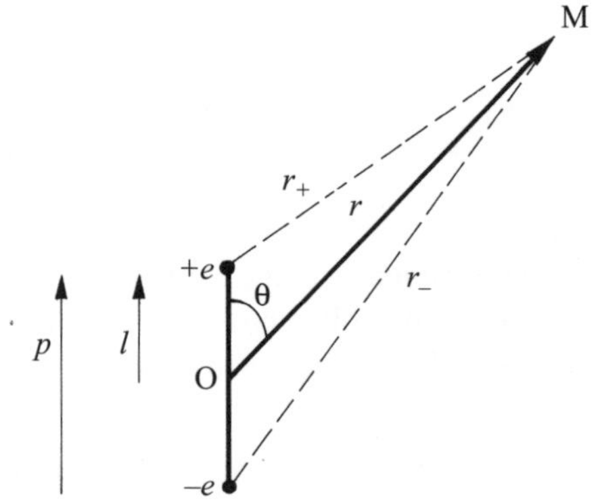

Fig. 3.14 Potential of a dipole at a point M in space.

3.3.4 Energy of ion-dipole interaction

If, in a homogeneous dielectric medium, we place an ion of a charge ze at a point M, the interaction energy between the ion and a dipole will be

$$w_{\text{ion-dipole}} = \int_0^{ze} V \mathrm{d}q = \int_0^{ze} \frac{\boldsymbol{p}\cdot\boldsymbol{r}}{4\pi\,\varepsilon_0\varepsilon_\mathrm{r}\,r^3}\mathrm{d}q = \frac{(ze)p\cos\theta}{4\pi\,\varepsilon_0\varepsilon_\mathrm{r}\,r^2} \tag{3.34}$$

This equation shows that for a cation, the ion-dipole interaction energy will be negative and maximum if the dipole is aligned with the vector of the electric field generated by the ion itself, and conversely, the interaction will be positive and maximum if the dipole is pointed towards the ion. In the case of water, the cation-water interaction is very strong for small ions such as lithium.

EXAMPLE

Let's calculate the interaction energy between a monovalent ion, such as sodium, and a single water molecule considered as a dipole with a moment of 1.85 Debye in the gas phase and 2.6 Debye in the liquid phase (1 Debye = $3.336{\cdot}10^{-30}$ C·m). We shall calculate this energy in the liquid phase and in the gas phase, taking 100 and 150 pm for the radii of the ion and the water respectively.
If we consider that the relative permittivity of water is 78, then we have

$$w = \frac{(ze)p\cos\theta N_\mathrm{A}}{4\pi\,\varepsilon_0\varepsilon_\mathrm{r}\,r^2} = \frac{1.6\cdot10^{-19}(\mathrm{C})\cdot2.6\cdot3.336\cdot10^{-30}(\mathrm{C\cdot m})\cdot6.02\cdot10^{23}(\mathrm{mol^{-1}})}{1.11\cdot10^{-10}(\mathrm{J^{-1}\cdot C^2\cdot m^{-1}})\cdot78\cdot250^2\cdot10^{-24}(\mathrm{m^2})}$$
$$= 1.5\ \mathrm{kJ\cdot mol^{-1}}$$

If we consider that the relative permittivity in the first hydration shell is 2, then we have a value of 60 kJ·mol^{-1}.
For the gas phase, taking the permittivity of vacuum, the numeric value resulting from equation (3.34) is then 85 kJ·mol^{-1}, which comes near to experimentally measured values (see Table 3.4).

3.3.5 Average ion-dipole interaction energy

In a polar solvent, the thermal agitation leads to a free rotation of the dipoles. The presence of an ion orients the dipoles in contact with it, but also influences the orientation of the dipoles outside the first solvation layer. To quantify this effect, we can calculate the average interaction energy between an ion and the solvent molecules by using the Boltzmann distribution function

$$< w_{\text{ion-dipole}} > = \frac{\int_0^{\pi}\int_0^{2\pi} w_{\text{ion-dipole}}(r,\theta,\phi)\exp^{-w_{\text{ion-dipole}}(r,\theta,\phi)/kT}\sin\theta\mathrm{d}\theta\mathrm{d}\phi}{\int_0^{\pi}\int_0^{2\pi}\exp^{-w_{\text{ion-dipole}}(r,\theta,\phi)/kT}\sin\theta\mathrm{d}\theta\mathrm{d}\phi} \tag{3.35}$$

By substituting expression (3.34) for the ion-dipole interaction, we have

$$< w_{\text{ion-dipole}} > \ = \ \frac{\displaystyle\int_0^{\pi}\int_0^{2\pi} \frac{(ze)p}{4\pi\,\varepsilon_0\varepsilon_r\,r^2} \exp^{-\frac{(ze)p\cos\theta}{4\pi\,\varepsilon_0\varepsilon_r\,r^2 kT}} \cos\theta \sin\theta \mathrm{d}\theta \mathrm{d}\phi}{\displaystyle\int_0^{\pi}\int_0^{2\pi} \exp^{-\frac{(ze)p\cos\theta}{4\pi\,\varepsilon_0\varepsilon_r\,r^2 kT}} \sin\theta \mathrm{d}\theta \mathrm{d}\phi} \tag{3.36}$$

To calculate these two integrals, we define the variables

$$x \ = \ -\frac{(ze)p\cos\theta}{4\pi\,\varepsilon_0\varepsilon_r\,r^2} \tag{3.37}$$

and

$$a \ = \ \frac{(ze)p}{4\pi\,\varepsilon_0\varepsilon_r\,r^2} \tag{3.38}$$

We have thus

$$\begin{aligned} < w_{\text{ion-dipole}} > \ &= \ \frac{-\int_{-a}^{a} x \exp^{x/kT} \mathrm{d}x}{\int_{-a}^{a} \exp^{x/kT} \mathrm{d}x} \ = \ \frac{-\left[xkT\exp^{x/kT} - k^2T^2\exp^{x/kT}\right]_{-a}^{a}}{\left[kT\exp^{x/kT}\right]_{-a}^{a}} \\ &= \ -a\coth\left(\frac{a}{kT}\right) + kT \ = \ -a\,L\left(\frac{a}{kT}\right) \end{aligned} \tag{3.39}$$

where $L(a)$ is the Langevin function as represented in Figure 3.15.

For small values of a, i.e. for quite large values of the ion-dipole distance r, a series expansion of the Langevin function gives

$$L\left(\frac{a}{kT}\right) \ \cong \ \frac{a}{3kT} \tag{3.40}$$

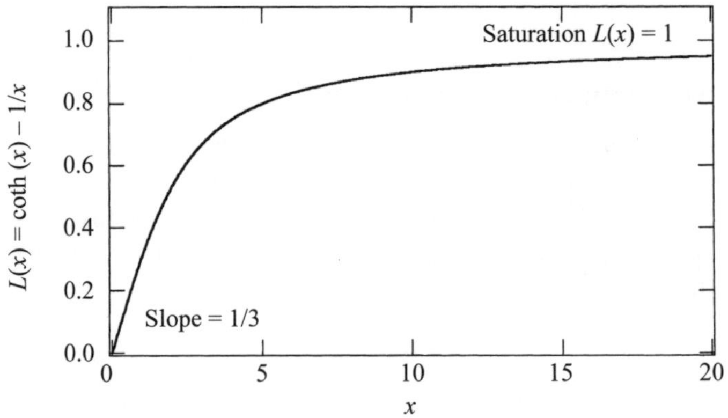

Fig. 3. 15 Langevin function

We finally obtain

$$< w_{\text{ion-dipole}} > \;=\; -\frac{a^2}{3kT} \;=\; -\frac{1}{3kT}\left[\frac{(ze)p}{4\pi\,\varepsilon_0\varepsilon_{\mathrm{r}}\,r^2}\right]^2 \qquad (3.41)$$

This relationship shows that the average interaction energy at large distances is inversely proportional to the temperature and depends on the distance as $1/r^4$.

EXAMPLE

Let's calculate the average interaction energy between a sodium ion and a water molecule at a distance of 250 pm and at 2.5 nm.

$$\begin{aligned} < w > &= \frac{N_{\mathrm{A}}}{3kT}\left[\frac{(ze)p}{4\pi\,\varepsilon_0\varepsilon_{\mathrm{r}}\,r^2}\right]^2 \\ &= \frac{6.02\cdot 10^{23}(\mathrm{mol}^{-1})}{1.38\cdot 10^{-23}(\mathrm{J}\cdot\mathrm{K}^{-1})\cdot 298(\mathrm{K})}\left[\frac{1.6\cdot 10^{-19}(\mathrm{C})\cdot 2.6\cdot 3.336\cdot 10^{-30}(\mathrm{C}\cdot\mathrm{m})}{1.11\cdot 10^{-10}(\mathrm{J}^{-1}\cdot\mathrm{C}^2\cdot\mathrm{m}^{-1})\cdot 78\cdot 250^2\cdot 10^{-24}(\mathrm{m}^2)}\right] \\ &= 1.0\ \mathrm{kJ}\cdot\mathrm{mol}^{-1} \end{aligned}$$

When the distance is 2.5 nm, we have

$< w > = 0.01\ \mathrm{kJ{\cdot}mol^{-1}}$

Meaning that, as soon as the second solvation layer is reached, the ion-dipole interactions become weak.

3.3.6 Comments

It is difficult to go further in the modelling of the structural aspects of solvation, because the degree of complexity increases rapidly as soon as we try to go further than these few energetic considerations of ion-dipole interactions.

3.4 ION-ION INTERACTIONS

3.4.1 Electrostatic interaction energy and the activity coefficient

In solution, the ions interact mainly by coulombic forces, either repulsive or attractive according to the signs. These electrostatic interactions increase when the concentration increases, as the distance between the ions diminishes.

If we consider that an ideally dilute solution is a solution in which the solutes do not interact, then an electrolyte solution looses its ideal behaviour at very low concentrations (micromolar range in water) whereas a non-ionic solution looses its ideal behaviour at higher concentrations (millimolar range in water). This is due to the fact that electrostatic forces are longer range than those involved in molecular interactions.

For an ideally dilute solution, the dilution work from a concentration c_1 to a concentration c_2 is purely osmotic and is given at constant pressure and temperature by

$$w_{osm} \quad = \quad nRT\ln\left(\frac{c_2}{c_1}\right) \tag{3.42}$$

where n is the total number of moles of solutes. Equation (3.42) shows that during a dilution $(c_2 < c_1)$, the reaction is exergonic ($w_{osm} < 0$).

For a real solution for which the activity coefficients cannot be neglected, the total dilution work is expressed more generally by

$$w_{dilut} \quad = \quad nRT\ln\left(\frac{a_2}{a_1}\right) \tag{3.43}$$

where a_2 and a_1 are the activities, e.g. on the concentration scale. If we assume that the solutions are sufficiently dilute that the deviation from the ideal behaviour is caused mainly by the electrostatic interactions, the dilution work of an electrolyte solution is then written as

$$w_{dilut} \quad = \quad nRT\ln\left(\frac{a_2}{a_1}\right) \quad = \quad nRT\ln\left(\frac{c_2}{c_1}\right) + nRT\ln\left(\frac{\gamma_2}{\gamma_1}\right) \quad = \quad w_{osm} + w_{elec} \tag{3.44}$$

The dilution work therefore comprises two terms, one osmotic and the other electrostatic

$$w_{elec} \quad = \quad nRT\ln\left(\frac{\gamma_2}{\gamma_1}\right) \tag{3.45}$$

Given that the activity coefficient γ tends to unity when the concentration decreases, the result is that the electric component of the dilution work w_{elec} is positive. Indeed, the reaction that consists of breaking the ion-ion interactions is endergonic. In order to calculate this work of ion-ion interaction, we can calculate the variation in Gibbs energy ΔG associated to the passage from an ideal solution formed of discharged ions to a real solution formed of charged ions where ion-ion interactions prevail. This is the gist of the Debye-Hückel theory presented below.

3.4.2 The Debye-Hückel theory (1923)

The objective of the Debye-Hückel theory which is based on the ionic atmosphere model, is to calculate the activity coefficient γ of an electrolyte as a function of is concentration in order to determine the work of ionic interaction given by equation (3.45).

Around a central ion, chosen arbitrarily inside a solution and taken henceforth as a reference, there are statistically always more ions of opposite charge than ions of the same charge; there is therefore a distribution of charges whose temporal average is well-defined in the moving frame which accompanies the central ion. This ionic cloud, also called the ***ionic atmosphere***, contains ions of both signs, but with a majority of ions having a charge opposite to that of the central ion. The total charge of this ionic atmosphere is equal and opposite to that of the central ion, so that the electroneutrality of the solution is ensured. To calculate the interaction energy, the Debye-Hückel theory is based on the following hypotheses

1. Only coulombic forces are taken into consideration; molecular interactions such as the Van der Waals forces occuring at smaller distances are ignored;
2. At all concentrations, the electrolyte is completely dissociated;
3. The relative permittivity of the solution is that of the pure solvent, variations wih the concentration are therefore ignored;
4. The ions are considered as rigid spheres, i.e. unpolarisable, whose charge causes a symmetrical spherical electric field;
5. The electrostatic interaction energy is weak compared to the energy due to the thermal agitation in the solution.

Before calculating the ion-ion interaction energy in solution, we have to establish the average electric potential distribution around a central ion, chosen arbitrarily inside the solution and considered as the origin of a reference frame. The quantities relative to the central ion carry the index *c,* and those concerning the positive or negative ions of the ionic atmosphere an index *i*.

In the following calculations, all the quantities are temporal averages, in order not to have to take into account the molecular agitation.

Let's consider that the average volumic density of the ions *i* carrying a charge $z_i e$ is N_i^∞. First of all, we shall assume that the ions are distributed throughout the solution according to a Boltzmann statistics. Thus, the density $N_i(r)$ of ions *i* in an elementary volume at a distance *r* from the the central ion is given by

$$N_i(r) = N_i^\infty \, e^{-z_i e[\phi(r)-\phi(\infty)]/kT} \tag{3.46}$$

where $\phi(r)$ is the electric potential at the distance *r* from the central ion. The term $z_i e[\phi(r) - \phi(\infty)]$ is the electrostatic work to bring an ion *i* from infinity to a distance *r* from the central ion.

Taking the bulk potential inside the solution $\phi(\infty)$ as zero, equation (3.46) reduces to

$$N_i(r) = N_i^\infty \, e^{-z_i e\phi(r)/kT} \tag{3.47}$$

The electric charge density $\rho(r)$ is then obtained by the summation of the charges of all the ions present

$$\rho(r) = \sum_i z_i e\, N_i(r) = \sum_i z_i e\, N_i^\infty e^{-z_i e\phi(r)/kT} \tag{3.48}$$

A hypothesis of this theory considers that the electrostatic interaction is weak compared to the thermal agitation ($z_i e\phi(r) << kT$), allowing the linearisation of expression (3.48).

NUMERICAL APPLICATION

Let's consider numerically the conditions imposed by $z_i e\phi(r) << kT$. Knowing that: $e = 1.6 \cdot 10^{-19}$ C, $4\pi\varepsilon_0 = 1.11 \cdot 10^{-10}$ $J^{-1} \cdot C^{-2} \cdot m^{-1}$, $k = 1.38 \cdot 10^{-23}$ $J \cdot K^{-1}$, we have kT/e = 25.7 mV.

In the absence of an ionic atmosphere, the potential generated by a monovalent cation is $V(r) = e/4\pi\varepsilon_0\varepsilon_r r$.
In a first approximation, we shall assume that $\phi(r) = V(r)$, and the resulting numerical values of the potential for different distances are tabulated below

r / nm	$\phi(r)$ / mV	$e\phi(r)/kT$	$\exp(-e\phi(r)/kT)$	$1 - e\phi(r)/kT$
0.1	185	7.21	$7.40\ 10^{-4}$	–6.21
1	18.5	0.721	0.486	0.279
10	1.85	0.0721	0.930	0.928
3	6.18	0.240	0.787	0.760

This brief calculation shows that we need a distance of at least a few nm for the approximation $z_i e\phi(r) << kT$ to be valid.
Now let's calculate the average distance between the ions in solution for different concentrations.
A 1 M solution of a monovalent salt contains $2000N_A = 12 \cdot 10^{26}$ ions per m^3.
If the volume of each ion is $4\pi a^3/3$, the radius of each co-sphere is

$$a = \sqrt[3]{\frac{3}{8000\pi\, N_A\, c}} = 6 \cdot 10^{-10} \cdot c^{-1/3}$$

which numerically reads

c / M	a / nm
1	0.6
0.1	1.3
0.01	2.8
0.001	6

This brief calculation, based on an oversimplified sphere packing model, shows that the condition $z_i e\phi(r) << kT$ requires that the concentration must be less than 0.01 M.

With the approximation $z_i e\phi(r) << kT$, a first order series expansion of the exponential in equation (3.48) yields

$$\rho(r) \cong \sum_i z_i e N_i^{\infty} \left[1 - \frac{z_i e\phi(r)}{kT} \right] \tag{3.49}$$

The electroneutrality of the solution implies that

$$\sum_i z_i e N_i^{\infty} = 0 \tag{3.50}$$

from which we obtain an expression of the charge density that reads

$$\rho(r) = -\sum_i \frac{z_i^2 e^2 N_i^\infty \phi(r)}{kT} = -\frac{e^2}{kT}\left[\sum_i N_i^\infty z_i^2\right]\phi(r) \tag{3.51}$$

The term in brackets being a constant, equation (3.51) indicates that there is a direct linear relationship between the charge density and the electric potential.

Calculation of the electric potential $\phi(r)$

Poisson's equation defined by equation (1.37) expresses in a general manner the relationship between the charge density and the electric potential

$$\mathrm{div}\boldsymbol{E} = -\mathrm{div}\,\mathbf{grad}\phi = -\nabla^2\phi = \frac{\rho}{\varepsilon_0\varepsilon_\mathrm{r}} \tag{3.52}$$

In spherical coordinates centered on the central ion, Poisson's equation can be written as

$$\frac{\partial^2\phi(r)}{\partial r^2} + \frac{2}{r}\frac{\partial\phi(r)}{\partial r} = -\frac{\rho(r)}{\varepsilon_0\varepsilon_\mathrm{r}} \tag{3.53}$$

which, by substitution of equation (3.51), reads

$$\frac{\partial^2\phi(r)}{\partial r^2} + \frac{2}{r}\frac{\partial\phi(r)}{\partial r} = \frac{e^2}{\varepsilon_0\varepsilon_\mathrm{r}\,kT}\left[\sum_i N_i^\infty z_i^2\right]\phi(r) = \kappa^2\,\phi(r) \tag{3.54}$$

with

$$\kappa^2 = \frac{e^2}{\varepsilon_0\varepsilon_\mathrm{r}\,kT}\sum_i N_i^\infty z_i^2 \tag{3.55}$$

κ is a constant with the dimension of a reciprocal length, and its precise physical significance will become clear later.

The integration of the differential equation (3.54) is facilitated by using the following identity

$$\frac{\partial^2[r\phi(r)]}{\partial r^2} = \frac{\partial}{\partial r}\left(\frac{\partial[r\phi(r)]}{\partial r}\right) = \frac{\partial}{\partial r}\left(r\frac{\partial\phi(r)}{\partial r} + \phi(r)\right) = r\frac{\partial^2\phi(r)}{\partial r^2} + 2\frac{\partial\phi(r)}{\partial r} \tag{3.56}$$

which gives

$$\frac{\partial^2[r\phi(r)]}{\partial r^2} = \kappa^2\,[r\phi(r)] \tag{3.57}$$

By putting $y = r\,\phi(r)$, we see that we have quite a simple differential equation of the type

$$\frac{\partial^2 y}{\partial x^2} = \kappa^2 y \tag{3.58}$$

for which the general solution is

$$y = C_1\,e^{-\kappa x} + C_2\,e^{\kappa x} \tag{3.59}$$

In the present case, the solution of equation (3.57) is therefore

$$r\phi(r) \quad = \quad C_1\, e^{-\kappa r} + C_2\, e^{\kappa r} \tag{3.60}$$

for which it remains to calculate the two integration constants C_1 and C_2. To do this, we shall make use of two properties of the system, namely that the potential becomes constant far from the central ion and that the solution is electroneutral.

In the bulk ($r \to \infty$), the potential is taken equal to zero by convention, therefore $C_2 = 0$. Moreover, the electroneutrality of the solution requires that the total charge of the ionic atmosphere should compensate that of the central ion

$$\int_a^\infty 4\pi r^2 \rho(r)\mathrm{d}r \quad = \quad -z_c e \tag{3.61}$$

where a is the minimum approach distance to the central ion. By combining equations (3.51), (3.55) and (3.60), the charge density around an ion reads

$$\rho(r) \quad = \quad -\phi(r)\,\varepsilon_0\varepsilon_\mathrm{r}\,\kappa^2 \quad = \quad -\frac{C_1\varepsilon_0\varepsilon_\mathrm{r}\kappa^2}{r}e^{-\kappa r} \tag{3.62}$$

By substitution of equation (3.62) into equation (3.61), we have

$$4\pi\, C_1\, \varepsilon_0\varepsilon_\mathrm{r}\kappa^2 \quad \int_a^\infty e^{-\kappa r} r\mathrm{d}r \quad = \quad z_c e \tag{3.63}$$

Knowing that

$$\int e^{-\kappa r}\, r\,\mathrm{d}r \quad = \quad \left[-\frac{re^{-\kappa r}}{\kappa}\right] - \int -\frac{1}{\kappa}e^{-\kappa r}\,\mathrm{d}r \quad = \quad \left[-\frac{(1+\kappa r)\,e^{-\kappa r}}{\kappa^2}\right] \tag{3.64}$$

we obtain the constant C_1 as being equal to

$$C_1 \quad = \quad \frac{z_c e}{4\pi\varepsilon_0\varepsilon_\mathrm{r}}\left[\frac{e^{\kappa a}}{1+\kappa a}\right] \tag{3.65}$$

The electrical potential distribution around the central ion is finally given by

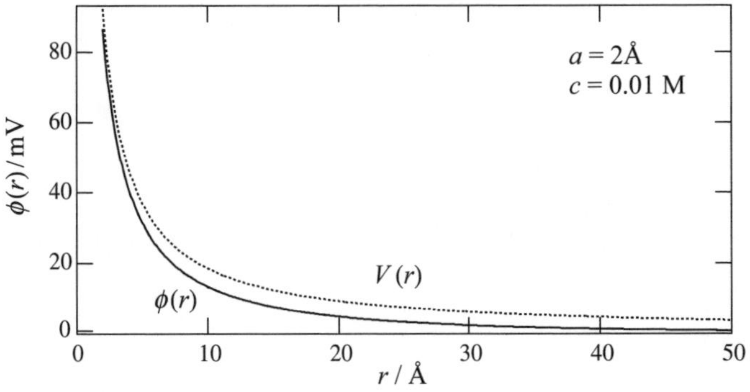

Fig. 3.16 Radial variation of the electrostatic potential in the presence and absence of an ionic atmosphere.

$$\phi(r) = \frac{z_c e}{4\pi\varepsilon_0\varepsilon_r}\left[\frac{e^{\kappa a}}{1+\kappa a}\right]\frac{e^{-\kappa r}}{r} \tag{3.66}$$

where a can in fact be considered as an average inter-ionic distance

$$a = r_c + r_i \tag{3.67}$$

Figure 3.16 illustrates the difference between $\phi(r)$ the electrostatic potential around the central ion in the presence of the ionic cloud, and $V(r)$ the electrostatic potential generated only by the central ion in the absence of other ions ($V(r) = z_c e / 4\pi\varepsilon_0\varepsilon_r r$). We see that very close to the central ion the two functions merge as the influence of the ionic atmosphere diminishes.

Ionic atmosphere

The density of the charge in the ionic atmosphere is written as

$$\rho(r) = \frac{-z_c e\kappa^2}{4\pi}\frac{e^{\kappa a}}{1+\kappa a}\frac{e^{-\kappa r}}{r} \tag{3.68}$$

In fact, this regularly decreasing function gives the density of the excess charge around the central ion. In order to understand better the distribution of ions in the ionic atmosphere, let's consider the infinitesimal charge $q_{\text{sph}}(r)$ carried by each shell with a radius r and thickness dr

$$q_{\text{sph}}(r) = 4\pi r^2\,\rho(r)\,\mathrm{d}r \tag{3.69}$$

or by substitution of equation (3.68)

$$q_{\text{sph}}(r) = -z_c e\kappa^2\frac{e^{\kappa a}}{1+\kappa a}\,re^{-\kappa r}\mathrm{d}r \tag{3.70}$$

The function $q_{\text{sph}}(r)$, shown in Figure 3.17, has a minimum; in effect, its derivative

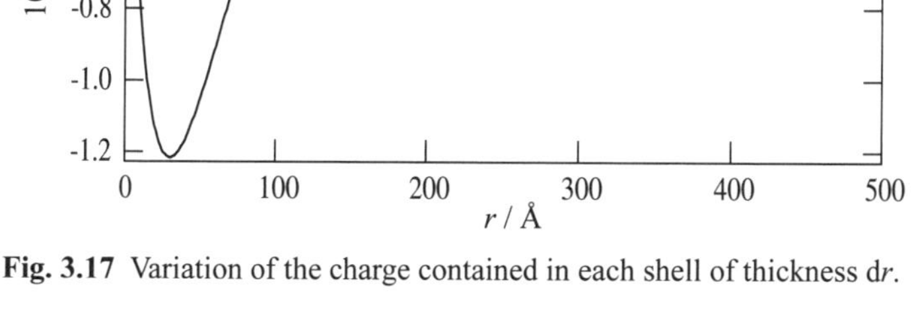

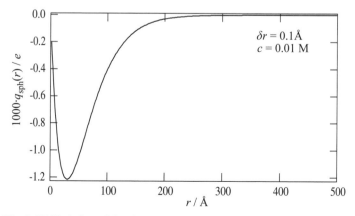

Fig. 3.17 Variation of the charge contained in each shell of thickness dr.

$$\frac{\mathrm{d}q_{\mathrm{sph}}(r)}{\mathrm{d}r} = -\frac{z_j e \kappa^2 e^{\kappa a}}{1+\kappa a} e^{-\kappa r}(1-\kappa r) \tag{3.71}$$

is equal to zero when $r_{\max} = 1/\kappa$. At this distance, the excess charge of opposite sign around the central ion is the largest. Thus, it appears that the constant $1/\kappa$ represents the most likely cation-anion distance and κ is often called the ***reciprocal Debye length***.

Actually, the reciprocal Debye length is often considered as the average radius of the ionic atmosphere. In effect, $q_{\mathrm{sph}}(r)/z_c e$ represents the distribution function of the counter-ions of the central ion, and the average radius is defined by

$$< r_{\mathrm{atm}} > = \int_a^\infty r \frac{q_{\mathrm{sph}}(r)}{z_c e} \mathrm{d}r = -\frac{\kappa^2 e^{\kappa a}}{1+\kappa a}\int_a^\infty r^2 e^{-\kappa r}\mathrm{d}r = \frac{1+(1+\kappa a)^2}{\kappa(1+\kappa a)} \tag{3.72}$$

This expression reduces to κ^{-1} if $\kappa a << 1$. Figure 3.18 shows in a schematic illustration of the ionic atmosphere.

By bringing the molar ionic concentration c_i in mol·l^{-1}

$$c_i = N_i^\infty / 1000\, N_{\mathrm{A}} \tag{3.73}$$

into equation (3.55) which gives κ^2, we have

$$\kappa^2 = \frac{1000\, e^2 N_{\mathrm{A}}}{\varepsilon_0 \varepsilon_{\mathrm{r}}\, kT}\sum_i c_i z_i^2 = \frac{2000\, F^2}{\varepsilon_0 \varepsilon_{\mathrm{r}}\, RT} I_c \tag{3.74}$$

where I_c is the ionic strength in the molarity scale

$$I_c = \tfrac{1}{2}\sum_i c_i z_i^2 \tag{3.75}$$

When the concentration increases, the average radius of the ionic atmosphere decreases parabolically (see Figure 3.19).

The average radius of the ionic atmosphere has an important physical significance, since it represents the ***screening distance*** of a charge. Effectively, any other charge situated at any distance greater $1/\kappa$ from a central charge 'will hardly see' and will not be able to interact with this charge, since seen from outside, the central charge is practically screened off by its cloud.

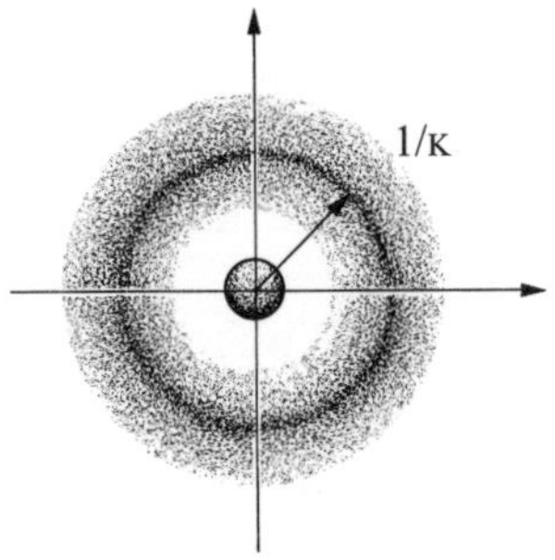

Fig. 3.18 Schematic view of the ionic atmosphere around a central ion.

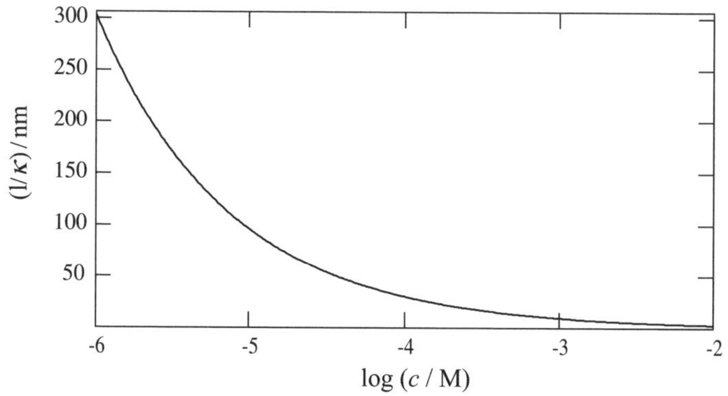

Fig. 3.19 Variation of the reciprocal Debye distance with the concentration of a monovalent salt.

When we compare the numerical values of the reciprocal Debye distance as a function of the concentration with those of the radii of the co-spheres (see the numerical application page 108), it becomes apparent that both are of the same order of magnitude. This implies that when a salt is in solution, the ions spread out more or less homogeneously in order to occupy all the space.

Calculation of the ionic activity coefficient

The potential $\phi(r)$ results as much from the charge of the central ion as from the charge of the ionic atmosphere. The central ion alone would give the potential

$$V(r) \quad = \quad \frac{z_c e}{4\pi\varepsilon_0\varepsilon_r}\frac{1}{r} \tag{3.76}$$

Using the principle of the superposition of potentials, we deduce that the part of the electric potential due to the charge of the ionic atmosphere is the difference:

$$\phi_{\text{atm}}(r) \quad = \quad \phi(r) - V(r) \quad = \quad \frac{z_c e}{4\pi\varepsilon_0\varepsilon_r}\left[\frac{e^{\kappa a}}{1+\kappa a}\frac{e^{-\kappa r}}{r} - \frac{1}{r}\right] \tag{3.77}$$

The potential $\phi_{\text{atm}}(r)$ is illustrated in Figure 3.20.

At a distance a from the central ion, the value of the potential produced by the ionic atmosphere is therefore

$$\phi_{\text{atm}}(a) \quad = \quad \frac{z_c e}{4\pi\varepsilon_0\varepsilon_r}\left[\frac{e^{\kappa a}}{1+\kappa a}\frac{e^{-\kappa a}}{a} - \frac{1}{a}\right] \quad = \quad -\frac{z_c e}{4\pi\varepsilon_0\varepsilon_r}\frac{\kappa}{1+\kappa a} \tag{3.78}$$

the distance a being the minimum distance to which an ion can approach the central ion.

The interaction between the central ion and its atmosphere corresponds to the work of charging this ion from zero to $z_c e$ as in the Born model, the difference being

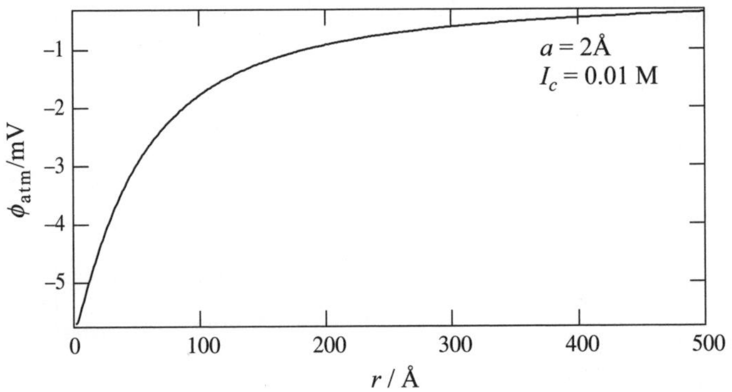

Fig. 3.20 Variation of the ionic atmosphere potential as a function of the distance relative to the central ion.

that the potential to consider is not the potential generated by the ion $V(r_{ion})$ as in equation (3.9) but the potential generated by the ionic atmosphere $\phi_{atm}(a)$ when the central ion carries the charge q

$$w_{ion-ion} = N_A \int_0^{z_c e} \phi_{atm}(a)\, dq = -\frac{N_A}{4\pi\varepsilon_0\varepsilon_r}\frac{\kappa}{1+\kappa a}\int_0^{z_c e} q\, dq = -\frac{N_A}{8\pi\varepsilon_0\varepsilon_r}\frac{\kappa(z_c e)^2}{1+\kappa a} \tag{3.79}$$

We should notice that we have taken r_{ion} as being equal to a, which is not strictly correct. However, this approximation is of little consequence (see below equation (3.89)), and is completely justified for dilute solutions where $\kappa a \ll 1$.

By combining this result with equation (3.45), we can calculate the electrical work of dilution from an ionic solution of a given concentration to an infinitely dilute solution as the opposite of the ion-ion interaction energy and thus obtain an expression for the activity coefficient on the scale of concentrations

$$\ln\gamma_i^c = -\frac{w_{ion-ion}}{RT} = -\frac{(z_i e)^2}{8\pi\varepsilon_0\varepsilon_r kT}\frac{\kappa}{1+\kappa a} \tag{3.80}$$

Also, by replacing κ by expression (3.74) and using common logarithms, we get

$$\log_{10}\gamma_i^c = -z_i^2\frac{A\sqrt{I_c}}{1+aB\sqrt{I_c}} \tag{3.81}$$

where A and B are constants. B is simply defined from equation (3.74) as being equal to

$$B = \frac{\kappa}{\sqrt{I_c}} = F\sqrt{\frac{2000}{\varepsilon_0\varepsilon_r RT}} \tag{3.82}$$

and A can be written as

$$A = \frac{e^2 B}{8\pi\varepsilon_0\varepsilon_r kT\ln(10)} \tag{3.83}$$

If I_c is expressed in mol·l^{-1} and a in metres, the values for A and B for water at 25°C are respectively 0.509 mol$^{-1/2}$·l$^{1/2}$ and 3.29·10^9 mol$^{1/2}$·l$^{-1/2}$·m^{-1}.

Calculation of the average ionic activity coefficient of a salt

Let's consider the case of a binary electrolyte $C_{\nu^+}A_{\nu^-}$. The chemical potential of a salt is a linear combination of the ionic electrochemical potentials (see equation (1.64)).

$$\mu_{\text{salt}} = \nu_+\tilde{\mu}_+ + \nu_-\tilde{\mu}_- = \nu_+\mu_+ + \nu_-\mu_- = \mu_{\text{salt}}^{ideal} + RT\ln\left(\gamma_+^{\nu_+}\gamma_-^{\nu_-}\right) = \mu_{\text{salt}}^{ideal} + RT\ln\gamma_{\text{salt}} \quad (3.84)$$

since the electroneutrality of the solution requires that

$$\nu_+ z_+ F\phi \quad + \quad \nu_- z_- F\phi \quad = \quad 0 \quad (3.85)$$

Given that we cannot *a priori* distinguish in an experimental manner the contribution of the cation and that of the anion when we observe the deviation from the ideal behaviour of an electrolytre solution upon increase of concentration, we define an average ionic activity coefficient $\gamma_\pm$

$$\gamma_\pm \quad = \quad \left(\gamma_+^{\nu_+}\,\gamma_-^{\nu_-}\right)^{1/\nu} \quad = \quad \gamma_{\text{salt}}^{1/\nu} \quad (3.86)$$

This value can be calculated from equations (3.81) and (3.84)

$$\log_{10}\gamma_\pm^c \quad = \quad -\left(\frac{\nu_+ z_+^2}{\nu} + \frac{\nu_- z_-^2}{\nu}\right)\frac{A\sqrt{I_c}}{1 + a_\pm B\sqrt{I_c}} \quad (3.87)$$

The electroneutrality condition allows us to write

$$\frac{\nu_+ z_+^2 + \nu_- z_-^2}{\nu} \quad = \quad \frac{(-\nu_- z_-)z_+ + (-\nu_+ z_+)z_-}{\nu} \quad = \quad -z_- z_+ \quad \frac{\nu_- + \nu_+}{\nu} \quad = \quad -z_- z_+ \quad (3.88)$$

from which eventually we obtain

$$\log_{10}\gamma_\pm^c \quad = \quad z_+ z_- \quad \frac{A\sqrt{I_c}}{1 + a_\pm B\sqrt{I_c}} \quad (3.89)$$

where $a_\pm$ is often considered as the minimum approach distance between the cation and the anion. For dilute solutions, equation (3.89) reduces to

$$\log_{10}\gamma_\pm^c \quad = \quad z_+ z_- \, A\sqrt{I_c} \quad (3.90)$$

which, for a 1:1 electrolyte, reads

$$\log\gamma_\pm^c \quad = \quad -A\sqrt{c_{\text{salt}}} \quad (3.91)$$

Equations (3.91) and (3.89) are shown on Figure 3.21 for solutions of NaCl, $CaCl_2$ and $LaCl_3$. In the linear approximation as given by equation (3.90), the slopes are respectively A, $2A\sqrt{3}$, $3A\sqrt{6}$, because of the different stoichiometries. The $a_\pm$ coefficients are respectively 400, 475 and 570 pm. Values of $a_\pm$ coefficients for other electrolytes are given in Table 3.5.

Table 3.5 Values of the $a_\pm$ parameter.

Salt	$a_\pm$ / pm
HCl	450
HBr	520
LiCl	430
NaCl	400
KCl	360

The comparison of activity coefficients on the molarity scale to values in the literature published on the molality scale can be a source of error and confusion for concentrated solutions. It is for this reason that in this paragraph, the scale used is carefully indicated using a superscript. Applying equation (1.27), we have

$$\gamma_\pm^c = \gamma_\pm^m \frac{d_0 m_{salt}}{c_{salt}} \tag{3.92}$$

where d_0 is the density of the pure solvent expressed in $kg{\cdot}l^{-1}$. Remembering that

$$m_{salt} = \frac{c_{salt}}{d - c_{salt} M_{salt}} \tag{3.93}$$

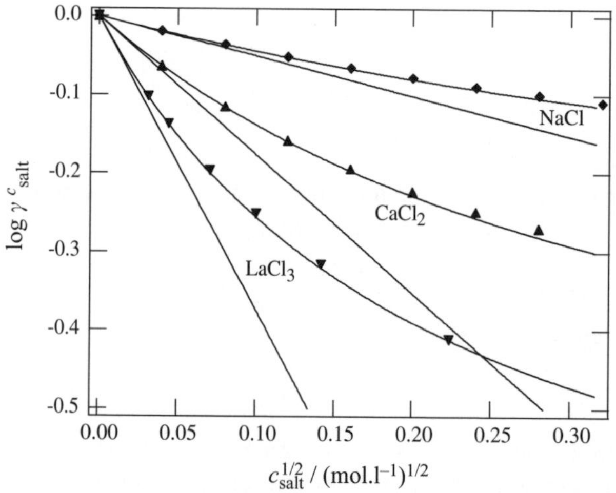

Fig 3.21 Experimental verification of the linear approximation (3.90) and the Debye-Hückel law (3.89). The experimental values (markers) are taken from: R.A. Robinson & R.H. Stokes, *Electrolyte Solutions*, Butterworth, London 1959.

where M_{salt} is the molar mass of the salt in kg·mol^{-1}. For example, in the case of NaCl and using experimental values given in the appendix (see page 131), we can express the molar concentration and the density in kg·l^{-1} as a function of the molality with the aid of a second order polynomial expression

$$c_{\text{NaCl}} = 0.99921 \cdot m_{\text{NaCl}} - 0.018859 \cdot m_{\text{NaCl}}^2 \tag{3.94}$$

and

$$d_{\text{NaCl}} = 1 + 0.039243 \cdot m_{\text{NaCl}} - 0.0010746 \cdot m_{\text{NaCl}}^2 \tag{3.95}$$

For dilute solutions where the density of the solution changes negligibly, the two scales overlap.

Average ionic activity coefficient of a concentrated salt

We can see in Figure 3.22 that at high solute concentrations, the activity coefficient attains a minimum before rising again. This is due to the fact that at such salt concentrations, a certain quantity of the solvent molecules are fixed in the hydration spheres of the ions.

If n_T is the total number of moles of solvent molecules (S) and n_H the total number of moles of hydration solvent molecules linked directly to the ions, the total Gibbs energy for a solution containing n moles of salt, is either written as

$$G = n_T \mu_S + n\, \mu_{salt} \tag{3.96}$$

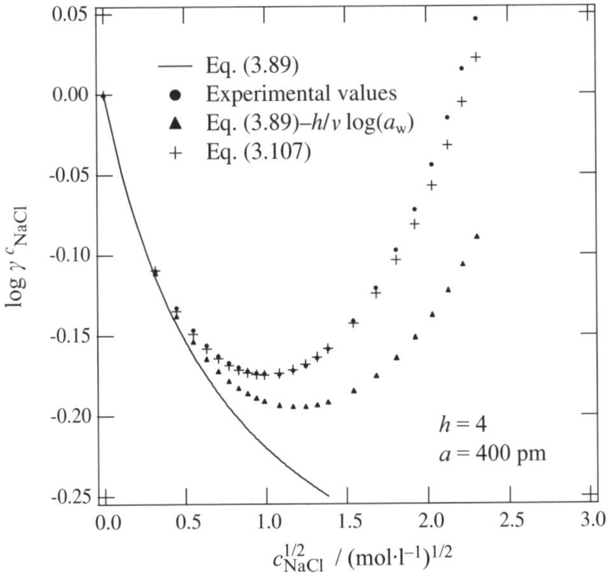

Fig 3.22 Verification of equation (3.107) for NaCl in water. The experimental values (points) are taken from: R.A. Robinson & R.H.Stokes, *Electrolyte Solutions*, Butterworth, London 1959.

or as

$$G = (n_{\rm T} - n_{\rm H})\,\mu_{\rm S} + n\,\mu^{h}_{\rm salt} \tag{3.97}$$

where $\mu^{h}_{\rm salt}$ is the chemical potential of the hydrated salt. Because these two expressions are equal, we can write

$$0 = n_{\rm H}\,\mu_{\rm S} + n\left(\mu_{\rm salt} - \mu^{h}_{\rm salt}\right) \tag{3.98}$$

To develop the chemical potential expressions, it is more convenient to work with the mole fraction scale rather than the concentration scale (molality or molarity). In this way, we can write

$$n_{\rm H}\mu^{\ominus}_{\rm S} + n\left(\mu^{\ominus}_{\rm salt} - \mu^{\ominus,h}_{\rm salt}\right) = -\,n_{\rm H}RT\ln a_{\rm S} - nRT\ln\left(\frac{x_{\rm salt}\,\gamma_{\rm salt}}{x^{h}_{\rm salt}\gamma^{h}_{\rm salt}}\right) \tag{3.99}$$

The mole fractions of the salt and the hydrated salt are respectively

$$x_{\rm salt} = \frac{n}{n_{\rm T}+n} \quad \text{and} \quad x^{h}_{\rm salt} = \frac{n}{\left(n_{\rm T}-n_{\rm H}\right)+n} \tag{3.100}$$

such that their ratio is

$$\frac{x_{\rm salt}}{x^{h}_{\rm salt}} = \frac{\left(n_{\rm T}-n_{\rm H}\right)+n}{n_{\rm T}+n} \tag{3.101}$$

For very dilute solutions, the solvent activity tends towards unity and so do the activity coefficients of the salt $\gamma^{h}_{\rm salt}$ and $\gamma^{h}_{\rm salt}$. Since the ratio $x_{\rm salt}/x^{h}_{\rm salt}$ also tends towards unity for very dilute solutions, we see that equation (3.99) is equal to zero for dilute solutions. This means that the linear combination of the standard terms (left hand side of equation (3.99)) is independent of the concentration and is equal to zero.

By introducing the average ionic activity coefficients on the mole fraction scale (superscript X), and taking into account equation (3.86), we therefore find that

$$\nu n\ln\gamma^{\rm X}_{\pm} = \nu n\ln\gamma^{h{\rm X}}_{\pm} - n_{\rm H}\ln a_{\rm S} - n\ln\left(\frac{n_{\rm T}-n_{\rm H}+n}{n_{\rm T}+n}\right) \tag{3.102}$$

In order to be able to make comparisons with the data of Figure 3.22 on the scale of concentrations, we need to calculate $\gamma^{c}_{\pm}$. To do that, we have to convert $\gamma^{h{\rm X}}_{\pm}$ into $\gamma^{hc}_{\pm}$. Applying equation (1.27), and expressing the mole fraction of the solvent

$$x_{\rm S} = \frac{n_{\rm T}}{n_{\rm T}+n_{\rm salt}} = \frac{d - c_{\rm salt}M_{\rm salt}}{d + c_{\rm salt}\left(M_{\rm S}-M_{\rm salt}\right)} \tag{3.103}$$

we have

$$\gamma^{h{\rm X}}_{\pm} = \frac{d - c_{\rm salt}M_{\rm salt}}{x_{\rm S}d_0}\,\gamma^{hc}_{\pm} = \frac{d + c_{\rm salt}\left(M_{\rm S}-M_{\rm salt}\right)}{{\rm d}_0}\,\gamma^{hc}_{\pm} \tag{3.104}$$

Next, we need to calculate the total number of moles of water

$$n_{\mathrm{T}} = \left[\frac{d}{c_{\mathrm{salt}} M_{\mathrm{S}}} - \frac{M_{\mathrm{salt}}}{M_{\mathrm{S}}}\right] \cdot n \qquad (3.105)$$

Putting $n_{\mathrm{H}} = hn$, where h is the number of moles of water linked per mole of salt, the last term of equation (3.102) is written as

$$\ln\left(\frac{n_{\mathrm{T}} - n_{\mathrm{H}} + n}{n_{\mathrm{T}} + n}\right) = \ln\left(\frac{d - c_{\mathrm{salt}}\left[M_{\mathrm{salt}} + M_{\mathrm{S}}(h-1)\right]}{d + c_{\mathrm{salt}}\left(M_{\mathrm{S}} - M_{\mathrm{salt}}\right)}\right) \qquad (3.106)$$

Substituting in equation (3.102), we then find

$$\ln \gamma_{\pm}^{c} = \ln \gamma_{\pm}^{hc} - \frac{h}{\nu} \ln a_{\mathrm{s}} - \ln\left(\frac{d - c_{\mathrm{salt}}\left[M_{\mathrm{salt}} + M_{\mathrm{S}}(h-1)\right]}{d_0}\right)^{1/\nu} \qquad (3.107)$$

The term $\gamma_{\pm}^{hc}$ corresponds to the interaction energy of the hydrated ions and is given by the Debye-Hückel expression (3.89), and it is possible to calculate an expression for $\gamma_{\pm}^{c}$ that depends on n_{H} and on the solvent activity a_{S}.

The solvent activity can be calculated from the equation

$$n_{\mathrm{T}} \mathrm{d}\mu_{\mathrm{S}} + n \mathrm{d}\mu_{\mathrm{salt}} = 0 \qquad (3.108)$$

which can be developed to yield

$$\mathrm{d} \ln a_{\mathrm{S}} = -\frac{n}{n_{\mathrm{T}}} \mathrm{d} \ln(\gamma_{\pm}^{c} c_{\mathrm{salt}}) \qquad (3.109)$$

Otherwise, the water activity can be measured and for example, for NaCl in water, the following polynomial can be used

$$a_{\mathrm{w}} = 1 - 0.030931 \cdot c_{\mathrm{NaCl}} - 0.0015192 \cdot c_{\mathrm{NaCl}}^{2} \qquad (3.110)$$

The second term in equation (3.107) corresponds to the work due to the disappearance of free water molecules into the solvation layers, and the last term to the work due to the increase in the effective salt concentration.

Figure 3.23 shows the experimental activity coefficients of alkali metal halides, which illustrate that the more the cation is hydrated, the more important the last two terms of equation (3.107) become. Table 3.6 gives values of the coefficients h and $a_{\pm}$ for different salts

Table 3.6 Experimental values of h and $a_{\pm}$.

	LiCl	NaCl	KCl	RbCl	HCl	$CaCl_2$	$MgCl_2$
h	7	3.5	2	1	8	12	14
$a_{\pm}$ / pm	430	400	360	350	450	475	500

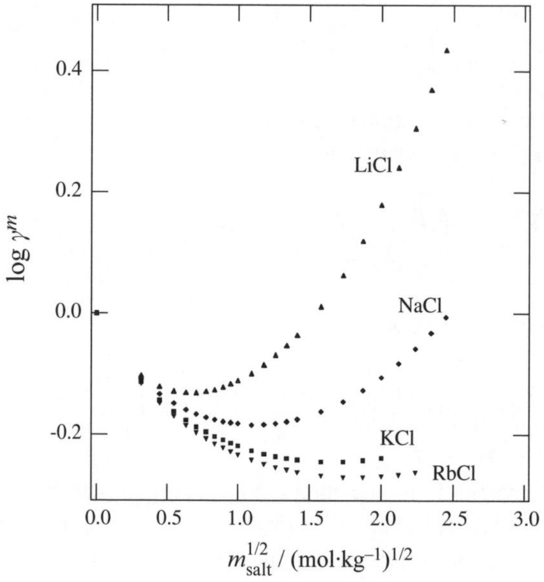

Fig 3.23 Influence of hydration on the activity coefficients of alkali metal halides. The experimental values (markers) are taken from : R.A. Robinson & R.H. Stokes, *Electrolyte Solutions*, Butterworth, London 1959.

3.4.3 Electrochemical measurement of activity coefficients

Measuring the potential of cells such as those studied in §2.1.4

$$\mathrm{Cu^I} \mid \mathrm{Pt} \mid \tfrac{1}{2}\mathrm{H_2}(f = p^{\ominus}),\ \ \mathrm{HCl} \mid \tfrac{1}{2}\mathrm{Hg_2Cl_2} \mid \mathrm{Hg} \mid \mathrm{Cu^{II}}$$

allows us to determine the average activity coefficient of the electrolyte, here hydrochloric acid

$$E = \left[E^{\ominus}_{\frac{1}{2}\mathrm{Hg_2Cl_2/Hg}}\right]_{\mathrm{SHE}} - \frac{RT}{F}\ln\left[\gamma^m_{\mathrm{H^+}}\gamma^m_{\mathrm{Cl^-}}\right] - \frac{2RT}{F}\ln\left[m_{\mathrm{HCl}}\right] \tag{3.111}$$

EXAMPLE

Let's take the numerical values from example §2.1.4, and calculate the standard redox potential of the calomel electrode given by equation (3.111).
Given that we are talking about dilute solutions, the scales of molarity and molality overlap, and so expressing the activity coefficients by equation (3.90) , we have

$$E = \left[E^{\ominus}_{\frac{1}{2}\mathrm{Hg_2Cl_2/Hg}}\right]_{\mathrm{SHE}} + \frac{2RT\ln 10}{F}A\sqrt{m_{\mathrm{HCl}}} - \frac{2RT}{F}\ln\left[m_{\mathrm{HCl}}\right]$$

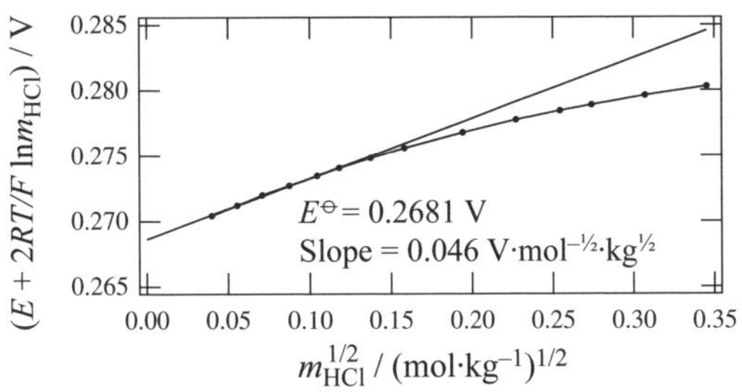

Thus we can plot

$$E + \frac{2RT}{F}\ln[m_{HCl}]$$

as a function of $\sqrt{m_{HCl}}$ and obtain the standard redox potential as the ordinate at the origin.

Thus, taking into account the activity coefficients we have a much better extrapolation that allows us to determine the standard redox potential to a tenth of a millivolt. The slope should be $2ART\ln 10/F$ which is 0.06 V/(mol·kg)$^{1/2}$ but in fact the linear extrapolation gives us a lower slope that indicates that equation (3.90) is not completely respected.

Actually, this example also illustrates another experimental method for measuring activity coefficients based on measuring the cell potential. Knowing the standard redox potential, from equation (3.111) we can measure the cell potential and from there determine the activity coefficents. This equation gives

$$\log \gamma_{\pm HCl} = \frac{F}{2RT\ln 10}\left(\left[E^{\ominus}_{\frac{1}{2}Hg_2Cl_2/Hg}\right]_{SHE} - E\right) - \log[m_{HCl}]$$

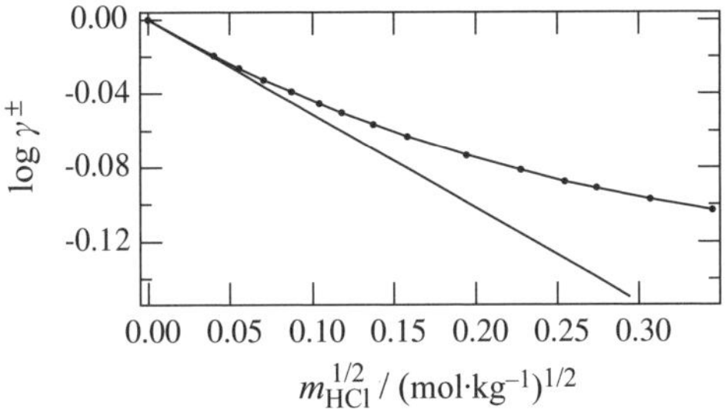

This graph where the standard redox potential taken is 0.2681 V confirms that equation (3.90) is not completely respected. Therefore, we must refine the evaluation of the standard potential by successive iterations. The final value obtained is 0.2679 V.

3.4.4 Kinetics of reactions between ions

It has been experimentally observed that second order reaction rate constants between charged species vary with the ionic strength (see Figure 3.24). When the ionic reactants have a similar charge, the speed increases with the ionic strength, whilst when they have opposite charges, the opposite happens.

Brønsted suggested that the rate of a reaction between two species A and B could be written as

$$k = k^{\ominus}\frac{\gamma_A \gamma_B}{\gamma^{\#}} \tag{3.112}$$

where $k^{\ominus}$ is the standard rate constant, and $\gamma^{\#}$ is the activity coefficient of the activated ionic complex, i.e. the transition state.

By applying the Debye-Hückel equation (3.81) in the linear approximation to express the activity coefficients of the ions, we obtain

$$\log_{10} k = \log_{10} k^{\ominus} - A\left[z_A^2 + z_B^2 - (z_A + z_B)^2\right] I^{1/2} = \log_{10} k^{\ominus} + 2Az_A z_B I^{1/2} \tag{3.113}$$

This equation has been tested on several systems, and generally describes the phenomena quite well. The results of Figure 3.24 illustrate the screening effect well. It is clear that for two ions of the same sign to be able to react, it is necessary to screen them, as much as possible to favour the reaction. Thus, an increase in the ionic

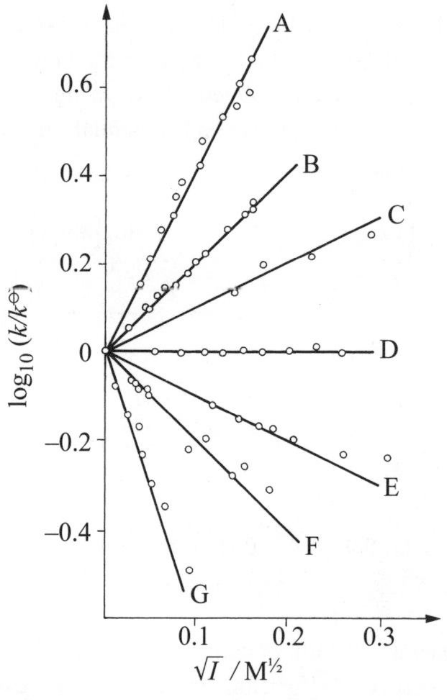

Fig. 3.24 Variation of the rate constant for reactions between ions as a function of the square root of the ionic strength (Laidler, *Chemical Kinetics*, Harper & Row, 1987, New York, USA).

A	$Co(NH_3)_5Br^{2+} + Hg^{2+}$	$z_A z_B = 4$
B	$S_2O_8^{2-} + I^-$	$z_A z_B = 2$
C	$CO(OC_2H_5)N{:}NO_2^- + OH^-$	$z_A z_B = 1$
D	$[Cr(urea)_6]^{3+} + H_2O$	$z_A z_B = 0$
	$CH_3COOC_2H_5 + OH^-$	$z_A z_B = 0$
E	$H^+ + Br^- + H_2O$	$z_A z_B = -1$
F	$Co(NH_3)_5Br^{2+} + OH^-$	$z_A z_B = -2$
G	$Fe^{2+} + Co(C_2O_4)^{3-}$	$z_A z_B = -6$

strength reduces the screening distance and consequently increases the rate of the reactions between two ions of the same sign. Inversely, two ions of opposite sign will be attracted to each other at longer distances in dilute solutions, and in this case, an increase in the ionic strength will slow down the reaction.

3.4.5 Comments

The Debye-Hückel theory based on electrostatics predicts very well the average ionic activity coefficents, in spite of the hypothesis that the electrolyte solution is considered as a homogeneous medium having a relative permittivity equal to that of the pure solvent. Thus, in spite of the complexity of the system, this statistical mechanics model provides a simple equation (3.89) with a single adjustable parameter $a_{\pm}$ that accounts reasonably well for the behaviour of electrolytic solutions up to concentrations of the order of 0.1 M.

A very important conceptual aspect of this theory is the notion of reciprocal distance κ which shows that the electrostatic interactions are screened by the presence of the electrolyte.

3.5 ION PAIRS

The Debye-Hückel theory considers that ions are isolated charges that are surrounded by an ionic atmosphere of the opposite charge. Nevertheless, it is possible that ions of opposite charge approach each other close enough for the coulombic attraction to become stronger than the kinetic energy of thermal agitation. When that happens, they form an ion pair that can be represented as a fluctuating dipole. The question is to know what is the concentration of the ion pairs.

3.5.1 The Bjerrum theory (1923)

Take a spherical shell of radius r and thickness $\mathrm{d}r$ around a central reference cation. Noting, that near to the central ion, the potential is mainly that created by the central ion itself (see Figure 3.16)

$$\phi(a) \gg \phi_{\mathrm{atm}}(a) \quad \text{or} \quad \phi(a) \approx V(a) \tag{3.114}$$

the number of negative charges $\mathrm{d}n_a(r)$ in this shell is defined from a Boltzmann distribution

$$\mathrm{d}n_a(r) = 4\pi r^2\, N_a(r)\,\mathrm{d}r = 4\pi r^2 \mathrm{d}r\, N_a^{\infty}\; e^{-\frac{z_+ z_- e^2}{4\pi\varepsilon_0\varepsilon_\mathrm{r}\, r\, kT}} \tag{3.115}$$

where $N_a(r)$ is the density of anions around the central cation.

If we plot the function $\mathrm{d}n_a(r)$ as in Figure (3.25), we observe that it is very high at a short distance, passes through a minimum and again becomes very high at longer distances as the volume of the shell becomes greater.

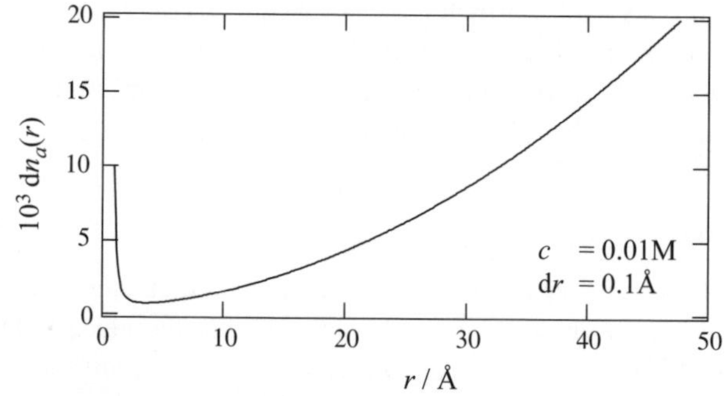

Fig. 3.25 Variation with distance of the number of counter-ions in shells of thickness dr.

This minimum can be found by differentiating equation (3.115)

$$\frac{\mathrm{d}n_a(r)}{\mathrm{d}r} = 8\pi r N_a^\infty\, e^{-\frac{z_+z_-e^2}{4\pi\varepsilon_0\varepsilon_r\, r\, kT}} + \frac{4\pi r^2 N_a^\infty z_+z_-e^2}{4\pi\varepsilon_0\varepsilon_r\, kT\, r^2}\, e^{-\frac{z_+z_-e^2}{4\pi\varepsilon_0\varepsilon_r\, r\, kT}} = 0 \qquad (3.116)$$

Thus, the minimum takes place when

$$r = q = -\frac{z_+z_-e^2}{8\pi\varepsilon_0\varepsilon_r kT} \qquad (3.117)$$

The gist of Bjerrum's model is to define an ion pair as existing when the distance between the centre of two ions of opposite sign is less than q (see Figure 3.26). In water ($\varepsilon_r = 78.5$), q is 357 pm while in organic solvents such as 1,2-dichloroethane ($\varepsilon_r = 10$), q is 2.8 nm.

The radii for alkali metals found by crystallography vary from 60 pm for lithium to 169 pm for cesium and those for halogens vary from 136 pm for fluorine to 216 pm

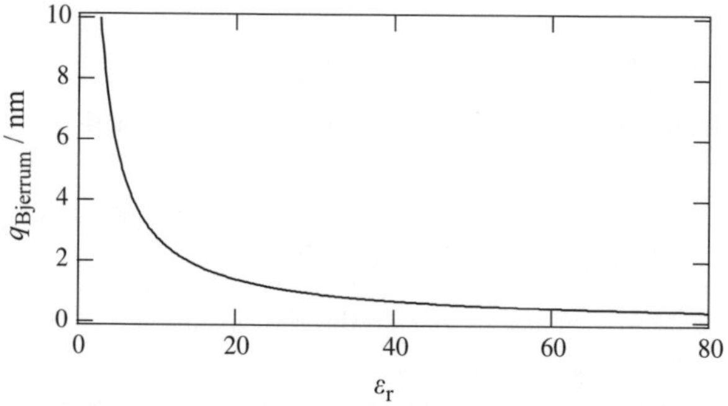

Fig. 3.26 Maximum formation distance for ion pairs as a function of the relative permittivity of the solvent.

for iodine (see Table 3.3). Comparison with q shows that cesium iodide cannot form ion pairs in water, at least in the sense of Bjerrum's theory.

This choice of q as the maximum separation distance to is not accidental. Firstly, at $r = q$ the electrostatic interaction energy $-z_+z_-e^2/4\pi\varepsilon_0\varepsilon_r r$ is equal to 2 kT, kT being the thermal agitation energy. Secondly, the number of anions in the sphere of radius q is

$$\theta \;=\; \int_a^q \mathrm{d}n_a(r)\,\mathrm{d}r \;=\; 4\pi N_a^{\infty} \int_a^q r^2\, e^{2q/r}\,\mathrm{d}r \tag{3.118}$$

where a is the minimum approach distance. By definition, θ must take a value between 0 and 1, i.e. the fraction of counter-ions around the central ion. Thus, by choosing the minimum of the function $\mathrm{d}n_a(r)$ as the upper limit of the integral, this integral converges and can be calculated numerically.

Because the density of free counter-ions around the ion N_a^{∞} depends both on the salt concentration, and on the fraction of ion pairs, the best way to calculate θ is by iteration.

3.5.2 Association constant

The equilibrium between the dissociated ions and the ion pairs can be written as

$$C^+ + A^- \rightleftarrows IP$$

The constant of this equilibrium is written as

$$K_A \;=\; \frac{a_{IP}}{a_{C^+}a_{A^-}} \tag{3.119}$$

If θ is the fraction of ion pairs, we have

$$K_A \;=\; \frac{\theta\, c\gamma_{IP}}{(1-\theta)c\gamma_{C^+}\;\;(1-\theta)c\gamma_{A^-}} \;=\; \frac{\theta}{(1-\theta)^2 c}\;\frac{\gamma_{IP}}{\gamma_{\pm}{}^2} \tag{3.120}$$

where c is the salt concentration, γ_{IP} the activity coefficient of the ion pair and $\gamma_{\pm}$ the average activity coefficient of the salt $\gamma_{\pm}{}^2 = \gamma_{C^+}\gamma_{A^-}$. From the fact that IP is a neutral species, and that we attribute the deviations of the activity coefficients from unity to electrostatic effects only, the value of γ_{IP} is taken to be 1.

Given that the association constant K_A is the same for all concentrations, the most judicious method is to calculate it for very dilute solutions, for which the activity coefficients can be considered as unity.

$$K_A \;=\; \frac{\theta}{(1-\theta)^2 c} \;\approx\; \frac{\theta}{c} \tag{3.121}$$

By replacing θ by equation (3.118) and expressing the ionic density N_a^{∞} (number of ions per m^3) in terms of molar concentrations and Avogadro's constant, we have

$$K_A \;=\; 4000\,\pi\, N_A \int_a^q r^2\, e^{2q/r}\,\mathrm{d}r \tag{3.122}$$

This shows that association constants can be calculated numerically and that the association constant is quite independent of the concentration.

To calculate θ in concentrated solutions, we can work iteratively in the following manner :

- calculate K_A
- assume that $\gamma_\pm$ is unity and calculate θ
- having obtained θ, calculate $\gamma_\pm$ from the Debye-Hückel formula modified in such a way that the ionic strength takes into account ion pairs. For 1:1 electrolytes, this is written as

$$\log \gamma_\pm = \frac{A\, z_+ z_- \sqrt{(1-\theta)c}}{1 + Bq\sqrt{(1-\theta)c}} \tag{3.123}$$

The choice of q as the value of the a parameter is rather arbitrary, but can be justified by the fact that this distance is the interaction distance of the ions when they form a pair in the Bjerrum sense.

- substitute the value obtained in K_A and recalculate
- And so on . . .

EXAMPLE

Let's calculate the proportion of ion pairs in a 10^{-3} M solution of tetrabutylammonium bromide in 1,2-dichloroethane, knowing that $\varepsilon_r = 10$, $r_{TBA^+} = 383$ pm, and $r_{Br^-} = 196$ pm.

First of all calculate the Bjerrum distance q (3.117):

$$q = -\frac{z_+ z_- e^2}{8\pi\varepsilon_0\varepsilon_r kT} = \frac{(1.6\cdot 10^{-19})^2}{8\times 3.14\times 8.85\cdot 10^{-12}\times 10\times 1.38\cdot 10^{-23}\times 298} = 2.8\ \text{nm}$$

and the equilibrium constant with (3.122). To do this, we can use a calculation software that can do numerical integrations. By taking the sum of the radii for a we then obtain :

$$K_A = 5222\ \text{M}^{-1}$$

The first value of θ is then the root of the quadratic equation (3.120) taking $\gamma_\pm = 1$. Next, we calculate the average activity coefficient, taking into account the fact that the constants A and B in equation (3.89) depend on the solvent, and here have values of 11.20 $\text{M}^{-1/2}$ and $9.21\cdot 10^9\ \text{M}^{1/2}\cdot\text{m}^{-1}$

The results obtained for 4 iterations are:

$\gamma_\pm$	θ
1	0.648
0.722	0.550
0.702	0.541
0.701	0.541

3.5.3 The Fuoss theory

The main defects in Bjerrum's theory are related to the arbitrary character of the distance q which defines the formation of ion pairs, and also the fact that the ions do not need to be in physical contact for an ion pair to exist. In the Fuoss theory, we only consider as ion pairs those ions that are of opposite charge, and in contact for an arbitrary length of time.

Let's consider a solution of a monovalent salt containing free ions and ion pairs. The number of free anions n_a^f is equal to the number of free cations n_c^f, that is to say to the total number of anions or cations n^t respectively, less the number of ion pairs existing n_p

$$n_a^f = n_a^t - n_p = n_c^f = n_c^t - n_p = n^f \tag{3.124}$$

If we add dn ions of each species to this solution, one part will form pairs and the other part will remain free. The variation in the number of free anions will be proportional to dn and the fraction of the volume not occupied by the cations will be

$$\mathrm{d}n_a^f = \frac{V - n_c^t V_c}{V}\,\mathrm{d}n \tag{3.125}$$

where V is the volume of the solution, and V_c the volume occupied by a cation. For dilute solutions, it is possible to neglect the volume occupied by the cations with respect to the volume of the solution, and equation (3.125) reduces to

$$\mathrm{d}n^f \cong \mathrm{d}n \tag{3.126}$$

The variation in the number of ion pairs is also proportional to dn, but this time proportional to the fraction of the volume occupied by the free ions, the whole being corrected by a Boltzmann distribution factor which brings in the cation-anion interaction

$$\mathrm{d}n_p = \frac{n_c^f V_c + n_a^f V_a}{V}\, e^{-U(r)/kT}\,\mathrm{d}n \tag{3.127}$$

Combining these two expressions, we get

$$\mathrm{d}n_p = \frac{n^f \left(V_c + V_a\right) e^{-U(r)/kT}}{V}\,\mathrm{d}n^f \tag{3.128}$$

Then by integration we obtain

$$n_p = \frac{\left(n^f\right)^2 \left(V_c + V_a\right) e^{-U(r)/kT}}{2\,V} \tag{3.129}$$

The interaction energy then corresponds to the coulombic interaction energy between an ion and a central ion when they are in contact. Knowing the expression for the electric potential around a central ion (see equation (3.66)), we then have

$$U(a) = \frac{z_+ z_- e^2}{4\pi\varepsilon_0\varepsilon_r\, a}\left[\frac{1}{1+\kappa a}\right] \tag{3.130}$$

In conclusion, the concentration of ion pairs can be calculated directly. By making the extra approximation that the average exclusion volume of an ion is $4/3\pi a^3$, we have

$$\frac{n_p}{V} = \left(\frac{n^f}{V}\right)^2 \frac{4\pi a^3}{3} e^{\frac{-z_+z_-e^2}{4\pi\varepsilon_0\varepsilon_r\, a\, kT}\left[\frac{1}{1+\kappa a}\right]} \tag{3.131}$$

By defining the fraction of ion pairs θ as

$$\frac{n_p}{V} = \theta\, 1000c\, N_A \tag{3.132}$$

where c is given in mol·l^{-1} , V in m^3, and the fraction of free ions by

$$\frac{n^f}{V} = (1-\theta)\, 1000c\, N_A \tag{3.133}$$

this fraction is obtained as the solution of the quadratic equation

$$\theta = 1000c\, N_A (1-\theta)^2 \frac{4\pi a^3}{3} e^{\frac{-z_+z_-e^2}{4\pi\varepsilon_0\varepsilon_r\, a\, kT}\left[\frac{1}{1+\kappa a}\right]} \tag{3.134}$$

Thus the equilibrium constant for the dilute solutions is written as

$$K_A = \frac{\theta}{(1-\theta)^2 c} = 1000 N_A \left(\frac{4\pi a^3}{3}\right) e^{\frac{-z_+z_-e^2}{4\pi\varepsilon_0\varepsilon_r\, a\, kT}\left[\frac{1}{1+\kappa a}\right]} \tag{3.135}$$

The dependence of K_A as a function of the relative permittivity (Figure 3.27) clearly shows that the formation of ion pairs is an important phenomenon for all weakly polar and non-polar solvents.

Experimental evidence corroborates fairly well both the Bjerrum and Fuoss theories. The association constants can be found from measuring the conductivity of solutions (see §4.2).

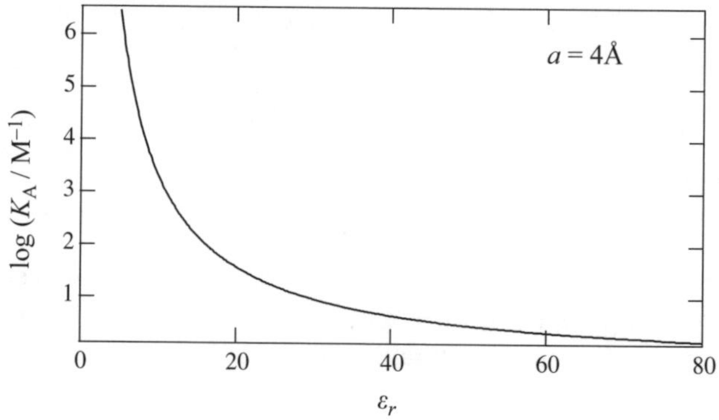

Fig. 3.27 Logarithm of the association constant for an 1:1 electrolyte as a function of the relative permittivity.

EXAMPLE

Let's calculate, this time by the Fuoss method, the proportion of ion pairs for a 10^{-3} M solution of tetrabutylammonium bromide in 1,2-dichloroethane.
By taking a as the sum of the radii, we have $K_A = 1895\ M^{-1}$ and $\theta = 0.49$. These values are of the same order of magnitude as those obtained previously using the Bjerrum method.

3.6 COMPUTATIONAL METHODS

3.6.1 The Monte-Carlo method

The Monte-Carlo method consists of considering a group of N+1 particles in a cubic box with a volume of $(N+1)L^3$. The initial configuration of this group is completely random. The configuration energy $U(q)$ is calculated as the sum of all the interaction energies between all the particles, which is

$$U = \sum_{i=1}^{N}\sum_{j=1}^{N} u_{ij} \tag{3.136}$$

The method consists of choosing a particle at random and moving it in a random and incremental manner, and calculating the energy of each new configuration. At each movement, the new configuration is accepted if its energy $U(t+1)$ is smaller than that of the preceding step $U(t)$. If it is not the case, the new configuration can be accepted only if the quantity $\exp[-U(t+1)/U(t)]$ is smaller than a number chosen at random between 0 and 1. These calculations stop when the configuration energy converges towards a minimum value. The calculations are in general repeated for various initial positions, in order to verify that the optimal configuration attained by convergence really is a property of the system.

When modelling an electrolyte solution, the ions are often considered as rigid spheres as in the Debye-Hückel theory, and the interaction energy between two ions is purely of a coulombic nature, i.e.

$$u_{ij} = \frac{z_i z_j e^2}{4\pi\varepsilon_0\varepsilon_r r_{ij}^2} \quad \text{for } r_{ij} \geq a \tag{3.137}$$

$$u_{ij} = \infty \quad \text{for } r_{ij} \leq a \tag{3.138}$$

From the optimal configuration, it is possible to calculate the correlation functions of pairs formed by ions of the same sign g and by ions of opposite sign $g_{+/-}$.

The correlation functions obtained using the Monte-Carlo method corroborate quite well those predicted by the Debye-Hückel method using equations (3.46) and (3.66).

$$g_{\text{D-H}} = \exp^{-ze\phi(r)/kT} = \exp^{\pm\frac{z^2e^2}{4\pi\varepsilon_0\varepsilon_r\, kT}\left[\frac{e^{\kappa a}}{1+\kappa a}\right]\frac{e^{-\kappa r}}{r}} \tag{3.139}$$

In a completely general way, the mean value of a property P of the canonical ensemble of particles can be calculated by taking a Boltzmann distribution function of the energy of the system

$$< P > = \frac{\int P(q)\exp^{-U(q)/kT}\,\mathrm{d}q}{\int \exp^{-U(q)/kT}\,\mathrm{d}q} \tag{3.140}$$

3.6.2 Molecular dynamics

More and more, molecular dynamics is being used for the study of the physical properties of liquids and electrolytic solutions.

In a nutshell, the principle of molecular dynamics consists of studying the behaviour of a system containing N particles, which can represent different species, by making the hypothesis that the system obeys the laws of classical mechanics.

What differentiates molecular dynamics from the Monte-Carlo methods comes from the fact that in molecular dynamics, the initial configuration contains not only a spatial distribution of the particles, but also a distribution of their velocity vectors. The system is allowed to evolve at constant kinetic energy (3/2 NkT) by calculating the displacements of the particles in steps of a few femtoseconds. This approach has the advantage of allowing the study of the trajectories of the different particles and obtaining information about the mode of transport of the species. The results of these calculations are enormously dependent on the interaction potentials between species, but there is no doubt that the increase in computational power of new computers has made molecular dynamics a standard tool in the study of solutions.

APPENDIX

Table 3.6 Physico-chemical data for NaCl solutions.
R.A. Robinson & R.H. Stokes, *Electrolyte Solutions*, Butterworth, London 1959.

m_{NaCl} /mol·kg^{-1}	c_{NaCl} / mol·l^{-1}	d_{NaCl} / kg·l^{-1}	a_W	γ^c_{NaCl}	n_T
0	0	1	1	1	55.5556
0.1	0.100126	1.00391	0.996888	0.777019	55.4478
0.2	0.199481	1.00781	0.993769	0.736911	55.3414
0.3	0.298459	1.01168	0.990633	0.713665	55.2351
0.4	0.39706	1.01553	0.987479	0.698131	55.1287
0.5	0.495284	1.01935	0.984308	0.687484	55.0224
0.6	0.593131	1.02316	0.981119	0.680794	54.9161
0.7	0.6906	1.02694	0.977915	0.676079	54.8099
0.8	0.787692	1.03071	0.974693	0.672344	54.7037
0.9	0.884407	1.03445	0.971456	0.670619	54.5975
1	0.980745	1.03817	0.968203	0.669899	54.4913
1.2	1.17229	1.04554	0.961652	0.66946	54.2791
1.4	1.36232	1.05283	0.955042	0.673114	54.067
1.6	1.55085	1.06004	0.948377	0.677822	53.855
1.8	1.73787	1.06716	0.941658	0.685667	53.6432
2	1.92338	1.07419	0.934888	0.694611	53.4315
2.5	2.38055	1.09139	0.917758	0.722522	52.9027
3	2.82829	1.10806	0.900366	0.757347	52.3747
3.5	3.26661	1.12419	0.88275	0.799301	51.8474
4	3.69549	1.13978	0.864948	0.84752	51.3209
4.5	4.11494	1.15483	0.846996	0.903293	50.7952
5	4.52497	1.16935	0.828932	0.965753	50.2703
5.5	4.92556	1.18333	0.81079	1.03623	49.7461
6	5.31673	1.19677	0.792604	1.1271	49.2228

CHAPTER 4

TRANSPORT IN SOLUTION

4.1 TRANSPORT IN ELECTROLYTE SOLUTIONS

Two types of mass transport are usually considered in electrolyte solutions:

- convection due to thermal or mechanical agitation
- transport due to a gradient of Gibbs energy.

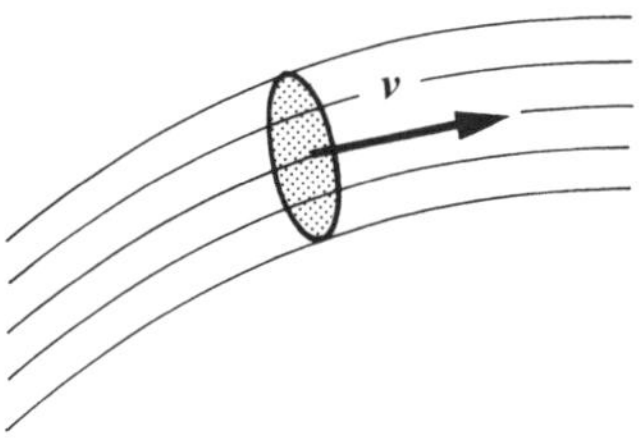

Fig. 4.1 Flux lines across a surface element of area d*S*.

Convection involves the movement of neutral or ionic species *i* with the solution that surrounds them as shown in Figure 4.1. The flux of these species through an element of area d*S* is a vector written as

$$\boldsymbol{J}_i \;=\; c_i\,\boldsymbol{v} \tag{4.1}$$

where c is the molar concentration (mol·m^{-3}) and $\boldsymbol{v}$ the velocity vector. The norm of this flux vector is a number of moles per unit time and area ($\text{mol·m}^{-2}\text{·s}^{-1}$).

4.1.1 Gradient of electrochemical potential

Ionic species in solution can also move under the effect of an electrochemical potential gradient, such that the driving force causing the displacement of an ion is

$$\boldsymbol{F} \;=\; -\frac{1}{N_\text{A}}\,\textbf{grad}\,\tilde{\mu} \tag{4.2}$$

where $\tilde{\mu}$ is the electrochemical potential for a mole of ions given by equation (1.60) and N_A is Avogadro's constant. The movement of the ion in a viscous solvent is

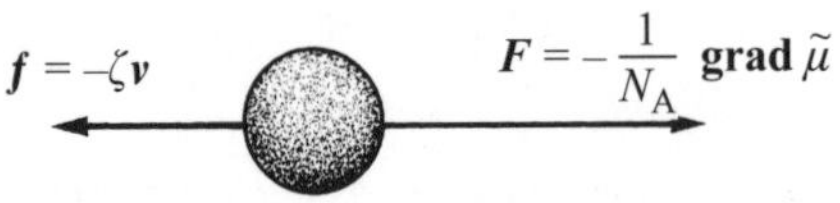

Fig. 4.2 Forces acting on a species in a viscous medium.

limited by a viscous friction force ***f*** proportional to the velocity vector of the ion as illustrated in Figure 4.2.

$$f = -\zeta \boldsymbol{v} \tag{4.3}$$

The resolution of Newton's equation

$$f + \boldsymbol{F} = m\boldsymbol{a} \tag{4.4}$$

shows that the velocity tends rapidly towards a steady state value (as for the free fall of a ball in a viscous liquid, see example below).

EXAMPLE

Let's calculate the velocity of a ball falling in a viscous liquid under the action of gravity in a laminar manner.
From equation (4.4), the differential equation to solve is :

$$f + \boldsymbol{F} = -\zeta v + mg = m\frac{dv}{dt}$$

The solution of the homogeneous equation is

$$v = \exp^{-\frac{\zeta t}{m}}$$

and considering the fact that the velocity is zero at the time $t = 0$, the equation for the velocity is written as

$$v = \frac{mg}{\zeta}\left[1 - \exp^{-\frac{\zeta t}{m}}\right]$$

We can thus see that the velocity tends towards a steady state value equal to mg/ζ.

In the case of the movement of a species down an electrochemical gradient, the steady state velocity is

$$\boldsymbol{v} = -\frac{1}{N_A \zeta} \textbf{grad}\, \tilde{\mu} \tag{4.5}$$

By combining equations (4.1) & (4.4), the flux of the species i is then written as

$$\boldsymbol{J}_i = c_i \boldsymbol{v} = -\frac{c_i}{N_A \zeta} \textbf{grad}\, \tilde{\mu}_i = -c_i \tilde{u}_i \textbf{ grad}\, \tilde{\mu}_i \tag{4.6}$$

where $\tilde{u}_i$ is called the ***electrochemical mobility***. A mobility is generally defined as the ratio of the velocity to the driving force. The units here are mol·s·kg^{-1} or mol·m^{2}·s^{-1}·J^{-1}. ***By definition, the electrochemical mobility is always positive both for cations and anions.***

If we consider that the ion is a sphere of radius r, hydrodynamics (see appendix, page 171) tells us that the friction constant ζ can be calculated as being equal to

$$\zeta = 6\pi\eta r \tag{4.7}$$

This expression is valid in the case of solid spheres where the velocity of the solvent molecules at the surface of the sphere is the same as that of the sphere (the *stick* condition), η being the viscosity of the solvent (kg·m^{-1}·s^{-1}). On the other hand, in the case of a fluid sphere, e.g. air bubbles rising in water, the surface velocity differs (the *slip* condition), and the friction constant is given by

$$\zeta = 4\pi\eta r \tag{4.8}$$

It is interesting to note in passing that this hydrodynamic theory is based on a macroscopic hydrodynamic approach that considers the liquid as a homogeneous viscous medium, and that equation (4.7) holds here at the molecular level for the movement of ions in a solvent, which is itself made up of molecules with translation, rotation and vibration modes.

By developing the expression for the electrochemical potential of a species i in an ideally dilute solution, taking into account the osmotic term linked to the pressure, we have

$$\tilde{\mu}_i = \mu_i^{\ominus} + RT\ln c_i + \overline{V}_i p + z_i F\phi \tag{4.9}$$

where $\overline{V}_i$ is the partial molar volume. Thence, we obtain the general expression for the flux under the electrochemical potential gradient

$$\boldsymbol{J}_i = -c_i\tilde{u}_i\mathbf{grad}\mu_i - z_iFc_i\tilde{u}_i\mathbf{grad}\phi - c_i\tilde{u}_i\overline{V}_i\mathbf{grad}p \tag{4.10}$$

This phenomenological equation is the general equation for the transport of a species i when gradients of chemical potential (i.e. concentration), of electrical potential and of pressure occur. Other gradients such as those linked to gravitational or centrifugal forces can also be included in a similar manner. In the following paragraphs, we shall look at how the general law (4.10) is related to the empirical laws observed for diffusion, migration and osmosis.

4.1.2 Fick's law for diffusion

Fick's first law states that diffusion is a process driven by concentration gradients ($\mathbf{grad}\phi = \mathbf{grad}\ p = 0$), and that the diffusion flux is directly proportional to this concentration gradient. In the case of one-dimensional gradients, the diffusion flux (i.e the norm of the diffusion flux vector) is therefore given by

$$J_i = -D_i \left(\frac{\partial c_i}{\partial x} \right)_{\phi,p} \tag{4.11}$$

where D_i is the ***diffusion coefficient*** expressed in $m^2 \cdot s^{-1}$, and more familiarly in $cm^2 \cdot s^{-1}$.

It is important to stress that diffusion phenomena in solution can only be observed if the thermal agitation, and or mechanical agitation, do not homogenise the concentration of the solutes in the solution. Diffusion phenomena in solution occur in particular close to solid walls. In effect, at the solid|liquid interface there is a 'stagnant' layer of solution, called the ***diffusion layer*** which has a thickness of the order of micrometres (see chapter 7). Thus, diffusion phenomena also occur in porous systems (a glass frit membrane, polyacrylamide gel, dialysis membranes, etc...) where convection phenomena are negligible.

Fick's first law is easily verified by measuring the flux, for example of dyes between two containers with different concentrations separated by a porous membrane. It can be observed that the initial flux is proportional to the difference in concentrations and inversely proportional to the thickness of the membrane. The dye concentration in each container is uniform because of the thermal agitation.

By comparison with the phenomenological equation (4.10), we get

$$D_i = RT\,\tilde{u}_i = \frac{kT}{6\pi\eta r} \tag{4.12}$$

because for dilute solutions for which we can neglect the activity coefficients, we have

$$\textbf{grad}\,\mu_i = \frac{RT}{c_i}\,\textbf{grad}\,c_i \tag{4.13}$$

The relation between the diffusion coefficient and the electrochemical mobility is called ***Einstein's law***, and the relation between the diffusion coefficient and the viscosity is called the ***Stokes-Einstein*** equation. We shall see in §4.6 that diffusion can be explained by a statistical approach.

4.1.3 Ohm's law for migration

An electric current density $\boldsymbol{j}$ is defined as a flow of positive charges through a surface element as shown in Figure 4.3. For metallic conductors, this flow is equal to the opposite of the flow of electrons through this surface. The current is equal to the current density multiplied by the cross-sectional area of the conductor; the electrical current is expressed in amperes ($A = C \cdot s^{-1}$).

Ohm's Law is an empirical law of proportionality between the current flowing between two equipotential surfaces A and B and the potential difference $V_A - V_B$.

$$R\,I_{A \to B} = V_A - V_B \tag{4.14}$$

where R is the resistance expressed in ohms ($\Omega = V \cdot A^{-1}$).

In ionic conductors, a similar law is also observed for each ionic species and the current density $\boldsymbol{j}$ which is a flux of charge is related to the flux $\boldsymbol{J}$ of the species i by

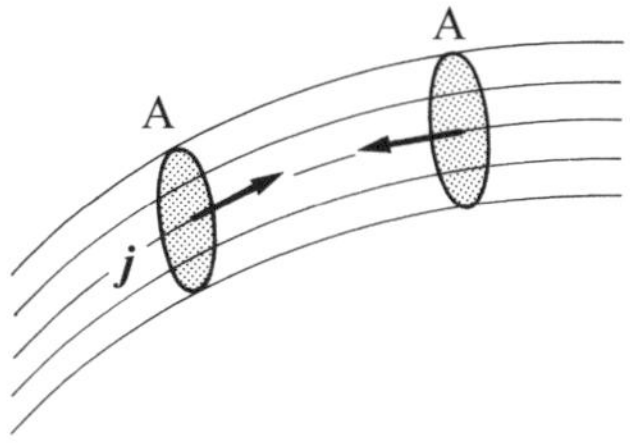

Fig. 4.3 Current distribution in a conductor.

$$\boldsymbol{j}_i = z_i F \boldsymbol{J}_i = -\sigma_i \,\mathbf{grad}\phi = \sigma_i \boldsymbol{E} \tag{4.15}$$

where σ_i is defined as a proportionality coefficient called ***ionic conductivity*** with the dimension $\Omega^{-1}\cdot m^{-1}$ or $S\cdot m^{-1}$, a Siemens being defined as $S = \Omega^{-1}$. Conductivity should not be confused with conductance G expressed in Siemens, the latter being defined as the inverse of the resistance.

A comparison with the phenomenological equation (4.10), when the only driving force is the electric field ($\mathbf{grad}c = \mathbf{grad}p = \mathbf{0}$) shows us that

$$\sigma_i = z_i^2 F^2 c_i \tilde{u}_i = z_i F c_i u_i = \lambda_i c_i = z_i^2 F^2 \frac{D_i c_i}{RT} \tag{4.16}$$

where u is called the ***electric mobility*** or sometimes the ***electrophoretic mobility*** ($m^2\cdot V^{-1}\cdot s^{-1}$). This mobility is related to the electrochemical mobility by

$$u_i = z_i F \tilde{u}_i \tag{4.17}$$

The electric mobility is defined as the ratio of the velocity to the electric field

$$\boldsymbol{v}_i = -z_i F \tilde{u}_i \,\mathbf{grad}\phi = z_i F \tilde{u}_i \boldsymbol{E} = u_i \boldsymbol{E} \tag{4.18}$$

By definition, the electric mobility is positive for cations and negative for anions.
λ_i is called the ***molar ionic conductivity*** and is defined as:

$$\lambda_i = z_i F u_i = \frac{\sigma_i}{c_i} \tag{4.19}$$

having a unit of $S\cdot mol^{-1}\cdot m^2$ or more usually $S\cdot mol^{-1}\cdot cm^2$.

This part of physical chemistry is encumbered with a multifarious vocabulary that is often rather old-fashioned. This is due to the fact that experimental studies preceded theoretical ones and so this domain progressed from empirical laws with their concomitant vocabulary. In particular, one must be wary of the old definition of electric mobility defined as the absolute value of the ratio of the velocity to the electric field. In this case, the ionic conductivity is the product of the absolute value of the charge, the electric mobility and the Faraday constant. Furthermore, certain old publications talk about the equivalent ionic conductivity (an equivalent being defined as a mole of monovalent cations); this equivalent ionic conductivity was defined as $\lambda_i / |z_i|$.

4.1.4 Osmotic flow

The flux of a species i generated by differences in pressure e.g. as in filtration by reverse osmosis, is written as

$$\boldsymbol{J}_i = -c_i\tilde{u}_i\,\overline{V}_i\,\mathbf{grad}p \tag{4.20}$$

4.2 CONDUCTIVITY OF ELECTROLYTE SOLUTIONS

4.2.1 Limiting molar conductivity

We have defined above the concept of molar ionic conductivity λ_i for a single species. In this paragraph, we shall look at the relation between the current density and the potential difference in an electrolyte solution containing several ionic species.

For an electrolyte $C_{\nu+}A_{\nu-}$, the total current density is

$$j = j_+ + j_- = -(\sigma_+ + \sigma_-)\,\mathbf{grad}\phi = -\sigma\,\mathbf{grad}\phi = \sigma\boldsymbol{E} \tag{4.21}$$

Comparison with Ohm's law shows us that the conductivity σ (sometimes written κ) is the inverse of the resistivity

$$\sigma = 1/\rho \tag{4.22}$$

We define the ***molar conductivity*** of an electrolyte as the ratio of the conductivity to the concentration

$$\Lambda_{\mathrm{m}} = \frac{\sigma}{c} = \frac{\sigma_+ + \sigma_-}{c} = \frac{\nu_+\sigma_+}{c_+} + \frac{\nu_-\sigma_-}{c_-} = \nu_+\lambda_+ + \nu_-\lambda_- \tag{4.23}$$

Its dimension is usually expressed in $\mathrm{S{\cdot}mol^{-1}{\cdot}cm^2}$. (It is wise to watch the units; if the conductivity is given in $\mathrm{S{\cdot}cm^{-1}}$ and if the concentration is given in $\mathrm{mol{\cdot}l^{-1}}$, then we have $\Lambda_{\mathrm{m}} = 1000\,\sigma/c$)

In the case of very dilute electrolyte solutions where the ion-ion interactions can be neglected, the molar ionic conductivity tends to a limiting value λ_i^{o} called the ***limiting molar ionic conductivity***. We then define the ***limiting molar conductivity*** of the electrolyte as the linear combination of the limiting molar ionic conductivities.

$$\Lambda_{\mathrm{m}}^{\mathrm{o}} = \nu_+\lambda_+^{\mathrm{o}} + \nu_-\lambda_-^{\mathrm{o}} \tag{4.24}$$

For example, the limiting molar conductivity of $ZnCl_2$ is

$$\Lambda_{\mathrm{ZnCl_2}}^{\mathrm{o}} = \lambda_{\mathrm{Zn^{2+}}}^{\mathrm{o}} + 2\,\lambda_{\mathrm{Cl^-}}^{\mathrm{o}} = 105.6 + 2\times 76.3 = 258.2 \quad \mathrm{S\cdot cm^2\cdot mol^{-1}}$$

From a purely electrical point of view, we can consider an electrolyte as a system of resistances in parallel, each resistance corresponding to the transport of the current by an ion, as illustrated schematically in Figure 4.4.

For a circuit of resistors in series, the total resistance is the sum of the resistances comprising the circuit, whereas when they are in parallel, it is easier to talk about

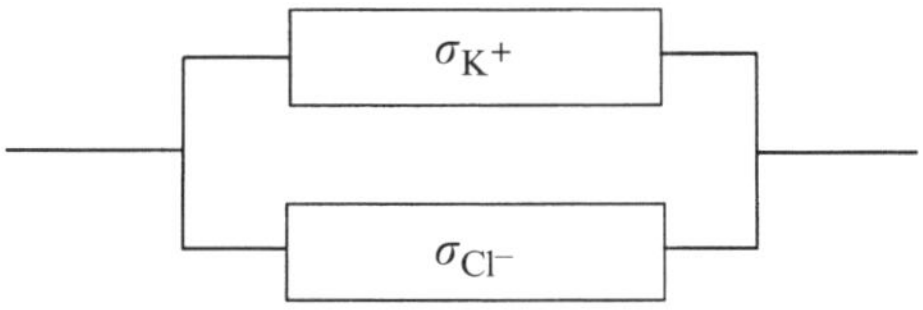

Fig. 4.4 Equivalent electric circuit for migration in a KCl solution.

conductance, because the total conductance of a circuit of resistors in parallel is the sum of the conductances comprising the circuit. This explains why, in the case of electrolyte solutions, we speak more easily in terms of conductance than resistance.

For dilute solutions, each ion transports current in its branch of the circuit without interacting with the movement of the other ions. For the electrolyte $C_{\nu+}A_{\nu-}$, the cation's transport number is given by

$$t_+ = \frac{j_+}{j} = \frac{\sigma_+}{\sigma} = \frac{\nu_+\lambda_+}{\Lambda_m} \tag{4.25}$$

For KCl, the limiting molar ionic conductivities are almost equal as shown in Table 4.1. This is equivalent to two roughly equal resistances in parallel, the current being transported nearly equally by the two ions ($t_{K^+} = 0.49$, $t_{Cl^-} = 0.51$). If we replace the potassium ion by a proton that has a much larger limiting molar ionic conductivity, the majority of the current (82%) is then carried by the proton and only 18% by the

Table 4.1 Limiting molar ionic conductivity at 25°C,
R.A. Robinson & R.H. Stokes, *Electrolyte Solutions*, Butterworth, London 1959.

Cation	λ_i^o /cm^2·Ω^{-1}·mol^{-1}	Anion	λ_i^o /cm^2·Ω^{-1}·mol^{-1}
H^+	349.8	OH^-	198.3
Li^+	38.7	F^-	55.4
Na^+	50.1	Cl^-	76.4
K^+	73.5	Br^-	78.1
Rb^+	77.8	I^-	76.4
Cs^+	77.3	NO_3^-	71.5
NH_4^+	73.6	ClO_4^-	67.4
Me_4N^+	44.9	SCN^-	66.0
Mg^{2+}	106.1	SO_4^{2-}	160.0
Ca^{2+}	119.0	$Fe(CN)_6^{3-}$	302.7
Fe^{3+}	204.0	$Fe(CN)_6^{4-}$	442.0

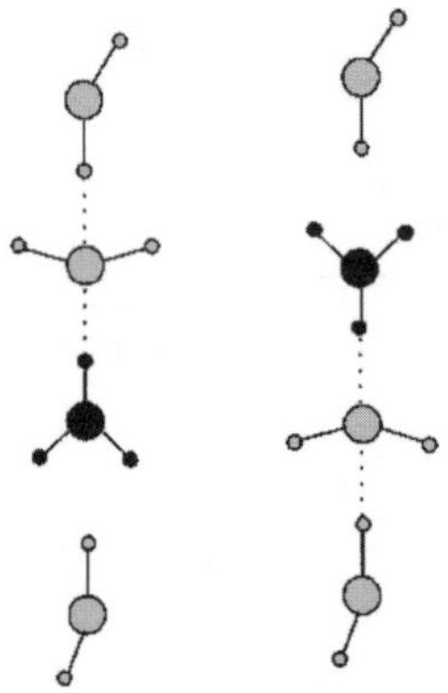

Fig. 4.5 Schematic illustration of proton mobility by 'structural diffusion' (Adapted from A.A. Kornyshev et al, *J. Phys. Chem. B.*, 107 (2003) 3351).

chloride ion ($t_{H^+} = 0.82$, $t_{Cl^-} = 0.18$). For $ZnCl_2$, the transport numbers are respectively $t_{Zn^{2+}} = 0.41$, $t_{Cl^-} = 0.59$.

The reason the limiting ionic molar conductivity of the proton is so high stems from the proton mobility mechanism that differs from that of the other ions which move by successive jumps, taking with them part of their hydration layer. In 1806, de Grotthus proposed a mechanism which considers that protons move rather in 'chain reactions', breaking covalent bonds and reforming in their place hydrogen bonds as illustrated in Figure 4.5. This transport mechanim is also called ***structural diffusion***.

Furthermore, we must take care when we write the proton in the chemical oxonium form H_3O^+, because this is already a major simplification. In fact, because the proton H^+ cannot exist in isolation in solution, certain theoretical work tends to prove that a stable complex of the aqueous proton is $H_9O_4^+$, i.e. an H_3O^+ core strongly hydrogen-bonded to three H_2O molecules sometimes called 'Eigen cation'. Another proposed structure is one where a proton is shared by two H_2O molecules to form the 'Zundel cation' $H_5O_2^+$. Anyway, the topic of proton mobility is a rather open debate, and recent works suggest that the proton influence can be spread over several hydrogen bonds.

4.2.2 Measurement of conductivities

In the same way that the electric resistance of a material depends on its resistivity and on its geometric dimension, the resistance of an electrolytic solution will depend on its conductivity and the geometrical characteristics of the measuring cell. In a general manner, Ohm's law can be written as the ***circulation*** (or path integral) of the electric field vector (see Annex A), which for a linear conductor reads

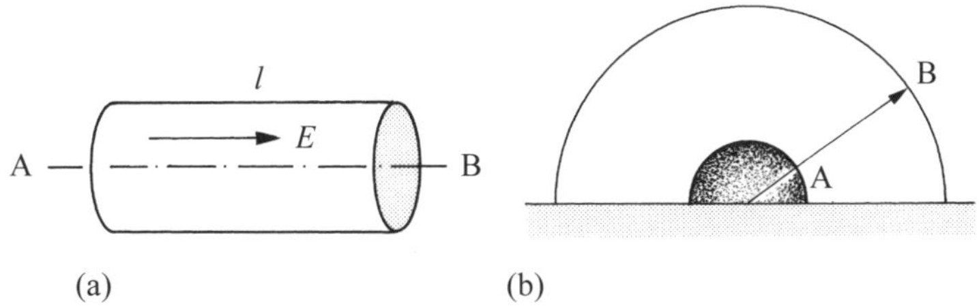

Fig. 4.6 Linear conduction (a) and hemispherical conduction (b).

$$V_A - V_B = \int_A^B \boldsymbol{E} \cdot d\boldsymbol{l} = \int_A^B \rho \, \boldsymbol{j} \cdot d\boldsymbol{l} = \rho j l \tag{4.26}$$

and the current as the flux through a cross-section is

$$I = \iint_S \boldsymbol{j} \cdot \boldsymbol{n} \, dS = jS \tag{4.27}$$

Thus, the resistance of a linear conductor (Figure 4.6a) is written, using equation (4.14), as

$$R = \frac{V_A - V_B}{I} = \frac{\rho l}{S} \tag{4.28}$$

For a current I between two concentric hemispheric electrodes (Figure 4.6b), the resistance can be calculated using the same method

$$V_A - V_B = \int_A^B \rho \boldsymbol{j} \cdot d\boldsymbol{r} = \int_A^B \rho \frac{I}{2\pi r^2} \cdot dr = \frac{\rho}{2\pi}\left(\frac{1}{r_A} - \frac{1}{r_B}\right) I = RI \tag{4.29}$$

A cell to measure the conductivity of electrolyte solutions usually comprises two parallel electrodes usually face to face. Given that the exact geometry of such a measuring cell, made up of two planar and parallel electrodes, is difficult to determine accurately, and that the distribution of the current lines is not uniform as illustrated in Figure 4.7, we call ***cell constant*** k_{cell}, the geometric factor equivalent of the ratio l/S, the units being m^{-1}. Thus, for each measuring cell, it is easier to determine the cell constant by calibration, i.e. by measuring the resistance of solutions with a well-known conductivity (e.g. KCl)

$$k_{cell} = \sigma_{KCl} \cdot R_{Measured} \tag{4.30}$$

From a practical point of view, the conductivity of a solution is measured by applying an AC potential difference between the two parallel electrodes in order to avoid changes in the chemical composition of the solution by oxidation at the anode and reduction at the cathode, that one would get with an applied DC voltage. With an AC applied potential difference, it is important to reduce the capacitive effects, and for this we use electrodes with large specific surfaces such as platinum black electrodes and operate at frequencies of the order of a few hundred Hz.

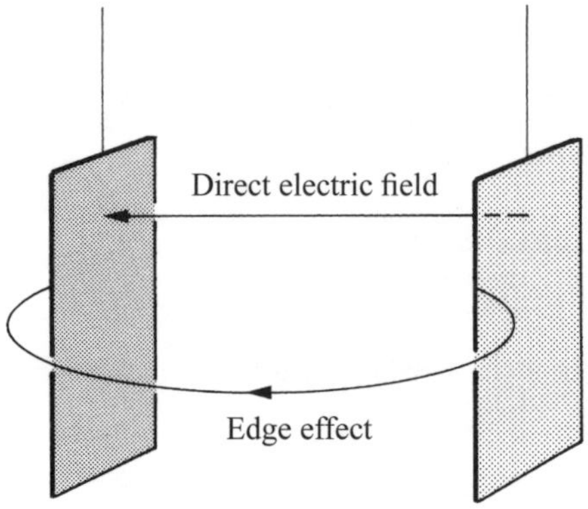

Fig. 4.7 Field lines between two parallel electrodes.

EXAMPLE

A measurement with a conductivity-measuring cell at 25°C shows a resistance of 747.5 Ω for a 0.01 M solution of KCl, and a value of 876 Ω when the electrolyte is a 0.005 M solution of $CaCl_2$. Calculate the conductivity and the molar conductivity of the solution of calcium chloride. Compare this value with the limiting molar conductivity using the data of Table 4.1.

First of all, we calculate the cell constant with the aid of equations (4.24) & (4.30) using the values from Table 4.1.

$$\begin{aligned} k_{\text{cell}} &= R_{\text{KCl}} \cdot \sigma_{\text{KCl}} = R_{\text{KCl}} \cdot \Lambda^{\text{o}}_{\text{KCl}} \cdot c_{\text{KCl}} \\ &= 747.5(\Omega) \cdot 149.9(\text{cm}^2 \cdot \Omega^{-1} \cdot \text{mol}^{-1}) \cdot 10^{-5}(\text{mol} \cdot \text{cm}^{-3}) = 1.1205\,\text{cm}^{-1} \end{aligned}$$

In fact, the molar conductivity of a 10^{-2} M solution of KCl is 141.3 ($\text{cm}^2 \cdot \Omega^{-1} \cdot \text{mol}^{-1}$), but by simplification we consider it dilute enough to take the limiting values of the molar ionic conductivity. Experimentally, it is clear that we must measure the cell constant for various dilutions and extrapolate to zero concentrations.

With this value, we can calculate the conductivity of the calcium chloride solution.

$$\sigma_{\text{CaCl}_2} = \frac{k_{\text{cell}}}{R_{\text{CaCl}_2}} = \frac{1.1205}{876} = 1.28 \cdot 10^{-3}\,\Omega^{-1} \cdot \text{cm}^{-1}$$

and thus the molar conductivity

$$\Lambda_{\text{CaCl}_2} = \frac{\sigma_{\text{CaCl}_2}}{c_{\text{CaCl}_2}} = \frac{1.28 \cdot 10^{-3}(\Omega^{-1} \cdot \text{cm}^{-1})}{5 \cdot 10^{-6}(\text{mol} \cdot \text{cm}^{-3})} = 255.8\,\Omega^{-1} \cdot \text{cm}^2 \cdot \text{mol}^{-1}$$

This value is less than the limiting molar conductivity value of $CaCl_2$

$$\Lambda^{\text{o}}_{\text{CaCl}_2} = \Lambda^{\text{o}}_{\text{Ca}^{2+}} + 2\Lambda^{\text{o}}_{\text{Cl}^-} = 119 + 2 \cdot 76.4 = 271.8\,\Omega^{-1} \cdot \text{cm}^2 \cdot \text{mol}^{-1}$$

This is due to ion-ion interactions as described in §4.3.1.

4.2.3 Measurement of acidity constants

For many years, conductometric measurements were frequently used for measuring the acidity constants of weak acids. Indeed, the very notions of weak and strong acids have their origin in their conductivity characteristics. In effect, a strong acid is by definition an acid whose molar conductivity follows Kohlrausch's law (see §4.3). A weak acid is therefore an acid whose molar conductivity is small at medium concentrations, but which increases rapidly when the concentration is reduced.

Let's look at the dissociation of the acid AH

$$AH + H_2O \rightleftarrows A^- + H_3O^+$$

The corresponding disssociation constant is then (neglecting the activity coefficients)

$$K_a \cong \frac{c_{H_3O^+}c_{A^-}}{c_{AH}} = \frac{\alpha^2}{1-\alpha}c_{total} \tag{4.31}$$

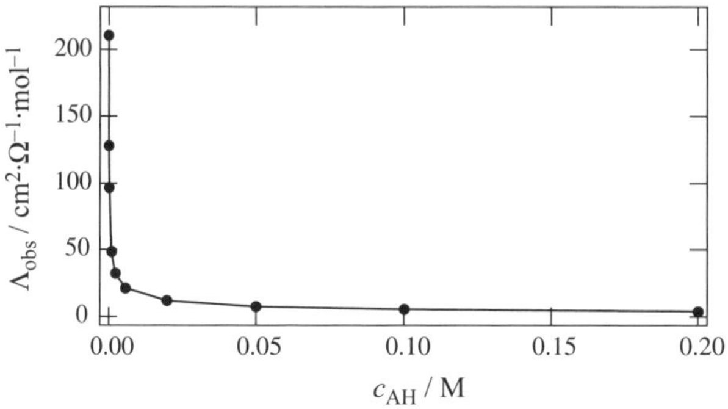

Fig. 4.8 Molar conductivity of acetic acid in solution. Λ_m^o = 390.7 cm^2·Ω^{-1}·mol^{-1} (Numerical values: R. A. Robinson & R. H. Stokes, *Electrolyte Solutions*, Butterworth, London 1959).

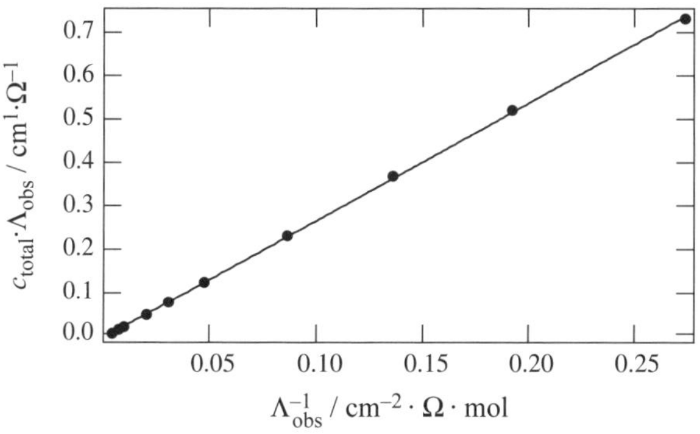

Fig. 4.9 Graph of equation (4.34).

where α is the degree of dissociation of the acid and c_{total} is the total acid concentration ($c_{\text{total}} = c_{\text{A}^-} + c_{\text{AH}}$). For low concentrations, we can make the hypothesis that the degree of dissociation becomes equal to the ratio of the molar conductivity to the limiting molar conductivity.

$$\alpha \quad = \quad \Lambda_{\text{m}} / \Lambda_{\text{m}}^{\text{o}} \tag{4.32}$$

The data in Figure 4.8 show values of the molar conductivity of acetic acid as a function of concentration. The problem is that it is experimentally difficult to obtain a precise value for the limiting molar conductivity. To circumvent this, we use the additive properties of molar conductivities, as for example

$$\Lambda_{\text{m(AH)}}^{\text{o}} \quad = \quad \Lambda_{\text{m(HCl)}}^{\text{o}} + \Lambda_{\text{m(NaA)}}^{\text{o}} - \Lambda_{\text{m(NaCl)}}^{\text{o}} \tag{4.33}$$

Thus by combining equations (4.31) & (4.32), we obtain Ostwald's law of dilution

$$c_{\text{total}} \Lambda_{\text{m}} \quad = \quad \frac{\left(\Lambda_{\text{m}}^{\text{o}}\right)^2 K_{\text{a}}}{\Lambda_{\text{m}}} \quad - \quad K_{\text{a}} \Lambda_{\text{m}}^{\text{o}} \tag{4.34}$$

that allows the determination of the dissociation constant by plotting $c_{\text{total}} \Lambda_{\text{m}}$ against Λ_{m}^{-1}. From the data in Figure 4.8 carried forward to Figure 4.9, we obtain a dissociation constant of $1.77 \cdot 10^{-5}$ M.

EXAMPLE

In 1894, Kohlrausch & Heydweiller measured the conductivity of water as $6.2 \cdot 10^{-6}$ $\text{S} \cdot \text{m}^{-1}$. From the values in Table 4.1, let's estimate the ionisation constant of water.

$$\Lambda_{\text{m(H}_2\text{O)}}^{\text{o}} = \lambda_{\text{H}^+}^{\text{o}} + \lambda_{\text{OH}^-}^{\text{o}} = 548.1 \, \text{cm}^2 \cdot \Omega^{-1} \cdot \text{mol}^{-1}$$

And so, applying equation (4.23) we can calculate the concentration c of dissociated water and make the hypothesis that the molar conductivity is equal to the limiting molar conductivity

$$c = \sigma / \Lambda_{\text{m(H}_2\text{O)}}^{\text{o}} = 6.2 \cdot 10^{-6} (\text{m}^{-1} \cdot \Omega^{-1}) / 548.1 \, (\text{cm}^2 \cdot \Omega^{-1} \cdot \text{mol}^{-1})$$
$$= 1.13 \cdot 10^{-7} \text{M}$$

which gives

$$K_{\text{H}_2\text{O}} = c^2 = 1.28 \cdot 10^{-14} \text{M}^2 .$$

4.3 INFLUENCE OF CONCENTRATION ON CONDUCTIVITY

4.3.1 Kohlrausch's law

Kohlrausch's law is an empirical law that expresses the dependence of the molar conductivity of an electrolyte as a function of its concentration for dilute solutions, as illustrated in Figure 4.10 for sodium chloride and potassium chloride

$$\Lambda_{\mathrm{m}} = \Lambda_{\mathrm{m}}^{\mathrm{o}} - K\sqrt{c} \tag{4.35}$$

Kohlrausch's law is a consequence of the ion-ion interactions when the ions are subjected to an electric field, and can be justified either simply by a thermodynamic approach or more phenomenologically by a microscopic modelling of the ion transport in solution.

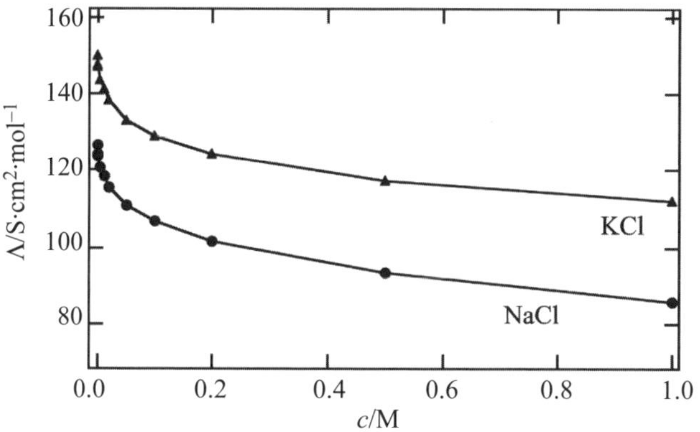

Fig. 4.10 Molar conductivities of aqueous solutions of NaCl and KCl.

4.3.2 Thermodynamic approach

Actually, considering ion-ion interactions means that the solution cannot be treated as ideal, and therefore we can no longer write that

$$\mathbf{grad}\,\mu = \frac{RT}{c}\,\mathbf{grad}\,c \tag{4.36}$$

Taking into account the activity coefficients, we have

$$\mathbf{grad}\,\mu = RT\,\mathbf{grad}\ln a = RT\left[\mathbf{grad}\ln\gamma + \mathbf{grad}\ln c\right] \tag{4.37}$$

Thus in one-dimensional systems, we see that

$$J_i = -RT\,c_i\tilde{u}_i\left[\frac{\partial \ln\gamma_i}{\partial x} + \frac{1}{c_i}\frac{\partial c_i}{\partial x}\right] = -D_i\frac{\partial c_i}{\partial x} \tag{4.38}$$

which yields a concentration dependence of the diffusion coefficient

$$D_i = RT\,\tilde{u}_i\left[1+c_i\left(\frac{\partial \ln \gamma_i}{\partial x}\right)\left(\frac{\partial x}{\partial c_i}\right)\right] \tag{4.39}$$

By substituting Einstein's law (4.12), we finally get

$$D_i = D_i^{\text{o}}\left[1+c_i\left(\frac{\partial \ln \gamma_i}{\partial c_i}\right)\right] \tag{4.40}$$

Using the Debye-Hückel law for 1:1 electrolytes given by equation (3.91)

$$\log \gamma_i^c = -z_i^2\,A\sqrt{I_c} = -A\sqrt{c} \tag{4.41}$$

we obtain

$$D_i = D_i^{\text{o}}\left[1-\frac{1}{2}A'\sqrt{c}\right] \tag{4.42}$$

To demonstrate Kohlrausch's law, we can make use of the additive property of molar ionic conductivities (4.23) and the definition of the electric mobility (4.17)

$$\begin{aligned}\Lambda_{\text{m}} &= \nu_+\lambda_+ + \nu_-\lambda_- = F(\nu_+ z_+ u_+ + \nu_- z_- u_-)\\ &= F^2(\nu_+ z_+^2\tilde{u}_+ + \nu_- z_-^2\tilde{u}_-) = \frac{F^2}{RT}(\nu_+ z_+^2 D_+ + \nu_- z_-^2 D_-)\end{aligned} \tag{4.43}$$

By substituting the concentration dependence of the diffusion coefficient given by equation (4.42), the concentration dependence of the molar conductivity for an 1:1 electrolyte reads

$$\Lambda_{\text{m}} = \frac{F^2}{RT}(\nu_+ z_+^2 D_+{}^{\text{o}} + \nu_- z_-^2 D_-{}^{\text{o}})\left[1-\frac{A'}{2}\sqrt{c}\right] \tag{4.44}$$

which reduces to the empirical Kohlrausch's law

$$\Lambda_{\text{m}} = \Lambda_{\text{m}}^{\text{o}} - K\sqrt{c} \tag{4.45}$$

4.3.3 Electrophoretic effect

When an electric field is applied, the central ion moves in the opposite direction with respect to its ionic atmosphere, as shown schematically in Figure 4.11. The central ion and the ionic atmosphere drag solvent molecules into their motion by friction. This causes a reduction in the conductivity.

The electrophoretic velocity of the ionic atmosphere can be calculated as that which would affect a sphere of radius $1/\kappa$ (the average radius of the ionic atmosphere defined by equation (3.72)) and whose charge would be the opposite of that of the central ion, equal to $z_c e$. In a steady state regime, the force of the electric field on the atmosphere is equal to the viscous friction force given by equations (4.3) & (4.7), from which we deduce that

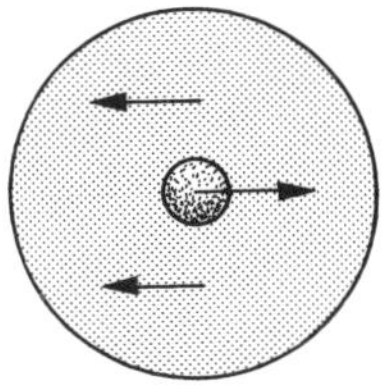

Fig. 4.11 The electrophoretic effect.

$$v_E = -\frac{z_c e E}{\zeta} = -\frac{z_c e \kappa E}{6\pi\eta} \tag{4.46}$$

This equation, obtained rather simplistically, is in fact the first degree approximation of the Onsager & Fuoss theory.

Onsager & Fuoss's approach consists of going from an equilibrium of forces $\boldsymbol{k}_S$ acting partly on the solvent molecules, and partly on the cations and anions in a solution at rest.

$$n_S \boldsymbol{k}_S = -\left(n_+ z_+ + n_- z_-\right) e\boldsymbol{E} \tag{4.47}$$

where n_S, n_+ and n_- are respectively the concentrations of solvent molecules, cations and anions in the solution. The sum of the forces df acting on a spherical shell of thickness dr around a central ion is written as

$$\mathrm{d}f = \left[\left(n_+^{\text{shell}} z_+ + n_-^{\text{shell}} z_-\right) e\boldsymbol{E} + n_S \boldsymbol{k}_S\right] 4\pi r^2 \mathrm{d}r \tag{4.48}$$

making the hypothesis that the solvent concentration is uniform. If we consider that each shell is subject to a friction force such as that described by equations (4.3) & (4.7), by elimination it follows that

$$v_E = \int_{r=a}^{\infty} \frac{e\boldsymbol{E}}{6\pi r\eta}\left[\left(n_+^{\text{shell}} - n_+\right) z_+ + \left(n_-^{\text{shell}} - n_-\right) z_-\right] 4\pi r^2 \mathrm{d}r \tag{4.49}$$

Assuming that the distribution of ions in the spherical shell obeys a Boltzmann distribution (see equation (3.47))

$$n_\pm^{\text{shell}} = n_\pm \exp^{-z_\pm e\phi(r)/kT} \tag{4.50}$$

which we can linearise for dilute solutions as we did for the Debye-Hückel theory, and assuming that the electric potential around the central ion is given by equation (3.66)

$$\phi(r) = \frac{z_c e}{4\pi\varepsilon_0\varepsilon_r}\left[\frac{e^{\kappa a}}{1+\kappa a}\right]\frac{e^{-\kappa r}}{r} \tag{4.51}$$

equation (4.49) becomes

$$
\begin{aligned}
v_E &= \frac{2e^2\boldsymbol{E}}{3\eta kT}\left[n_+z_+^2+n_-z_-^2\right]\int_{r=a}^{\infty}\phi(r)r\,\mathrm{d}r \\
&= \frac{z_c e^3\boldsymbol{E}}{6\pi\varepsilon_0\varepsilon_r\eta kT}\left[\frac{e^{\kappa a}}{1+\kappa a}\right]\left[n_+z_+^2+n_-z_-^2\right]\int_{r=a}^{\infty}e^{-\kappa r} \\
&= -\frac{z_c e^3\boldsymbol{E}}{6\pi\varepsilon_0\varepsilon_r\eta kT}\left[\frac{1}{1+\kappa a}\right]\left[n_+z_+^2+n_-z_-^2\right]
\end{aligned}
\tag{4.52}
$$

The last term in brackets is twice the ionic strength defined by equation (3.75), and can be expressed as a function of the reciprocal Debye length defined by equation (3.55)

$$
\left[n_+z_+^2+n_-z_-^2\right] = \kappa^2\frac{\varepsilon_0\varepsilon_r kT}{e^2} \tag{4.53}
$$

By substitution, we obtain

$$
v_E = -\frac{z_c e\boldsymbol{E}}{6\pi\eta}\left(\frac{\kappa}{1+\kappa a}\right) \tag{4.54}
$$

The term $1+\kappa a$ is linked to the consideration of the size of the ions. If we neglect this term, we find ourselves again with equation (4.46).

4.3.4 Relaxation effect of the ionic atmosphere

This effect is due to the relaxation time of the ionic atmosphere that varies between 1 μs and 1 ns according to the concentration. As a first approximation, the relaxation time can be estimated from the Einstein-Smoluchowski equation (4.162).

$$
\tau_R = \frac{(1/\kappa)^2}{2D} \tag{4.55}
$$

In effect, when there is a movement of the central ion and its atmosphere under the action of the electric field, there is a break in the spherical symmetry. This translates to a restraining force of the ionic atmosphere on the central ion, and τ_R represents a characteristic time for the central ion to move with respect to its atmosphere.

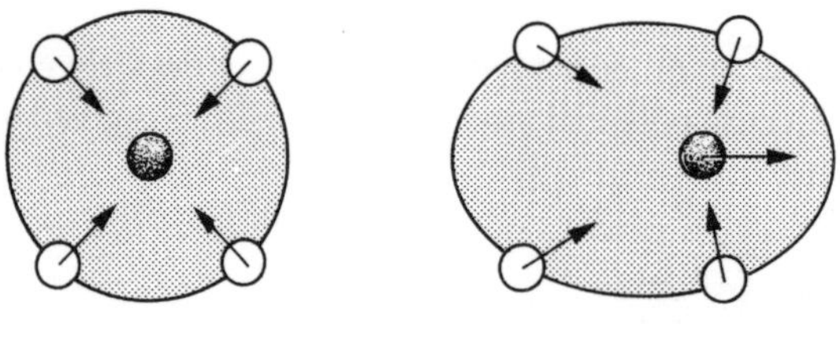

Fig. 4.12 Relaxation of the ionic atmosphere.

The Onsager-Falkenhagen theory that allows the modelling of ionic atmosphere relaxation phenomena involves a mathematical complexity that is beyond the scope of this book. However, the principal concept of this theory is to consider that the electric field $\boldsymbol{E}$ is partially compensated by a 'relaxation field' $\Delta\boldsymbol{E}$ such that

$$\frac{\Delta \boldsymbol{E}}{\boldsymbol{E}} = \frac{z_+ z_- e^2}{12\pi\varepsilon_0\varepsilon_r kT}\left(\frac{q}{1+\sqrt{q}}\right)\left(\frac{\kappa}{1+\kappa a}\right) \tag{4.56}$$

with

$$q = \frac{-z_+ z_-}{z_+ - z_-}\left(\frac{\lambda_+^o + \lambda_-^o}{z_+\lambda_-^o - z_-\lambda_+^o}\right) \tag{4.57}$$

The parameter q reduces to a factor of 1/2 for symmetrical electrolytes.

Thus the force exerted on a shell of thickness $\mathrm{d}r$ around a central ion is written as

$$\left[\left(n_+^{\text{shell}} z_+ + n_-^{\text{shell}} z_-\right) e(\boldsymbol{E}+\Delta\boldsymbol{E}) + n_\mathrm{S}\boldsymbol{k}_\mathrm{S}\right] 4\pi r^2 \mathrm{d}r \tag{4.58}$$

and a calculation similar to the previous one gives us an expression similar to equation (4.54) i.e.

$$\boldsymbol{v}_E = -\frac{z_c e(\boldsymbol{E}+\Delta\boldsymbol{E})}{6\pi\eta}\left(\frac{\kappa}{1+\kappa a}\right) \tag{4.59}$$

So the velocity of the central ion slowed down by the relaxation of the atmosphere is

$$\boldsymbol{v}_R = \tilde{u}_c^o z_c F(\boldsymbol{E}+\Delta\boldsymbol{E}) \tag{4.60}$$

By combining the electrophoretic effect and the relaxation effect of the ionic atmosphere, we can calculate the molar conductivity of an electrolyte as follows

$$\boldsymbol{v} = \boldsymbol{v}_R + \boldsymbol{v}_E = \tilde{u}_c^o z_c F(\boldsymbol{E}+\Delta\boldsymbol{E}) - \frac{z_c e(\boldsymbol{E}+\Delta\boldsymbol{E})}{6\pi\eta}\left(\frac{\kappa}{1+\kappa a}\right) \tag{4.61}$$

The resulting electric mobility is then written as

$$u_\text{app} = \frac{\boldsymbol{v}}{\boldsymbol{E}} = \tilde{u}_c^o z_c F\left(1 - \frac{1}{6\pi\eta\tilde{u}_c^o N_\mathrm{A}}\left(\frac{\kappa}{1+\kappa a}\right)\right)\left(1+\frac{\Delta\boldsymbol{E}}{\boldsymbol{E}}\right) \tag{4.62}$$

Thus, the molar ionic conductivities defined by equation (4.19) are given by

$$\begin{aligned} \lambda_+ &= z_+ F u_{\text{app}+} = \tilde{u}_+^o z_+^2 F^2\left(1 - \frac{1}{6\pi\tilde{u}_+^o\eta N_\mathrm{A}}\left(\frac{\kappa}{1+\kappa a}\right)\right)\left(1+\frac{\Delta\boldsymbol{E}}{\boldsymbol{E}}\right) \\ &= \lambda_+^o\left(1 - \frac{z_+^2 F^2}{6\pi\eta\lambda_+^o N_\mathrm{A}}\left(\frac{\kappa}{1+\kappa a}\right)\right)\left(1+\frac{\Delta\boldsymbol{E}}{\boldsymbol{E}}\right) \end{aligned} \tag{4.63}$$

$$\begin{aligned}\lambda_- &= z_- F u_{\text{app}-} = \tilde{u}_-^{\text{o}} z_-^2 F^2 \left(1 - \frac{1}{6\pi \tilde{u}_-^{\text{o}} \eta N_{\text{A}}} \left(\frac{\kappa}{1+\kappa a}\right)\right)\left(1 + \frac{\Delta E}{E}\right) \\ &= \lambda_-^{\text{o}} \left(1 - \frac{z_-^2 F^2}{6\pi \lambda_-^{\text{o}} \eta N_{\text{A}}} \left(\frac{\kappa}{1+\kappa a}\right)\right)\left(1 + \frac{\Delta E}{E}\right)\end{aligned} \tag{4.64}$$

The molar conductivity of the solution is then obtained by a classic linear combination of the molar ionic conductivities.

$$\begin{aligned}\Lambda_{\text{m}} &= \nu_+ \lambda_+ + \nu_- \lambda_- \\ &= \left(\Lambda_{\text{m}}^{\text{o}} - \frac{\left(\nu_+ z_+^2 + \nu_- z_-^2\right) F^2}{6\pi \eta N_{\text{A}}} \left(\frac{\kappa}{1+\kappa a}\right)\right)\left(1 + \frac{\Delta E}{E}\right)\end{aligned} \tag{4.65}$$

Neglecting the second order terms and the term $1+\kappa a$ for the size of the ion, we obtain

$$\Lambda_{\text{m}} = \Lambda_{\text{m}}^{\text{o}} - \left(\frac{z_+ z_- e^2 \Lambda_{\text{m}}^{\text{o}}}{12\pi \varepsilon_0 \varepsilon_{\text{r}} kT} \left(\frac{q}{1+\sqrt{q}}\right) + \frac{\left(\nu_+ z_+^2 + \nu_- z_-^2\right) F^2}{6\pi \eta N_{\text{A}}}\right)\kappa \tag{4.66}$$

Given that the reciprocal Debye length varies with the square root of the ionic force, we arrive again at Kohlrausch's empirical law (4.45).

EXAMPLE

Let's develop equation (4.66) for NaCl, $CaCl_2$ and $LaCl_3$ and compare the results to the experimental values.

	$\lambda_+^{\text{o}}/\text{cm}^2{\cdot}\Omega^{-1}{\cdot}\text{mol}^{-1}$	$\lambda_-^{\text{o}}/\text{cm}^2{\cdot}\Omega^{-1}{\cdot}\text{mol}^{-1}$	q	$\dfrac{q}{1+\sqrt{q}}$
NaCl	50.1	76.4	0.500	0.2929
$CaCl_2$	119	76.4	0.464	0.2760
$LaCl_3$	209.1	76.4	0.384	0.2371

For the reciprocal Debye length, we can use equation (3.82).

$\kappa = B\sqrt{I_c} = 3.29 \cdot 10^9 \text{ m}^{-1}$

Thus the variation in molar conductivity of sodium chloride is given by:

$$\Lambda_{\text{NaCl}} = \Lambda_{\text{NaCl}}^{\text{o}} - \left[\frac{e^2 B \Lambda_{\text{NaCl}}^{\text{o}}}{12\pi \varepsilon_0 \varepsilon_{\text{r}} kT} \left(\frac{q_{\text{NaCl}}}{1+\sqrt{q_{\text{NaCl}}}}\right) + \frac{2F^2 B}{6\pi \eta N_{\text{A}}}\right]\sqrt{c_{\text{NaCl}}}$$

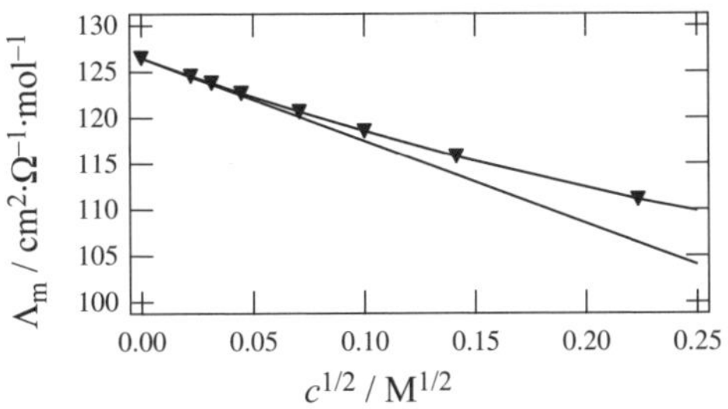

For calcium chloride :

$$\Lambda_{CaCl_2} = \Lambda^{o}_{CaCl_2} - \left[\frac{2e^2 B\sqrt{3}\Lambda^{o}_{CaCl_2}}{12\pi\varepsilon_0\varepsilon_r kT}\left(\frac{q_{CaCl_2}}{1+\sqrt{q_{CaCl_2}}}\right)+\frac{6F^2 B\sqrt{3}}{6\pi\eta N_A}\right]\sqrt{c_{CaCl_2}}$$

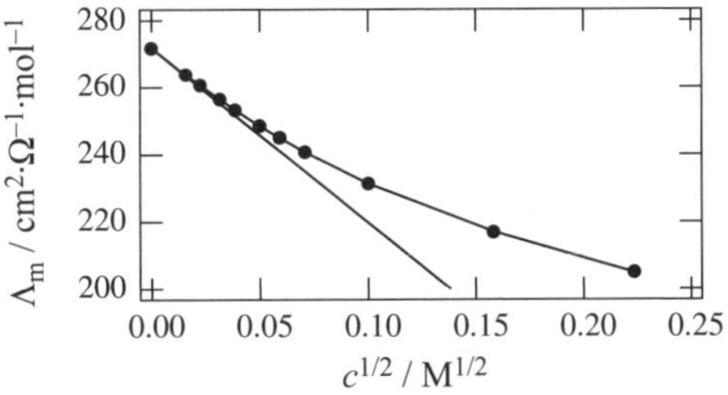

and for lanthanum chloride

$$\Lambda_{LaCl_3} = \Lambda^{o}_{LaCl_3} - \left[\frac{3e^2 B\sqrt{6}\Lambda^{o}_{LaCl_3}}{12\pi\varepsilon_0\varepsilon_r kT}\left(\frac{q_{LaCl_3}}{1+\sqrt{q_{LaCl_3}}}\right)+\frac{4F^2 B\sqrt{6}}{6\pi\eta N_A}\right]\sqrt{c_{LaCl_3}}$$

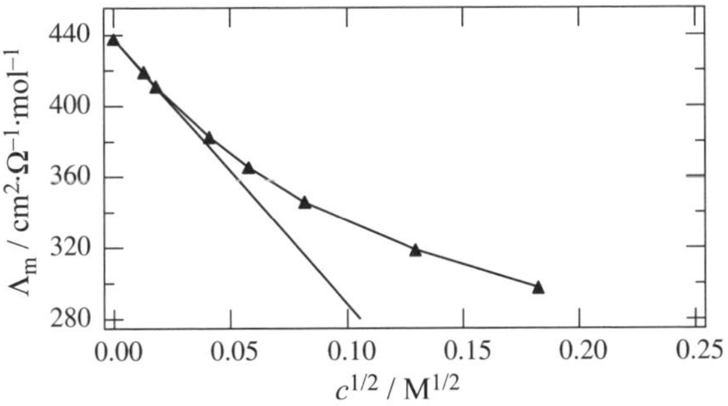

The experimental values for these three graphs shown as markers are taken from: R. A. Robinson & R. H. Stokes, *Electrolyte Solutions*, Butterworths, London 1959. The straight lines represent the respective equations.

These results validate the modelling of the Kohlrausch's law based on the combined action of the electrophoretic effect and relaxation on the ionic atmosphere. We can thus conclude that the experimental Kohlrausch's law is caused by ion-ion interactions.

4.3.5 Relaxation effect at high frequencies

The relaxation time of the ionic atmosphere is of the order of 10 to 100 ns, and so if we measure the conductivity of a solution using a high frequency potential source, in the radio frequency band, e.g. 10 MHz, the ionic atmosphere can no longer follow the speed of movement of the ions, and so the conductivities measured are higher. This effect, predicted by Debye and Falkenhagen, was verified experimentally in 1928 by Sack, and so takes the name of the ***Debye-Falkenhagen effect*** or the ***Sack effect***.

Another effect can be observed by placing an electrolytic solution in an electric field strong enough so that the central ion passes through its ionic cloud so quickly that the latter cannot follow. The ion, no longer being retarded by an electrophoretic effect, has a greater mobility, and the conductivities measured are higher. This second effect is called the ***Wien effect***.

4.4 DIELECTRIC FRICTION

In the latter section, we looked at the reduction in conductivity induced by ion-ion interactions. In this section, we are going to look at the reduction in conductivity in polar solvents induced by ion-dipole interactions. Before launching into this question of dielectric friction of the solvent on the ion movement, it will be useful to complete our knowledge of ion-dipole interactions and consider the electric field induced by a dipole.

4.4.1 Electric field created by a dipole

With the electric field being defined as the gradient of the electric potential

$$\boldsymbol{E} \;=\; -\,\mathbf{grad}\,V \tag{4.67}$$

the expression for the potential generated by a dipole (equation 3.33) is written as

$$4\pi\varepsilon_0\,\boldsymbol{E} \;=\; -\,\mathbf{grad}\left(\frac{\boldsymbol{p}\cdot\boldsymbol{r}}{r^3}\right) \;=\; -(\boldsymbol{p}\cdot\boldsymbol{r})\,\mathbf{grad}\left(\frac{1}{r^3}\right) \;-\; \frac{1}{r^3}\,\mathbf{grad}\,(\boldsymbol{p}\cdot\boldsymbol{r}) \tag{4.68}$$

and is by the definition of the vector gradient equal to $\boldsymbol{p}$. In effect, we have

$$\begin{aligned}\mathbf{grad}\,(\boldsymbol{p}\cdot\boldsymbol{r}) \;&=\; \boldsymbol{i}\frac{\partial}{\partial x}(p_x\,x)+\boldsymbol{j}\frac{\partial}{\partial y}(p_y\,y)+\boldsymbol{k}\,\frac{\partial}{\partial z}(p_z\,z)\\ &=\; \boldsymbol{i}\,p_x+\boldsymbol{j}\,p_y+\boldsymbol{k}\,p_z \;=\; \boldsymbol{p}\end{aligned} \tag{4.69}$$

Also, we have

$$\mathbf{grad}\left(\frac{1}{r^3}\right) \;=\; -\frac{3}{r^4}\,\mathbf{grad}\,r \;=\; -\frac{3\,\boldsymbol{r}}{r^5} \tag{4.70}$$

And, if $r^2 = x^2 + y^2 + z^2$,

$$\mathbf{grad}\, r = \boldsymbol{i}\frac{\partial r}{\partial x} + \boldsymbol{j}\frac{\partial r}{\partial y} + \boldsymbol{k}\frac{\partial r}{\partial z} = \frac{\boldsymbol{r}}{r} = \hat{\boldsymbol{r}} \tag{4.71}$$

since

$$\frac{\partial r}{\partial x} = \frac{2x}{2\sqrt{x^2+y^2+z^2}} = \frac{x}{r} \tag{4.72}$$

Thus, the field is written as

$$\boldsymbol{E} = \frac{3\,\boldsymbol{p}\cdot\boldsymbol{r}}{4\pi\,\varepsilon_0\,r^5}\,\boldsymbol{r} - \frac{\boldsymbol{p}}{4\pi\,\varepsilon_0\,r^3} \tag{4.73}$$

By developing the scalar products, we obtain

$$4\pi\,\varepsilon_0 E_x = \frac{3\,x^2 - r^2}{r^5}\,p_x + \frac{3\,xy}{r^5}\,p_y + \frac{3\,xz}{r^5}\,p_z \tag{4.74}$$

$$4\pi\,\varepsilon_0 E_y = \frac{3\,yx}{r^5}\,p_x + \frac{3\,y^2 - r^2}{r^5}\,p_y + \frac{3\,yz}{r^5}\,p_z \tag{4.75}$$

$$4\pi\,\varepsilon_0 E_z = \frac{3\,zx}{r^5}\,p_x + \frac{3\,zy}{r^5}\,p_y + \frac{3\,z^2 - r^2}{r^5}\,p_z \tag{4.76}$$

which, in tensor notation, reads

$$\boldsymbol{E} = -\boldsymbol{T}\cdot\boldsymbol{p} \tag{4.77}$$

where the negative sign is due to a convention, and where the tensor $\boldsymbol{T}$ is given by

$$\boldsymbol{T} = \frac{4\pi\,\varepsilon_0}{r^3}\begin{pmatrix} 1-\frac{3x^2}{r^2} & -\frac{3xy}{r^2} & -\frac{3xz}{r^2} \\ -\frac{3yx}{r^2} & 1-\frac{3y^2}{r^2} & -\frac{3yz}{r^2} \\ -\frac{3zx}{r^2} & -\frac{3zy}{r^2} & 1-\frac{3z^2}{r^2} \end{pmatrix} \tag{4.78}$$

In tensor notation, we would write

$$\boldsymbol{T} = \frac{4\pi\,\varepsilon_0}{r^3}\left(\boldsymbol{I} - \frac{3\,\boldsymbol{r}\cdot\boldsymbol{r}}{r^2}\right) = \frac{4\pi\,\varepsilon_0}{r^3}\left(\boldsymbol{I} - 3\,\hat{\boldsymbol{r}}\cdot\hat{\boldsymbol{r}}\right) \tag{4.79}$$

where $\boldsymbol{I}$ is the unity tensor and $\boldsymbol{T}$ is called the tensor of a dipolar field.

4.4.2 Relaxation of the dielectrics

From a macroscopic point of view, polar solvents can be considered as homogeneous dielectric media (see §1.1.3). We have already seen that at high frequencies, dielectrics have a certain inertia as illustrated by the diminution of the relative permittivity when the frequency of the perturbation increases. Thus for time dependent systems, their response time to a perturbation has to be taken into account. In order to understand dielectric friction phenomena, it is necessary to study briefly the principles linked to the relaxation of dielectric media following perturbations.

We defined by equation (1.52) the electric displacement vector as being proportional to the electric field, the proportionality constant being the permittivity of the medium. This definition is valid for steady state or harmonic systems. For transitory systems, the inertia of the dipoles will cause a retarded response of $\boldsymbol{D}$ with respect to $\boldsymbol{E}$. If we apply an electric field pulse to a dielectric at a time t_0 for a duration $\mathrm{d}t$, the orientation inertia of the dipoles means that the response of the displacement vector will relax back to zero, but with a slight delay as shown in Figure 4.13. In general, part of the dielectric reacts instantaneously (electronic polarisation) and so the response of the electric displacement vector to the impulse of an electric field pulse is written as

$$\begin{aligned} \boldsymbol{D}(t-t_0) &= \varepsilon_0\varepsilon_\infty\,\boldsymbol{E}(t_0)+\boldsymbol{E}(t_0)\alpha(t_0)\mathrm{d}t & \text{for } t_0<t<t_0+\mathrm{d}t \\ \boldsymbol{D}(t-t_0) &= \boldsymbol{E}(t_0)\alpha(t-t_0)\mathrm{d}t & \text{for } t>t_0+\mathrm{d}t \end{aligned} \qquad (4.80)$$

where ε_∞ is the relative permittivity at frequencies tending to infinity, also sometimes called the ***optical relative permittivity*** for frequencies corresponding to those of light. The function $\alpha(t-t_0)$ is a damping function, also sometimes called a memory function, which we assume to be constant during the time interval $\mathrm{d}t$.

If we now apply, at time $t=0$, an electric field $\boldsymbol{E}(t)$, by the principle of superposition of responses given by equation (4.80) for intervals of time $\mathrm{d}t$, the displacement vector is then written in the form of an integral

$$\boldsymbol{D}(t) = \varepsilon_0\varepsilon_\infty\,\boldsymbol{E}(t)+\int_0^t \boldsymbol{E}(u)\,\alpha(t-u)\,\mathrm{d}u \qquad (4.81)$$

We can take as a first approximation an exponential damping function

$$\alpha(t) = \alpha(0)\exp^{-t/\tau} \qquad (4.82)$$

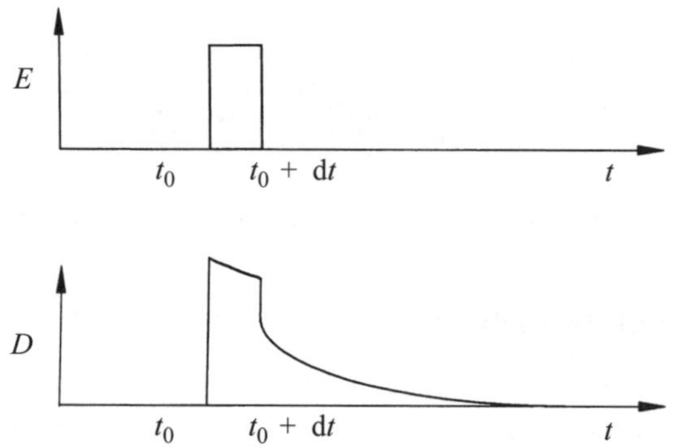

Fig. 4.13 Response of an electric displacement vector to the impulse of an electric field.

Knowing that

$$\frac{\partial}{\partial t}\left[\int_a^t F(u)G(t-u)\mathrm{d}u\right] = F(t)G(0)+\int_a^t F(u)\left(\frac{\partial G(t-u)}{\partial t}\right)\mathrm{d}u \tag{4.83}$$

we have, by differentiating equation (4.81) and multiplying by τ

$$\tau\frac{\partial \boldsymbol{D}(t)}{\partial t} = \varepsilon_0\varepsilon_\infty\tau\frac{\partial \boldsymbol{E}(t)}{\partial t} + \tau\alpha(0)\boldsymbol{E}(t)-\int_0^t \boldsymbol{E}(u)\,\alpha(t-u)\,\mathrm{d}u \tag{4.84}$$

By substituting equation (4.81) in equation (4.84), we get

$$\tau\frac{\partial}{\partial t}\left[\boldsymbol{D}(t)-\varepsilon_0\varepsilon_\infty\boldsymbol{E}(t)\right] + \left[\boldsymbol{D}(t)-\varepsilon_0\varepsilon_\infty\boldsymbol{E}(t)\right] = \tau\alpha(0)\boldsymbol{E}(t) \tag{4.85}$$

To eliminate $\alpha(0)$, let's look at the static case where $\boldsymbol{D}=\varepsilon_0\varepsilon_r\boldsymbol{E}$. Then we have:

$$\frac{\partial}{\partial t}\left[\boldsymbol{D}(t)-\varepsilon_0\varepsilon_\infty\boldsymbol{E}(t)\right] = 0 \tag{4.86}$$

and thus

$$\tau\alpha(0) = \varepsilon_0\left(\varepsilon_\mathrm{r}-\varepsilon_\infty\right) \tag{4.87}$$

The damping function is then written as

$$\alpha(t) = \frac{\varepsilon_0\left(\varepsilon_\mathrm{r}-\varepsilon_\infty\right)}{\tau}\exp^{-t/\tau} \tag{4.88}$$

Let's look at the relaxation of the dielectric in a flat capacitator following a graded variation of the charge. The approach to equilibrium in these conditions is described by

$$\frac{\partial \boldsymbol{D}(t)}{\partial t} = 0 \tag{4.89}$$

and $\boldsymbol{D}(t)=\boldsymbol{D}_0$ since a constant charge on the capacitor implies a constant polarisation. Equation (4.85) becomes

$$\frac{\tau\varepsilon_\infty}{\varepsilon_\mathrm{r}}\frac{\partial \boldsymbol{E}(t)}{\partial t} + \boldsymbol{E}(t) = \frac{\boldsymbol{D}_0}{\varepsilon_0\varepsilon_\mathrm{r}} \tag{4.90}$$

which integrates to give

$$\boldsymbol{E} = \frac{\boldsymbol{D}_0}{\varepsilon_0\varepsilon_\mathrm{r}}\left[1-\exp^{-\varepsilon_\mathrm{r}\tau\, t/\varepsilon_\infty}\right] = \frac{\boldsymbol{D}_0}{\varepsilon_0\varepsilon_\mathrm{r}}\left[1-\exp^{-t/\tau_\mathrm{L}}\right] \tag{4.91}$$

τ_L is called the ***longitudinal relaxation time***. Now, if we look at the relaxation of a dielectric following a graded variation in potential, the approach to equilibrium is in these conditions given by

$$\frac{\mathrm{d}\boldsymbol{E}(t)}{\mathrm{d}t} = 0 \tag{4.92}$$

and $\boldsymbol{E}(t) = \boldsymbol{E}_0$ since a constant potential difference implies a constant electric field. Equation (4.85) becomes

$$\tau\frac{\mathrm{d}\boldsymbol{D}(t)}{\mathrm{d}t} + \boldsymbol{D}(t) = \varepsilon_0\varepsilon_\mathrm{r}\boldsymbol{E}_0 \tag{4.93}$$

which integrates to give

$$\boldsymbol{D} = \varepsilon_0\varepsilon_\mathrm{r}\left[1-\exp^{-t/\tau}\right] \tag{4.94}$$

The relaxation time for a variation in potential is called the ***Debye relaxation time*** τ_D. For the two kinds of perturbations, we have an exponential approach to equilibrium. The two relaxation times defined above are linked by

$$\tau_\mathrm{L} = \frac{\varepsilon_\infty}{\varepsilon_\mathrm{r}}\tau_\mathrm{D} \tag{4.95}$$

For water, τ_L is 25 ps. This corresponds to a wavelength of about 1 cm which is in the range of microwaves.

4.4.3 Dielectric friction coefficient

In order to model the mechanisms of dielectric friction, let's first of all look at an immobile ion in a solvent. The molecules of polar solvents around the ion are subjected to an orientation polarisation, but by symmetry, no resultant force is exercised in return on the ion. On the other hand, if the ion is moved by the action of an external force, while the polarisation has remained 'frozen', the resultant broken symmetry will produce a force corresponding to a restoring force as illustrated in Figure 4.14.

For small displacements, we can make the hypothesis that the restoring force is, like in the case of a spring, proportional to the displacement, and thus write

$$\boldsymbol{F}^\mathrm{d} = \left[\frac{\partial \boldsymbol{F}^\mathrm{d}}{\partial \boldsymbol{r}}\right]\cdot\boldsymbol{r} \tag{4.96}$$

where the exponent d relates to the restoring force of the dielectric.

If the ion is displaced by the effect of an external force $\boldsymbol{F}^\mathrm{e}$, we can write that in a steady state regime equation (4.4) has an extra term due to the dielectric friction

$$\boldsymbol{F}^\mathrm{e} = \boldsymbol{F}^\mathrm{d} + \boldsymbol{f}^\mathrm{visc} \tag{4.97}$$

where $\boldsymbol{f}^\mathrm{visc}$ is the force of the viscous friction proportional to the velocity defined by equation (4.5). This equation may be rewritten in tensorial notation by using the unity tensor $\boldsymbol{I}$

$$\boldsymbol{f}^\mathrm{visc} = \zeta^\mathrm{o}\,\boldsymbol{I}\cdot\boldsymbol{v} \tag{4.98}$$

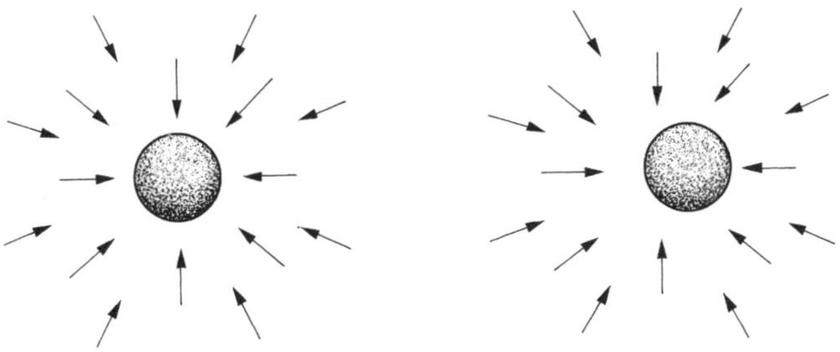

Fig. 4.14 Restoring forces of dipoles 'frozen' during the displacement of an ion

We recall that in the case of 'stick' friction, the friction coefficient is written as

$$\zeta^{o} = 6\pi\eta a \tag{4.99}$$

where a is the radius of the ion and η the viscosity of the solvent (see appendix page 171).

To link the displacement to the velocity, let's look at the displacement of an ion under an external force, as illustrated in Figure 4.14. If the ion moves rapidly with respect to the solvent that we can consider it as 'frozen', the ion displacement will happen up to the moment where the external force is compensated for by the dielectric restoring force. At that moment, the ion stops and the solvent can relax with a relaxation time characteristic of the longiditudinal relaxation time τ_L of the dielectric. By considering the movement of an ion as a series of elementary displacements dl followed by relaxation, the speed of the ion is dl/τ_L, and thus the elementary distance travelled is therefore $v\tau_L$. The friction force can then be written as

$$\boldsymbol{F}^{d} = \left[\frac{\partial \boldsymbol{F}^{d}}{\partial \boldsymbol{r}}\right] \cdot \boldsymbol{v}\, \tau_L \tag{4.100}$$

and equation (4.97) in tensorial notation gives

$$\boldsymbol{F}^{e} = \left[\frac{\partial \boldsymbol{F}^{d}}{\partial \boldsymbol{r}}\right] \cdot \boldsymbol{v}\, \tau_L + \zeta^{o}\, \boldsymbol{I} \cdot \boldsymbol{v} = \zeta\, \boldsymbol{I} \cdot \boldsymbol{v} \tag{4.101}$$

where ζ is the effective friction coefficient such that

$$\zeta\, \boldsymbol{I} = \zeta^{o}\, \boldsymbol{I} + \left[\frac{\partial \boldsymbol{F}^{d}}{\partial \boldsymbol{r}}\right] \tau_L \tag{4.102}$$

Now, we can try to evaluate the restoring force. The electric field due to the ion at the distance r is

$$\boldsymbol{E}(\boldsymbol{r}) = \frac{q}{4\,\pi\,\varepsilon_0 \varepsilon_r\, r^2}\, \hat{\boldsymbol{r}} \tag{4.103}$$

The polarisation of an element with a volume $d^3\boldsymbol{r}$ at the point r due to the molecular orientation caused by the presence of the charge q is then written as

$$\boldsymbol{p}(\boldsymbol{r})\,\mathrm{d}^3\boldsymbol{r} \;=\; (\varepsilon_\mathrm{r}-\varepsilon_\infty)\,\varepsilon_0\,\boldsymbol{E}(\boldsymbol{r})\,\mathrm{d}^3\boldsymbol{r} \;=\; \left(\frac{\varepsilon_\mathrm{r}-\varepsilon_\infty}{4\,\pi}\right)\frac{q}{\varepsilon_\mathrm{r}r^2}\,\hat{\boldsymbol{r}}\,\mathrm{d}^3\boldsymbol{r} \tag{4.104}$$

Note that we must subtract from the total polarisation ($\boldsymbol{P}_\mathrm{total} = \chi\varepsilon_0\boldsymbol{E} = (\varepsilon_\mathrm{r}-1)\varepsilon_0\boldsymbol{E}$) the contribution due to the electronic polarisation which is quasi-instantaneous ($\boldsymbol{P}_\infty = (\varepsilon_\infty - 1)\varepsilon_0\boldsymbol{E}$).

The force exerted in return on the ion by the polarisation induced by this volumic element is

$$\boldsymbol{f}(\boldsymbol{r})\,\mathrm{d}^3\boldsymbol{r} \;=\; \frac{q\boldsymbol{E}_\mathrm{pol}}{4\pi\varepsilon_0\varepsilon_\infty}\,\mathrm{d}^3\boldsymbol{r} \;=\; -\frac{q}{4\pi\varepsilon_0\varepsilon_\infty}\,\boldsymbol{T}(\boldsymbol{r})\cdot\boldsymbol{p}(\boldsymbol{r})\,\mathrm{d}^3\boldsymbol{r} \tag{4.105}$$

since all the dipoles are considered as 'frozen' and the relative permittivity of the medium is thus ε_∞. $\boldsymbol{T}(\boldsymbol{r})$ is the tensor of the dipolar field previously defined by equation (4.79).

In order to make this vectorial calculation more comprehensible, we will develop it step by step.

$$\begin{pmatrix} f_x \\ f_y \\ f_z \end{pmatrix} = -\frac{(\varepsilon_\mathrm{r}-\varepsilon_\infty)q^2}{16\pi^2\varepsilon_0\varepsilon_\mathrm{r}\varepsilon_\infty r^8}\begin{pmatrix} 3x^2-r^2 & 3xy & 3xz \\ 3yx & 3y^2-r^2 & 3yz \\ 3zx & 3zy & 3z^2-r^2 \end{pmatrix}\begin{pmatrix} x \\ y \\ z \end{pmatrix} = -\frac{(\varepsilon_\mathrm{r}-\varepsilon_\infty)q^2}{16\pi^2\varepsilon_0\varepsilon_\mathrm{r}\varepsilon_\infty r^8}\begin{pmatrix} 2xr^2 \\ 2yr^2 \\ 2zr^2 \end{pmatrix} \tag{4.106}$$

which, in vectorial notation is

$$\boldsymbol{f}(\boldsymbol{r}) \;=\; -\frac{q^2\,(\varepsilon_\mathrm{r}-\varepsilon_\infty)}{8\,\pi^2\,\varepsilon_0\varepsilon_\mathrm{r}\varepsilon_\infty}\,\frac{\hat{\boldsymbol{r}}}{r^5} \tag{4.107}$$

If the ion is displaced by the distance $\mathrm{d}\boldsymbol{r} = \mathrm{d}x\boldsymbol{i} + \mathrm{d}y\boldsymbol{j} + \mathrm{d}z\boldsymbol{k}$, the variation in the force can be expressed by the Jacobian force matrix defined by

$$\mathrm{d}\boldsymbol{f}(\boldsymbol{r}) \;=\; \begin{pmatrix} \frac{\partial}{\partial x}f_x & \frac{\partial}{\partial y}f_x & \frac{\partial}{\partial z}f_x \\ \frac{\partial}{\partial x}f_y & \frac{\partial}{\partial y}f_y & \frac{\partial}{\partial z}f_y \\ \frac{\partial}{\partial x}f_z & \frac{\partial}{\partial y}f_z & \frac{\partial}{\partial z}f_z \end{pmatrix}\begin{pmatrix} \mathrm{d}x \\ \mathrm{d}y \\ \mathrm{d}z \end{pmatrix} \tag{4.108}$$

By developing each term in the matrix, we arrive at

$$\mathrm{d}\boldsymbol{f}(\boldsymbol{r}) = -\frac{(\varepsilon_\mathrm{r}-\varepsilon_\infty)q^2}{8\pi^2\varepsilon_0\varepsilon_\mathrm{r}\varepsilon_\infty}\begin{pmatrix} \frac{\partial}{\partial x}\left(\frac{x}{\left(x^2+y^2+z^2\right)^3}\right) & \frac{\partial}{\partial y}\left(\frac{x}{\left(x^2+y^2+z^2\right)^3}\right) & \frac{\partial}{\partial z}\left(\frac{x}{\left(x^2+y^2+z^2\right)^3}\right) \\ \frac{\partial}{\partial x}\left(\frac{y}{\left(x^2+y^2+z^2\right)^3}\right) & \frac{\partial}{\partial y}\left(\frac{y}{\left(x^2+y^2+z^2\right)^3}\right) & \frac{\partial}{\partial z}\left(\frac{y}{\left(x^2+y^2+z^2\right)^3}\right) \\ \frac{\partial}{\partial x}\left(\frac{z}{\left(x^2+y^2+z^2\right)^3}\right) & \frac{\partial}{\partial y}\left(\frac{z}{\left(x^2+y^2+z^2\right)^3}\right) & \frac{\partial}{\partial z}\left(\frac{z}{\left(x^2+y^2+z^2\right)^3}\right) \end{pmatrix}\begin{pmatrix} \mathrm{d}x \\ \mathrm{d}y \\ \mathrm{d}z \end{pmatrix} \tag{4.109}$$

which simplifies to

$$\mathrm{d}\boldsymbol{f}(\boldsymbol{r}) = -\frac{(\varepsilon_\mathrm{r}-\varepsilon_\infty)q^2}{8\pi^2\,\varepsilon_0\varepsilon_\mathrm{r}\varepsilon_\infty}\begin{pmatrix} \frac{r^2-6x^2}{(x^2+y^2+z^2)^4} & \frac{-6xy}{(x^2+y^2+z^2)^4} & \frac{-6xz}{(x^2+y^2+z^2)^4} \\ \frac{-6xy}{(x^2+y^2+z^2)^4} & \frac{r^2-6y^2}{(x^2+y^2+z^2)^4} & \frac{-6yz}{(x^2+y^2+z^2)^4} \\ \frac{-6xz}{(x^2+y^2+z^2)^4} & \frac{-6yz}{(x^2+y^2+z^2)^4} & \frac{r^2-6z^2}{(x^2+y^2+z^2)^4} \end{pmatrix}\begin{pmatrix}\mathrm{d}x\\ \mathrm{d}y\\ \mathrm{d}z\end{pmatrix} \tag{4.110}$$

which is again

$$\mathrm{d}\,\boldsymbol{f}(\boldsymbol{r}) \quad = \quad -\frac{q^2\,(\varepsilon_\mathrm{r}-\varepsilon_\infty)}{8\,\pi^2\,\varepsilon_0\varepsilon_\mathrm{r}\varepsilon_\infty}\;\frac{1}{r^6}\;(\boldsymbol{I}-6\,\hat{\boldsymbol{r}}\cdot\hat{\boldsymbol{r}})\,\mathrm{d}\boldsymbol{r} \tag{4.111}$$

Hence, we get

$$\left[\frac{\partial \boldsymbol{F}^\mathrm{d}}{\partial \boldsymbol{r}}\right] \;=\; \iiint\left[\frac{\partial \boldsymbol{f}(\boldsymbol{r})}{\partial \boldsymbol{r}}\right]\mathrm{d}^3\boldsymbol{r} \;=\; -\frac{(\varepsilon_\mathrm{r}-\varepsilon_\infty)q^2}{8\pi^2\varepsilon_0\varepsilon_\mathrm{r}\varepsilon_\infty}\iiint\frac{1}{r^6}\;(\boldsymbol{I}-6\,\hat{\boldsymbol{r}}\cdot\hat{\boldsymbol{r}})\,\mathrm{d}^3\boldsymbol{r} \tag{4.112}$$

The integral of the tensor can be calculated by changing to spherical coordinates

$$\iiint\frac{1}{r^6}\quad(\boldsymbol{I}-6\,\hat{\boldsymbol{r}}\cdot\hat{\boldsymbol{r}})\,\mathrm{d}^3\boldsymbol{r}$$

$$= \int_0^{2\pi}\int_0^{\pi}\int_a^{\infty}\begin{pmatrix} \frac{\sin\phi-6\cos^2\theta\sin^3\phi}{r^4} & \frac{-6\cos\theta\sin\theta\sin^3\phi}{r^4} & \frac{-6\cos\theta\cos\phi\sin^2\phi}{r^4} \\ \frac{-6\cos\theta\sin\theta\sin^3\phi}{r^4} & \frac{\sin\phi-6\sin^2\theta\sin^3\phi}{r^4} & \frac{-6\sin\theta\cos\phi\sin^2\phi}{r^4} \\ \frac{-6\cos\theta\cos\phi\sin^2\phi}{r^4} & \frac{-6\sin\theta\cos\phi\sin^2\phi}{r^4} & \frac{\sin\phi-6\cos^2\phi\sin\phi}{r^4} \end{pmatrix}\mathrm{d}\theta\,\mathrm{d}\phi\,\mathrm{d}r \tag{4.113}$$

Knowing that

$$\int_0^{\pi}\sin^3\phi\,\mathrm{d}\phi=\frac{4}{3},\quad \int_0^{2\pi}\cos^2\theta\,\mathrm{d}\theta=\pi\text{, and that }\int_a^{\infty}\frac{1}{r^4}\,\mathrm{d}r=\frac{1}{3a^3},$$

term (1,1) equals $-4\pi/3a^3$. In the same way,

$\int_0^{2\pi}\sin^2\theta\,\mathrm{d}\theta=\pi$, and so term (2,2) equals $-4\pi/3a^3$. Finally,

$\int_0^{\pi}\cos^2\phi\sin\phi\,\mathrm{d}\phi=\frac{2}{3}$, and therefore term (3,3) also equals $-4\pi/3a^3$.

Furthermore, for the non-diagonal terms, knowing that

$$\int_0^{2\pi}\cos\theta\sin\theta\,\mathrm{d}\theta=0\text{, and }\int_0^{\pi}\cos\phi\sin^2\phi\,\mathrm{d}\phi=0$$

all these non-diagonal terms are equal to zero.
And so we have

$$\left[\frac{\partial \boldsymbol{F}^\mathrm{d}}{\partial \boldsymbol{r}}\right] \;=\; \frac{(\varepsilon_\mathrm{r}-\varepsilon_\infty)q^2}{8\pi^2\varepsilon_0\varepsilon_\mathrm{r}\varepsilon_\infty}\,\frac{4\pi}{3a^3}\,\boldsymbol{I} \tag{4.114}$$

The coefficient of dielectric friction finally reads

$$\zeta^{\mathrm{d}} = \frac{q^2(\varepsilon_{\mathrm{r}}-\varepsilon_\infty)}{6\pi\varepsilon_0\varepsilon_{\mathrm{r}}\varepsilon_\infty\, a^3}\,\tau_{\mathrm{L}} = \frac{q^2(\varepsilon_{\mathrm{r}}-\varepsilon_\infty)}{6\pi\varepsilon_0\varepsilon_{\mathrm{r}}^2\, a^3}\,\tau_{\mathrm{D}} \tag{4.115}$$

From a practical point of view, it is clear that the smaller the ion and the higher the relative permittivity of the solvent, the more the effect is non-negligible.

This calculation is presented as an example of the type of vectorial calculation one is confronted with when investigating ion-dipole interactions.

4.5 THERMODYNAMICS OF IRREVERSIBLE SYSTEMS

4.5.1 Onsager's phenomenological equations

To understand transport phenomena from a thermodynamic point of view, it is important to be able to link the flux that represents directionality in time of an evolution with the driving forces. The transport phenomena are therefore characterised by the fact that the system is not at equilibrium and produces entropy by continually dissipating energy when an external driving force is applied. This increase in entropy is expressed from the definition of the ***dissipation function*** ϕ,

$$\phi = T\frac{\mathrm{d}S}{\mathrm{d}t} = \sum_i \boldsymbol{J}_i\boldsymbol{X}_i \tag{4.116}$$

where $\boldsymbol{J}_i$ is the flux of the species i and $\boldsymbol{X}_i$ the driving force acting on that species. If the systems are linear, i.e. if the flux is proportional to the driving forces, we have

$$\boldsymbol{J}_i = \sum_j L_{ij}\boldsymbol{X}_j \tag{4.117}$$

Onsager's theory shows that the coefficients of symmetrical coupling are equal

$$L_{ij} = L_{ji} \tag{4.118}$$

In the case of an electrolyte we have

$$\boldsymbol{J}_{\mathrm{C}^+} = L_{\mathrm{C}^+\mathrm{C}^+}\boldsymbol{X}_{\mathrm{C}^+} + L_{\mathrm{C}^+\mathrm{A}^-}\boldsymbol{X}_{\mathrm{A}^-} \tag{4.119}$$

$$\boldsymbol{J}_{\mathrm{A}^-} = L_{\mathrm{A}^-\mathrm{C}^+}\boldsymbol{X}_{\mathrm{C}^+} + L_{\mathrm{A}^-\mathrm{A}^-}\boldsymbol{X}_{\mathrm{A}^-} \tag{4.120}$$

4.5.2 Diffusion potential

An interesting phenomenon is the potential difference established in solution when a salt diffuses. For example, if we place in contact a concentrated NaCl solution and a dilute NaCl solution, the salt will diffuse from the region where its concentration is higher to the region where its concentration is lower. However, the ionic mobilities of the two ions are not equal (see Table 4.1) and consequently their diffusion coefficients are not equal. As it happens, chloride ions are faster than the sodium ions. To maintain electroneutrality, an electric field is created during the diffusion process of species of different mobilities. In the case of NaCl, the electric field acts as to slow down the chloride ions and to speed up the sodium cations.

To quantify this phenomenon, let's consider a volume element of thickness dx crossed by a concentration gradient of a salt C^+A^-. If the cation and the anion have different mobilities, their fluxes can be expressed from equations (4.119-4.120).

The electroneutrality condition for the volume element is written as

$$z_{C^+}F\boldsymbol{J}_{C^+} \quad + \quad z_{A^-}F\boldsymbol{J}_{A^-} \quad = \quad 0 \tag{4.121}$$

By substitution, we have

$$\left[z_{C^+}FL_{C^+C^+} + z_{A^-}FL_{A^-C^+}\right]\boldsymbol{X}_{C^+} \quad + \quad \left[z_{C^+}FL_{C^+A^-} + z_{A^-}FL_{A^-A^-}\right]\boldsymbol{X}_{A^-} \quad = \quad 0 \tag{4.122}$$

For a salt, we shall limit ourselves to the transport by diffusion-migration, and the driving force is simply

$$\boldsymbol{X}_i \quad = \quad -\,\mathbf{grad}\,\tilde{\mu}_i \quad = \quad -\,\mathbf{grad}\mu_i \quad -\,z_iF\mathbf{grad}\phi \tag{4.123}$$

Thus, taking

$$\alpha \quad = \quad z_{C^+}FL_{C^+C^+} + z_{A^-}FL_{A^-C^+} \tag{4.124}$$

and

$$\beta \quad = \quad z_{C^+}FL_{C^+A^-} + z_{A^-}FL_{A^-A^-} \tag{4.125}$$

we obtain by substitution a relation which shows that a potential difference is established by the diffusion fluxes

$$-F\mathbf{grad}\phi \quad = \quad \frac{\alpha}{\alpha z_{C^+} + \beta z_{A^-}}\mathbf{grad}\mu_{C^+} \quad + \quad \frac{\beta}{\alpha z_{C^+} + \beta z_{A^-}}\mathbf{grad}\mu_{A^-} \tag{4.126}$$

The resulting potential difference is called the ***diffusion potential***.

We can show that

$$\frac{\alpha}{\alpha z_{C^+} + \beta z_{A^-}} \quad = \quad \frac{t_{C^+}}{z_{C^+}} \tag{4.127}$$

Actually,

$$\frac{\alpha}{\alpha z_{C^+} + \beta z_{A^-}} \quad = \quad \frac{z_{C^+}FL_{C^+C^+} + z_{A^-}FL_{A^-C^+}}{z_{C^+}\left[z_{C^+}FL_{C^+C^+} + z_{A^-}FL_{A^-C^+}\right] + z_{A^-}\left[z_{C^+}FL_{C^+A^-} + z_{A^-}FL_{A^-A^-}\right]} \tag{4.128}$$

and by multiplying the denominator and the numerator by $\mathbf{grad}\phi$, we get

$$\frac{\alpha}{\alpha z_{C^+} + \beta z_{A^-}} = \left[\frac{J_{C^+}}{z_{C^+}J_{C^+} + z_{A^-}J_{A^-}}\right]_{\mathbf{grad}\mu=0} = \frac{1}{z_{C^+}}\left[\frac{j_{C^+}}{j_{C^+} + j_{A^-}}\right]_{\mathbf{grad}\mu=0} = \frac{t_{C^+}}{z_{C^+}} \tag{4.129}$$

In the same way, knowing that $t_{C+} + t_{A-} = 1$, it follows that

$$\frac{\beta}{\alpha z_{C^+} + \beta z_{A^-}} = \frac{t_{A^-}}{z_{A^-}} \tag{4.130}$$

From this we deduce an equation for the diffusion potential as a function of the transport numbers of the ions.

$$-\mathbf{grad}\phi = \sum_i \frac{t_i}{z_i F} \mathbf{grad}\mu_i \tag{4.131}$$

This relationship indicates that the greater the difference in transport number (e.g. HCl), the greater is the diffusion potential, whilst it is small for salts for which the transport numbers of the cation and anion are similar, e.g. KCl.

4.5.3 The Planck-Henderson equation

Consider a liquid junction such as that illustrated in Figure 2.17 and let's calculate the diffusion potential across the glass frit. To do this, we need to integrate equation (4.131) between $x = 0$ and $x = l$, where l is the thickness of the junction.

$$\Delta\phi = \sum_i \frac{t_i}{z_i F} \int_0^l \frac{\mathrm{d}\mu_i}{\mathrm{d}x} \mathrm{d}x \tag{4.132}$$

If we make the hypothesis that the solutions are ideal, equation (4.132) integrates to give

$$\Delta\phi = \sum_i \frac{RTt_i}{z_i F} \int_0^l \frac{\ln c_i}{\mathrm{d}x} \mathrm{d}x = \sum_i \frac{RT}{F} \frac{t_i}{z_i} \ln\left[\frac{c_i(x=l)}{c_i(x=0)}\right] \tag{4.133}$$

For an 1:1 electrolyte, we have

$$\Delta\phi = \frac{RT}{F}(t_{C^+} - t_{A^-}) \ln\left[\frac{c(l)}{c(0)}\right] = \frac{RT}{F}(2t_{C^+} - 1) \ln\left[\frac{c(l)}{c(0)}\right] \tag{4.134}$$

If the salt concentration (e.g. KCl) in the reference electrode is in excess with respect to that of the solution in which the reference electrode is submerged, then the transport number of the ions from the solution diffusing through the glass frit will be negligible. Equation (4.134) thus applies mainly to the salt from the internal compartment of the reference electrode diffusing through the glass frit to the solution. If this salt is KCl, the values in Table 4.1 show us that $t_{C+} \cong t_{A-}$ and therefore the potential of the liquid junction is very small.

4.6 STATISTICAL ASPECTS OF DIFFUSION

We have seen for macroscopic systems that diffusion corresponds to a flux of species in a concentration gradient. In this section, we shall treat this type of mass transfer statistically.

4.6.1 Random motion in one dimension

Let's consider a particle moving along an axis by a series of jumps of length l. Each jump, either forwards or backwards, has a probability of 1/2. After N jumps from the position $x = 0$, the particle may find itself at any position between $-Nl$ and Nl

$-Nl \quad -(N-1)l \quad \cdots 0 \cdots \quad (N-1)l \quad Nl$

Now let's try to calculate the probability $W(m,N)$ that the particle will arrive at a point m after N jumps.

The probability for any sequence of N jumps is $(1/2)^N$ if all the jumps have the same probability. Therefore, the probability $W(m,N)$ is $(1/2)^N$ times the number of distinct sequences which lead to m after N jumps. To arrive at the point m, $(N+m)/2$ jumps were in the right direction, and $(N-m)/2$ jumps were in the opposite direction. Of course, if N is even, m must also be even and the converse if N is odd.

The number of distinct sequences is

$$\frac{N!}{\left(\frac{N+m}{2}\right)!\left(\frac{N-m}{2}\right)!}$$

and there are thus 4 ways to get to position 2 in 4 jumps.

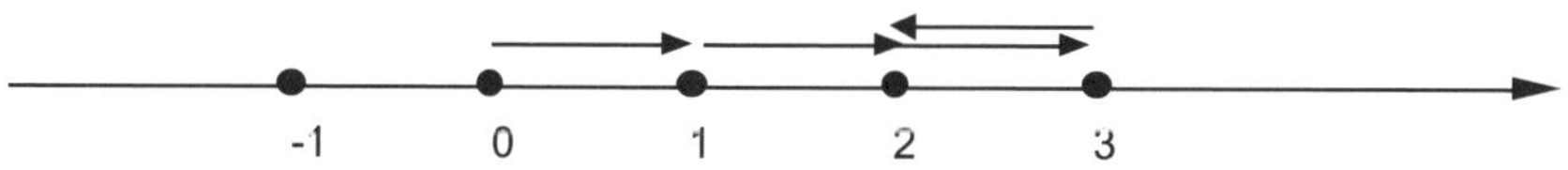

Fig. 4.15 Diffusion on an axis. Example of displacement to position 2 in 4 jumps.

The probability $W(m,N)$ is given by

$$W(m,N) \quad = \quad \left(\frac{1}{2}\right)^N \frac{N!}{\left(\frac{N+m}{2}\right)!\left(\frac{N-m}{2}\right)!} \tag{4.135}$$

The probability of going to position 2 in 4 jumps is therefore 0.25.

For the case where N is large and/or $m << N$, we can apply the Stirling approximation to this equation

$$\ln N! \;=\; \frac{1}{2}\ln(2\pi) \;+\; \left(N+\frac{1}{2}\right)\ln N \;-\; N \tag{4.136}$$

and thus obtain

$$\begin{aligned}\ln\left(W(m,N)\right) \;=\;& N\ln\frac{1}{2} \;+\; \left[\frac{1}{2}\ln(2\pi)+\left(N+\frac{1}{2}\right)\ln N-N\right]\\ &-\left[\frac{1}{2}\ln(2\pi)+\left(\frac{N+m+1}{2}\right)\ln\left(\frac{N+m}{2}\right)-\left(\frac{N+m}{2}\right)\right]\\ &-\left[\frac{1}{2}\ln(2\pi)+\left(\frac{N-m+1}{2}\right)\ln\left(\frac{N-m}{2}\right)-\left(\frac{N-m}{2}\right)\right]\\ =\;& -N\ln 2 \;-\; \frac{1}{2}\ln(2\pi) \;+\; \left(N+\frac{1}{2}\right)\ln N\\ &-\;\left(\frac{N+m+1}{2}\right)\ln\left[\frac{N}{2}\left(1+\frac{m}{N}\right)\right] \;-\; \left(\frac{N-m+1}{2}\right)\ln\left[\frac{N}{2}\left(1-\frac{m}{N}\right)\right]\end{aligned} \tag{4.137}$$

By developing to the second order the terms $\ln\left(1+\frac{m}{N}\right)$ and $\ln\left(1-\frac{m}{N}\right)$ following the series expansion

$$\ln(1+\varepsilon) \;\cong\; \varepsilon-\frac{\varepsilon^2}{2}$$

we get

$$\ln\left(W(m,N)\right) \;=\; -\frac{1}{2}\ln N+\ln 2-\frac{1}{2}\ln 2\pi-\frac{m^2}{2N} \tag{4.138}$$

Thus, for large values of N, the probability that the particle gets to the point m is

$$W(m,N) \;=\; \sqrt{\frac{2}{\pi N}}\;\exp^{-(m^2/2N)} \tag{4.139}$$

The probability that the particle gets to the space between $x = ml$, where l is the length of a jump, and $x + \Delta x$ (with $\Delta x >> l$) after N jumps is such that

$$W(x,N)\,\Delta x \;=\; W(m,N)\left(\frac{\Delta x}{2l}\right) \tag{4.140}$$

the factor 1/2 being due to the respect for the common parity between m and N. Actually, m can only take even or odd values if the number of jumps is even or odd from which comes the probability 1/2. Thus,

$$W(x,N) \;=\; \frac{1}{\sqrt{2\pi Nl^2}}\;\exp^{-(x^2/2Nl^2)} \tag{4.141}$$

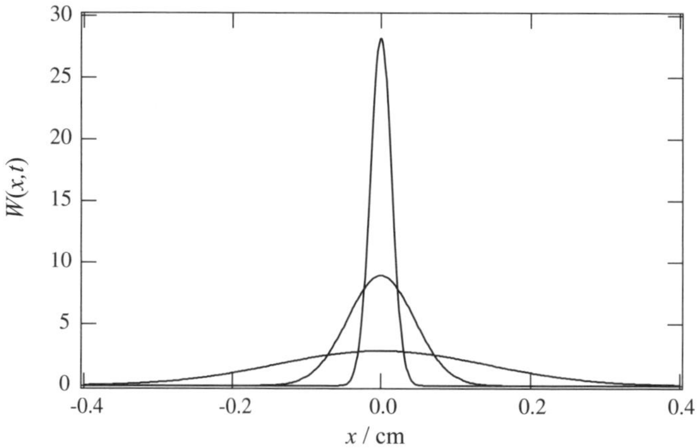

Fig. 4.16 Concentration profile with unidirectional diffusion.

Supposing that the particle makes n jumps per unit time, the probability that it will be at a position between x and $x + \Delta x$ is then

$$W(x,t)\,\Delta x \quad = \quad \frac{1}{\sqrt{4\pi Dt}} \quad \exp^{-(x^2/4Dt)}\,\Delta x \tag{4.142}$$

putting

$$D \quad = \quad \frac{1}{2} \quad nl^2 \tag{4.143}$$

Equation (4.142) is a gaussian, as shown in Figure 4.16, that describes the enlargement of an initial Dirac distribution by linear diffusion such as you can observe in a chromatographic column.

4.6.2 One-dimensional random motion with an absorbent wall

Consider an absorbent wall at a distance $m - m_1$. The first point to consider is the probability that a particle can get to the wall in N jumps. The second point is the average speed of the particles arriving at the wall.

In the first case, it is clear that to calculate the probability $W\,(m,N;m_1)$ of going to position m knowing that there is an absorbent wall at position m_1, we should take into account the fact that the particles cannot get past the absorbent wall or even touch it. Thus, compared to the previous calculation, there are forbidden trajectories which must be subtracted from $W\,(m,N)$, i.e. those which arrive in the absence of the absorbent wall at the image of point m with respect to the wall, i.e. at $2\,m_1 - m$. We have then

$$W\,(m,\,N;\,m_1) \quad = \quad W\,(m,\,N) - W\,(2\,m_1 - m,\,N) \tag{4.144}$$

In the example of Figure 4.17, there are 5 ways of getting to $m = 3$ in 5 jumps. However, one of the trajectories goes by the absorbent wall and is therefore forbidden.

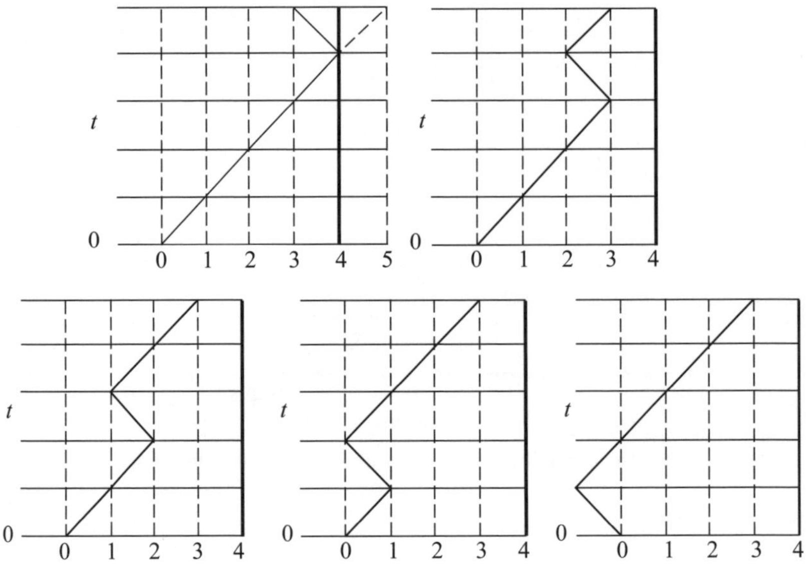

Fig. 4.17 Number of ways of arriving at $m = 3$ in 5 jumps. Absorbing wall at $m_1 = 4$.

In this example, the forbidden trajectory goes by the plane located at $2\,m_1 - m = 8 - 3 = 5$, the image of m with respect to the absorbing path.

When N is large, by using equation (4.139) we then get

$$W(m,N;m_1) \quad = \quad \sqrt{\frac{2}{\pi\, N}}\left[\exp^{-m^2/2N} - \exp^{-(2m_1-m)^2/2N}\right] \tag{4.145}$$

or again

$$W(x,t;x_1) \quad = \quad \frac{1}{\sqrt{4\,\pi Dt}}\left[\exp^{-x^2/4Dt} - \exp^{-(2x_1-x)^2/4Dt}\right] \tag{4.146}$$

Note in passing that

$$W\,(x_1,\,t;\,x_1) \quad = \quad 0 \tag{4.147}$$

and that N and m_1 must be of the same parity.

Concerning the velocity of arrival at the wall, we can say that the number of permitted ways of arriving at m_1 in N jumps is the total number of ways of arriving at m_1 in N jumps in the absence of the wall, less twice the number of ways of arriving at m_1+1 in $N-1$ jumps, still in the absence of the wall.

The factor 2 takes account of the particles which, in the absence of the wall, have actually attained the position m_1+1 as well as particles which have attained m_1-1 in $N-1$ jumps, but whose trajectories have crossed the absorbent wall. Given that the number of trajectories arriving at $m_1 - 1$ in $N-1$ jumps having crossed the wall is the same as the number of trajectories attaining the image of $m_1 - 1$ at $2\,m_1 - (m_1 - 1) = m_1+1$ still in $N - 1$ jumps, we can see that the arrival speed is

$$\frac{N!}{\left(\frac{N+m_1}{2}\right)!\left(\frac{N-m_1}{2}\right)!} - 2\frac{(N-1)!}{\left(\frac{N+m_1}{2}\right)!\left(\frac{N-m_1-2}{2}\right)!}$$

$$= \frac{N!}{\left(\frac{N-m_1}{2}\right)!\left(\frac{N+m_1}{2}\right)!}\left(1-\frac{N-m_1}{N}\right) = \frac{m_1}{N}W(m_1,N) \tag{4.148}$$

For the case where N is large, equation (4.139) gives us an arrival speed of

$$\frac{m_1}{N}\sqrt{\frac{2}{\pi N}}\exp^{-m_1^2/2N} \tag{4.149}$$

Thus, the probability that a particle arrives at x_1 during the time interval between t and $t + \Delta t$ is

$$q(x_1,t) = \frac{x_1}{t}\frac{1}{2\sqrt{\pi Dt}}\exp^{-x_1^2/4Dt} = -D\left(\frac{\partial W(x,t;x_1)}{\partial x}\right)_{x=x_1} \tag{4.150}$$

And so we find again a diffusion flux as defined empirically by equation (4.11).

These two examples in one and two dimensions illustrate how diffusion can be treated in a statistical manner. Naturally, the method may be extended to a two- or a three-dimensional system.

4.6.3 Brownian Motion

Brownian motion was discovered in 1827 by the biologist Robert Brown who studied the movement of pollen in suspension on water. It was only in 1905 that Einstein succeeded in explaining Brownian motion by combining the stochastic principle of the ***random walk*** with the distribution concept of the Maxwell-Boltzmann statisitics. The main idea of this theory is the following:

In a non-viscous system, the collisions cause variations in velocity. On the other hand, in a viscous system, the variations in velocity are rapidly dissipated and, in fact, collisions only lead to a change in direction. The principal hypothesis is that the collisions cause jumps in random directions. The trajectory of a particle is then a random walk.

Langevin's equation

Langevin's approach considers the movement of a colloid particle in suspension, making the hypothesis that the particle is submitted to a friction force and a fluctuating random force $F(t)$, which is a characteristic of Brownian motion.

$$m\frac{\mathrm{d}^2x(t)}{\mathrm{d}t}+\zeta\frac{\mathrm{d}x(t)}{\mathrm{d}t} = F(t) \tag{4.151}$$

As a first approximation, we can consider that the friction force obeys Stokes' law

$$\zeta = 6\pi\eta r \tag{4.152}$$

The fluctuating force has two specific characteristics:

- $F(t)$ is independent of the velocity of the particle
- $F(t)$ varies in an infinitely rapid way when compared to the displacement of the particle.

By expressing the Langevin equation as a function of the velocities on the x-axis, v_x, by multiplying by x we get

$$mx\frac{\mathrm{d}v_x}{\mathrm{d}t} = -\zeta x\, v_x + F(t)\, x \tag{4.153}$$

By using the two identities

$$x\, v_x = \frac{1}{2}\frac{\mathrm{d}x^2}{\mathrm{d}t} \tag{4.154}$$

and

$$x\frac{\mathrm{d}v_x}{\mathrm{d}t} = \frac{1}{2}\frac{\mathrm{d}}{\mathrm{d}t}\left(\frac{\mathrm{d}x^2}{\mathrm{d}t}\right) - v_x^2 \tag{4.155}$$

we get

$$\frac{m}{2}\frac{\mathrm{d}}{\mathrm{d}t}\left(\frac{\mathrm{d}x^2}{\mathrm{d}t}\right) - mv_x^2 = -\zeta\frac{\mathrm{d}x^2}{\mathrm{d}t} + F\, x \tag{4.156}$$

Taking the average of this equation, the fluctuating term Fx disappears and we have

$$\frac{m}{2}\frac{\mathrm{d}}{\mathrm{d}t}<\frac{\mathrm{d}x^2}{\mathrm{d}t}> - m<v_x^2> = -\frac{\zeta}{2}<\frac{\mathrm{d}x^2}{\mathrm{d}t}> \tag{4.157}$$

Using the energy equipartition theorem

$$\frac{1}{2}m<v_x^2> = \frac{1}{2}kT \tag{4.158}$$

we get

$$\frac{m}{2}\frac{\mathrm{d}}{\mathrm{d}t}<\frac{\mathrm{d}x^2}{\mathrm{d}t}> + \frac{\zeta}{2}<\frac{\mathrm{d}x^2}{\mathrm{d}t}> = kT \tag{4.159}$$

Putting

$$u = <\frac{\mathrm{d}x^2}{\mathrm{d}t}> = \frac{\mathrm{d}<x^2>}{\mathrm{d}t} \tag{4.160}$$

the differential equation is written as

$$\frac{m}{2}\frac{\mathrm{d}u}{\mathrm{d}t} + \frac{\zeta}{2}u = kT \tag{4.161}$$

The solution of this equation is of the type

$$u = \frac{2kT}{\zeta} + A\exp^{-\zeta\, t\, /m} \tag{4.162}$$

where A is an integration constant.

If the ratio $\zeta t/m$ is large, that is to say for particles of small mass in suspension in viscous media, the second term of this equation becomes negligible. If this is the case, then by integration of equation (4.160) we get

$$\left[< x^2 >\right]_0^{\Delta x} = \frac{2kT}{\zeta}\,[t]_0^{\tau} \tag{4.163}$$

or again

$$< \Delta x^2 > = \frac{2kT}{\zeta}\,\tau = 2\,D\,\tau \tag{4.164}$$

Einstein's diffusion equation in one dimension

Another approach, still for modelling the diffusion, is based on the probability of transition. Consider a system containing f particles at a time t in an elementary volume contained between x and $x + \mathrm{d}x$. After a period τ, consider the volume element situated at x' . During the period τ , particles have gone into and come out of this volume. The first hypothesis that we will make is to consider that the probability of a particle entering into the volume element only depends on the distance $x' - x$ and the period τ. This is $\phi(x' - x, \tau)$ a probability density sometimes called the transition probability. Thus, the density at the time τ is

$$f(x', t+\tau) = \int_{-\infty}^{\infty} f(x,t)\,\phi(x'-x,\tau)\,\mathrm{d}x \tag{4.165}$$

By putting $X = x - x'$ and fixing x', equation (4.163) becomes

$$f(x', t+\tau) = \int_{-\infty}^{\infty} f(x'+X,t)\,\phi(X,\tau)\,\mathrm{d}X \tag{4.166}$$

Note in passing that the function ϕ is even in as much as the probability of a movement to the right is the same as that of a movement to the left, and therefore that $\phi(X,\tau) = \phi(-X,\tau)$.

By developing equation (4.166) in a series expansion, we get

$$f(x',t) + \tau\frac{\partial f}{\partial t} + \ldots = \int_{-\infty}^{\infty}\left\{ f(x',t) + X\frac{\partial f}{\partial x} + \frac{X^2}{2!}\frac{\partial^2 f}{\partial x^2} + \ldots\right\}\phi(X,\tau)\,\mathrm{d}X \tag{4.167}$$

$$= f(x',t)\int_{-\infty}^{\infty}\phi(X,\tau)\,\mathrm{d}X + \frac{\partial f}{\partial x}\int_{-\infty}^{\infty} X\phi(X,\tau)\,\mathrm{d}X + \frac{1}{2!}\frac{\partial^2 f}{\partial x^2}\int_{-\infty}^{\infty} X^2\,\phi(X,\tau)\,\mathrm{d}X + \ldots$$

$\phi(X,r)$ being a normalised even distribution function, the following properties are verified:

$$\int_{-\infty}^{\infty} \phi(X,\tau)\,\mathrm{d}X = 1 \tag{4.168}$$

$$\int_{-\infty}^{\infty} X\,\phi(X,\tau)\,\mathrm{d}X = 0 \tag{4.169}$$

$$\int_{-\infty}^{\infty} X^2\,\phi(X,\tau)\,\mathrm{d}X = <X^2> \tag{4.170}$$

By neglecting the higher order terms, we thus have

$$\tau\frac{\partial f}{\partial t} = \frac{1}{2}\frac{\partial^2 f}{\partial x^2}<X^2> \tag{4.171}$$

f being a density of particles (in other words a concentration), a comparison with Fick's first law

$$\frac{\partial f}{\partial t} = D\,\frac{\partial^2 f}{\partial x^2} \tag{4.172}$$

leads to

$$D = \frac{<x^2>}{2\,\tau} \tag{4.173}$$

And we thus arrive again at equation (4.164). This equation is very important, since it gives access to the time scale of diffusion phenomena.

The solution of differential equation (4.172) is

$$f(x,t) = \frac{1}{\sqrt{4\pi Dt}}\,e^{-x^2/4Dt} \tag{4.174}$$

We find again that the distribution law is gaussian as described previously by equation (4.142) and the graph in Figure 4.16.

APPENDIX: ELEMENTS OF FLUID MECHANICS

Since the transport phenomena treated in this chapter take place in liquids, let's take a basic look at fluid mechanics.

Continuity equation

For a liquid, the divergence of the mass flux density (here $\rho\boldsymbol{v}$) across a surface containing a volume is equal to the variation over time of the volumic mass, which is

$$\mathrm{div}(\rho\boldsymbol{v}) = \nabla\cdot(\rho\boldsymbol{v}) = -\frac{\partial\rho}{\partial t} \tag{4.A1}$$

[This equation is parallel to the equation of conservation of charge in electro-magnetism

$$\frac{\partial\rho}{\partial t} + \mathrm{div}\boldsymbol{j} = 0 \tag{4.A2}$$

where in this case ρ is the volumic charge density and $\boldsymbol{j}$ the current density.]

By developing equation (4.A1), we obtain the relation

$$\mathrm{div}(\rho\boldsymbol{v}) = \rho\mathrm{div}\boldsymbol{v} + \boldsymbol{v}\cdot\mathbf{grad}\rho = -\frac{\partial\rho}{\partial t} \tag{4.A3}$$

which by regrouping gives

$$\left[\frac{\partial\rho}{\partial t} + \boldsymbol{v}\cdot\mathbf{grad}\rho\right] + \rho\mathrm{div}\boldsymbol{v} = \frac{\mathrm{d}\rho}{\mathrm{d}t} + \rho\mathrm{div}\boldsymbol{v} = 0 \tag{4.A4}$$

The term $\mathrm{d}\rho/\mathrm{d}t$ represents the variation in volumic mass over time for a fluid element which we follow during the flow.

If the liquid is incompressible, its density is constant and the hydrodynamic continuity equation reduces to

$$\mathrm{div}\boldsymbol{v} = \nabla\cdot\boldsymbol{v} = 0 \tag{4.A5}$$

Tensor constraint for a viscous fluid

The tensor constraint is a mathematical tool that allows us to treat the deformations of an elastic body. Take a volume element of which one element of the surface is subjected to a force $\boldsymbol{F}$. We will call S_{ix} the ratio of the projection of the force acting on the surface element perpendicular to the x-axis to the surface area

$$S_{ix} = \left.\frac{F_{ix}}{\partial y\partial z}\right|_{i=x,y,z} \tag{4.A6}$$

The first index relates to the direction of the force component, and the second index refers to the normal direction at the surface (see Figure 4A.1).

The tensor constraint $\boldsymbol{S}$ is defined such that the force $\boldsymbol{S}_\mathrm{n}$ acting on any surface element perpendicular to the unity vector $\hat{\boldsymbol{n}}$

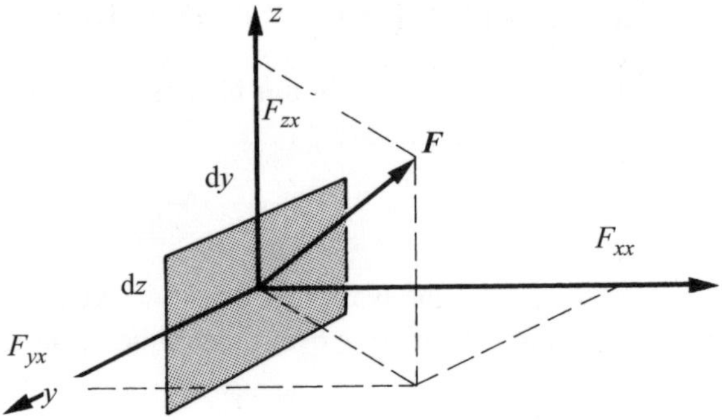

Fig. 4.A1 The force $\boldsymbol{F}$ on a surface element dydz perpendicular to the x axis decomposes into 3 components.

$$S_{in} = \sum_{j=x,y,z} S_{ij} n_j \tag{4.A7}$$

In a fluid at rest, the only force acting on a surface element, and perpendicularly to it, is the pressure. In this case the tensor constraint is a diagonal tensor since there is no shearing force on the volume element

$$S_{ij} = -p\delta_{ij} \tag{4.A8}$$

where p is the hydrostatic pressure. The negative sign expresses the fact that the volume is compressed by the fluid.

In a moving fluid, the friction forces due to the viscosity of the solution are exerted tangentially to the element. In addition to the pressure term, the tensor constraint also contains elements linked to the viscous friction, which for incompressible fluids gives

$$S_{ij} = \eta\left(\frac{\partial v_i}{\partial x_j} + \frac{\partial v_j}{\partial x_i}\right) - p\delta_{ij} = S'_{ij} - p\delta_{ij} \tag{4.A9}$$

where η is the viscosity and $\boldsymbol{v}$ the velocity vector. $\boldsymbol{S}'$, sometimes called the viscosity tensor, is a symmetrical tensor.

In the absence of external forces exerted on a volume element, Newton's equation applied to it is generally written as

$$\frac{\mathrm{d}}{\mathrm{d}t}\left[\iiint_V \rho \boldsymbol{v}\,\mathrm{d}v\right] = \iint_S \boldsymbol{S}\cdot\hat{\boldsymbol{n}}\,\mathrm{d}s \tag{4.A10}$$

This equation translates that the variation over time of the momentum is equal to the forces exerted on the surface of the volume. By applying the Green-Ostrogradski theorem, we get

$$\rho\frac{\mathrm{d}\boldsymbol{v}}{\mathrm{d}t} = \mathrm{div}\boldsymbol{S} \tag{4.A11}$$

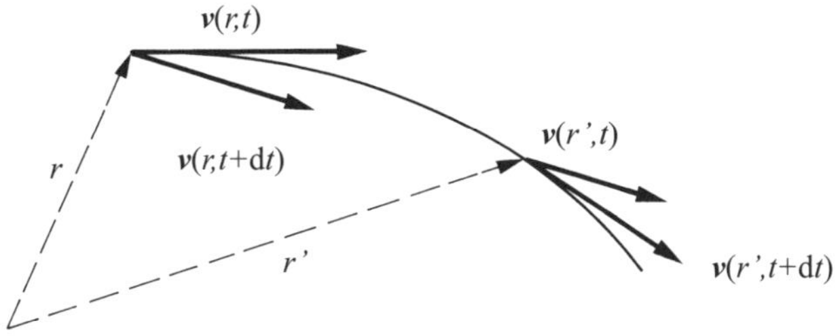

Fig. 4.A2 Acceleration in a fluid medium

The term d$\boldsymbol{v}$/dt represents the acceleration, i.e. the variation in velocity of the particles forming the fluid during their movement in space, and not the variation in velocity with respect to a fixed starting point.

$$\mathrm{d}\boldsymbol{v} \;=\; \boldsymbol{v}(r',t+\mathrm{d}t)-\boldsymbol{v}(r,t) \;=\; \frac{\partial \boldsymbol{v}}{\partial t}\mathrm{d}t+\frac{\partial \boldsymbol{v}}{\partial x}\mathrm{d}x+\frac{\partial \boldsymbol{v}}{\partial y}\mathrm{d}y+\frac{\partial \boldsymbol{v}}{\partial z}\mathrm{d}z \qquad (4.\mathrm{A}12)$$

The acceleration is therefore made up of two terms

$$\frac{\mathrm{d}\boldsymbol{v}}{\mathrm{d}t} \;=\; \frac{\partial \boldsymbol{v}}{\partial t}+\left(\frac{\partial \boldsymbol{v}}{\partial x}\right)\frac{\mathrm{d}x}{\mathrm{d}t}+\left(\frac{\partial \boldsymbol{v}}{\partial y}\right)\frac{\mathrm{d}y}{\mathrm{d}t}+\left(\frac{\partial \boldsymbol{v}}{\partial z}\right)\frac{\mathrm{d}z}{\mathrm{d}t} \;=\; \frac{\partial \boldsymbol{v}}{\partial t}+(\boldsymbol{v}\cdot\mathbf{grad}v) \qquad (4.\mathrm{A}13)$$

The first is the derivative $\partial\boldsymbol{v}/\partial t$ at a position x,y,z fixed in space. The second term is relative to the difference in velocities at the same instant at two points in space separated by a distance d$\boldsymbol{r}$, where d$\boldsymbol{r}$ is the distance covered by the fluid during the time dt.

Thus by substituting in equation (4.A11) we have

$$\rho\frac{\mathrm{d}\boldsymbol{v}}{\mathrm{d}t} \;=\; \rho\left[\frac{\partial \boldsymbol{v}}{\partial t}+(\boldsymbol{v}\cdot\nabla)\boldsymbol{v}\right] \;=\; \mathrm{div}\boldsymbol{S} \;=\; \eta\,\mathrm{div}(\mathbf{grad}\boldsymbol{v})-\mathbf{grad}p \qquad (4.\mathrm{A}14)$$

This equation is called the Navier-Stokes equation if there are no external forces. For incompressible fluids with weak velocities, the term $(\boldsymbol{v}\cdot\nabla)\boldsymbol{v}$ is negligible (small Reynolds number) and the Navier-Stokes equation reduces to the following linear form

$$\rho\frac{\partial \boldsymbol{v}}{\partial t} \;=\; \eta\nabla^2\boldsymbol{v} \;-\; \nabla p \qquad (4.\mathrm{A}15)$$

Flow friction of a liquid around a sphere

Consider the movement of a sphere in a fluid and let's calculate the friction coefficient acting on the sphere of radius R. For this, we need to consider an immobile sphere in a liquid flux, of uniform speed $\boldsymbol{u}$ when far away from the sphere (Figure 4A.3). The problem is to resolve equation (4.A15) for a steady state regime, which is

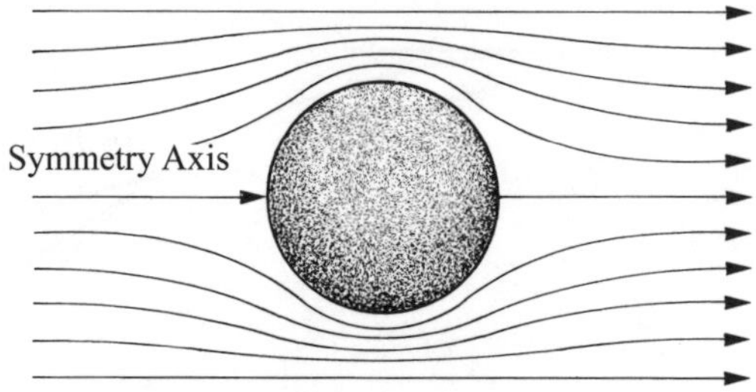

Fig. 4.A3 Viscous flow around a sphere.

$$\nabla p \quad - \quad \eta \nabla^2 \boldsymbol{v} \quad = \quad 0 \tag{4.A16}$$

with the boundary conditions $\boldsymbol{v} = 0$ for $r = R$ and $\boldsymbol{v} = \boldsymbol{u}$ when r tends towards infinity. We will take a system of spherical coordinates such that the axis corresponds to the velocity of the fluid ($\theta = 0$).

The solution of this equation is

$$\boldsymbol{v} \quad = \quad -\frac{3R}{4}\left[\frac{\boldsymbol{u}+(\boldsymbol{u}\cdot\hat{\boldsymbol{n}})\hat{\boldsymbol{n}}}{r}\right]-\frac{R^3}{4}\left[\frac{\boldsymbol{u}-3(\boldsymbol{u}\cdot\hat{\boldsymbol{n}})\hat{\boldsymbol{n}}}{r}\right]+\boldsymbol{u} \tag{4.A17}$$

We thus verify that $\boldsymbol{v} = \boldsymbol{u}$ when r tends towards infinity.

The expression of the velocity in spherical coordinates obtained from equation (4.A17) is

$$v_n \quad = \quad u\cos\theta\left[1-\frac{3R}{2r}+\frac{R^3}{2r^3}\right] \tag{4.A18}$$

$$v_\theta \quad = \quad -u\sin\theta\left[1-\frac{3R}{4r}-\frac{R^3}{4r^3}\right] \tag{4.A19}$$

We verify that at the surface of the sphere $r = R$, the two velocity components are zero.

The friction force is then defined as the resultant of the forces exerted on the surface of the sphere S

$$\boldsymbol{F} \quad = \quad \iint_S \left[-p\cos\theta + S'_{nn}\cos\theta - S'_{n\theta}\sin\theta\right] \mathrm{d}S \tag{4.A20}$$

where θ is the angle between $\boldsymbol{u}$ and $\boldsymbol{r}$.

The components linked to the viscosity of the constraint tensor in spherical coordinates are

$$S'_{nn} \quad = \quad 2\eta\frac{\partial v_n}{\partial r} \tag{4.A21}$$

$$S'_{n\theta} \quad = \quad \eta\left[\frac{1}{r}\frac{\partial v_n}{\partial \theta}+\frac{\partial v_\theta}{\partial r}-\frac{v_\theta}{r}\right] \tag{4.A22}$$

Putting into these equations the velocity vector components (4.A18) & (4.A19), we have

$$S'_{nn} = 0 \tag{4.A23}$$

$$S'_{n\theta} = -\frac{3\eta}{2R} u \sin\theta \tag{4.A24}$$

The pressure at the surface of the sphere is not uniform in the moving fluid. It is given by

$$p = p_\infty - \frac{3\eta R}{2r^2} \boldsymbol{u} \cdot \hat{\boldsymbol{n}} \tag{4.A25}$$

Developing in spherical coordinates, this becomes

$$p = p_\infty - \frac{3\eta R}{2r^2} u \cos\theta \tag{4.A26}$$

The pressure at the surface of the sphere due to the movement of the fluid is then

$$p = -\frac{3\eta}{2R} u \cos\theta \tag{4.A27}$$

The integration of equation (4.A20) reduces to

$$\boldsymbol{F} = -\frac{3\eta u}{2R} \iint \mathrm{d}S = -\frac{3\eta u}{2R} 4\pi R^2 \tag{4.A28}$$

which gives

$$\boldsymbol{F} = -6\pi\eta R\, \boldsymbol{u} \tag{4.A29}$$

We can therefore justify equation (4.7).

CHAPTER 5

ELECTRIFIED INTERFACES

We saw in chapter 1 that the bulk of all phases is neutral, and that all the excess charges are distributed at the interfaces. The presence of these charges generates interfacial electric fields that can be modified by changing the experimental conditions, in order, for example, to control charge transfer reactions. But before considering the structure and distribution of charges at electrified interfaces, we need to remind ourselves a little about interfacial thermodynamics.

5.1 INTERFACIAL TENSION

5.1.1 Surface Gibbs energy

Consider a liquid in contact with its vapour. At equilibrium, there is a dynamic exchange of molecules between the two phases. In order to understand how such an equilibrium is attained, let's take a vessel containing a liquid and air, and evacuate the air using a vacuum pump. Equilibrium is obtained when the rate of evaporation is equal to the rate of condensation. The molecules of the vapour phase have a greater potential energy than those in the liquid phase that are stabilised by isotropic molecular interactions. The molecules at the interface are submitted to anisotropic intermolecular forces which result in a force of attraction towards the centre of the phase as shown in Figure 5.1 and have thus a potential energy value which lies between the two bulk values. The difference between the average potential energy of a molecule in the vapour and in the liquid phase defines the molar vapourisation enthalpy $\Delta H_{\text{vap,m}}$ (= 40.7 kJ·mol^{-1} for water)

$$\Delta < \varepsilon > \;=\; \Delta H_{\text{vap,m}} \,/\, N_{\text{A}} \tag{5.1}$$

where N_{A} is the Avogadro constant.

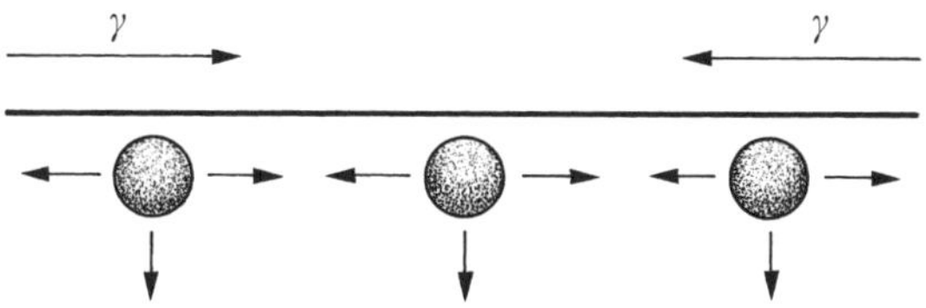

Fig. 5.1 Intermolecular forces at a liquid | gas surface.

The energy transfer from the surroundings to the system to increase the area of the interface by an elementary quantity dA, is called the ***surface Gibbs energy density***. If $\Delta < \varepsilon >$ is the energy required to break the interactions between neighbouring molecules, the surface Gibbs energy density is the energy required to break some of these interactions. This can be expressed empirically as a function of the vapourisation enthalpy

$$\gamma = 0.3 \frac{\Delta H_{\text{vap,m}}}{N_A} \left[\frac{N_A \rho}{M}\right]^{2/3} \tag{5.2}$$

where ρ is the mass density, M the molar mass of the liquid. The last term represents the number of molecules per unit surface area. For rare gases, this equation predicts the experimental values quite well, as shown in Table 5.1.

In the framework of a thermodynamic description, which has the goal of establishing a relation between observable experimental quantities, it appears that the work that the surroundings has to do on the system to increase the area of the interface by an elementary quantity dA is directly proportional to this increase in the surface area

$$\mathrm{d}W = \gamma\, \mathrm{d}A \tag{5.3}$$

The surface Gibbs energy density γ is expressed in N·m^{-1} and appears as a force parallel to the surface, trying to reduce the area of the interface, in the same way as pressure appears as a perpendicular force to an interface, trying to increase the volume of a gas phase. In the absence of gravity, the equilibrium geometry for a liquid phase is a sphere, i.e. the smallest surface per unit volume.

Table 5.1 Surface tension calculated using equation (5.2) and the corresponding values measured experimentally, (D. Tabor : *Gases, liquids and solids*, Cambridge University Press,1969, Cambridge).

	$\gamma_{\text{calculated}}$ / mN·m^{-1}	γ_{observed} / mN·m^{-1}
Argon	14	13
Neon	4	5.5
Nitrogen	11	10.5
Oxygen	13	18
Mercury	630	600
Benzene	110	40
Water		78

5.1.2 How is the surface Gibbs energy density linked to the mechanical properties of the interface?

From a mechanical point of view, there are two ways to modify a surface: stretching and shearing. We define a ***line tension*** T_{ij} (force per unit length) as the force in the direction j per unit of length of edge of a normal surface in the direction i .

To increase a surface element with an area of $\mathrm{d}A = \mathrm{d}x\mathrm{d}y$ by stretching, the work required is

$$\delta W \quad = \quad (T_{xx}\mathrm{d}y)\,\delta \mathrm{d}x \quad = \quad \delta(\gamma\,\mathrm{d}A) \quad = \quad \delta\gamma\,\mathrm{d}A \quad + \quad \gamma\,\delta \mathrm{d}A \tag{5.4}$$

where $\delta\mathrm{dA}$ (= $\delta\mathrm{d}x\,\mathrm{d}y$) is the variation in area of the surface element. We then have

$$T_{xx} \quad = \quad \gamma \quad + \quad \left[\frac{\partial\gamma}{\partial\varepsilon_{xx}}\right]_{\text{other deformations}} \tag{5.5}$$

with

$$\delta\varepsilon_{xx} \quad = \quad \frac{\delta \mathrm{d}x}{\partial\varepsilon_{yx}} \tag{5.6}$$

for the stretching on the x-axis which follows. For a shear force, we have

$$\delta W \quad = \quad (T_{yx}\mathrm{d}x)\,\delta \mathrm{d}x \quad = \quad \delta\gamma\,\mathrm{d}A \tag{5.7}$$

because $\delta\mathrm{dA} = 0$ (no change in area), from which we get

$$T_{yx} \quad = \quad \left[\frac{\partial\gamma}{\partial\varepsilon_{yx}}\right]_{\text{other deformations}} \tag{5.8}$$

with

$$\delta\varepsilon_{yx} \quad = \quad \frac{\delta \mathrm{d}x}{\mathrm{d}y} \tag{5.9}$$

This brief introduction shows that we can define a surface force tensor whose elements are written as

$$T_{ij} \quad = \quad \gamma\,\delta_{ij} \quad + \quad \left(\frac{\partial\gamma}{\partial\varepsilon_{ij}}\right)_{\text{other deformations}} \tag{5.10}$$

δ_{ij} being the Kronecker symbol (= 1 if $i = j$ and = 0 if $i \neq j$).

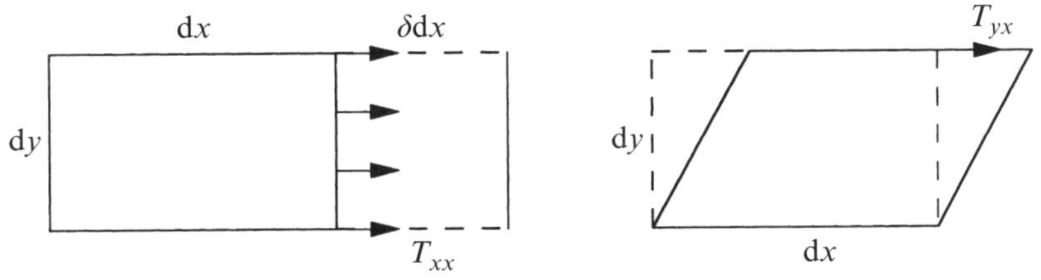

Fig. 5.2 Stretching and shearing forces on a surface element dxdy.

If the surface is isotropic and homogeneous as in the case of a liquid surface, the stretching forces are independent of the direction and there are no shear force. Thus for liquid | liquid and liquid | gas surfaces, the surface force tensor reduces to the diagonal elements

$$T_{ii} = \gamma + \left(\frac{\partial \gamma}{\partial \varepsilon}\right) = \gamma + A\left(\frac{\partial \gamma}{\partial A}\right) \tag{5.11}$$

When the molecules at the centre of a phase can move rapidly towards the surface when its area increases, the system reaches a thermodynamic equilibrium. The term $\partial\gamma/\partial A$ is then zero, and the surface Gibbs energy density γ is equal to the line tension.

In these cases, the surface Gibbs energy density is called the ***interfacial tension*** for liquid | liquid interfaces, or ***surface tension*** for liquid | gas interfaces

5.2 INTERFACIAL THERMODYNAMICS

5.2.1 The Gibbs adsorption equation

There are two thermodynamic approaches for treating interfaces. The oldest was put forward by Willard Gibbs, and the more recent by Edward Guggenheim.

The fact that it is not possible to determine a physical boundary of separation between two phases is a dilemma for treating an interface: there is always a region of space where the two phases mix. We shall adopt here Guggenheim's approach, based on the definition of an interphase.

An interphase is defined as a phase that contains all the discontinuities. Its boundaries are arbitrary as long as the phases separated by the interphase are homogeneous.

Consider two phases α and β separated by an interphase σ (see Figure 5.3).

Even though the boundaries $\alpha|\sigma$ and $\sigma|\beta$ are arbitrary, the total volume V remains the sum of the volumes of the three phases

$$V = V^{\alpha} + V^{\sigma} + V^{\beta} \tag{5.12}$$

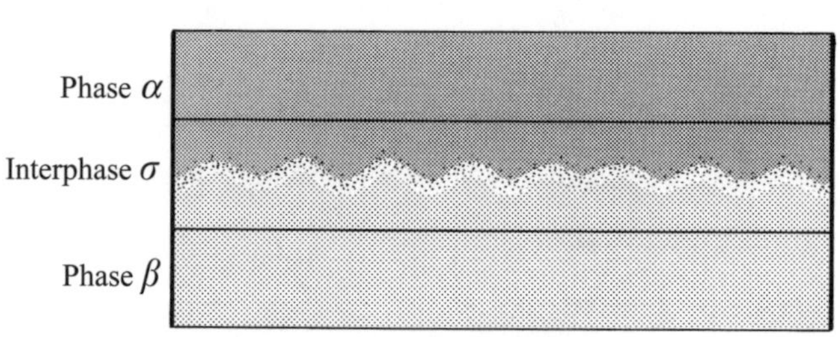

Fig. 5.3 Schematic diagram of an interphase σ between two phases α and β.

In the same way, for all the extensive variables X of the system (n_i, U, S, etc.), we will have

$$X = X^{\alpha} + X^{\sigma} + X^{\beta} \tag{5.13}$$

At equilibrium, all the intensive variables (T, p, μ_i, etc.) must be equal in the three phases. Consequently, we can write the variations in internal energy of each phase as

$$\mathrm{d}U^{\alpha} = T\mathrm{d}S^{\alpha} - p\mathrm{d}V^{\alpha} + \sum_i \mu_i \mathrm{d}n_i^{\alpha} \tag{5.14}$$

$$\mathrm{d}U^{\sigma} = T\mathrm{d}S^{\sigma} - p\mathrm{d}V^{\sigma} + \gamma\,\mathrm{d}A + \sum_i \mu_i \mathrm{d}n_i^{\sigma} \tag{5.15}$$

$$\mathrm{d}U^{\beta} = T\mathrm{d}S^{\beta} - p\mathrm{d}V^{\beta} + \sum_i \mu_i \mathrm{d}n_i^{\beta} \tag{5.16}$$

The internal energy being a homogeneous first order state function, we can apply Euler's theorem of integration with constant intensive variables and write

$$U = \sum_j \left(\frac{\partial U}{\partial X_j}\right)_{k \neq j} X_j \tag{5.17}$$

which, in this instance, reads

$$U^{\alpha} = TS^{\alpha} - pV^{\alpha} + \sum_i \mu_i n_i^{\alpha} \tag{5.18}$$

$$U^{\sigma} = TS^{\sigma} - pV^{\sigma} + \gamma\,A + \sum_i \mu_i n_i^{\sigma} \tag{5.19}$$

$$U^{\beta} = TS^{\beta} - pV^{\beta} + \sum_i \mu_i n_i^{\beta} \tag{5.20}$$

By differentiating these equations and subtracting respectively equations (5.14) to (5.16), we get

$$S^{\alpha}\mathrm{d}T - V^{\alpha}\mathrm{d}p + \sum_i n_i^{\alpha}\mathrm{d}\mu_i = 0 \tag{5.21}$$

$$S^{\sigma}\mathrm{d}T - V^{\sigma}\mathrm{d}p + A\,\mathrm{d}\gamma + \sum_i n_i^{\sigma}\mathrm{d}\mu_i = 0 \tag{5.22}$$

$$S^{\beta}\mathrm{d}T - V^{\beta}\mathrm{d}p + \sum_i n_i^{\beta}\mathrm{d}\mu_i = 0 \tag{5.23}$$

Equations (5.21) and (5.23) defined for the phases α and β are the Gibbs-Duhem equations for these phases, which show that we cannot independently vary the temperature, the pressure and the composition of a phase.

The Gibbs-Duhem equation for the interphase is called the ***Gibbs adsorption equation***

$$-A\,\mathrm{d}\gamma = S^{\sigma}\mathrm{d}T - V^{\sigma}\mathrm{d}p + \sum_i n_i^{\sigma}\mathrm{d}\mu_i \tag{5.24}$$

Equation (5.24) provides a thermodynamic definition of

$$\gamma = \left(\frac{\partial U^{\sigma}}{\partial A}\right)_{S,V,n_i} \tag{5.25}$$

but this is not a practical definition since U^{σ} depends on the arbitrary boundaries.

5.2.2 Surface excess concentration

The Gibbs adsorption equation is a very useful tool in surface studies, and the following paragraph is an example of how it can be used.

Consider a water | air interface in equilibrium at constant temperature and pressure. To keep the system simple, we make the hypothesis that the gas phase only contains a neutral gas (G) and water vapour (S). In the same way, we shall limit ourselves to a liquid phase containing only water as the solvent (S) and a neutral non-volatile solute (N).

In this very simple case, the Gibbs adsorption equation (5.24) with T & p constant reduces to

$$-A\,\mathrm{d}\gamma = n_{\mathrm{N}}^{\sigma}\,\mathrm{d}\mu_{\mathrm{N}} + n_{\mathrm{S}}^{\sigma}\,\mathrm{d}\mu_{\mathrm{S}} + n_{\mathrm{G}}^{\sigma}\,\mathrm{d}\mu_{\mathrm{G}} \tag{5.26}$$

We define the ***surface concentration*** as the number of moles of a given species in the interphase divided by the geometric area of the interface, i.e. $\Gamma = n^{\sigma}/A$. By definition, the surface concentration depends on the arbitrary choice of the boundaries $\alpha|\sigma$ and $\sigma|\beta$ and the arbitrary thickness of the interphase. Equation (5.26) can then be written as

$$-\mathrm{d}\gamma = \Gamma_{\mathrm{N}}\,\mathrm{d}\mu_{\mathrm{N}} + \Gamma_{\mathrm{S}}\,\mathrm{d}\mu_{\mathrm{S}} + \Gamma_{\mathrm{G}}\,\mathrm{d}\mu_{\mathrm{G}} \tag{5.27}$$

The Gibbs-Duhem equation for the liquid phase is written as

$$n_{\mathrm{N}}^{\mathrm{l}}\,\mathrm{d}\mu_{\mathrm{N}} + n_{\mathrm{S}}^{\mathrm{l}}\,\mathrm{d}\mu_{\mathrm{S}} = 0 \qquad \text{if} \qquad n_{\mathrm{G}}^{\mathrm{l}} \approx 0 \tag{5.28}$$

Similarly, the Gibbs-Duhem equation for the gas phase becomes

$$n_{\mathrm{S}}^{\mathrm{g}}\,\mathrm{d}\mu_{S} + n_{\mathrm{G}}^{\mathrm{g}}\,\mathrm{d}\mu_{G} = 0 \qquad \text{if} \qquad n_{\mathrm{N}}^{\mathrm{g}} \approx 0 \tag{5.29}$$

These two equations establish a link between the variations in chemical potential of the three components S, N and G.

$$\mathrm{d}\mu_{\mathrm{S}} = -\frac{n_{\mathrm{N}}^{\mathrm{l}}}{n_{\mathrm{S}}^{\mathrm{l}}}\,\mathrm{d}\mu_{\mathrm{N}} \tag{5.30}$$

and

$$\mathrm{d}\mu_{\mathrm{G}} = -\frac{n_{\mathrm{S}}^{\mathrm{g}}}{n_{\mathrm{G}}^{\mathrm{g}}}\,\mathrm{d}\mu_{\mathrm{S}} \tag{5.31}$$

By substituting these expressions into the Gibbs adsorption equation (5.26), we get

$$-\mathrm{d}\gamma = \left[\Gamma_{\mathrm{N}} - \frac{n_{\mathrm{N}}^{\mathrm{l}}}{n_{\mathrm{S}}^{\mathrm{l}}}\Gamma_{S} + \frac{n_{\mathrm{S}}^{\mathrm{g}}\,n_{\mathrm{N}}^{\mathrm{l}}}{n_{\mathrm{G}}^{\mathrm{g}}\,n_{\mathrm{S}}^{\mathrm{l}}}\Gamma_{\mathrm{G}}\right]\mathrm{d}\mu_{\mathrm{N}} \tag{5.32}$$

The bracketed term $\Gamma_N^{(S)}$ is called the ***surface excess concentration***. Neglecting the second order terms ($n_S^g << n_G^g$) and ($n_N^l << n_S^l$), the surface excess concentration can simply be expressed as

$$\Gamma_N^{(S)} = \Gamma_N - \frac{n_N^l}{n_S^l}\Gamma_S \tag{5.33}$$

The surface excess concentration of N with respect to S is the number of moles of N per unit surface area (surface concentration) minus the number of molecules of N which one would have if the bulk concentration ratio n_N^l / n_S^l was maintained down to the interfacial region where the surface concentration of the solvent is Γ_S as shown in Figure 5.4.

Whilst Γ_N is a function of the arbitrary choice of the boundaries $\alpha|\sigma$ and $\sigma|\beta$, the surface excess concentration $\Gamma_N^{(S)}$ is independent of this arbitrary choice, as long as this choice guarantees that the adjacent phases α and β are homogeneous. This justifies *a posteriori* the Guggenheim approach and the arbitrary definition of the interphase.

$\Gamma_N^{(S)}$ is experimentally accessible by measuring the dependence of the interfacial tension as a function of the concentration of N in the aqueous phase. For dilute solutions, we have

$$\Gamma_N^{(S)} = -\frac{\partial\gamma}{RT\partial\ln a_N} \cong -\frac{\partial\gamma}{RT\partial\ln c_N} \tag{5.34}$$

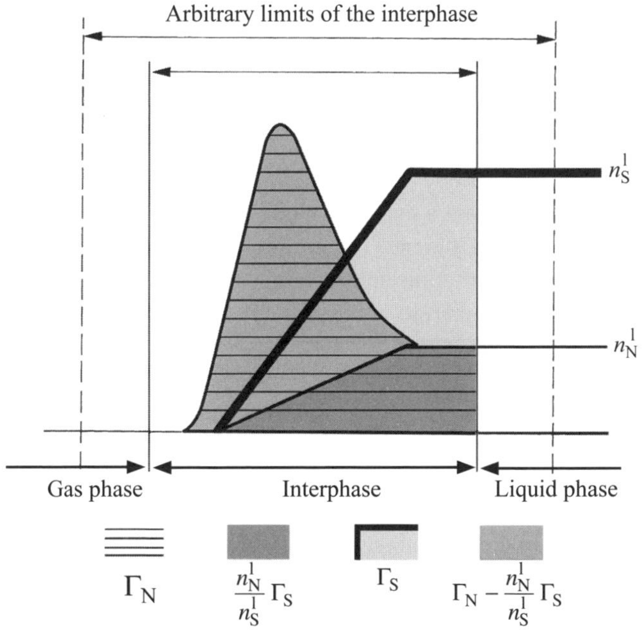

Fig. 5.4 Schematic definition of a surface excess concentration.

EXAMPLE

Let's demonstrate that the surface excess defined by equation (5.33) does not depend on the arbitrary boundaries.

To do this, we will consider the displacement of the right-hand boundary from the solid line to the dotted line as illustrated in Figure 5.4. Doing this, the quantity of neutral solute N in the interphase becomes $n_N^\sigma + dn_N$ and that of the solvent $n_S^\sigma + dn_S$, whilst in the liquid phase these quantities become respectively $n_N^1 - dn_N$ and $n_S^1 - dn_S$. Nonetheless, if the volume of the liquid phase is large enough, we can neglect these reductions. Taking into account these variations, equation (5.33) becomes

$$\Gamma_N^{(S)} = \left[\Gamma_N + \frac{dn_N^\sigma}{A}\right] - \frac{n_N^1}{n_S^1}\left[\Gamma_S + \frac{dn_S^\sigma}{A}\right]$$

In the volume added at the time of the displacement of the boundary, we have

$$\frac{dn_N^\sigma}{dn_S^\sigma} = \frac{n_N^1}{n_S^1}$$

since the solution is homogeneous to the right of the boundary marked with a solid line. By substitution, the expression above reduces to equation (5.33) in spite of the displacement of the boundary.

5.3 THERMODYNAMICS OF ELECTRIFIED INTERFACES

There are various types of electrified interfaces between conducting phases that can be metallic, semi-conductors, ionic crystals (e.g. AgI), molten salts, electrolyte solutions etc.

It is convenient to distinguish between polarisable and unpolarisable interfaces. A ***polarisable interface*** is one where we can apply a Galvani potential difference without causing noticeable changes in the chemical composition of the phases in contact, in other words without the passage of any noticeable faradaic current. So a polarisable interface is, from a thermodynamic point of view, always in equilibrium, and the Galvani potential difference is therefore an adjustable parameter of the system in addition to the classical adjustable parameters that are the concentrations of the different species, the temperature or the pressure.

5.3.1 Mercury | electrolyte interface

Consider a mercury electrode in an aqueous solution of NaCl. This interface is polarisable over a large range of electrode potentials. Effectively, if we polarise the interface with negative electrode potentials, no current will be observed until we reach a potential where a reduction takes place and where a noticeable cathodic current starts to flow. Given the present species, the first reduction will be that of the proton. For kinetic reasons explained later in chapter 7, this reduction does not take place around the standard redox potential of 0 V, but at much more negative electrode

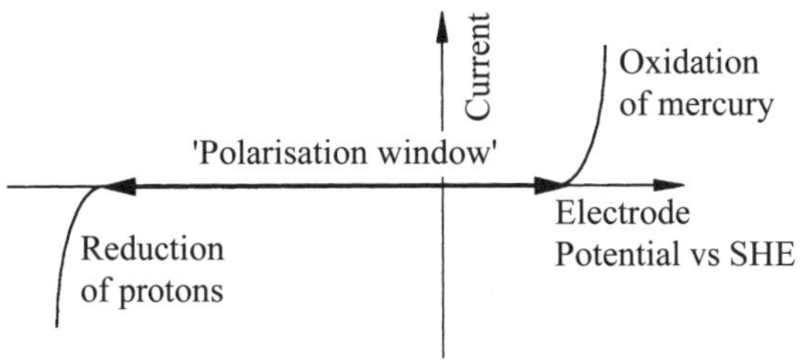

Fig 5.5 Polarisation window for a mercury drop electrode in a solution of NaCl.

potential values between –1.5 V and –2 V depending on the pH of the solution. If the interface is polarised with positive electrode potentials, no current will be observed until potentials where an oxidation takes place and where a noticeable anodic current flows. For the system under consideration, we shall oxidise mercury at potentials close to the standard redox potential of the $\frac{1}{2}Hg_2^{2+}/Hg$ couple, that is, about 0.2 V in the presence of chloride. Between these two limits, we have what is commonly called a ***polarisation window*** as illustrated in Figure 5.5, where no noticeable current flows when the electrode potential is varied.

In this paragraph, we shall present the thermodynamic aspects of the polarised mercury | electrolyte interface by way of example. Let's choose the example of a cadmium-mercury amalgam in contact with a solution of $MgCl_2$ and HCl. The polarisation window for this system is limited on the side of positive electrode potentials by the oxidation of cadmium that takes place at electrode potentials of about – 0.5 V as shown in Figure 5.6.

The Gibbs adsorption equation whose general formulation for an interface containing ions can be written either as a function of the chemical potentials of the neutral species (metal, salt etc.) or as a function of the electrochemical potentials of the corresponding charged species (ions, electrons etc.). To illustrate this second approach, we can write

$$\frac{S^\sigma \mathrm{d}T}{A} - \frac{V^\sigma \mathrm{d}p}{A} + \sum_i \Gamma_i \mathrm{d}\tilde{\mu}_i = -\mathrm{d}\gamma \tag{5.35}$$

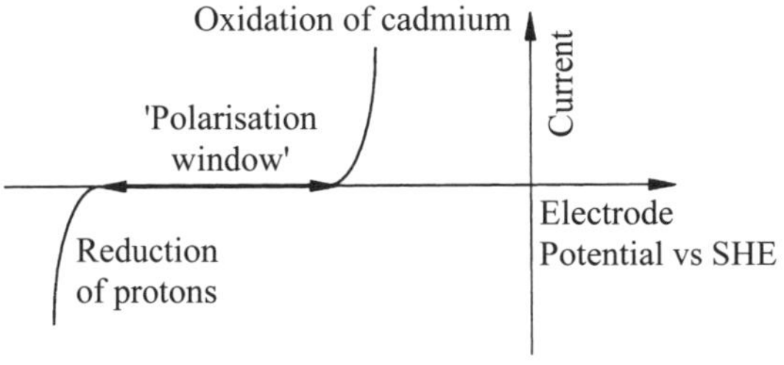

Fig 5.6 Polarisation window for a mercury-cadmium amalgam drop electrode in a solution of $MgCl_2$+HCl.

which, at constant temperature and pressure, in the present case reads

$$\begin{aligned}\Gamma_{Cd^{2+}}d\tilde{\mu}_{Cd^{2+}} + \Gamma_{Hg^{+}}d\tilde{\mu}_{Hg^{+}} + \Gamma_{e^{-}}d\tilde{\mu}_{e^{-}} + \Gamma_{Mg^{2+}}d\tilde{\mu}_{Mg^{2+}} \\ + \Gamma_{H^{+}}d\tilde{\mu}_{H^{+}} + \Gamma_{Cl^{-}}d\tilde{\mu}_{Cl^{-}} + \Gamma_{H_2O}d\mu_{H_2O} &= -d\gamma\end{aligned} \tag{5.36}$$

Since this equation contains a large number of variables, before we go any further, we need to calculate the ***degree of freedom*** or the ***variance*** of the system – i.e. the number of independent variables.

By definition, the variance is the sum of the number of intensive variables of each of the adjacent phases minus the number of relations that link them.

Considering the charged species, we can define 5 intensive variables for the metallic phase

$$T,\, p,\, \tilde{\mu}_{Cd^{2+}},\, \tilde{\mu}_{Hg^{+}} \text{ and } \tilde{\mu}_{e-}$$

Similarly, we can define 6 intensive variables for the aqueous phase

$$T,\, p,\, \tilde{\mu}_{Mg^{2+}},\, \tilde{\mu}_{H^{+}},\, \tilde{\mu}_{Cl^{-}} \text{ and } \mu_{H_2O}$$

There are 5 equilibrium relations between the two adjacent phases

- the pressure equilibrium
- the thermal equilibrium
- a Gibbs-Duhem equation for each bulk phase
- the electroneutrality of the system

The variance of the system is therefore 6 , which reduces to 4 at constant T & p. This means that we can independently vary four variables in the system, which in fact fixes all the other variables.

The Gibbs adsorption equation (5.36) contains 8 variables; we therefore need to make use of relationships among these 8 variables to reduce their number to only 4 independent variables, which should preferably be experimentally accessible.

To start with, we can consider the ionic equilibria

$$Mg^{2+} + 2\,Cl^{-} \rightleftarrows MgCl_2$$

and

$$H^{+} + Cl^{-} \rightleftarrows HCl$$

and apply equation (1.64) which in this case reads

$$d\mu_{MgCl_2} = d\tilde{\mu}_{Mg^{2+}} + 2\,d\tilde{\mu}_{Cl^{-}} \tag{5.37}$$

and

$$d\mu_{HCl} = d\tilde{\mu}_{H^{+}} + d\tilde{\mu}_{Cl^{-}} \tag{5.38}$$

Thus, the sum in equation (5.36) involving the ionic species $\Gamma_{Mg^{2+}}\,d\tilde{\mu}_{Mg^{2+}} + \Gamma_{H^+}\,d\tilde{\mu}_{H^+} + \Gamma_{Cl^-}\,d\tilde{\mu}_{Cl^-}$ can be re-written as a a linear combination of the chemical potential of the salts

$$\Gamma_{Mg^{2+}}\,d\tilde{\mu}_{Mg^{2+}} + \Gamma_{H^+}\,d\tilde{\mu}_{H^+} + \Gamma_{Cl^-}\,d\tilde{\mu}_{Cl^-} = \Gamma_{Mg^{2+}}\,d\mu_{MgCl_2} + \Gamma_{H^+}\,d\mu_{HCl} - (2\Gamma_{Mg^{2+}} + \Gamma_{H^+} - \Gamma_{Cl^-})\,d\tilde{\mu}_{Cl^-} \quad (5.39)$$

The term in brackets in equation (5.39) represents the surface density of ionic charges in solution at the interphase

$$\sigma^E = (2\Gamma_{Mg^{2+}} + \Gamma_{H^+} - \Gamma_{Cl^-})\,F \quad (5.40)$$

In the same way, the sum $\Gamma_{Cd^{2+}}\,d\tilde{\mu}_{Cd^{2+}} + \Gamma_{Hg^+}\,d\tilde{\mu}_{Hg^+} + \Gamma_{e^-}\,d\tilde{\mu}_{e^-}$ can be written as a linear combination of the chemical potentials of the metals and of the electrochemical potential of the electron

$$\Gamma_{Cd^{2+}}\,d\tilde{\mu}_{Cd^{2+}} + \Gamma_{Hg^+}\,d\tilde{\mu}_{Hg^+} + \Gamma_{e^-}\,d\tilde{\mu}_{e^-} = \Gamma_{Cd^{2+}}\,d\mu_{Cd} + \Gamma_{Hg^+}\,d\mu_{Hg} - (2\Gamma_{Cd^{2+}} + \Gamma_{Hg^+} - \Gamma_{e^-})\,d\tilde{\mu}_{e^-} \quad (5.41)$$

where the term in brackets represents the surface charge density on the metal σ^M

$$\sigma^M = (2\Gamma_{Cd^{2+}} + \Gamma_{Hg^+} - \Gamma_{e^-})\,F \quad (5.42)$$

The electroneutrality of the interphase imposes the condition $\sigma^E = -\sigma^M$, and the Gibbs adsorption equation (5.36) now reduces to

$$\Gamma_{Mg^{2+}} d\mu_{MgCl_2} + \Gamma_{H^+} d\mu_{HCl} + \Gamma_{Cd^{2+}} d\mu_{Cd} + \Gamma_{Hg^+} d\mu_{Hg} + \Gamma_{H_2O} d\mu_{H_2O} + \frac{\sigma^M}{F}(d\tilde{\mu}_{Cl^-} - d\tilde{\mu}_{e^-}) = -d\gamma \quad (5.43)$$

The Gibbs-Duhem equation for the aqueous phase at constant T & p can be written considering either the ionic species or the salts. In the latter case, we have

$$n^E_{H_2O}\,d\mu_{H_2O} + n^E_{MgCl_2}\,d\mu_{MgCl_2} + n^E_{HCl}\,d\mu_{HCl} = 0 \quad (5.44)$$

Substituting for $d\mu_{H_2O}$ into equation (5.43) provides a definition of the surface excess concentration of aqueous cations

$$\Gamma^{(H_2O)}_{Mg^{2+}} = \Gamma_{Mg^{2+}} - \frac{n^E_{MgCl_2}}{n^E_{H_2O}}\Gamma_{H_2O} \quad (5.45)$$

and

$$\Gamma^{(H_2O)}_{H^+} = \Gamma_{H^+} - \frac{n^E_{HCl}}{n^E_{H_2O}}\Gamma_{H_2O} \quad (5.46)$$

For the amalgam, the Gibbs-Duhem equation can also be written considering either the ionic species or the metals. In the latter case, we have

$$n_{\text{Cd}}^{\text{M}}\,\text{d}\mu_{\text{Cd}} \;+\; n_{\text{Hg}}^{\text{M}}\,\text{d}\mu_{\text{Hg}} \;=\; 0 \tag{5.47}$$

and similarly substituting for $\text{d}\mu_{\text{Hg}}$ into equation (5.43) provides a definition of the surface excess concentration of the cadmium ions at the surface of the amalgam

$$\Gamma_{\text{Cd}^{2+}}^{(\text{Hg})} \;=\; \Gamma_{\text{Cd}^{2+}} - \frac{n_{\text{Cd}}^{\text{M}}}{n_{\text{Hg}}^{\text{M}}}\,\Gamma_{\text{Hg}^{+}} \tag{5.48}$$

We have now eliminated two more variables and the Gibbs adsorption equation (5.43) thus reduces to

$$\Gamma_{\text{Cd}^{2+}}^{(\text{Hg})}\,\text{d}\mu_{\text{Cd}} + \Gamma_{\text{Mg}^{2+}}^{(\text{H}_2\text{O})}\,\text{d}\mu_{\text{MgCl}_2} + \Gamma_{\text{H}^{+}}^{(\text{H}_2\text{O})}\,\text{d}\mu_{\text{HCl}} + \frac{\sigma^{\text{M}}}{F}(\text{d}\tilde{\mu}_{\text{Cl}^-} - \text{d}\tilde{\mu}_{e^-}) \;=\; -\text{d}\gamma \tag{5.49}$$

Since the variance of the system is equal to 4, equation (5.49) describes the variation of the interfacial tension as a function of 4 independent variables. Written in this way, the term $(\text{d}\tilde{\mu}_{\text{Cl}^-} - \text{d}\tilde{\mu}_{e^-})$ is not very explicit. To explain it, let's consider the electrochemical cell as a whole

$$\text{Cu}^{\text{I}} \mid \text{Ag} \mid \text{AgCl} \mid \text{MgCl}_2 + \text{HCl} \;\|\; \text{Cd} + \text{Hg} \mid \text{Cu}^{\text{II}}$$

Considering the equilibria

$$\text{Ag}^+ + e^- \leftrightarrows \text{Ag}$$

and

$$\text{Ag}^+ + \text{Cl}^- \leftrightarrows \text{AgCl}$$

And applying equation (1.64) we can first write

$$\text{d}\tilde{\mu}_{\text{Ag}^+} + \text{d}\tilde{\mu}_{e^-}^{\text{Ag}} \;=\; \text{d}\mu_{\text{Ag}} \;=\; 0 \tag{5.50}$$

and

$$\text{d}\tilde{\mu}_{\text{Ag}^+} + \text{d}\tilde{\mu}_{\text{Cl}^-} \;=\; \text{d}\mu_{\text{AgCl}} \;=\; 0 \tag{5.51}$$

since silver and silver chloride are pure substances ($a_{\text{Ag}} = a_{\text{AgCl}} = 1$). As we have an electronic equilibrium between Cu^{I} and Ag ($\text{d}\tilde{\mu}_{e^-}^{\text{Cu}^{\text{I}}} = \text{d}\tilde{\mu}_{e^-}^{\text{Ag}}$), by substituting equations (5.50) & (5.51) we therefore obtain a direct relationship between $\text{d}\tilde{\mu}_{e^-}^{\text{Cu}^{\text{I}}}$ and $\text{d}\tilde{\mu}_{\text{Cl}^-}$

$$\text{d}\tilde{\mu}_{e^-}^{\text{Cu}^{\text{I}}} \;=\; \text{d}\tilde{\mu}_{e^-}^{\text{Ag}} \;=\; -\text{d}\tilde{\mu}_{\text{Ag}^+} \;=\; \text{d}\tilde{\mu}_{\text{Cl}^-} \tag{5.52}$$

The variation in electrical potential at the terminals of the cell dE is thus

$$\text{d}E \;=\; \text{d}(\phi^{\text{Cu}^{\text{II}}} - \phi^{\text{Cu}^{\text{I}}}) \;=\; -(\text{d}\tilde{\mu}_{e-}^{\text{Cu}^{\text{II}}} - \text{d}\tilde{\mu}_{e-}^{\text{Cu}^{\text{I}}})/F \;=\; (\text{d}\tilde{\mu}_{\text{Cl}^-} - \text{d}\tilde{\mu}_{e-})/F \tag{5.53}$$

taking into account the electronic equilibrium between Cu^{II} and the amalgam ($\text{d}\tilde{\mu}_{e^-}^{\text{Cu}^{\text{II}}} = \text{d}\tilde{\mu}_{e^-}^{\text{Cd+Hg}} = \text{d}\tilde{\mu}_{e^-}$). By substituting equation (5.53), the Gibbs adsorption equation (5.49) finally reads

$$\Gamma_{Cd^{2+}}^{(Hg)}\, d\mu_{Cd} \;+\; \Gamma_{Mg^{2+}}^{(H_2O)}\, d\mu_{MgCl_2} \;+\; \Gamma_{H^+}^{(H_2O)}\, d\mu_{HCl} \;+\; \sigma^M\, dE \;=\; -d\gamma \quad (5.54)$$

This equation tells us that by fixing experimentally the concentration of cadmium in the amalgam (μ_{Cd}), the concentration of magnesium chloride in solution (μ_{MgCl_2}), the concentration of hydrochloric acid in solution (μ_{HCl}), and the potential difference E applied to the terminals of the cell, then the interfacial tension will be determined unequivocally by the choice of these four independent experimental values.

Equation (5.54) also tells us how to determine experimentally the surface excess concentrations. For example, we can measure the interfacial tension for different amounts of cadmium in the amalgam, and obtain the surface excess concentration of cadmium by plotting γ as a function of $\ln a_{Cd}$ keeping all the other variables constant, and by taking the slope of this graph to obtain

$$\Gamma_{Cd^{2+}}^{(Hg)} \;=\; -\left(\frac{\partial \gamma}{\partial \mu_{Cd}}\right)_{\text{salt, acid, applied potential}} \quad (5.55)$$

Similarly, we can determine experimentally the surface excess concentrations of Mg^{2+} and that of H^+, by plotting γ as a function of $\ln a_{MgCl_2}$ and γ as a function of $\ln a_{HCl}$ respectively. The surface excess concentration of Cl^- can be determined indirectly by measuring the interfacial charge density $\sigma^E = -\sigma^M$. Indeed, it is worth pointing out that the interfacial charge density is in fact a surface excess quantity that we can also write as

$$\sigma^E \;=\; (2\Gamma_{Mg^{2+}} + \Gamma_{H^+} - \Gamma_{Cl^-})F \;=\; (2\Gamma_{Mg^{2+}}^{(H_2O)} + \Gamma_{H^+}^{(H_2O)} - \Gamma_{Cl^-}^{(H_2O)})F \quad (5.56)$$

The surface charge density σ^E is therefore, from a thermodynamic point of view, a ***surface excess charge***. It can be experimentally determined for a given system by measuring the interfacial tension as a function of the applied potential difference and plotting what is called an ***electro-capillary curve***, and by determining the slope according to

$$\sigma^M \;=\; -\left(\frac{\partial \gamma}{\partial E}\right)_{T,p,\mu_i} \quad (5.57)$$

Electrocapillary phenomena were investigated as early as 1875, and equation (5.57) is called the ***Lippmann equation***.

By differentiating equation (5.57), we can define the ***differential capacity*** of the interface C_d as

$$C_d \;=\; \frac{\partial \sigma^M}{\partial E} \;=\; -\left(\frac{\partial^2 \gamma}{\partial E^2}\right)_{T,p,\mu_i} \quad (5.58)$$

This thermodynamic approach developed above can be used more generally to investigate how a mercury drop electrode behaves when it is polarised in an electrolyte solution. Figure 5.7 shows the results originally reported by Grahame (1947) for a mercury electrode polarised in different electrolyte solutions.

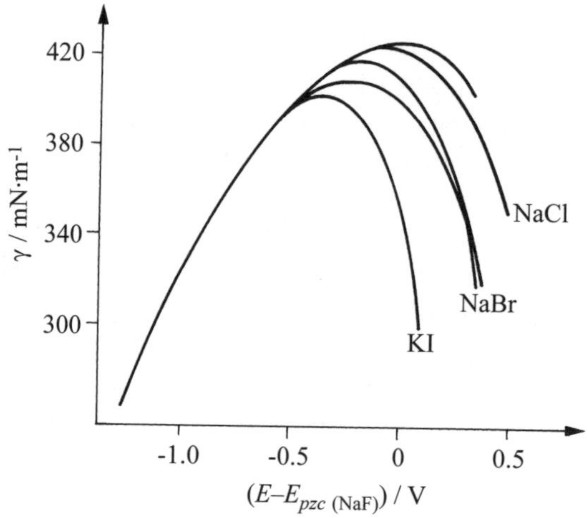

Fig. 5.7 Electrocapillary curves on a mercury electrode (Adapted from David C. Grahame, *Chem. Rev.*, 41 (1947) 441).

At negative electrode potentials, the slope of the graph is positive and equation (5.57) indicates that the surface excess charge density on the metal is then negative. In solution, the cations are more numerous at the interface than the anions, giving an excess of positive charges in the solution. The fact that the interfacial tension is the same for Na^+ and K^+ indicates at first sight the absence of specific interactions between the cations and the mercury. When the interfacial tension is at its maximum, the excess charge is zero both on the metal and in solution. The corresponding potential is called the ***potential of zero charge*** *(pzc)*. At positive electrode potentials, the anions are more numerous than the cations, and it can be observed that the interfacial tension varies with the nature of the anion. The stronger the short-range interactions between the mercury and the anions, the more the interfacial tension diminishes.

Therefore, for metal | solution interfaces, there is a physical separation of the charges when the interface is polarised, with electrons on one side and ions on the other. The interaction between these excess charges produces a reduction of the potential energy of the charged species at the interface, and thus a reduction of the interfacial tension. A classic mistake is to think that the interfacial tension is reduced because of the repulsion of ions of the same sign accumulating on the aqueous side of the interface. In fact, the interfacial tension schematically pictured in Figure 5.1 is defined as a force that appears parallel to the interface but that is in fact due to the unbalanced forces perpendicular to the interface and oriented towards the inside of the respective phases. Thus, the reduction of the interfacial tension with the applied electrode potential is due to the compensation of these unbalanced forces by the coulombic interaction between the two excess charges (electronic on the metal and ionic in the electrolyte).

5.3.2 Electrolyte | electrolyte interface

Consider, for example, an interface separating a solution of tetrabutylammonium tetraphenylborate (TBA^+TPB^-) in 1,2-dichloroethane in contact with an aqueous solution of lithium chloride. This interface is also polarisable. In fact, if the aqueous phase is polarised positively with respect to the organic phase, no current crosses the interface before the noticeable transfer of either a cation from the aqueous phase towards the organic phase, or an anion from the organic phase to the aqueous phase. In the present case, TPB^- transfers first and limits the potential window at positive Galvani potential differences. Conversely, if the aqueous phase is polarised negatively with respect to the organic phase, no current crosses the interface before the noticeable transfer of either an anion from the aqueous phase to the organic phase, or a cation from the organic phase to the aqueous phase. For the system studied here, the chloride ion transfers first and limits the window at negative Galvani potential differences (see Figure 5.8).

The Gibbs adsorption equation (5.35) for this example (at constant temperature and pressure) is written as

$$\begin{aligned}\Gamma_{TBA^+}d\tilde{\mu}_{TBA^+} + \Gamma_{TPB^-}d\tilde{\mu}_{TPB^-} + \Gamma_{DCE}d\mu_{DCE} \\ + \Gamma_{Li^+}d\tilde{\mu}_{Li^+} + \Gamma_{Cl^-}d\tilde{\mu}_{Cl^-} + \Gamma_{H_2O}d\mu_{H_2O} &= -d\gamma\end{aligned} \tag{5.59}$$

Let's calculate the degree of freedom or variance of the system, i.e. the number of independent variables.

We can consider 5 intensive variables in the organic phase

$$T,\ p,\ \tilde{\mu}_{TBA^+},\ \tilde{\mu}_{TPB^-} \text{ and } \mu_{DCE}$$

and 5 intensive variables in the aqueous phase

$$T,\ p,\ \tilde{\mu}_{Li^+},\ \tilde{\mu}_{Cl^-} \text{ and } \mu_{H_2O}$$

There are then 5 equilibrium relations between the two phases:

- The pressure equilibrium and the thermal equilibrium
- A Gibbs-Duhem equation for each phase
- The electroneutrality of the system

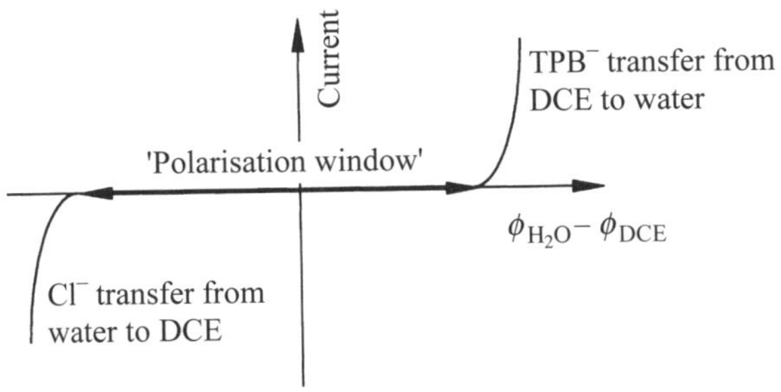

Fig. 5.8 Polarisation window for an interface between a TBATPB solution in 1,2-dichloroethane and an aqueous solution of LiCl.

The variance of the system is therefore 5, which reduces to 3 at constant T & p. This means that we can independently vary 3 variables of the system, which then fix all the other variables.

The Gibbs adsorption equation (5.59) contains 7 variables; as previously, we shall use relationships to reduce this number to only 3 independent variables, which should preferably be experimentally accessible.

To begin with, we consider the ionic equilibria in the organic phase

$$TBA^+ + TPB^- \rightleftarrows TBATPB$$

and in the aqueous phase

$$Li^+ + Cl^- \rightleftarrows LiCl$$

Thus, neglecting the formation of interfacial LiTPB or TBACl ion pairs in a first approximation, and by making use of equation (1.64) for the two salts, we have

$$\Gamma_{TBA^+}\, d\tilde{\mu}_{TBA^+} + \Gamma_{TPB^-}\, d\tilde{\mu}_{TPB^-} = \Gamma_{TBA^+}\, d\mu_{TBATPB} - (\Gamma_{TBA^+} - \Gamma_{TPB^-})\, d\tilde{\mu}_{TPB^-} \tag{5.60}$$

and

$$\Gamma_{Li^+}\, d\tilde{\mu}_{Li^+} + \Gamma_{Cl^-}\, d\tilde{\mu}_{Cl^-} = \Gamma_{Li^+}\, d\mu_{LiCl} - (\Gamma_{Li^+} - \Gamma_{Cl^-})\, d\tilde{\mu}_{Cl^-} \tag{5.61}$$

The terms in brackets in equations (5.60) & (5.61) represent the surface densities of ionic charges in respectively the organic solution and the aqueous solution

$$\sigma^{DCE} = (\Gamma_{TBA^+} - \Gamma_{TPB^-})\, F \tag{5.62}$$

and

$$\sigma^{H_2O} = (\Gamma_{Li^+} - \Gamma_{Cl^-})\, F \tag{5.63}$$

Taking into account the electroneutrality of the interphase $\sigma^{H_2O} = -\sigma^{DCE}$, the Gibbs adsorption equation (5.59) may now be written as

$$\begin{aligned} &\Gamma_{TBA^+} d\mu_{TBATPB} + \Gamma_{DCE} d\mu_{DCE} + \Gamma_{Li^+} d\mu_{LiCl} + \Gamma_{H_2O} d\mu_{H_2O} \\ &\qquad + \sigma^{H_2O}\left(d\tilde{\mu}_{TPB^-} - d\tilde{\mu}_{Cl^-}\right) = -d\gamma \end{aligned} \tag{5.64}$$

By neglecting the respective solubilities of water in dichloroethane and that of dichloroethane in water, the Gibbs-Duhem equations for the organic and aqueous phases can be written as

$$n_{DCE}^{DCE}\, d\mu_{DCE} + n_{TBATPB}^{DCE}\, d\mu_{TBATPB} = 0 \tag{5.65}$$

and

$$n_{H_2O}^{H_2O}\, d\mu_{H_2O} + n_{LiCl}^{H_2O}\, d\mu_{LiCl} = 0 \tag{5.66}$$

By substituting the terms relative to the solvents, we can define the surface excess concentration of the organic and aqueous cations respectively

$$\Gamma_{\mathrm{TBA^+}}^{(\mathrm{DCE})} = \Gamma_{\mathrm{TBA^+}} - \frac{n_{\mathrm{TBATPB}}^{\mathrm{DCE}}}{n_{\mathrm{DCE}}^{\mathrm{DCE}}}\,\Gamma_{\mathrm{DCE}} \tag{5.67}$$

and

$$\Gamma_{\mathrm{Li^+}}^{(\mathrm{H_2O})} = \Gamma_{\mathrm{Li^+}} - \frac{n_{\mathrm{LiCl}}^{\mathrm{H_2O}}}{n_{\mathrm{H_2O}}^{\mathrm{H_2O}}}\,\Gamma_{\mathrm{H_2O}} \tag{5.68}$$

We have in this way eliminated two variables and the Gibbs adsorption equation (5.64) reduces to

$$\Gamma_{\mathrm{TBA^+}}^{(\mathrm{DCE})}\mathrm{d}\mu_{\mathrm{TBATPB}} + \Gamma_{\mathrm{Li^+}}^{(\mathrm{H_2O})}\mathrm{d}\mu_{\mathrm{LiCl}} + \sigma^{\mathrm{H_2O}}\left(\mathrm{d}\tilde{\mu}_{\mathrm{TPB^-}} - \mathrm{d}\tilde{\mu}_{\mathrm{Cl^-}}\right) = -\mathrm{d}\gamma \tag{5.69}$$

We can now develop the term $(\mathrm{d}\tilde{\mu}_{\mathrm{TPB^-}} - \mathrm{d}\tilde{\mu}_{\mathrm{Cl^-}})$, considering the electrochemical cell as a whole

$$\mathrm{Cu^I} \mid \mathrm{Ag^I} \mid \mathrm{AgCl} \mid \mathrm{LiCl} \parallel \mathrm{TBATPB} \mid \mathrm{AgTPB} \mid \mathrm{Ag^{II}} \mid \mathrm{Cu^{II}}$$

In this cell, the reference electrode is an Ag | AgTPB electrode submerged directly in the organic solvent. From a practical viewpoint, the stability of this electrode is not ideal, but we shall use it as an example. With equilibrium across the whole cell, we can write by analogy with equation (5.52):

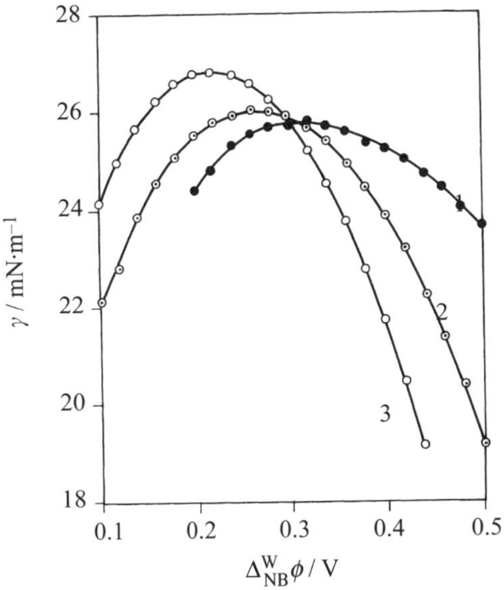

Fig. 5.9 Electrocapillary curves for an interface between a solution of tetrabutylammonium tetraphenylborate in nitrobenzene and a solution of: 1) 0.01M, 2) 0.1M & 3) 1M LiCl in water. [T. Kakiuchi & M. Senda, Bull. *Chem. Soc. Jpn*, 56 (1983) 1753, with permission from the Chemical Society of Japan.

$$\mathrm{d}\tilde{\mu}_{e^-}^{\mathrm{Cu}^{\mathrm{I}}} = \mathrm{d}\tilde{\mu}_{e^-}^{\mathrm{Ag}^{\mathrm{I}}} = -\mathrm{d}\tilde{\mu}_{\mathrm{Ag}^+} = \mathrm{d}\tilde{\mu}_{\mathrm{Cl}^-} \tag{5.70}$$

and also

$$\mathrm{d}\tilde{\mu}_{e^-}^{\mathrm{Cu}^{\mathrm{II}}} = \mathrm{d}\tilde{\mu}_{e^-}^{\mathrm{Ag}^{\mathrm{II}}} = -\mathrm{d}\tilde{\mu}_{\mathrm{Ag}^+} = \mathrm{d}\tilde{\mu}_{\mathrm{TPB}^-} \tag{5.71}$$

The potential difference at the cell terminals dE is

$$\mathrm{d}E = \mathrm{d}(\phi^{\mathrm{Cu}^{\mathrm{II}}} - \phi^{\mathrm{Cu}^{\mathrm{I}}}) = -(\mathrm{d}\tilde{\mu}_{e-}^{\mathrm{Cu}^{\mathrm{II}}} - \mathrm{d}\tilde{\mu}_{e-}^{\mathrm{Cu}^{\mathrm{I}}})/F = -(\mathrm{d}\tilde{\mu}_{\mathrm{TPB}^-} - \mathrm{d}\tilde{\mu}_{\mathrm{Cl}^-})/F \tag{5.72}$$

Thus, we finally have the Gibbs adsorption equation in the form

$$\Gamma_{\mathrm{TBA}^+}^{(\mathrm{DCE})}\mathrm{d}\mu_{\mathrm{TBATPB}} + \Gamma_{\mathrm{Li}^+}^{(\mathrm{H_2O})}\mathrm{d}\mu_{\mathrm{LiCl}} - \sigma^{\mathrm{H_2O}}\mathrm{d}E = -\mathrm{d}\gamma \tag{5.73}$$

Equation (5.73) indicates that we can determine experimentally the interfacial charge density by measuring electrocapillary curves such as those in Figure 5.9. As for the Lippmann's equation, we have

$$\sigma^{\mathrm{H_2O}} = \left(\frac{\partial \gamma}{\partial E}\right)_{\mu_{\mathrm{LiCl}},\mu_{\mathrm{TBATPB}}} \tag{5.74}$$

Similarly, we can measure the surface excess concentrations, using for example

$$\Gamma_{\mathrm{TBA}^+}^{(\mathrm{DCE})} = -\left(\frac{\partial \gamma}{\partial \mu_{\mathrm{TBATPB}}}\right)_{\mu_{\mathrm{LiCl}},\, E} \tag{5.75}$$

5.4 SPATIAL DISTRIBUTION OF POLARISATION CHARGES

Electrified interfaces are places where heterogeneous charge transfer reactions happen. And so it is important to know for the different types of interface the spatial distribution of charges and the distribution of the electric potential across the interphase. In this section, we shall review various electrified interfaces; model ones such as the mercury | electrolyte interface or others such as liquid | liquid interfaces or those between two semiconductors.

5.4.1 Metal | electrolyte interface

We have seen that excess charges can only exist at interfaces. The distribution of the excess electronic and ionic charge is therefore an important facet of the metal | electrolyte electrochemical interface. Since the excess of electronic charge is spread over a very small thickness of 100 to 200 pm maximum, most of the Galvani potential difference will take place in the electrolyte.

The distribution of the ions near the electrode then becomes a major point of interest.

The Gouy–Chapman theory

Gouy (1910) and Chapman (1913) independently proposed a model of charged metal | electrolyte interfaces. This model has common hypotheses with the Debye-Hückel theory, i.e.

- The interactions are purely electrostatic
- The ions are considered as point charges. The polarisability of ions is neglected
- The model also considers that the metal can be represented as a planar surface supporting a surface density of charge σ
- The distribution of ions follows Boltzmann's statistics

From the last condition, we can define the spatial distribution of ionic charges

$$N_i(x) \quad = \quad N_i^\infty \, e^{-z_i e \, (\phi(x) - \phi^\infty) \, / \, kT} \tag{5.76}$$

where ϕ^∞ is the inner potential of the solution when $x \to \infty$. $N_i(x)$ is the volumic density of ions in the elementary volume defined by the planes of abscissae x and $x + \mathrm{d}x$. The term $z_i e(\phi(x) - \phi^\infty)$ represents the work to bring a species i from the bulk to the interfacial region at a distance x from the metal.

For this volume element, the Laplacian of the potential is a function of the volumic charge density ρ as given by Poisson's equation (1.37)

$$\Delta\phi \quad = \quad -\frac{\rho}{\varepsilon_0 \varepsilon_\mathrm{r}} \tag{5.77}$$

where due to the present geometry, only the x-axis perpendicular to the surface of the electrode is considered, and equation (5.77) reduces to

$$\frac{\partial^2 \phi(x)}{\partial x^2} \quad = \quad -\frac{\rho(x)}{\varepsilon_0 \varepsilon_\mathrm{r}} \tag{5.78}$$

where ε_0 is the permittivity of vacuum and ε_r the relative permittivity of the electrolyte solution. The volumic charge density in solution is simply expressed as

$$\rho(x) \quad = \quad \sum_i z_i e \, N_i(x) \tag{5.79}$$

By substituting equation (5.79) in Poisson's equation (5.78), we have

$$\frac{\partial^2 \phi(x)}{\partial x^2} \quad = \quad \frac{-1}{\varepsilon_0 \varepsilon_\mathrm{r}} \sum_i z_i e \, N_i^\infty \, e^{-z_i e[\phi(x) - \phi^\infty]/kT} \tag{5.80}$$

This differential equation is known as the ***Poisson–Boltzmann equation***. The boundary conditions for $x \to \infty$ are

$$\phi(x) \to \phi^\infty \qquad \text{and} \qquad \frac{\partial \phi(x)}{\partial x} \to 0$$

Even though it is possible to solve this differential equation for different types of electrolytes, we shall content ourselves with studying the systems of 1:1 univalent electrolytes.

In this case, the Poisson-Boltzmann equation reduces to

$$\begin{aligned}\frac{\partial^2\phi(x)}{\partial x^2} &= \frac{eN^\infty}{\varepsilon_0\varepsilon_r}\left[e^{e[\phi(x)-\phi^\infty]/kT} - e^{-e[\phi(x)-\phi^\infty]/kT}\right] \\ &= \frac{2eN^\infty}{\varepsilon_0\varepsilon_r}\sinh\left[\frac{e[\phi(x)-\phi^\infty]}{kT}\right]\end{aligned} \tag{5.81}$$

since

$$2\sinh x = e^x - e^{-x} \tag{5.82}$$

By using the identity

$$2\left(\frac{\partial^2\phi}{\partial x^2}\right) = 2\frac{\partial}{\partial x}\left(\frac{\partial\phi}{\partial x}\right) = 2\left[\frac{\partial}{\partial\phi}\left(\frac{\partial\phi}{\partial x}\right)\right]\cdot\left(\frac{\partial\phi}{\partial x}\right) = \frac{\partial}{\partial\phi}\left[\left(\frac{\partial\phi}{\partial x}\right)^2\right] \tag{5.83}$$

the differential equation (5.81) becomes

$$\frac{\partial}{\partial\phi}\left[\left(\frac{\partial\phi}{\partial x}\right)^2\right] = \frac{4eN^\infty}{\varepsilon_0\varepsilon_r}\sinh\left[\frac{e}{kT}(\phi(x)-\phi^\infty)\right] \tag{5.84}$$

Integrating between x and ∞, we thus obtain

$$\begin{aligned}-\left(\frac{\partial\phi}{\partial x}\right)^2 &= \frac{4eN^\infty}{\varepsilon_0\varepsilon_r}\int_{\phi(x)}^{\phi^\infty}\sinh\left[\frac{e}{kT}(\phi(x)-\phi^\infty)\right]\mathrm{d}\phi \\ &= \frac{4N^\infty kT}{\varepsilon_0\varepsilon_r}\left[1-\cosh\frac{e}{kT}(\phi(x)-\phi^\infty)\right]\end{aligned} \tag{5.85}$$

Using the relation

$$1-\cosh 2x = -2(\sinh x)^2 \tag{5.86}$$

we get an expression of the electric field as a function of the distance from the electrode

$$\frac{\partial\phi}{\partial x} = -\left[\frac{8N^\infty kT}{\varepsilon_0\varepsilon_r}\right]^{1/2}\sinh\left[\frac{e}{2kT}(\phi(x)-\phi^\infty)\right] \tag{5.87}$$

The choice of sign, taking the square root of equation (5.85) is due to the fact that if the potential difference $\phi(x)-\phi^\infty$ is positive, the slope must be negative.

Equation (5.87) gives us the variation in the electric field as a function of the distance normal to the electrode. The interphase has to be overall electrically neutral. Thus, the electronic surface charge density on the metal is equal and opposite to the ionic surface charge density in solution.

$$\sigma^{\mathrm{M}} = -\int_0^\infty \rho(x)\mathrm{d}x \tag{5.88}$$

which, by substituting the Poisson equation (5.78), gives

$$\sigma^{\mathrm{M}} = \int_0^\infty \varepsilon_0\varepsilon_{\mathrm{r}}\left(\frac{\partial^2\phi(x)}{\partial x^2}\right)\mathrm{d}x = -\varepsilon_0\varepsilon_{\mathrm{r}}\left(\frac{\partial\phi(x)}{\partial x}\right)_{x=0} \tag{5.89}$$

Knowing the electric field at the interface given by equation (5.87), we then get

$$\sigma^{\mathrm{M}} = \left[8N^\infty kT\varepsilon_0\varepsilon_{\mathrm{r}}\right]^{1/2}\sinh\left[\frac{e}{2kT}(\phi(0)-\phi^\infty)\right] \tag{5.90}$$

This equation links the potential drop in the electrolyte $(\phi(0)-\phi^\infty)$ to the charge on the metal, and this relation is shown in Figure 5.10. Equation (5.89) also represents the Gauss equation (1.38) for the case of a metallic conductor in contact with a dielectric medium with a relative permitivity ε_{r}.

In order to find the potential distribution $\phi(x)$, we can integrate equation (5.87)

$$\int_{\phi(0)-\phi^\infty}^{\phi(x)-\phi^\infty}\left[\sinh\left(\frac{e(\phi(x)-\phi^\infty)}{2kT}\right)\right]^{-1}\mathrm{d}(\phi(x)-\phi^\infty) = -\left[\frac{8N^\infty kT}{\varepsilon_0\varepsilon_{\mathrm{r}}}\right]^{1/2}\int_0^x \mathrm{d}x \tag{5.91}$$

to obtain

$$\frac{2kT}{e}\ln\left[\frac{\tanh\left[e(\phi(x)-\phi^\infty)/4kT\right]}{\tanh\left[e(\phi(0)-\phi^\infty)/4kT\right]}\right] = -\left[\frac{8N^\infty kT}{\varepsilon_0\varepsilon_{\mathrm{r}}}\right]^{1/2}x \tag{5.92}$$

Introducing the Debye-Hückel reciprocal length defined by equation (3.55)

$$\kappa = \left[\frac{2N^\infty e^2}{\varepsilon_0\varepsilon_{\mathrm{r}}kT}\right]^{1/2} \tag{5.93}$$

we finally get

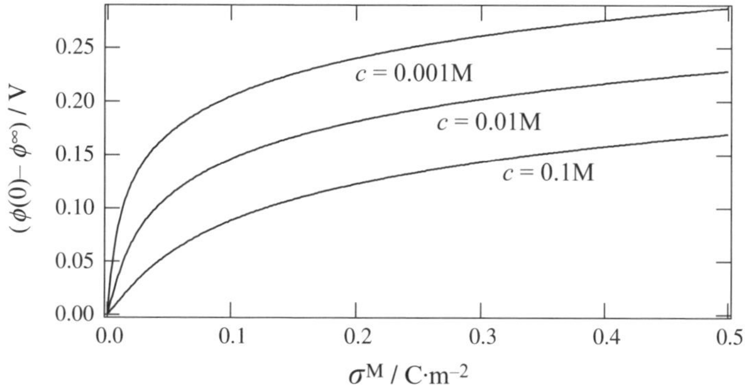

Fig. 5.10 Potential drop in the diffuse layer as a function of the charge on the electrode for different concentrations of an 1:1 electrolyte.

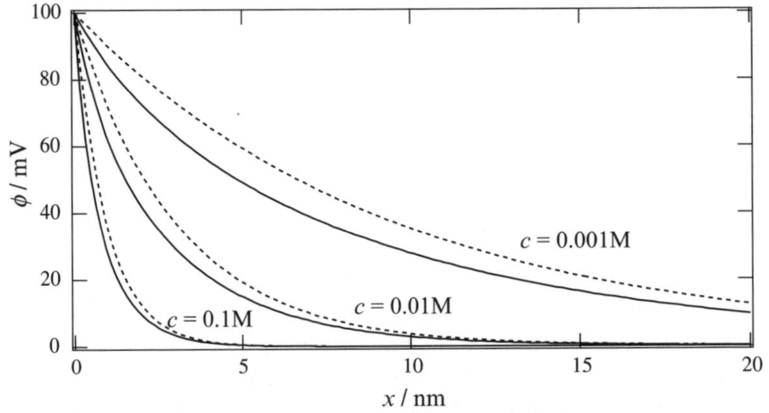

Fig. 5.11 Potential distribution for different electrolyte concentrations. Potential drop = 100 mV.

$$\tanh\left[\frac{e}{4kT}(\phi(x)-\phi^{\infty})\right] = e^{-\kappa x}\tanh\left[\frac{e}{4kT}(\phi(0)-\phi^{\infty})\right] \tag{5.94}$$

The potential distribution thus obtained is shown by the solid lines in Figure 5.11. The layer of electrolyte close to the electrode where the potential varies is where the ionic surface excess charge is distributed. As a result, it is called the ***space charged region*** or the ***diffuse layer***; the latter should not be confused with the diffusion layer discussed in chapters 4 & 7.

In fact, the potential distribution more or less resembles a function of the type

$$\phi = \left[\phi(0)-\phi^{\infty}\right] e^{-\kappa x} \tag{5.95}$$

as shown by the dotted curves in Figure 5.11. This is why the reciprocal Debye length is often taken as a measure of the diffuse layer thickness.

From the potential distribution, we can calculate the corresponding distribution of the cations and anions by applying Boltzmann's distribution law as given by equation (5.76). The concentration profiles thus obtained and illustrated in Figure 5.12 clearly show the asymmetry of the distribution between the accumulation of anions and the depletion of cations. Because the size of the ions is neglected, the concentration of anions at the surface of the electrode tends to very high and unrealistic values.

From equation (5.90), we can define the Gouy-Chapman capacity C_{GC} corresponding to the volumic distribution of ions predicted by the Gouy-Chapman theory by differentiation of the surface charge density with respect to the potential drop in solution

$$C_{GC} = \frac{\partial\sigma^{M}}{\partial(\phi(0)-\phi^{\infty})} = \left[\frac{2N^{\infty}e^{2}\varepsilon_{0}\varepsilon_{r}}{kT}\right]^{1/2}\cosh\left[\frac{e}{2kT}(\phi(0)-\phi^{\infty})\right] \tag{5.96}$$

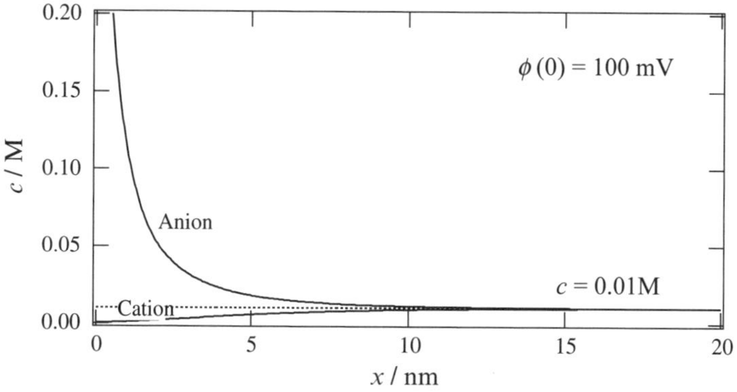

Fig. 5.12 Ionic concentration profile.

To test the validity of the Gouy-Chapman theory, it is common to compare the capacity measured experimentally from impedance measurements (see chapter 9) with that predicted by equation (5.96). The graphs in Figure 5.13 show these values for a mercury electrode in contact with an aqueous KF solution for different electrolyte concentrations.

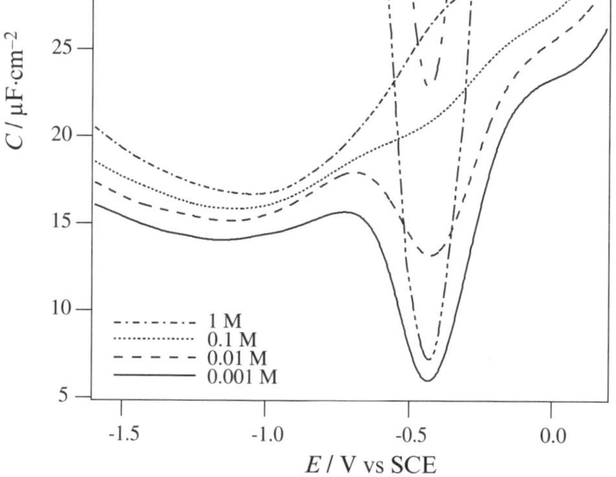

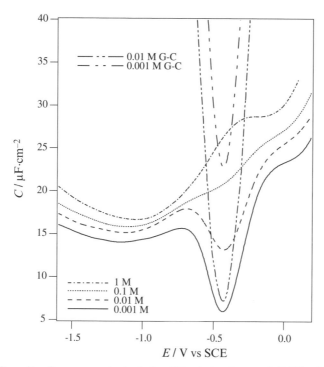

Fig. 5.13 Capacity of a mercury electrode in a KF solution ($pzc = -0.433$V) with the respective Gouy-Chapman capacity [Paolo Galleto, EPFL thesis].

We can see that at low electrolyte concentrations, the theory corroborates the experimental results only for potentials close to the *pzc*. At higher concentrations, the diffuse layer becomes thinner, and the experimental capacity becomes dominated by the presence of a water layer close to the electrode that is neglected in the Gouy-Chapman theory that considers ionic charges as point charges, not taking into account the ionic hydration layer.

Influence of the metal and the inner layer

The experimental curves in Figure 5.13 show clearly that at potentials far from the *pzc*, the Gouy-Chapman theory is not sufficient to model the capacity of the electrified interface. To alleviate this problem, we can consider, as a first approximation, that the capacities measured correspond in fact to two capacitors in series: one purely dielectric corresponding to the presence of a monolayer of water on the electrode as illustrated in Figure 5.14 and whose capacity is called the ***Helmholtz capacity*** C_H.

The second capacitor is the one associated with the charge distribution in the diffuse layer and whose capacity is the Gouy-Chapman capacity. Thus, the observed capacity can be written as

$$\frac{1}{C_{obs}} = \frac{1}{C_H} + \frac{1}{C_{GC}} \tag{5.97}$$

Figure 5.15 illustrates schematically that when two capacitors are in series, the total capacity is always dominated by the smaller one. Therefore, near the *pzc,*

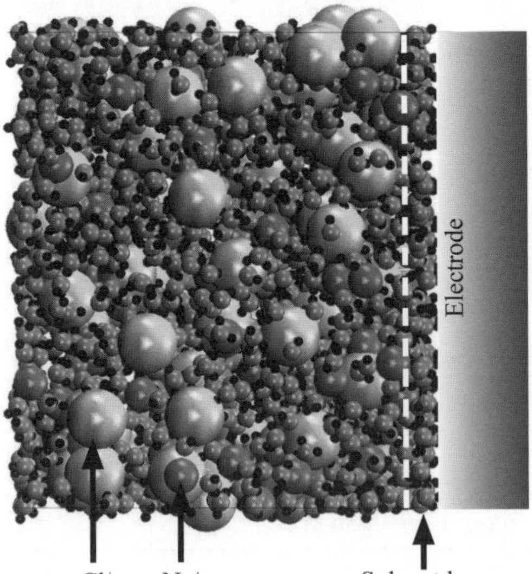

Fig. 5.14 Molecular dynamics illustration of the ionic double layer of a 3M NaCl solution. The charge on the electrode is $-1\mu C\cdot cm^{-2}$ (Personnal communication, Prof. Michael Philpott, Singapor University).

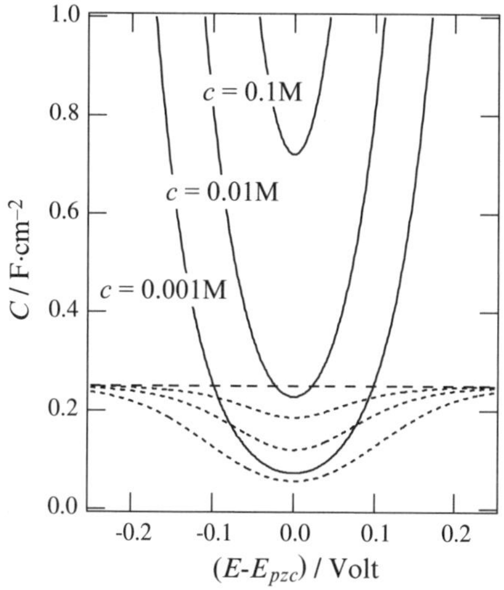

Fig. 5.15 Schematic illustration of two capacitors in series. The Gouy-Chapman capacities are the solid lines, calculated from equation (5.96); the Helmholtz capacity is taken constant at 0.25 $F{\cdot}m^{-2}$. The resultant capacities are shown as dotted lines.

the Gouy-Chapman capacity is dominant at low concentrations, whereas, when the surface charge density is large, the Helmholtz capacity is dominant.

In fact, C_H can, as a first approximation, be taken as equal to C_{obs} when the electrolyte concentration is large (>1M). By doing so, the solvent model allows a reasonably good prediction of the experimental results as shown in Figure 5.16. The most important conclusion from this model is that the potential distribution

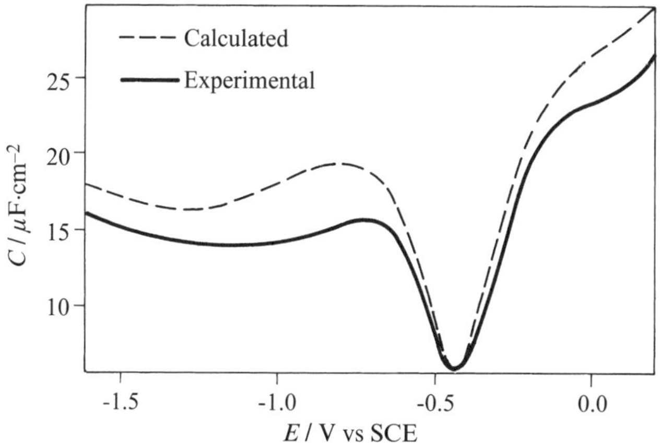

Fig. 5.16 Experimental capacity compared to that calculated using equation (5.97) for a mercury electrode in a 0.001M KF solution. C_H is the experimental value for a 1M solution of KF taken from Figure 5.13 (Paolo Galetto, EPFL thesis).

takes place principally in the solvent layer when the electrolyte concentration is high, or when the relative permittivity of the solvent is large. Thus, it is important to realise that a polarisation of 0.3V with respect to the *pzc* of a polarisable electrode submerged in a concentrated electrolyte generates very strong electric fields – of the order of gigavolts per metre (10^{-9} V·m^{-1}), given that the potential drop occurs over a monolayer of water whose molecule diameter is around 300 pm.

To estimate the Helmholtz capacity from experimental results, it is usual to trace a Parsons-Zobel plot at fixed charge density with on the *y*-axis the inverse of the experimental capacity per unit surface, and on the *x*-axis the inverse of the Gouy-Chapman capacity. It is clear that if equation (5.97) is to be verified, the slope must be unity and the intercept must correspond to the Helmholtz capacity. If the graph is not linear, other phenomena such as specific adsorption of ions have to be taken into account. Furthermore, if a straight line is obtained, this method can be useful for measuring the geometric surface area of the electrode, as in the case of mercury drops, for which measuring the exact area can prove difficult. In this case, we plot the inverse of the experimental capacity as a function of the inverse of the Gouy-Chapman capacity, and the slope gives the surface area.

For the case of KF and the values shown in Figure 5.13, the Helmholtz capacity obtained at the *pzc* is 29.5 μF·cm^{-2}. If the whole capacity were due to the solvent layer, this value would allow us to estimate the relative permittivity of the first water monolayer. Knowing that $C_H = \varepsilon_0 \varepsilon_r / d$, and taking as a first approximation that the thickness of the monolayer is 300 pm, we get a value of 10 for the relative permittivity. Taking up again the arguments of Figure 1.7, this value indicates that the molecules of water at the surface have libration modes.

We have assumed so far that the solvent layer is ion-free, as it is expected for

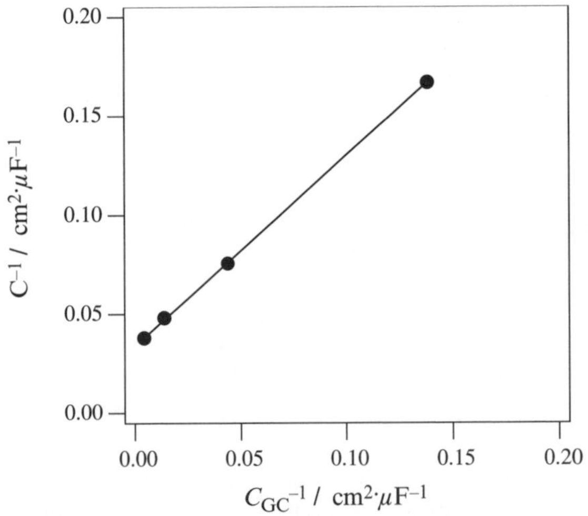

Fig. 5. 17 Parsons-Zobel plot at the *pzc* from the data in Figure 5.13.

the case of KF on mercury. Electrocapillary data such as those shown in Figure 5.7 suggest, however, that some anions, such as chloride, bromide or iodide, do specifically adsorb on mercury. In this case, we have to consider that the anions enter the solvent layer, loosing some hydration water molecules and interacting directly with the metal. The interfacial layer containing solvent molecules and specifically adsorbed ions is often called the ***Stern layer***. Because cations tend to be more strongly hydrated, specific adsorption of cations is more seldom. If some ions are specifically adsorbed, it is clear that the capacity C_H will depend on the presence of ions in the Stern layer.

In fact, C_H varies not only with the charge density on the metal, but also with the nature of the metal, and in the case of solid electrodes with the crystallographic face. This shows that the solvent model is incomplete, and that the metal itself plays a role. The Jellium model (see Figure 1.10) shows that the spillover of the electrons from the metal depends on the charge density. If the interaction between the electrons from the metal and the solvent are weak, we can then consider that the Helmholtz capacity is itself made up of two capacitors in series, one for the electronic spillover, and the other for the solvent monolayer.

$$\frac{1}{C_H} = \frac{1}{C_{metal}} + \frac{1}{C_{solution}} \tag{5.98}$$

The smallest possible value for $C_{solution}$ corresponds to the dielectric saturation of the solvent (molecules of water 'frozen') being $\varepsilon_r = 2$ which is about $6\mu F \cdot cm^{-2}$. This brief calculation shows that the contribution of the metal is more or less of the same order of magnitude as that of the solvent.

Potential of zero charge

The potential of zero charge of a metal is linked in a certain way to the work function of the metal, as illustrated in Figure 5.18.

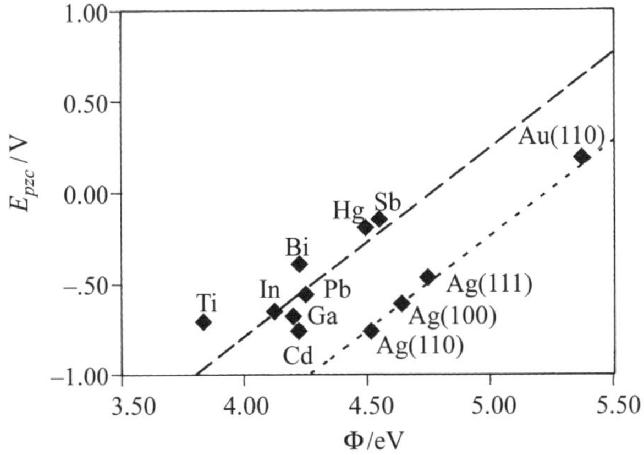

Fig. 5.18 Potentiel of zero charge versus the work function (Adapted from W. Schmickler, *Interfacial Electrochemistry*, Oxford University Press, 1985).

Effectively, for a metal M, the *pzc* measured with respect to, say, a standard hydrogen reference electrode with the cell

$$\mathrm{Cu^I} \mid \mathrm{Pt} \mid \mathrm{H^+},\ \tfrac{1}{2}\mathrm{H_2}... \parallel \text{Solution} \mid \mathrm{M} \mid \mathrm{Cu^{II}}$$

can be written as

$$\left[E_{pzc}\right]_{\mathrm{SHE}} = \left(\phi^{\mathrm{Cu^{II}}} - \phi^{\mathrm{Cu^{I}}}\right) = \left(\phi^{\mathrm{Cu^{II}}} - \phi^{\mathrm{M}}\right) + \left(\phi^{\mathrm{M}} - \phi^{\mathrm{S}}\right)_{pzc} + \left(\phi^{\mathrm{S_R}} - \phi^{\mathrm{Pt}}\right) + \left(\phi^{\mathrm{Pt}} - \phi^{\mathrm{Cu^{I}}}\right) \tag{5.99}$$

This equation implies that the potential difference across the liquid junction between the solution S and the acid solution of the reference electrode S_R is negligible. By using the definition of the electrochemical potential given by equation (1.60), the first term and the last term can be combined as follows

$$\left(\phi^{\mathrm{Cu^{II}}} - \phi^{\mathrm{M}}\right) + \left(\phi^{\mathrm{Pt}} - \phi^{\mathrm{Cu^{I}}}\right) = \left(\mu_{e^-}^{\mathrm{Pt}} - \mu_{e^-}^{\mathrm{M}}\right)/F \tag{5.100}$$

The term relative to the metal | solution interface can be written using equation (1.59)

$$\phi^{\mathrm{M}} - \phi^{\mathrm{S}} = \Delta_{\mathrm{S}}^{\mathrm{M}} g(\mathrm{dip}) + \Delta_{\mathrm{S}}^{\mathrm{M}} g(\mathrm{ion}) \tag{5.101}$$

and reduces to $\Delta_{\mathrm{S}}^{\mathrm{M}} g(\mathrm{dip})$ at the *pzc,* as by definition there is no potential distribution associated to ionic species at the *pzc*.

Finally, the term related to the reference electrode can be written using equation (2.37)

$$\left(\phi^{\mathrm{S_R}} - \phi^{\mathrm{Pt}}\right) = -\left[E^{\ominus}_{\mathrm{H^+/\frac{1}{2}H_2}}\right]_{\mathrm{abs}} - \left(\mu_{e^-}^{\mathrm{Pt}}/F\right) + \chi^{\mathrm{S}} \tag{5.102}$$

By substituting equations (5.100)-(5.102) into equation (5.98), we get

$$\begin{aligned}\left[E_{pzc}\right]_{\mathrm{SHE}} &= -\left(\mu_{e^-}^{\mathrm{M}}/F\right) - \left[E^{\ominus}_{\mathrm{H^+/\frac{1}{2}H_2}}\right]_{\mathrm{abs}} + \Delta_{\mathrm{S}}^{\mathrm{M}} g(\mathrm{dip}) + \chi^{\mathrm{S}} \\ &= \Phi^{\mathrm{M}}/F - \left[E^{\ominus}_{\mathrm{H^+/\frac{1}{2}H_2}}\right]_{\mathrm{abs}} + \Delta_{\mathrm{S}}^{\mathrm{M}} g(\mathrm{dip}) + \chi^{\mathrm{S}} - \chi^{\mathrm{N}}\end{aligned} \tag{5.103}$$

where Φ^{M} and χ^{M} are respectively, the work function and the surface potential of pure mercury.

The dipolar contribution $\Delta_{\mathrm{S}}^{\mathrm{M}} g(\mathrm{dip})$ to the Galvani potential difference can be split into a metallic dipolar contribution (Jellium type) and a solvent contribution. On the metal side, the electrons overspill less in the presence of water than at the metal | vacuum interface. In the case of mercury, the difference $g^{\mathrm{Hg}}(\mathrm{dip}) - \chi^{\mathrm{Hg}}$ is estimated to be of the order of 0.32 V. On the water side, the molecules of the mercury | solution interface are less aligned than at the air | water interface and the difference $g^{\mathrm{S}}(\mathrm{dip}) - \chi^{\mathrm{S}}$ is roughly –0.06 V. It has been proposed to use equation (5.103) to determine the absolute standard potential of the standard hydrogen electrode. The results thus obtained vary between 4.3 V and 4.8 V, and correspond quite well to the value calculated in §2.1.5 using thermodynamic data on hydrogen. It must be noted however, that the two approaches are not completely independent, and it is therefore not surprising to find the same values.

Streaming electrode

For a liquid metallic electrode such as mercury, a reasonably simple method for measuring the potential of zero charge consists of flowing the mercury into the electrolyte solution whilst spraying the jet. In this way, the surface of the electrode is constantly renewed, and therefore carries no excess charges. By doing this, and by measuring with a voltmeter the potential diffence between the mercury electrode and a reference electrode placed in the solution, the potential of zero charge can be measured directly.

5.4.2 Electrolyte | electrolyte interface

When two immiscible electrolytes are put in contact with each other, it is possible to polarise the interface. In the absence of specific adsorption, the excess charges are distributed in the two diffuse layers back to back. As a first approximation, we can apply the Gouy-Chapman theory.

The Gouy–Chapman theory

Consider an interface between an organic solution (o) and an aqueous solution (w) such that the interface is defined by the coordinate $x = 0$.

Using the same hypotheses as in §5.4.1, in the case of two 1:1 electrolytes in contact, the Poisson-Boltzmann equation to be solved is still equation (5.81) which reads

$$\frac{\partial^2\phi(x)}{\partial x^2} = \frac{2eN^\infty}{\varepsilon_0\varepsilon_\mathrm{r}}\sinh\left[\frac{e[\phi(x)-\phi^\infty]}{kT}\right] \tag{5.104}$$

The boundary conditions for $x \to -\infty$ are

$$\phi(x)\to\phi^\mathrm{o} \qquad \text{and} \qquad \frac{\partial\phi(x)}{\partial x}\to 0 \tag{5.105}$$

and for $x \to \infty$

$$\phi(x)\to\phi^\mathrm{w} \qquad \text{and} \qquad \frac{\partial\phi(x)}{\partial x}\to 0 \tag{5.106}$$

By integrating in the organic phase between x and $-\infty$, we get

$$\frac{\partial\phi(x)}{\partial x} = \left[\frac{8N^\mathrm{o}kT}{\varepsilon_0\varepsilon_\mathrm{o}}\right]^{1/2}\sinh\left[\frac{e}{2kT}(\phi(x)-\phi^\mathrm{o})\right] \tag{5.107}$$

and in the aqueous phase between x and ∞

$$\frac{\partial\phi(x)}{\partial x} = -\left[\frac{8N^\mathrm{w}kT}{\varepsilon_0\varepsilon_\mathrm{w}}\right]^{1/2}\sinh\left[\frac{e}{2kT}(\phi(x)-\phi^\mathrm{w})\right] \tag{5.108}$$

The choice of signs is dictated by the fact that if $(\phi^\mathrm{w} - \phi^\mathrm{o}) \geqslant 0$, then $(\phi(x) - \phi^\mathrm{o}) \geqslant 0$ and $(\phi(x) - \phi^\mathrm{w}) \leqslant 0$ and therefore the slope is positive.

The surface excess charge density in the organic solution is

$$\sigma^{\mathrm{o}} = \int_{-\infty}^{0} \rho(x)\mathrm{d}x = -\int_{-\infty}^{0} \varepsilon_0\varepsilon_{\mathrm{o}}\left(\frac{\partial^2\phi(x)}{\partial x^2}\right)\mathrm{d}x = -\varepsilon_0\varepsilon_{\mathrm{o}}\left(\frac{\partial\phi(x)}{\partial x}\right)_{x=0} \tag{5.109}$$

and in the aqueous solution

$$\sigma^{\mathrm{w}} = \int_{0}^{\infty} \rho(x)\mathrm{d}x = -\int_{0}^{\infty} \varepsilon_0\varepsilon_{\mathrm{w}}\left(\frac{\partial^2\phi(x)}{\partial x^2}\right)\mathrm{d}x = \varepsilon_0\varepsilon_{\mathrm{w}}\left(\frac{\partial\phi(x)}{\partial x}\right)_{x=0} \tag{5.110}$$

The electroneutrality of the interphase then gives us

$$\sinh\left[\frac{e}{2kT}(\phi(0)-\phi^{\mathrm{w}})\right] = -\sqrt{\frac{\varepsilon_{\mathrm{o}}N_{\mathrm{o}}}{\varepsilon_{\mathrm{w}}N_{\mathrm{w}}}}\sinh\left[\frac{e}{2kT}(\phi(0)-\phi^{\mathrm{o}})\right] \tag{5.111}$$

From this equation, we can calculate the potential difference in one phase in terms of the total potential difference

$$\tanh\left[\frac{e}{2kT}(\phi(0)-\phi^{\mathrm{o}})\right] = \frac{\sqrt{\dfrac{\varepsilon_{\mathrm{o}}N_{\mathrm{o}}}{\varepsilon_{\mathrm{w}}N_{\mathrm{w}}}}\sinh\left[\dfrac{e}{2kT}(\phi^{\mathrm{w}}-\phi^{\mathrm{o}})\right]}{1+\sqrt{\dfrac{\varepsilon_{\mathrm{o}}N_{\mathrm{o}}}{\varepsilon_{\mathrm{w}}N_{\mathrm{w}}}}\sinh\left[\dfrac{e}{2kT}(\phi^{\mathrm{w}}-\phi^{\mathrm{o}})\right]} \tag{5.112}$$

In the case of a water | 1,2-dichloroethane interface for which the relative permittivities ε_{w} and ε_{o} are respectively 78 and 10, Figure 5.19 shows that the applied potential difference is spread between the two phases according to the ratio of the concentrations. To find ϕ as a function of x, we can integrate equations (5.105) and (5.106) which gives for each phase

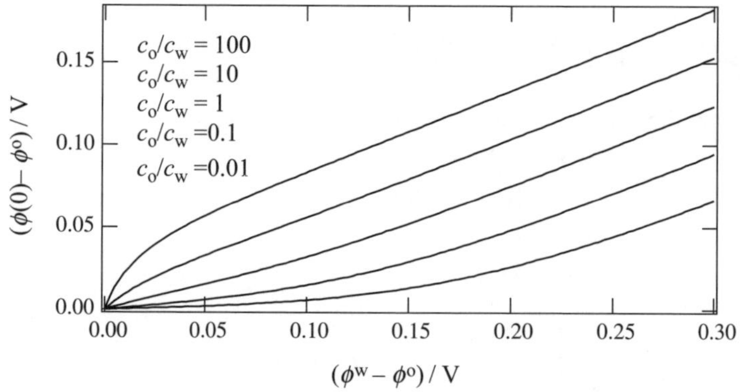

Fig. 5. 19 Potential drop in the organic phase as a function of the applied potential difference for different ratios of concentration. The top curve corresponds to a ratio of 100 and the bottomone to a ratio of 0.01.

$$\tanh\left[\frac{e}{4kT}(\phi(x)-\phi^{\infty})\right] = e^{-\kappa x}\tanh\left[\frac{e}{4kT}(\phi(0)-\phi^{\infty})\right] \tag{5.113}$$

The potential distribution thus obtained is shown in Figure 5.20

In a general manner, the Gouy–Chapman theory predicts the form of capacity curves, as shown in Figure 5.21. Contrary to metal electrodes, the capacities measured

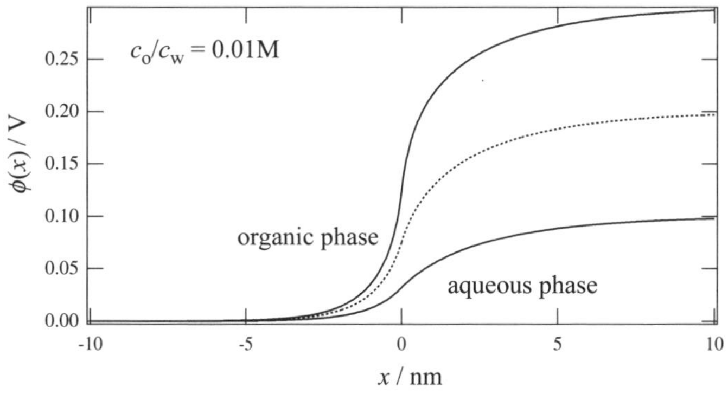

Fig. 5.20 Potential distribution at a 1,2-dichloroethane | water interface for different interfacial polarisations: 0.1, 0.2 and 0.3 V.

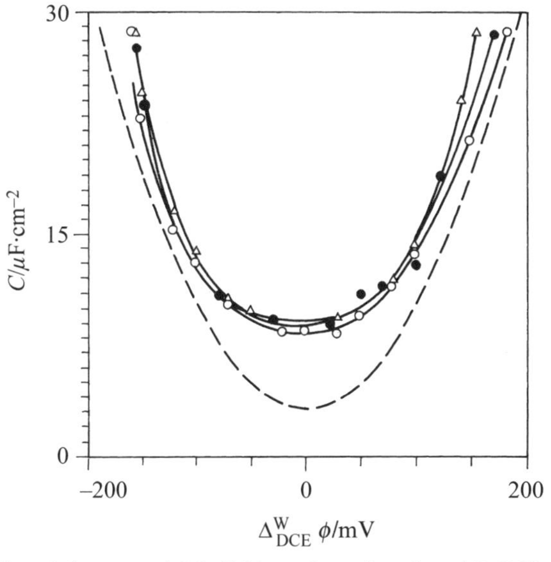

Fig. 5.21 Capacity of the water | 1,2-dichloroethane interface (Δ) KCl, (O)LiCl, (●) HCl. Ag | AgCl | 0.01M XCl | TBATPB 0.01M | TBACl 0.01M | AgCl | Ag. The dotted curve is the Gouy-Chapman capacity [C. Melo Pereira et al, *J. Chem. Soc. Faraday Trans.*,90 (1994) 143, with permission from the Royal Society of Chemistry).

experimentally at liquid | liquid interfaces are higher than those predicted by the Gouy-Chapman theory. This can be explained by considering either an inter-penetration of the diffuse layers (coulombic attraction) or a certain roughness of the interface due to the capillary waves present at all liquid interfaces.

Potential of zero charge

The potential of zero charge at liquid | liquid interfaces is a parameter which is easily measured with a streaming system similar to the one used for the mercury electrode. The Galvani potential difference at the *pzc* is simply $\Delta_o^w g$(dip). This quantity is probably small if the two solvents have little dipole-dipole interactions, and the Galvani potential difference at the *pzc* can be taken in a first approximation as the origin of this potential scale.

5.4.3 *p*|*n* junction

Intrinsic and extrinsic semiconductors

We saw in §1.2 that a solid could be considered as a macromolecule containing N atoms whose molecular orbitals of close energy levels form bands. For diamond ($1s^2$, $2s^2$, $2p^2$), the carbon atoms form four equivalent bonds with hybrid orbitals sp^3. Thus, instead of making a full $2s$ band and a partially-occupied $2p$ band which one would get if there was no hybridisation of the $2s$ and $2p$ orbitals, we have a $2s$-$2p$ bonding band containing $4N$ electrons, therefore completely filled, and also a completely empty anti-bonding band also able to contain $4N$ electrons. The energy difference between these two bands is about 5 eV, which is about 500 kJ·mol^{-1}.

For silicon ($1s^2$, $2s^2$, $2p^6$ $3s^2$, $3p^2$) and germanium, the crystal also has the tetravalent structure of diamond, but the energy differences between the bonding and anti-bonding bands are less (1.2 and 0.7 eV, respectively). Thus at ordinary temperatures, the Fermi-Dirac distribution function shows that certain electrons have access to energy levels in the anti-bonding band. When an electron goes from the

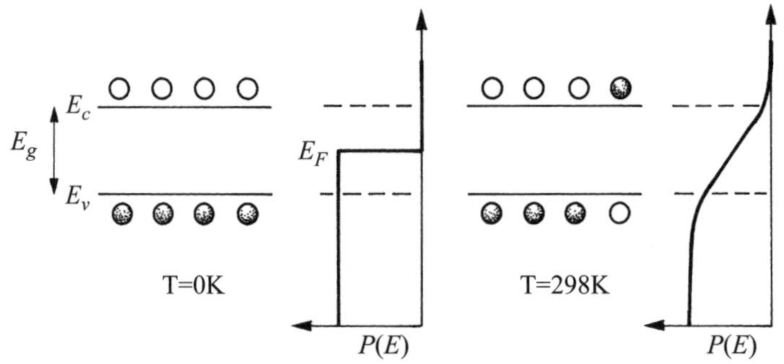

Fig. 5.22 Electron distribution for an intrinsic semiconductor.

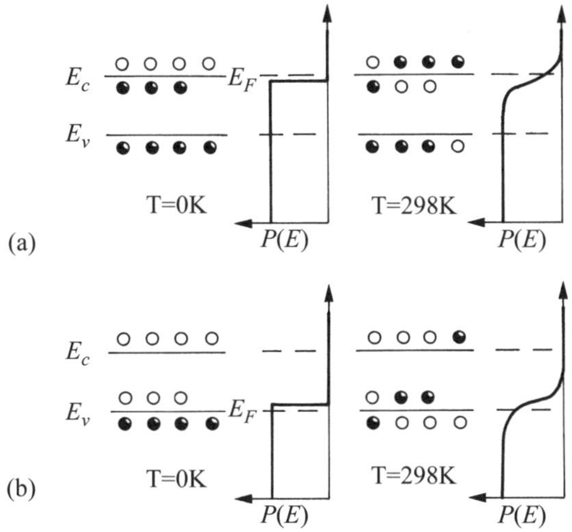

Fig. 5.23 Electron distribution in extrinsic semiconductors. (a) n-type semiconductor, (b) p-type semiconductor.

valence band to the conduction band, it leaves an unoccupied level called an ***electron hole***. This hole can migrate in the valence band and can therefore be considered as a positive charge carrier. A semiconductor containing electron-hole pairs is called ***intrinsic***. The number of charge carriers is of the order of 10^{10} to 10^{16} per cm^3 compared to 10^{22} for a metal (number of atoms/cm^3 of Ge , Si or GaAs $\approx 5 \times 10^{22}$).

In the case of an intrinsic semiconductor, the Fermi level is found in the middle of the band gap.

It is possible to modify the number of electrons and holes by adding an electron donor or acceptor. The semiconductor thus obtained is called ***doped or extrinsic***. It is known as ***n*-type** when it contains donors and ***p*-type** when it contains acceptors.

For silicon that is tetravalent, phosphorus (pentavalent) can serve as a donor, and boron (trivalent) can serve as an acceptor. The doping densities are of the order of 10^{15} to 10^{17} atoms/cm^3. The Fermi level of an *n* doped semiconductor is near the conduction band, and of a *p* doped semiconductor near the valence band.

Junction between a *p*-type semiconductor and a *n*-type semiconductor

When two doped semiconductors, one *n*-type, the other *p*-type, are put in contact, the electrons flow from the *n*-type semiconductor until the Fermi levels are equal in the two phases. At equilibrium, the *n*-type semiconductor is positively charged, whilst the *p*-type is negatively charged. The excess charge in the semiconductor is on the surface, but distributed in a surface layer with a certain thickness. The subsequent electric field, which forms in this region, leads to a curving of the valence and conduction bands. The curve of the bands rises at a positive charge and dips when the semiconductors are negatively charged, as shown in Figure 5.24.

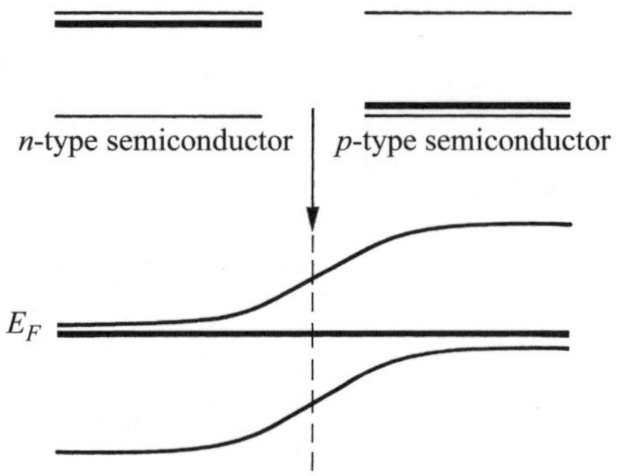

Fig. 5.24 Curvature of the bands at the interface with equalisation of the Fermi levels.

After the flux of electrons from the *n*-type to the *p*-type until the equalisation of the Fermi levels, the excess charges remaining on the *n*-side are those of the fixed positively-charged donor ions, and on the *p*-side, those of the negatively charged acceptor ions. A junction then happens between two depletion layers containing equal but opposite fixed charges (see Figure 5.25). The effective width and the capacity of this layer can be easily calculated by solving Poisson's equation (1.37), expressed in a one-dimensional axis corresponding to the normal of the interface.

$$\frac{\partial^2 \phi(x)}{\partial x^2} = -\frac{\rho(x)}{\varepsilon_0 \varepsilon_r} \tag{5.114}$$

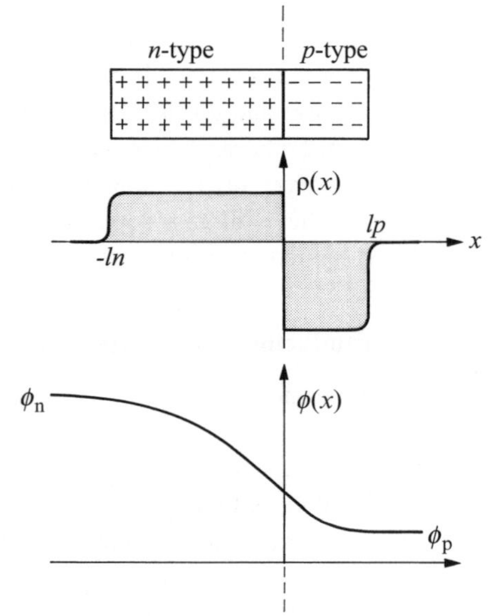

Fig. 5.25 'Very schematic' diagram of charge and potential distribution at a $p \mid n$ junction.

If the volumic density is constant (homogeneous doping), we can integrate directly on the n and p sides between x and the limits $-l_n$ & l_p respectively (see Figure 5.25).

$$\varepsilon_0\varepsilon_{\rm r}\left(\frac{\partial\phi}{\partial x}\right)_{-l_n<x\leq 0} = -\rho_{\rm d}\left(x+l_n\right) \tag{5.115}$$

and

$$\varepsilon_0\varepsilon_{\rm r}\left(\frac{\partial\phi}{\partial x}\right)_{0\geq x>l_p} = -\rho_{\rm a}\left(x-l_p\right) \tag{5.116}$$

Outside of the depletion layer, the field is equal to zero. It should be stressed that the notion of layer width is obviously a simplified view in as much as the volumic charge density $\rho(x)$ is not a true step function.

The electroneutrality condition at the interface gives

$$-\rho_{\rm d}\,l_n = \rho_{\rm a}\,l_p \qquad \text{or} \qquad N_{\rm d}\,l_n = N_{\rm a}\,l_p \tag{5.117}$$

with N the number of doping atoms per unit volume.

By integrating equation (5.115) & (5.116) between x=0 and x, we obtain the potential distribution in the adjacent layers:

$$\varepsilon_0\varepsilon_{\rm r}\phi(x)_{-l_n<x\leq 0} = -\rho_{\rm d}\left(x^2/2+l_nx\right)+\phi(0) \tag{5.118}$$

$$\varepsilon_0\varepsilon_{\rm r}\phi(x)_{0\geq x>l_p} = -\rho_{\rm a}\left(x^2/2-l_px\right)+\phi(0) \tag{5.119}$$

The potential drop on the n-side and p-side are then respectively equal to

$$\varepsilon_0\varepsilon_{\rm r}(\phi_n-\phi_0) = \rho_{\rm d}\,l_n^2/2 \tag{5.120}$$

and

$$\varepsilon_0\varepsilon_{\rm r}(\phi_0-\phi_p) = -\rho_{\rm a}\,l_p^2/2 \tag{5.121}$$

The total potential across the junction then simply reads

$$\Delta_p^n\phi = \phi_n-\phi_p = \left[\rho_{\rm d}l_n^2-\rho_{\rm a}l_p^2\right]/2\varepsilon_0\varepsilon_{\rm r} = e\left[N_{\rm d}l_n^2+N_{\rm a}l_p^2\right]/2\varepsilon_0\varepsilon_{\rm r} \tag{5.122}$$

The surface charge being

$$Q = e\,N_{\rm d}\,l_n \tag{5.123}$$

equation (5.122) can be written as

$$\Delta_p^n\phi = \left[\frac{Q^2}{2\varepsilon_0\varepsilon_{\rm r}e}\right]\left[\frac{N_{\rm a}+N_{\rm d}}{N_{\rm a}N_{\rm d}}\right] \tag{5.124}$$

from which we can express the capacity of the junction by

$$C = \frac{{\rm d}Q}{{\rm d}\Delta_p^n\phi} = \frac{1}{2}\sqrt{\left[\frac{2\varepsilon_0\varepsilon_{\rm r}e}{\Delta_p^n\phi}\right]\left[\frac{N_{\rm a}N_{\rm d}}{N_{\rm a}+N_{\rm d}}\right]} \tag{5.125}$$

We therefore note that the $p \mid n$ junction can be used as a capacitor whose capacity varies according to $\Delta_p^n\phi^{-1/2}$. This is the working principle of the variable-capacity diode or varactor, used in modulation circuits and frequency synchronisation.

We can calculate l_n from the equation (5.122) taking into account the electroneutrality condition (5.117)

$$N_{\mathrm{a}}\left(\frac{N_{\mathrm{d}}l_n}{N_{\mathrm{a}}}\right)^2 + N_{\mathrm{d}}l_n^2 = \frac{N_{\mathrm{d}}}{N_{\mathrm{a}}}\left(N_{\mathrm{d}}+N_{\mathrm{a}}\right)l_n^2 = \frac{2\varepsilon_0\varepsilon_{\mathrm{r}}\Delta\phi}{e} \tag{5.126}$$

from which we obtained

$$l_n = \sqrt{\frac{2\varepsilon_0\varepsilon_{\mathrm{r}}\Delta\phi N_{\mathrm{a}}}{eN_{\mathrm{d}}\left(N_{\mathrm{d}}+N_{\mathrm{a}}\right)}} \quad \text{and} \quad l_p = \frac{N_{\mathrm{d}}l_n}{N_{\mathrm{a}}} = \sqrt{\frac{2\varepsilon_0\varepsilon_{\mathrm{r}}\Delta_p^n\phi N_{\mathrm{d}}}{eN_{\mathrm{a}}\left(N_{\mathrm{d}}+N_{\mathrm{a}}\right)}} \tag{5.127}$$

The width of the $p \mid n$ junction is therefore given by

$$W = l_p + l_n = \left[\frac{2\varepsilon_0\varepsilon_{\mathrm{r}}\Delta_p^n\phi\,(N_{\mathrm{a}}+N_{\mathrm{d}})}{e\,N_{\mathrm{a}}N_{\mathrm{d}}}\right]^{1/2} \tag{5.128}$$

EXAMPLE

Knowing that generally we have $N_{\mathrm{a}} = N_{\mathrm{d}} = 10^{15}\ \mathrm{cm}^{-3}$ and that $\varepsilon_{\mathrm{r}} = 11.9$ for silicon, let's calculate the width of the $p \mid n$ junction.

If $\Delta\phi$ is 1 volt, we have :

$$W = \sqrt{\frac{2\,11.9\,8.85\,10^{-12}\,1\,2\,10^{21}}{1.6\,10^{-19}\,10^{42}}} = 1.6\ \mu\mathrm{m}$$

This example shows that the width of the charge distribution in doped semiconductors is of the order of micrometres, compared to a few nanometres for electrolytic solutions and a few picometres for metallic conductors.

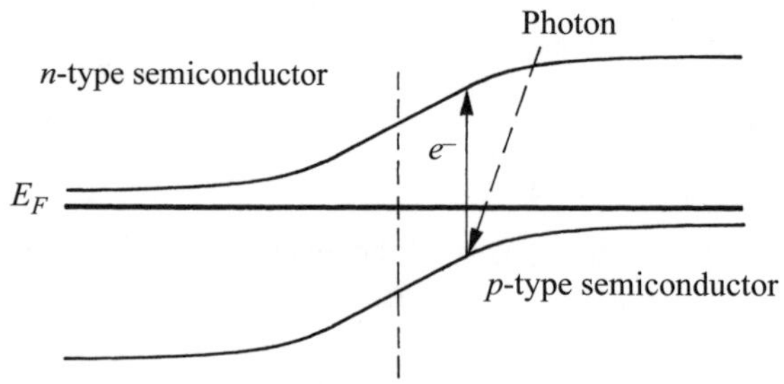

Fig 5.26 Photogeneration of an electron-hole pair at a $p \mid n$ junction.

One of the applications of the $p \mid n$ junction is the conversion of solar energy. Under illumination, the absorption of photons whose energy corresponds to the width of the band gap results in the creation of electron/hole pairs. These can either recombine or separate under the effect of the interfacial electric field, this separation causing the generation of an electric current.

5.4.4 Semiconductor | electrolyte interface.

In order to understand the structure of the semiconductor | electrolyte interface, we shall first study different types of charged regions in a semiconductor with capacitors of the type semiconductor | insulator | metal, where all charge transfers can be ignored.

Flat band case

Consider a semiconductor | insulator | metal capacitor such as Si | SiO_2 | M, used in field-effect transistors. At equilibrium, neither of the two phases is charged. The absence of charged regions in the semiconductor means an absence of electric field. The energy levels are therefore constant and this situation is known as ***flat bands.***

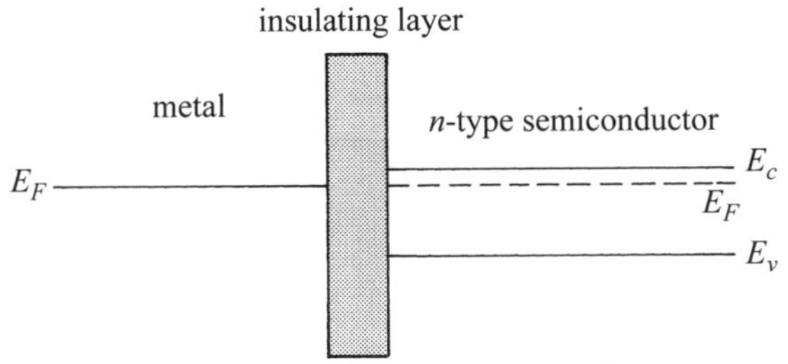

Fig. 5.27 Non-charged metal | insulator | semiconductor interface. The Fermi levels are arbitrarily taken as equal.

Accumulation, depletion and inversion layers

Consider, for example, an n-type semiconductor in the capacitor shown in Figure 5.27. If we charge the metal positively by lowering its Fermi level with respect to that of the semiconductor, we will create a region of negative charges in the semiconductor. To do this, we transport electrons, which are the majority carriers, from inside the crystal to the surface, where they will accumulate.

The ***accumulation layer*** is the region of the interphase where there is an accumulation of majority carriers. This accumulation of charges induces a curving of the bands downwards for a n-type semiconductor, and upwards for a p-type. Also note that there cannot be any charge transfer across the insulator and consequently the Fermi level in the bulk of the semiconductor remains constant.

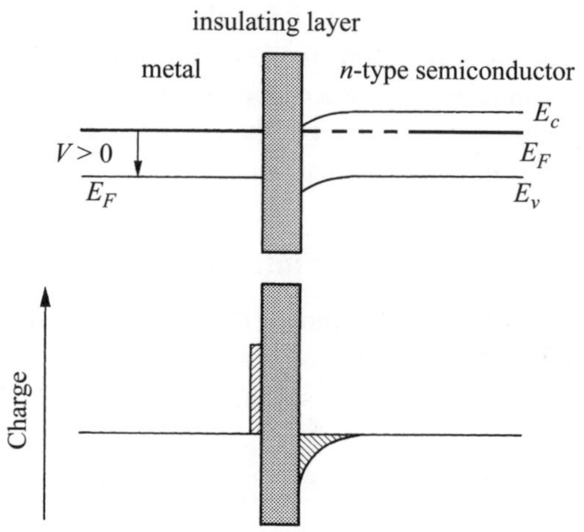

Fig. 5.28 Accumulation layer in a negatively charged *n*-type semiconductor.

Still looking at an *n*-type semiconductor, consider what happens if we charge the metal negatively by increasing its Fermi level with respect to that of the semiconductor. As previously, we will create in the semiconductor a positively charged region. The majority carriers being the electrons, they will move from the surface towards the centre of the crystal, leaving behind them an almost homogeneous region of donors. The ***depletion layer*** is the region of the interphase impoverished in majority carriers. This depletion causes a curving of the bands upwards for a *n*-type semiconductor, and downwards for a *p*-type semiconductor.

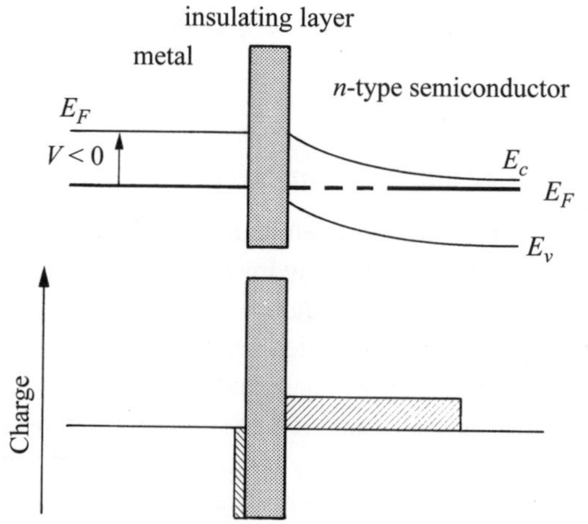

Fig. 5.29 Depletion layer in a positively-charged *n*-type semiconductor.

When there is a very large depletion, the minority carriers, in this case the holes, can contribute to the positive charge of the semiconductor, thus creating an ***inversion layer*** in which the accumulation of holes locally transforms the n-type semiconductor into a p type one. In this case, the Fermi level is near to the energy of the valence band.

Potential distribution in the semiconductor | electrolyte interphase

We have already studied the potential distribution in metal | electrolyte interfaces and shown that this potential is spread mainly in the electrolyte between a compact Helmholtz or Stern layer and a diffuse Gouy-Chapman layer.

In the case of semiconductor electrodes, the potential distribution in the curved region of the bands also needs to be considered. In the polarisation regions where an accumulation layer forms, the electrode behaves rather like a metal and the potential difference between the semiconductor and the solution is distributed mainly in the electrolyte. In the polarisation regions where a depletion layer forms, any variation in electrode potential is distributed mainly in the semiconductor. The result is that the position of the edges of the surface bands remains fixed when the electrode potential is varied. The edges of the surface bands are then said to be pinned.

5.4.5 Conclusion

In a general manner, we can conclude that if a phase is charged, the charge is distributed at the surface. The thickness of this charged region varies from a few picometres for electronic conductors, to a few nanometres for ionic conductors such as electrolyte solutions where the ions are mobile, and to a few micrometres for doped semiconductors where the charged sites are fixed in the solid matrix.

Additionally, the potential drop happens mainly in the phase with the lowest dielectric constant and/or the smallest concentration of charge carriers (ions or electrons).

It is good to remember that the bulk of each phase is globally electro-neutral.

5.5 STRUCTURE OF ELECTROCHEMICAL INTERFACES

5.5.1 Monocrystal | electrolyte interface

One of the major advances in electrochemistry at the end of the 20th century was the study of monocrystalline electrodes, where only one of the crystal faces is put in contact with the solution. The simplest way of making such an electrode is to melt the end of a wire and let it cool in an inert atmosphere. The ball of molten metal at the end of the wire crystallises to form a monocrystal with several faces. Note in passing that the surface of an almost spherical monocrystal behaves like that of a polycrystalline metal. After having determined the axes of symmetry of the monocrystal obtained, it can be trimmed and polished in such a way as to expose a particular crystallographic face. As shown below, the crystallographic faces have different topographies.

It is perhaps useful to recall the Miller principle of indices for crystals. From the three axes *x, y,* and *z*, we define the crystallographic planes according to the Miller indices (*h,k,l*) obtained as follows:

- Identify the intercept of the plane on the *x, y* and *z* axes as a function of the units of the lattice *a*, *b* and *c*
- Take the inverse of these figures
- Multiply them by the same factor in order to get only integer numbers
- If the intercept is negative, put a bar over the index

Example

Let's define the Miller indices for these three planes

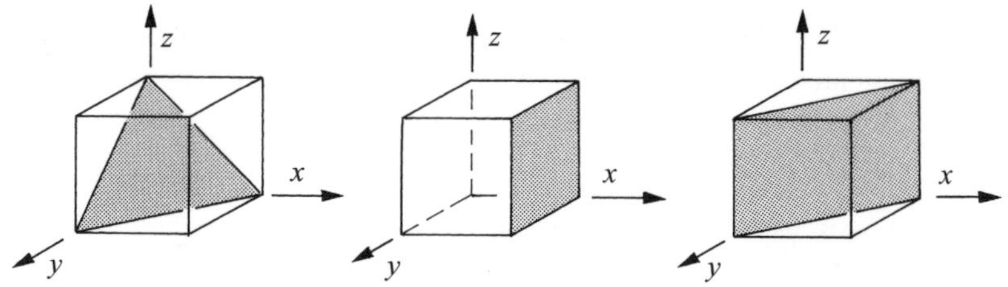

The first has intercepts : 1, 1, 1 and it is therefore the plane (111)
The second has intercepts : 1, ∞, ∞ and it is therefore the plane (100)
The third has intercepts : 1, 1, ∞ and it is therefore the plane (110)

For platinum and gold that crystallise in a cubic face-centred system, the various planes have therefore different surface densities of atoms, as shown in Figures 5.30 and 5.31.

In order to put only one crystal face in contact with the solution, a common method is to bring the face close to the solution until a meniscus forms. This technique

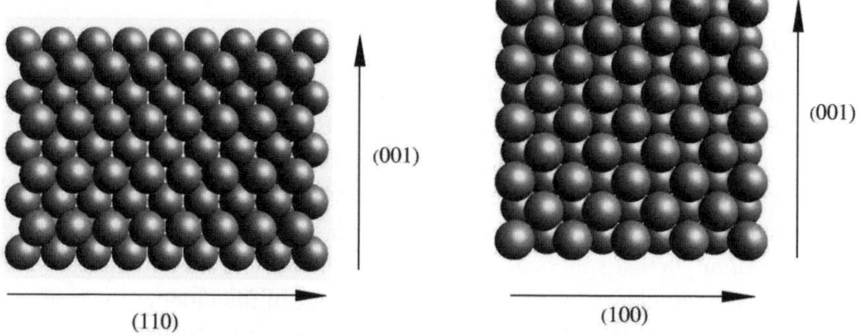

Fig. 5.30 Surface of a monocrystal (110) and (100).

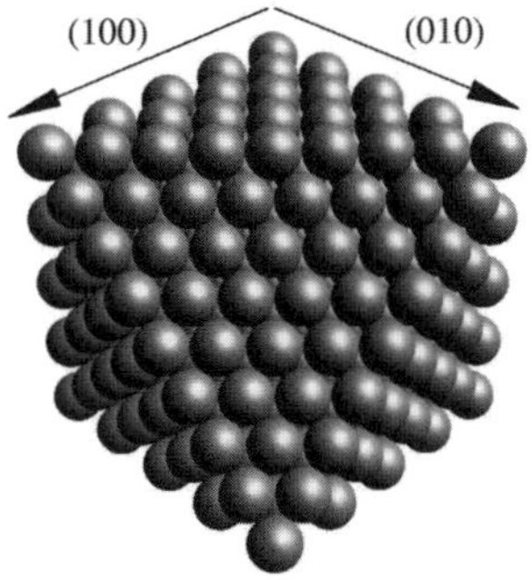

Fig. 5.31 Surface of a monocrystal (111).

allows the measurement of the capacity by cyclic voltammetry as shown in Figure 5.32 for a platinum (111) electrode in sulphuric acid.

The measuring principle is to apply, between the monocrystalline working electrode and a reference electrode in solution, a potential ramp with a sweep rate ν such that the potential of the working electrode is

$$E = E_i + \nu t \tag{5.129}$$

where E_i is an initial value of the electrode potential. Thus, in the absence of a faradaic reaction, the current measured is a capacitive current given by

$$I = \frac{dq}{dt} = C\frac{dE}{dt} = \nu C \tag{5.130}$$

where C is the capacity of the electrode. When the sweep rate is rather slow, the measured capacity is representative of surface phenomena such as adsorption, and the contribution of the diffuse layer, such as illustrated in Figure 5.13 is negligible. This electrochemical method of studying surfaces is called ***cyclic voltammetry*** (developed in more detail in chapter 10 for the study of faradaic reactions) and is a very sensitive method to study adsorbed monolayers using a rather simple equipment.

In the example in Figure 5.32 relating to a platinum (111) electrode in a solution of sulfuric acid, the current below 0.3 V corresponds to the states of adsorption of the hydrogen atoms on the platinum. The charge obtained by integration corresponds to one electron per atom of platinum at the surface. The current measured between 0.3 and 0.5 V is not observed in perchloric acid and the signal obtained is no doubt related to the adsorption of the anion HSO_4^-. Above 0.5 V, there is adsorption of Pt-OH hydroxides and then at even more positive potentials, formation of an oxide layer.

Recent developments of scanning microscopies (**STM: *Scanning tunnelling microscopy***, **AFM: *Atomic Force Microscopy***, etc.) have allowed the observation of the surface of the electrode *in situ* with atomic resolution. The basic principle of an STM is to scan the surface of the electrode with a point sharpened to the atomic level by etching, and to measure the current obtained by tunnelling when the distance

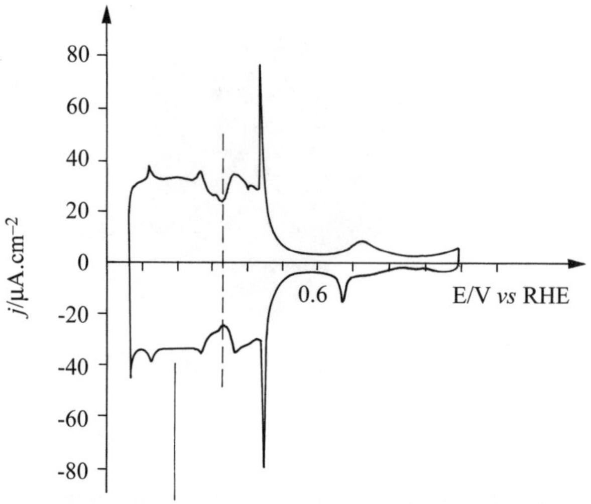

Fig. 5.32 Cyclic voltammogram of a Pt(111) electrode having just been polished, flame-treated and cooled in air. Solution: 0.5 M H_2SO_4. Sweep rate 50 $mV\cdot s^{-1}$ (J. Clavilier, R. Faure, G. Guinet and R. Durand, *J. Electroanal. Chem.*, 107 (1980) 205, with permission from Elsevier Science).

between the point and the surface is small. The basic principle of AFM is to scan the surface of the electrode with a point mounted on a cantilever and to measure the topography of the surface. These two methods thus allow atomic-resolution cartography that can be combined to electrochemical characterisation.

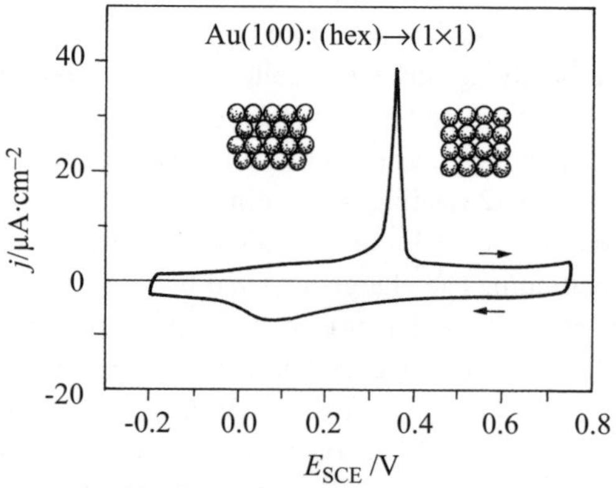

Fig. 5.33 Cyclic current-potential curve for Au(100) in 0.1 M H_2SO_4, starting with a freshly prepared reconstruction surface at –0.2 V vs. SCE. Scan rate: 50 $mV\cdot s^{-1}$. Lifting of the (hex) reconstruction during the positive scan is seen by a pronounced current peak. The subsequent scan in negative direction reflects the electrochemical behaviour of Au(100)-(1×1). [Dakkouri & Kolb, in *Interfacial Electrochemistry*, Marcel Dekker, 1999, with permission from Marcel Dekker].

These techniques of scanning microscopy have been used for example to investigate surface reconstruction phenomena. In fact, the surface atoms experience different forces to those inside the metal, and as for the molecules in Figure 5.1, these atoms have a higher potential energy. The first consequence of this effect is a slight contraction of the distance between the first and second atomic layers, followed by an expansion of the two following layers. This is called the ***surface relaxation effect***. The second consequence is the breaking and formation of covalent bonds between the atoms in the surface layer. These phenomena are observed during ***surface reconstruction***. To minimise their excess potential energy, the surface atoms move in order to create new surface structures. A well-characterised reconstruction is that of gold (111) which, in spite of its high density reconstructs into a $(\sqrt{3} \times 22)$ structure which is characterised by a lateral compression of 4.4% of the surface atoms in one of the three 110 directions such that the 23rd atom of a row is repositioned under the 22nd atom of the next structure below. These surface reconstructions can also be observed by cyclic voltammetry as in the case of gold (100) where the surface reconstructs in the compact hexagonal structure as shown in Figure 5.33.

5.5.2 Electrolyte | electrolyte interface

The electrolyte | electrolyte interface is a molecular interface with a certain dynamic. There are no experimental methods at the moment for studying the structure of this interface. A few preliminary results obtained using neutron reflectivity do not

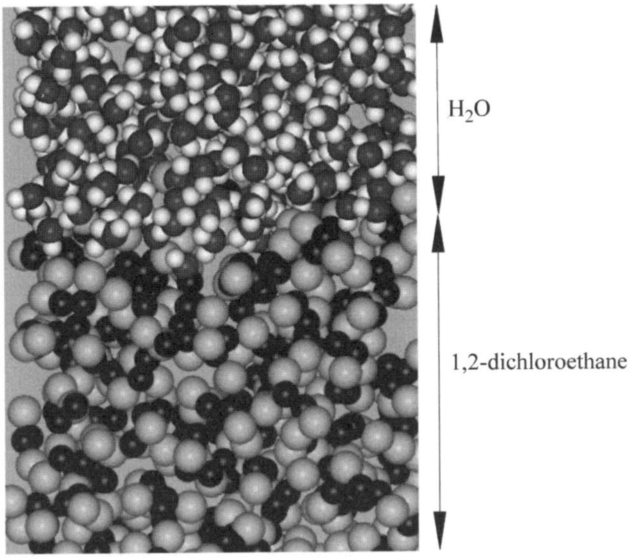

Fig. 5.34 Molecular dynamics image of the water|1,2-dichloroethane interface. (Personal communication from Prof. I. Benjamin, Santa Cruz, USA).

give clear information about the thickness of the interphase layer. The only interfacial images which we have, come from molecular dynamics simulations as shown in Figure 5.34.

Even if molecular dynamics produces beautiful images, we must bear in mind that they are only the visualisation of a model...

CHAPTER 6

ELECTROKINETIC PHENOMENA AND ELECTROCHEMICAL SEPARATION METHODS

6.1 ELECTROKINETIC PHENOMENA

6.1.1 Definitions

We saw in chapter 5 that at the interface between a charged solid phase and an electrolyte solution, there is a volumic distribution of excess charge in solution that can be modelled using the Gouy-Chapman theory. We have so far considered electrolyte solutions next to polarised metals or semi-conductors, but interfaces of interest also include for example:

- the surfaces of metal oxides in solution, such as those illustrated in Figure 6.1, for which the surface charges depend on the pH of the solution
- soft interfaces formed by self-assembled layers e.g. soap bubbles, lipid bio-membranes
- the surface of ion exchange resins, etc.

Electrokinetic phenomena are associated with the movement of an electrolyte solution near a charged solid, and in fact four types can be distinguished.

Electro-osmosis is the phenomenon associated to the movement of an electrolyte solution near a charged interface under the influence of an electric field parallel to the surface. Electro-osmosis has many applications, one of the most recent being electrophoresis in a silica capillary.

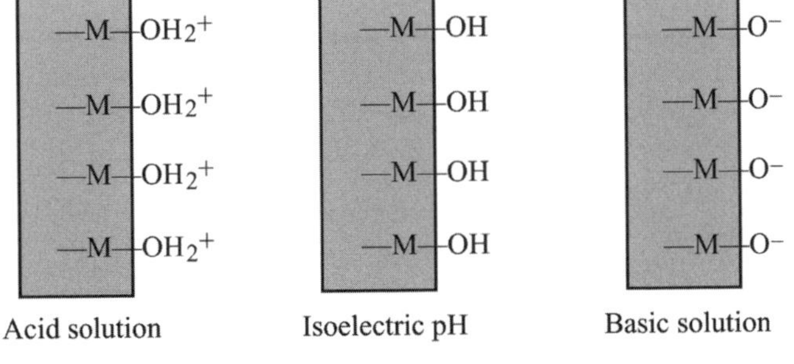

Fig. 6.1 Surface ionisation of metal oxides in aqueous solutions

Electrophoresis of charged colloids is the movement of colloids in a solution under the effect of an electric field, and the term is applied especially to solid particles whose radius is larger than the thickness of the diffuse layer.

Streaming potential corresponds to the forced streaming of a liquid electrolyte near a charged interface, which thus generates a potential difference opposing the flow of the solution. These potentials must be taken into account when considering the passage of electrolytes across porous materials by convection.

Sedimentation potential is the potential generated by sedimentation of charged particles in an electrolyte.

6.1.2 Electro-osmosis

Before we study the phenomena of electro-osmosis, it is perhaps useful to revise the macroscopic definition of viscosity. Consider two parallel plates separated by a distance L and a liquid phase. One plate is fixed and the other moves at a constant velocity $\boldsymbol{v}_y$ as shown in Figure 6.2.

At steady state, the moving plate is subject to a frictional force proportional to the velocity of the plate, but also inversely proportional to the separation distance between the two plates. The steady state velocity profile between the two plates, called the two-dimensional Couette flow, is a linear profile going from zero to $\boldsymbol{v}_y$

$$\frac{F_y}{S} = -\eta\frac{v_y}{L} = -\eta\frac{\partial v_y(x)}{\partial x} \tag{6.1}$$

Thus, the viscosity is defined as the proportionality coefficient between the frictional force which would act on a unit surface plate parallel to the direction of the moving liquid and the velocity of the liquid. The molecular explanation of viscosity is based on the transfer of momentum that takes place during the molecular collisions. The molecules that have a low velocity on the y-axis going towards a zone, where the molecules have a higher velocity, will slow them down, and vice-versa.

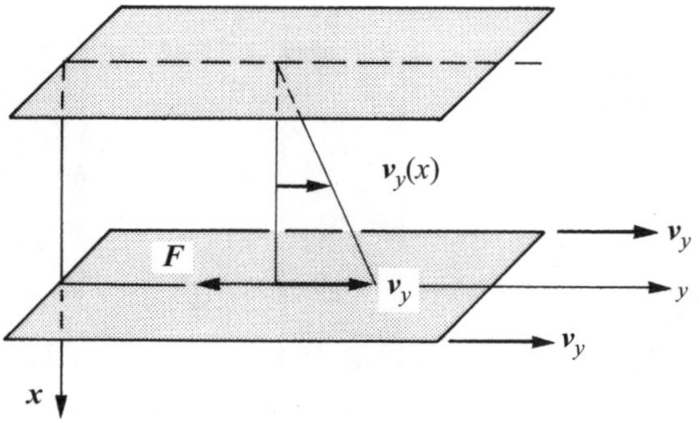

Fig. 6.2 Summary of forces on a moving plate in a viscous liquid.

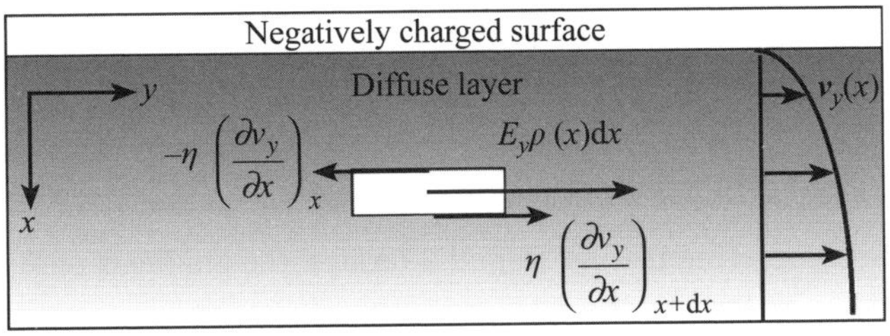

Fig. 6.3 Electric field parallel to the negatively charged surface of a solid phase and forces acting on the volume element.

Consider a charged interface, and suppose we apply an electric field parallel to the interface. The ions in the diffuse layer are then subjected to a coulombic force parallel to the interface and drag the surrounding solvent molecules with them by friction. Take a volume element of electrolyte solution $\mathrm{d}x\mathrm{d}y\mathrm{d}z$ where the volumic charge density is $\rho(x)$. At steady state (no acceleration), the coulombic force exerted on this volume element is equal to the sum of the friction forces.

$$E_y\rho(x)\mathrm{d}x = -\left(\eta\frac{\partial v_y(x)}{\partial x}\right)_{x+\mathrm{d}x} + \left(\eta\frac{\partial v_y(x)}{\partial x}\right)_x \tag{6.2}$$

where $v_y(x)$ is the velocity of the solution parallel to the wall.

If we take it that the viscosity is independent of the distance to the solid surface and therefore of the velocity gradient (Newtonian liquid), then expression (6.2) can be written as

$$E_y\rho(x)\mathrm{d}x = -\eta\frac{\partial^2 v_y(x)}{\partial x^2}\mathrm{d}x \tag{6.3}$$

If we replace $\rho(x)$ by the Poisson equation in one dimension (see equation (5.78)), we have

$$E_y\,\varepsilon_0\varepsilon_\mathrm{r}\frac{\partial^2\phi(x)}{\partial x^2}\mathrm{d}x = \eta\frac{\partial^2 v_y(x)}{\partial x^2}\mathrm{d}x \tag{6.4}$$

The velocity of the solution near a solid surface is zero (stick condition). Thus, we can integrate equation (6.4) with the following boundary conditions

$$v_y(x_s) = 0 \quad \text{for} \quad \phi(x_s) = \zeta \quad \text{at the shear plane}$$
$$v_y(x)\rightarrow v_E \quad \text{for} \quad \phi(x)\rightarrow 0 \quad \text{in the bulk}$$

where v_E is the electro-osmotic velocity of the liquid attained by the fluid outside the diffuse layer and where ζ is the potential at the shear plane. This potential is called the ***zeta potential*** or the ***electrokinetic potential***. Note that in Figure 6.3, ζ is negative since the wall is negatively charged.

Therefore, integrating between x and infinity, we have

$$\int_x^{\infty} E_y\,\varepsilon_0\varepsilon_{\rm r}\frac{\partial^2\phi(x)}{\partial x^2}{\rm d}x \;=\; \int_x^{\infty}\eta\frac{\partial^2 v_y(x)}{\partial x^2}{\rm d}x \;=\; \left[E_y\,\varepsilon_0\varepsilon_{\rm r}\frac{\partial\phi(x)}{\partial x}\right]_x^{\infty} \;=\; \left[\eta\frac{\partial v_y(x)}{\partial x}\right]_x^{\infty} \tag{6.5}$$

which yields, considering that far from the surface, the gradients on x of the electric potential and that of the velocity $v_y(x)$ are zero

$$E_y\,\varepsilon_0\varepsilon_{\rm r}\frac{\partial\phi(x)}{\partial x} \;=\; \eta\frac{\partial v_y(x)}{\partial x} \tag{6.6}$$

A second integration between the shear plane and x leads to

$$\int_{x_s}^{x} E_y\,\varepsilon_0\varepsilon_{\rm r}\frac{\partial\phi(x)}{\partial x}{\rm d}x \;=\; \int_{x_s}^{x}\eta\frac{\partial v_y(x)}{\partial x}{\rm d}x \;=\; \left[E_y\,\varepsilon_0\varepsilon_{\rm r}\phi(x)\right]_{x_s}^{x} \;=\; \left[\eta v_y(x)\right]_{x_s}^{x} \tag{6.7}$$

With the above boundary conditions, we get

$$v_y(x) \;=\; \frac{E_y\varepsilon_0\varepsilon_{\rm r}}{\eta}\left[\phi(x)-\zeta\right] \tag{6.8}$$

Furthermore, making the approximation (5.95) to estimate the potential profile $\phi(x)$ and considering that the shear plane coincides with the boundary separating the first solvent layer on the solid (either the Helmholtz or the Stern layer) from the diffuse layer (see Figure 5.14), we then have

$$\phi(x) \;=\; \left[\phi(x_s)-\phi^{\infty}\right]\,e^{-\kappa x} \;=\; \zeta e^{-\kappa x} \tag{6.9}$$

Substituting in equation (6.8), we obtain

$$v_y(x) \;=\; -\frac{E_y\varepsilon_0\varepsilon_{\rm r}\zeta}{\eta}\left[1-e^{-\kappa x}\right] \tag{6.10}$$

where κ is the reciprocal Debye length characteristic of the thickness of the diffuse layer. The negative sign shows that the velocity is positive, i.e. in the direction of the electric field when the zeta potential is negative.

Equation (6.10) also shows that the velocity of the solution reaches a constant value v_E outside the diffuse layer whose thickness depends on the ionic strength of the solution (see Figure 5.11). This velocity is then given by the ***Smoluchowski equation***, which is written as

$$\zeta \;=\; -\frac{\eta}{\varepsilon_0\varepsilon_{\rm r}E_y}\,v_y \tag{6.11}$$

It is often interesting to link the zeta potential to the surface charge density on the wall. The latter is defined by equation (5.88) that is, taking into account approximation (6.9)

$$\sigma_{\rm wall} \;=\; -\int_{x_s}^{\infty}\rho(x){\rm d}x \;\cong\; \int_{x_s}^{\infty}\varepsilon_0\varepsilon_{\rm r}\left(\frac{\partial^2\phi(x)}{\partial x^2}\right){\rm d}x \;=\; \varepsilon_0\varepsilon_{\rm r}\zeta\kappa \tag{6.12}$$

This equation shows that the zeta potential is directly proportional to the surface charge density and inversely proportional to the reciprocal Debye length.

6.1.3 Electrophoresis of charged colloids

In this section, we shall limit ourselves to the study of electrophoresis of non-conducting charged spheres. The Gouy-Chapman theory for spherical objects requires us to solve the Poisson equation (3.53) with spherical coordinates, i.e.

$$\nabla^2\phi = -\frac{\rho(r)}{\varepsilon_0\varepsilon_r} = \frac{\partial^2\phi(r)}{\partial r^2}+\frac{2}{r}\frac{\partial\phi(r)}{\partial r} \tag{6.13}$$

We have already seen the integration of this equation in §3.4.3 in the framework of the Debye-Hückel theory. The solution is therefore

$$\phi(r) = C_1\frac{e^{-\kappa r}}{r} \tag{6.14}$$

where C_1 is an integration constant. In the Debye-Hückel theory, this constant was determined using the electroneutrality of the solution as a boundary condition. The same approach can be used here if we take care not to forget that equation (3.66) was obtained using the hypothesis that the electrostatic interaction energy between the sphere and the solution is weak with respect to the thermal agitation, which implies that we must limit ourselves to cases of small zeta potential values. Thus, we have

$$\phi(r) = \frac{q_s e^{-\kappa(r-r_s)}}{4\pi\varepsilon_0\varepsilon_r r\,(1+\kappa r_s)} \tag{6.15}$$

where r_s can be considered equal to the radius of the charged sphere. At the surface of the sphere, this expression can be taken as the sum of the potential generated by the charges on the sphere and the potential generated by the charges in solution. So, equation (6.15) can be re-written as

$$\phi(r_s) = \frac{q_s}{4\pi\varepsilon_0\varepsilon_r r_s} - \frac{q_s}{4\pi\varepsilon_0\varepsilon_r(r_s+\kappa^{-1})} \tag{6.16}$$

Consider now the movement of a charged sphere in an electric field. For dilute solutions where κ^{-1} is large, the zeta potential is given directly by equation (6.16)

$$\zeta = \frac{q_s}{4\pi\varepsilon_0\varepsilon_r r_s} \tag{6.17}$$

The balance of forces exerted on the sphere including the electrostatic force $q\boldsymbol{E}$ and the friction force $-6\pi r_s\eta\boldsymbol{v}$ gives

$$\boldsymbol{v} = \frac{q_s\boldsymbol{E}}{6\pi\eta r_s} = \frac{2\varepsilon_0\varepsilon_r\zeta\boldsymbol{E}}{3\eta} \tag{6.18}$$

This equation, sometimes called the ***Hückel equation***, gives a direct relationship between the electrophoretic mobility defined by equation (4.18) and the zeta potential

$$u = \frac{2\varepsilon_0\varepsilon_r\zeta}{3\eta} \tag{6.19}$$

It is possible to calculate the electrophoretic mobility in solutions of high ionic strength. However, the calculations required are too extensive to be presented here. It is also possible to take into account the effects of the relaxation of the ionic atmosphere as we did in §4.3.4.

6.1.4 Streaming potential

The streaming potential is linked to the movement of ionic excess charges in the diffuse layer swept along when a liquid flows near a charged wall.

Consider a tube of length L and radius a, and a laminar flow generated by a pressure difference Δp between the ends of the tube (see Figure 6.4). The current due to the displacement of the ionic surface excesses near to the charged wall of the tube will generate a potential difference between the two ends of the tube. This potential will, in turn, generate a compensating current since it is clear that if, for example, we circulate a sample of phosphate buffer solution under pressure in a silica capillary, the ingoing and outgoing solutions will be electroneutral.

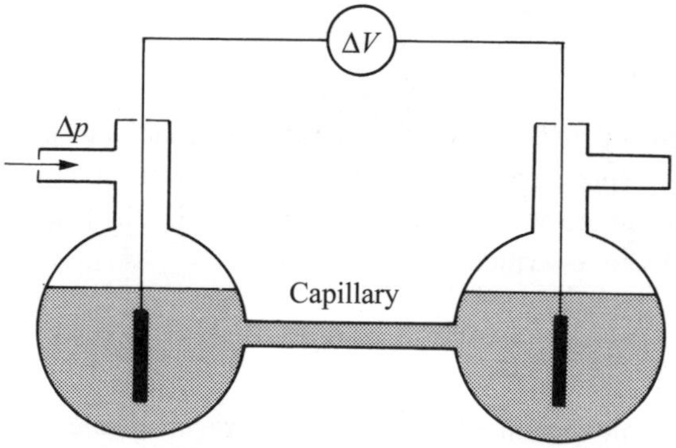

Fig. 6.4 Experimental set-up for measuring a streaming potential

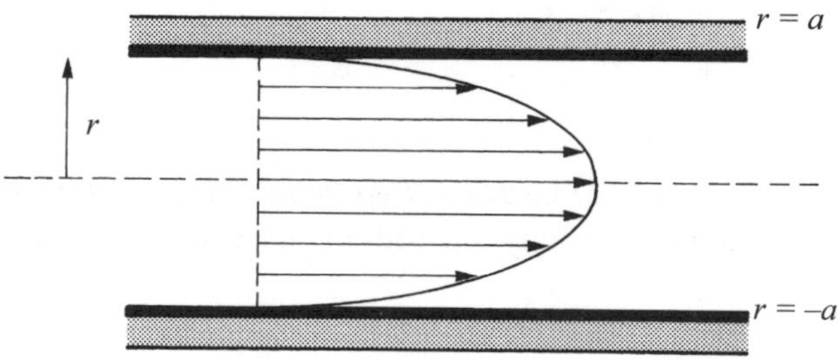

Fig. 6.5 Velocity profile in a cylindrical tube

In a first instance, we will calculate the current due to the flow of ions in the diffuse layer, and then the potential difference across the tube.

The velocity profile of a liquid in a tube of radius a and length L is a parabolic profile, as shown in Figure 6.5, given by the Poiseuille equation

$$v(r) = \frac{\Delta p}{4\eta L}\left(a^2 - r^2\right) \tag{6.20}$$

The electric current due to convection of the ions in the diffuse layer is written as

$$I = \int_0^a 2\pi r\rho(r)v(r)\mathrm{d}r \tag{6.21}$$

Since the excess charge is located close to the wall in the diffuse layer, it is clear that the resulting current will also have its field lines close to the wall. In this part of the tube, within a thickness of a few nanometres, it is possible to linearise equation (6.20) to get

$$v(r) = \frac{a\,\Delta p}{2\eta L}(a - r) \tag{6.22}$$

since when $r \to a$, then $a^2 - r^2 = (a+r)(a-r) \to 2a(a-r)$.

If the diameter of the capillary a is very much greater than the thickness of the diffuse layer ($\kappa^{-1} << a$), we can express the volumic charge density as a function of the Poisson equation in one dimension (see equation (5.78)). Strictly speaking, we should use cylindrical coordinates but this tends to complicate the calculation. So, combining equations (6.21) and (6.22), we have

$$I = -\frac{\pi\varepsilon_0\varepsilon_\mathrm{r}\,a\,\Delta p}{\eta L}\int_0^a r\left(\frac{\partial^2\phi(r)}{\partial r^2}\right)(a - r)\mathrm{d}r \tag{6.23}$$

By integrating twice by parts, and still using the approximation (5.95)

$$\phi(r) = \zeta e^{-\kappa(a-r)} \tag{6.24}$$

we get

$$I = -\frac{\pi\varepsilon_0\varepsilon_\mathrm{r}\,a\,\Delta p}{\eta L}\left(\left[\left(\frac{\partial\phi(r)}{\partial r}\right)r(a-r)\right]_0^a - \int_0^a\left(\frac{\partial\phi(r)}{\partial r}\right)(a - 2r)\mathrm{d}r\right)$$

$$I = \frac{\pi\varepsilon_0\varepsilon_\mathrm{r}\,a\,\Delta p}{\eta L}\left(\left[\phi(r)(a-2r)\right]_0^a + 2\int_0^a\phi(r)\mathrm{d}r\right) \tag{6.25}$$

$$I = \frac{\pi\varepsilon_0\varepsilon_\mathrm{r}\,a\,\Delta p}{\eta L}\left(-\zeta a + 2\zeta\left[\frac{1}{\kappa}e^{-\kappa(a-r)}\right]_0^a\right) = \frac{\pi\varepsilon_0\varepsilon_\mathrm{r}\,a\,\Delta p}{\eta L}\zeta a = \frac{\varepsilon_0\varepsilon_\mathrm{r}\,\pi a^2\,\Delta p\,\zeta}{\eta L}$$

Still using the hypothesis that $\kappa^{-1} << a$, the compensation current is defined by

$$I_c = \pi a^2\sigma E \tag{6.26}$$

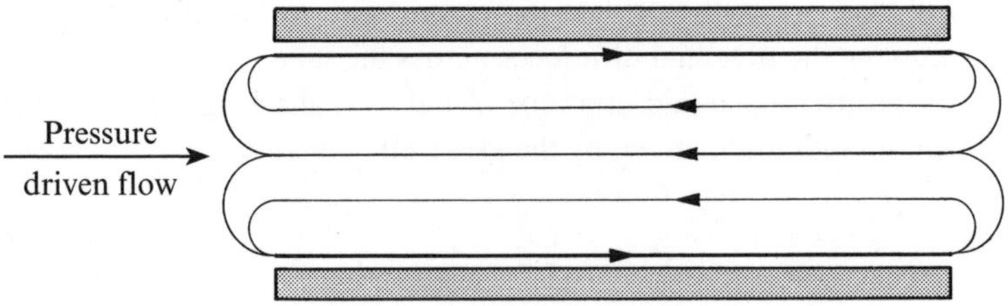

Fig. 6.6 Current loops during electrolyte streaming in a tube with negatively charged walls.

where E is the electric field induced and σ the conductivity of the solution.

Given that the total current across the tube is zero, we can write

$$E = \frac{\varepsilon_0 \varepsilon_r \, \Delta p \, \zeta}{\eta \, \sigma \, L} \tag{6.27}$$

and the potential difference between the two ends of the tube is

$$\Delta V = \frac{\varepsilon_0 \varepsilon_r \, \Delta p \, \zeta}{\eta \, \sigma} \tag{6.28}$$

Looking at the current lines inside the tube, we have loops of current that can be drawn schematically as in Figure 6.6.

It is interesting to compare the phenomena of electro-osmosis and streaming potential. The irreversible thermodynamic approach described by equation (4.117) shows that both the flow velocity of the solution in a tube, and the current can be written as a linear combination of the two forces resulting from the pressure gradient and the electric potential gradient.

$$J_i = \sum_i L_{ij} X_j = \begin{cases} \boldsymbol{v} = L_{11} \mathbf{grad} p + L_{12} \boldsymbol{E} \\ \boldsymbol{i} = L_{21} \mathbf{grad} p + L_{22} \boldsymbol{E} \end{cases} \tag{6.29}$$

The Onsager reciprocity relation ($L_{12} = L_{21}$) predicts that the electro-osmotic mobility u_{eo} is equal to the ratio of the electric current generated by a pressure driven flow to that pressure gradient.

$$u_{eo} = \left(\frac{v}{E} \right)_{\mathrm{grad}p=0} = \left(\frac{i}{\mathrm{grad}p} \right)_{E=0} \tag{6.30}$$

6.1.5 Sedimentation potential

Like the streaming potential, the sedimentation potential is generated by the movement of charged colloids in an electrolyte solution. Each sphere brings with it the diffuse layer surrounding it, and so, to maintain the electroneutrality, a compensating electric field is generated. The movement of the colloids can be induced by gravity or, in an ultra-centrifuge whose rotational speed may reach more than 75 000 r.p.m. The Onsager equations also show the link between electrophoresis and sedimentation potential.

6.2 CAPILLARY ELECTROPHORESIS

6.2.1 Capillary electro-osmotic flow

The principle of electrophoresis in a silica capillary (internal diameter of the order of 100 μm) is the establishment of an electro-osmotic plug flow, as shown in Figure 6.7). In effect, because the silica is negatively charged in a wide range of pH, the application of an electric field of about 10 to 100 kV·m^{-1} can generate a flow of a few nanoliters per second towards the cathode. The plug flow has a flat velocity profile compared to the parabolic profile of a pressure driven flow. This is due to the fact that the driving force of an electro-osmotic flow is located in the diffuse layer as explained in §6.1.2, and that there is no friction within the solution inside the capillary. As a result, the solution moves within the capillary with a uniform speed.

The rate of the electro-osmotic flow is simply obtained from the definition of the zeta potential given in equation (6.11) that is written as

$$v_{\mathrm{eo}} = -\frac{\varepsilon_0 \varepsilon_{\mathrm{r}}\, \zeta\, \boldsymbol{E}}{\eta} = u_{\mathrm{eo}} \boldsymbol{E} = u_{\mathrm{eo}} \frac{\Delta V}{L} \tag{6.31}$$

where u_{eo} is the electro-osmotic mobility and ΔV is the absolute value of the high voltage applied across the capillary of length L. The negative sign is due to the fact that the flow goes with the electric field when ζ is negative. It is important to note that equation (6.31) holds for capillaries of different shapes, such as a serpentine, and sizes, as long as the zeta potential can be considered constant and not too large over the whole capillary surface. As a consequence, the easiest way to calculate the velocity profile in a micro-structure is to calculate the electric field distribution that can be easily obtained by solving the Laplace equation ($\nabla^2 V = 0$).

In fact, strictly speaking, we should solve the Navier-Stokes equation (4.A14), taking into account the force exerted by the electric field on the volumic charge distribution. This resolution should be made in cylindrical coordinates if the diameter of the capillary is very small, or in unidimensional coordinates as in §6.1.1 if the thickness of the diffuse layer is negligible compared to the capillary diameter ($\kappa^{-1} << a$). However, in steady state conditions, the simplified approach presented here is sufficient to describe the electro-osmotic plug flow.

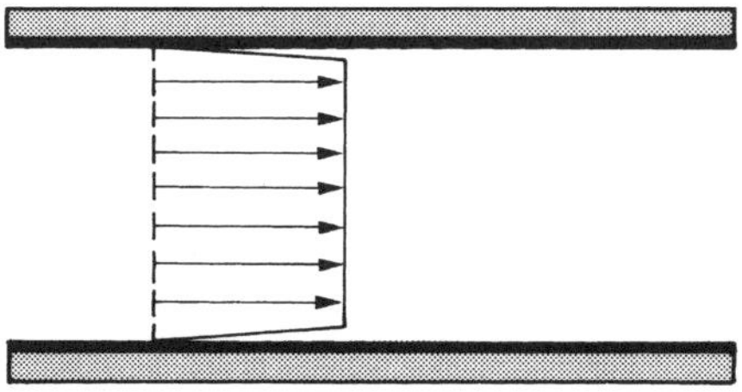

Fig. 6.7 Schematic plug flow in a capillary

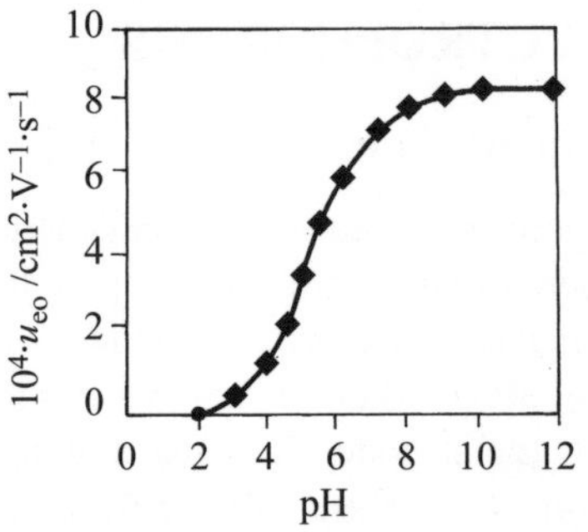

Fig. 6.8 Variation of the electro-osmotic mobility as a function of the pH for a silica capillary pre-washed in 0.1M NaOH (Adapted from P.D. Grossman & J.C. Colburn: *Capillary Electrophoresis*. Academic Press).

For a capillary of diameter a, the volumic flow rate is therefore

$$F_V = v_{eo} S = \frac{\pi a^2 u_{eo} \Delta V}{L} \tag{6.32}$$

To calculate the equivalent pressure we need to exert to have the same flow rate, we can apply the Poiseuille formula (6.20) which, by integration, gives

$$F_V = \int_0^a \frac{\Delta p}{4L\eta}(a^2 - r^2) 2\pi r \mathrm{d}r = \frac{\pi a^4 \Delta p}{8L\eta} \tag{6.33}$$

It is interesting to notice that for very narrow capillaries, electro-osmotic pumping becomes advantageous ($\Delta p = 8 u_{eo} \eta \Delta V / a^2$).

The zeta potential varies with the surface charge density, which in the case of silica depends on the pH as shown in Figure 6.8. This type of curve, in fact, represents an acid-base titration curve at the wall of the silica capillary. At high pH, the silica is negatively charged, and the mobility is at its maximum. At the lowest pH, the wall is neutralised and the electro-osmotic mobility decreases. At very acidic pHs, the wall becomes positively charged and the flow direction is inversed.

EXAMPLE

Consider that the charge on a silica capillary is approximately $-0.01 \mathrm{C \cdot m^{-2}}$ ($= -1 \mu\mathrm{C \cdot cm^{-2}}$). Let's calculate the zeta potential for a 0.01M solution of a 1:1 salt.

The calculation of equation (5.93) yields

$\kappa = [(2 \cdot 10 \cdot 6.02 \cdot 10^{23})/(8.85 \cdot 10^{-12} \cdot 78 \cdot 1.38 \cdot 10^{-23} \cdot 298)]^{1/2} \cdot 1.6 \cdot 10^{-19} = 3.3 \cdot 10^8 \mathrm{~m^{-1}}$

corresponding to a diffuse layer thickness of 3 nm as shown in Figure 5.11.

The zeta potential given by equation (6.12) is therefore about –43 mV. If the viscosity is 0.001 $\mathrm{N \cdot s \cdot m^{-2}}$, the electrophoretic mobility defined by equation (6.31) is then $3 \cdot 10^{-8} \mathrm{~m^2 \cdot V^{-1} \cdot s^{-1}}$. This can be compared to the experimental values in Figure 6.8 showing that the order of magnitude of the surface charge density of silica is about $0.01 \mathrm{C \cdot m^{-2}}$.

For this surface charge density, a field of 10 $\mathrm{kV \cdot m^{-1}}$ gives a volumic flow rate of

$2.4 \cdot 10^{-12}$ $m^3 \cdot s^{-1}$, i.e. 2.4 $nl \cdot s^{-1}$ for a capillary with an internal diameter of 50 μm.
The pressure gradient per linear metre of capillary that we would have to apply for the same flow is about 10^4 ($N \cdot m^{-3}$) that is about 0.1 $atm \cdot m^{-1}$.

Separation techniques using electro-kinetic flow in a glass capillary comprise three classes of methods:

- electrophoretic separation
- micellar electro-kinetic separation
- electrophoretic/chromatographic separation.

6.2.2 Zone electrophoresis in a capillary

The principle of this separation technique is simply based on the difference in electrophoretic mobility of the charged species. The velocity of a given species in the capillary is then the sum of the electro-osmotic velocity and the electrophoretic velocity. At neutral pH, silica is negatively charged and the cations then move forward faster than the solvent and the anions are retarded. In steady state conditions, the velocity of the ion is then

$$v = \left(u_{eo} + u\right)\frac{\Delta V}{L} \tag{6.34}$$

where u is the electrophoretic mobility defined by equation (4.17) and ΔV the absolute value of the voltage drop in the capillary. Note that often the electro-osmotic mobility is greater than the electrophoretic mobility, which allows the separation of both the cations and anions by a single injection.

The retention time, defined as the duration of the ion's journey in the capillary, is then simply given by

$$t_R = \frac{L}{v} = \frac{L^2}{\left(u_{eo} + u\right)\Delta V} \tag{6.35}$$

If we consider that the mass transfer is limited to the electro-osmotic flow of the solution and the diffusion-migration of the species injected into the capillary, the concentration distribution of the ionic species in the capillary is given by the differential equation

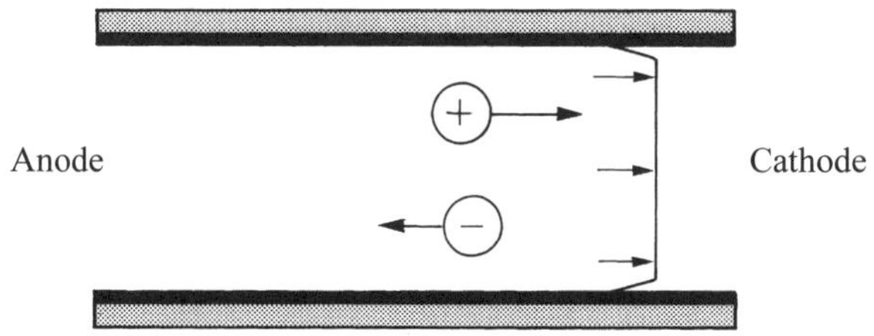

Fig. 6.9 Scheme of the movement of the solvent and ions in zone electrophoresis in a glass capillary.

$$\frac{\partial c_i}{\partial t} = D_i \frac{\partial^2 c_i}{\partial y^2} - v_y \frac{\partial c_i}{\partial y} \tag{6.36}$$

To solve this differential equation, it is useful to change the variable y to

$$Y = y - v_y t \tag{6.37}$$

in order to eliminate the term relative to the velocity. In effect, the partial derivatives with respect to y remain identical

$$\frac{\partial c_i}{\partial y} = \left(\frac{\partial c_i}{\partial Y}\right)\left(\frac{\partial Y}{\partial y}\right) = \frac{\partial c_i}{\partial Y} \tag{6.38}$$

and those with respect to time are found by considering the total differential

$$\begin{aligned} \mathrm{d}c_i &= \frac{\partial c_i(y,t)}{\partial y}\mathrm{d}y + \frac{\partial c_i(y,t)}{\partial t}\mathrm{d}t = \frac{\partial c_i(y,t)}{\partial y}\left[\mathrm{d}Y + v_y \mathrm{d}t\right] + \frac{\partial c_i(y,t)}{\partial t}\mathrm{d}t \\ &= \frac{\partial c_i(y,t)}{\partial y}\mathrm{d}Y + \left[\frac{\partial c_i(y,t)}{\partial t} + v_y \frac{\partial c_i(y,t)}{\partial y}\right]\mathrm{d}t = \frac{\partial c_i(Y,t)}{\partial Y}\mathrm{d}Y + \frac{\partial c_i(Y,t)}{\partial t}\mathrm{d}t \end{aligned} \tag{6.39}$$

Thus, equation (6.36) reduces to a Fick equation without a convection-migration term

$$\frac{\partial c_i(Y,t)}{\partial t} = D_i \frac{\partial^2 c_i(Y,t)}{\partial Y^2} \tag{6.40}$$

In the case of an injection of a Dirac impulse of samples, the Laplace transformation of the Fick equation (6.40) becomes (see §8.1.1)

$$D_i \frac{\partial^2 \bar{c}_i}{\partial Y^2} - s\bar{c}_i + c_0 \delta(Y) = 0 \tag{6.41}$$

The solution of this equation is

$$\bar{c}_i(Y) = -\text{Heaviside}(Y)\frac{c_0}{2\sqrt{Ds}}\left[\exp^{\sqrt{\frac{s}{D}}Y} - \exp^{-\sqrt{\frac{s}{D}}Y}\right] + A\exp^{\sqrt{\frac{s}{D}}Y} + B\exp^{-\sqrt{\frac{s}{D}}Y} \tag{6.42}$$

The concentration when Y tends towards minus infinity and plus infinity is zero. From this, we deduce that the constant B is also zero and that the constant A is

$$A = \frac{c_0}{2\sqrt{Ds}} \tag{6.43}$$

We now have in fact two functions according to the sign of Y

$$\bar{c}_i(Y) = \frac{c_0}{2\sqrt{Ds}}\left[\exp^{\sqrt{\frac{s}{D}}Y} - \text{Heaviside}(Y)\left[\exp^{\sqrt{\frac{s}{D}}Y} - \exp^{-\sqrt{\frac{s}{D}}Y}\right]\right] \tag{6.44}$$

Knowing the following Laplace transformation,

$$L\left[\frac{\exp^{-a^2/4t}}{\sqrt{\pi t}}\right] = \frac{\exp^{-a\sqrt{s}}}{\sqrt{s}} \tag{6.45}$$

the solution of equation (6.40) is a Gaussian distribution centred on Y given by

$$c_i(Y,t) = \frac{c_0}{2\sqrt{\pi D t}}\exp^{-\left[\frac{Y^2}{4Dt}\right]} = \frac{c_0}{2\sqrt{\pi D t}}\exp^{-\left[\frac{(y-vt)^2}{4Dt}\right]} \tag{6.46}$$

Thus, we find again the results in Figure (4.16) for linear diffusion.

The width of the peak represented by the variance is solely due to the axial diffusion

$$\sigma_{\mathrm{L}}^2 = 2Dt_{\mathrm{R}} = \frac{2DL^2}{(u_{\mathrm{eo}}+u)\Delta V} \tag{6.47}$$

and the number of theoretical equivalent plates is

$$N = \left(\frac{t_{\mathrm{R}}}{\sigma_{\mathrm{t}}}\right)^2 = \left(\frac{L}{\sigma_{\mathrm{L}}}\right)^2 = (u_{\mathrm{eo}}+u)\frac{\Delta V}{2D} \tag{6.48}$$

In electrophoretic separation, it is interesting to compare the friction coefficient with the mass of the species. For a sphere, the friction coefficient is $\zeta = 6\pi r\eta$ (see equation (4.7)) and is therefore proportional to $m^{1/3}$.

(N.B. Do not confuse the zeta potential and the friction coefficient, even if these two quantities are represented by the same symbol).

In the case of long charged polyelectrolyte chains, the electrophoretic friction coefficient is directly proportional to the mass. Therefore, we cannot separate DNA molecules of different lengths by zone electrophoresis in a capillary, since the charge (2 charges per base) and the friction coefficient are directly proportional to the mass. Nevertheless, we can separate small oligo-nucleotides (up to 20 or 30 bases). This remark for the separation of DNA also applies to proteins denatured by a detergent such as SDS (Sodium DodecylSulfate) where the charge is proportional to the size. For these biopolymers, a porous matrix, such as a gel, has to be used for the separation.

6.2.3 Micellar electrokinetic capillary chromatography

This approach is based on the distribution of species to be separated between the mobile phase (electro-osmotic flow towards the cathode) and a negatively charged micellar phase (e.g. sodium dodecylsulfate) travelling more slowly. The charged or neutral lipophilic molecules reside partially in the micellar phase (pseudo-stationary phase) migrating to the anode and are therefore held back relative to the hydrophilic molecules moving together with the electro-osmotic flow towards the cathode. The separation is at the same time electrophoretic due to the migration of the micelles under the influence of the electric field and chromatographic due to the distribution of the lipophilic molecules between the aqueous eluent and the organic pseudo-stationary phase.

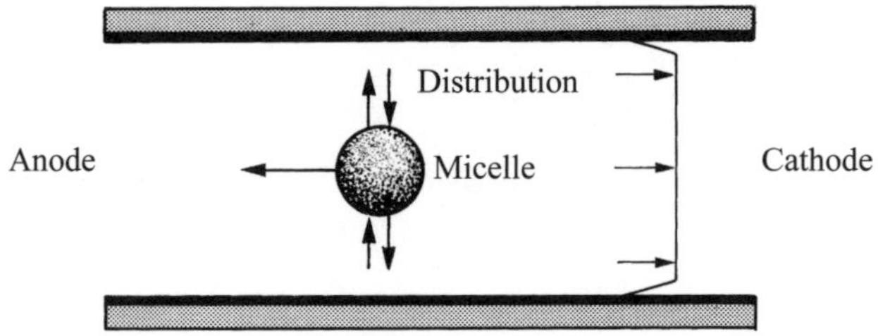

Fig. 6.10 Retarded movement of a micelle in an electro-osmotic flow.

6.2.4 Capillary electro-chromatography (CEC)

This method uses an electro-osmotic flow to circulate the mobile phase in a capillary column filled with particles forming a stationary phase. The plug flow is, in general, only slightly perturbed by the filling. The retention principle is the same as for HPLC chromatography.

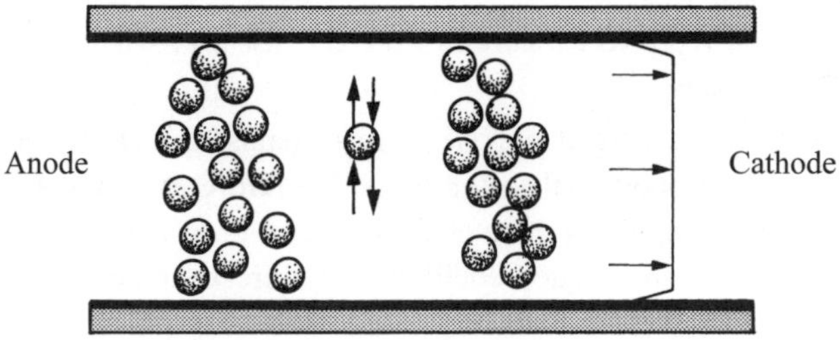

Fig. 6.11 Electro-chromatography with electro-osmotic pumping.

6.2.5 Capillary separation techniques not using electro-osmosis

Commercially available capillary electrophoresis equipment offers other separation techniques that do not use properties linked to the electro-osmotic flow, but those associated with the small diameter of the capillary such as the elimination of natural convection phenomena. These techniques are just adaptations of the classic electrophoresis methods described in §6.3, such as:

- electrophoresis in a gel
- isoelectric focusing
- isotachophoresis

For these techniques, the electro-osmotic flow has to be suppressed, for example by covering the inner wall of the silica capillary with a fine layer of a viscous polymer such as linear polyacrylamide. In effect, equation (6.8) shows us that to eliminate the electro-osmotic flow, all we need is a high viscosity in the few nanometres of the diffuse layer. This high surface viscosity slows down the movement of the ions in the diffuse layer, and thus inhibits the onset of the electro-osmosis flow of the solution.

6.2.6 Experimental methods

A capillary electrophoresis system consists of a high volatge power supply, two reservoirs of buffer solutions, a sampling unit, a separation capillary and a detector, as illustrated schematically in Figure 6.12.

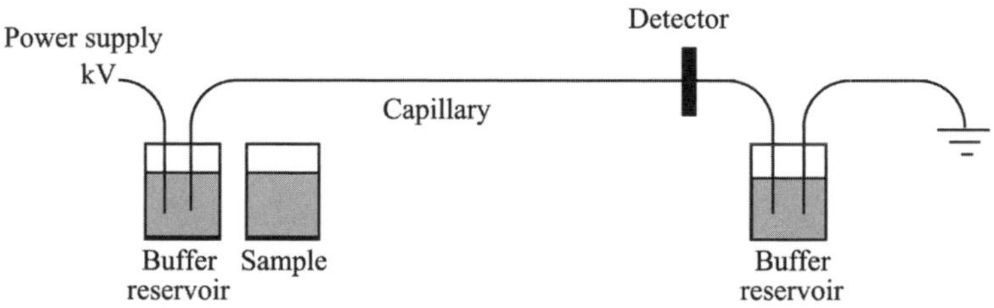

Fig. 6.12 Schematic representation of a capillary electrophoresis system.

Electrokinetic injection

Electrokinetic injection, or injection by electromigration, simply consists of placing the anodic end of the capillary and the high voltage electrode into the recipient containing the sample for a short duration, and then replacing these two again in the buffer reservoir.

The electrophoretic and electro-osmotic velocities of the sample may be calculated from their respective mobilities and from the injection voltage ΔV_{inj} imposed during the injection

$$v = v_{\text{ep}} + v_{\text{eo}} = \left(u + u_{\text{eo}}\right)\frac{\Delta V_{\text{inj}}}{L} \tag{6.49}$$

where L is the length of the column.

The length of the sample plug then depends on the duration of the injection

$$l = \left(v_{\text{ep}} + v_{\text{eo}}\right) t_{\text{inj}} \tag{6.50}$$

The weight of the sample injected is then

$$w = \pi r^2 lc = (u + u_{\text{eo}})\frac{\pi r^2\, c\, \Delta V_{\text{inj}}\, t_{\text{inj}}}{L} \tag{6.51}$$

where c is the mass concentration in kg·l^{-1}. Using the electrokinetic injection method, we control the amount injected by varying the applied potential or the injection duration. Also, this method allows the differentiation of the cations and anions from the neutral species as soon as the injection is made. In effect, the quantity of cations injected is larger than the quantity of neutral species that is in turn larger than the quantity of anions, which is not necessarily an advantage.

Injection under pressure

This injection method consists of putting the end of the capillary into a closed vessel containing the sample and putting the latter under pressure. The quantity of sample injected using this technique is

$$w = \frac{\Delta P\, \pi r^4\, c\, t_{\text{inj}}}{8\eta L} \tag{6.52}$$

Commercial equipment often uses either electrokinetic injection or injection under pressure. Nevertheless, other techniques exist, among which we might mention the following methods:

Gravity injection

This method consists of raising the sample reservoir to a height of Δh for a duration Δt. The injected volume is then simply given by

$$V_{\text{inj}} = \frac{\rho g \pi r^4 \Delta h \Delta t}{8\eta L} \tag{6.53}$$

Note that the amount of sample injected under pressure or by gravity does not depend on its electrophoretic mobility.

Electrokinetic sample loop

This injection method, applicable mainly to micro-manufactured capillaries in glass, silica or polymer plates, consists of using the law of addition of currents to drive the samples when they pass over the junction between two capillaries. As in Figure 6.13, we first make the sample circulate by electro-osmosis between the source and the drain (shaded zone). Then, we apply a potential difference between the entry and exit of

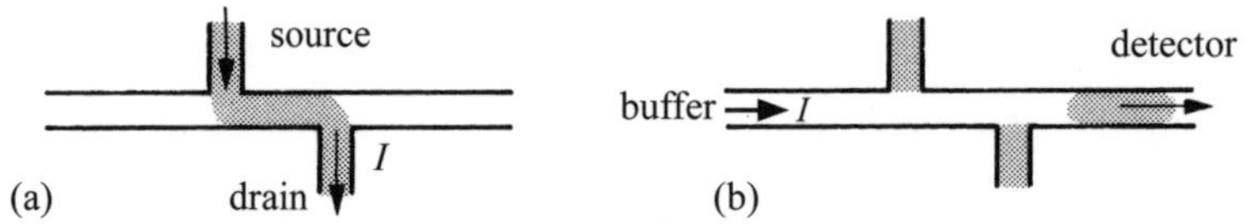

Fig. 6.13 Electrokinetic injection using a double 'T'. (a) The sample circulates in the part of the capillary between the source and the drain, (b) The potential is applied to the terminals of the capillary.

the separation capillary. The part of the sample in the separation capillary between the source and the drain is then sent along the length of the capillary by electro-osmosis.

Columns and the effect of temperature

The capillaries are in general made of hollow silica fibres with an internal diameter of less than 100 μm. This allows the application of high potential differences whilst having a good thermal dissipation of the Joule heating. The capillaries are usually covered in polyimide, which makes them much less fragile and easier to handle.

The electric current passing in the capillary causes heating by the Joule effect. Using equations (4.21) and (4.23), the electrical power to be dissipated per unit length is

$$\frac{P}{L} = \frac{I\Delta V}{L} = \frac{\Lambda_{\mathrm{m}} c\,\pi r^2\,\Delta V^2}{L^2} \tag{6.54}$$

where Λ_{m} is the molar conductivity of the solution.

A thermal balance on a silica capillary covered with a fine layer of plastic can be found by resolving the thermal conductivity equations in cylindrical coordinates.

$$\mathrm{div}\boldsymbol{q}_r = \frac{1}{r}\frac{\mathrm{d}}{\mathrm{d}r}(rq_r) = \frac{I^2}{\Lambda_{\mathrm{m}} c} \tag{6.55}$$

where $\boldsymbol{q}_r$ is the heat flux per unit length and the second term is the power generated per unit volume. A first integration gives

$$q_r = \frac{I^2}{2\Lambda_{\mathrm{m}} c} r \tag{6.56}$$

Fourier's law is written as

$$\boldsymbol{q}_r = -\lambda_t \mathbf{grad}T = -\lambda_t \frac{\mathrm{d}T}{\mathrm{d}r} \tag{6.57}$$

where λ_{t} is the thermal conductivity of the buffer solution. If it is a constant, then we have

$$T(r) - T_1 = \frac{I^2 r_1^2}{4\lambda_{\mathrm{t}}\Lambda_{\mathrm{m}} c}\left[1 - \left(\frac{r}{r_1}\right)^2\right] \tag{6.58}$$

This relation shows that the temperature profile inside a capillary is parabolic. In the same way, the temperature drop in the silica capillary is given by the integration of equation (6.57)

$$-\lambda_{\mathrm{s}}\frac{\mathrm{d}T}{\mathrm{d}r} = q = \frac{G}{2\pi r} = \frac{I^2}{\Lambda_{\mathrm{m}} c}\frac{\pi r_1^2}{2\pi r} \tag{6.59}$$

where G is the total quantity of heat lost per unit length. The temperature difference at the edges of the silica tube is therefore written as

$$T_2 - T_1 = \frac{I^2 r_1^2}{2\lambda_{\mathrm{s}}\Lambda_{\mathrm{m}} c}\ln\left(\frac{r_2}{r_1}\right) \tag{6.60}$$

whilst that in the polymer layer is

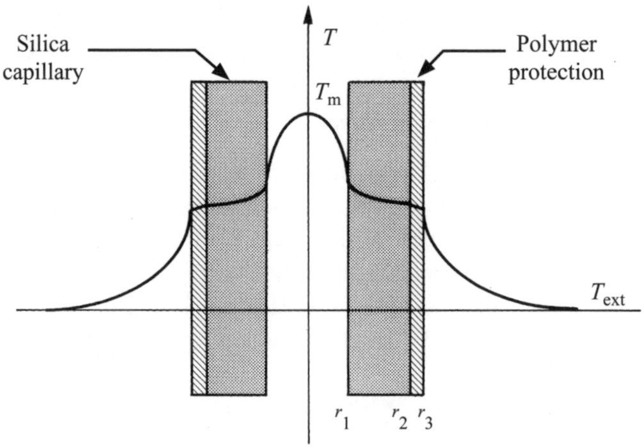

Fig. 6.14 Temperature profile during the electro-osmotic pumping in a capillary.

$$T_3 - T_2 = \frac{I^2 r_1^2}{2\lambda_p \Lambda_m c} \ln\left(\frac{r_3}{r_2}\right) \tag{6.61}$$

where λ_s and λ_p are respectively, the thermal conductivities of the silica and the polymer. The coefficient of heat transfer between the polymer and the cooling system (air or water according to the apparatus used) is such that

$$2\pi r_3\, h\left(T_3 - T_{ext}\right) = \frac{I^2\, \pi r_1^2}{\Lambda_m\, c} \tag{6.62}$$

where h is the heat transfer coefficient.

Thus, the temperature difference between the centre of the capillary and the outside is written as

$$T_0 - T_{ext} = \frac{I^2 r_1^2}{2\Lambda_m c}\left[\frac{1}{2\lambda_t} + \frac{1}{\lambda_s}\ln\left(\frac{r_2}{r_1}\right) + \frac{1}{\lambda_p}\ln\left(\frac{r_3}{r_2}\right) + \frac{1}{r_3 h}\right] \tag{6.63}$$

The last term dominates the temperature distribution.

This calculation clearly shows that controlling the temperature of the column is an essential factor in assuring reproducibility of retention times.

Detection methods

The small quantities of samples introduced into the separation capillary render the detection of peaks rather difficult. We can distinguish five classes of detectors:

- UV VIS
- fluorescence
- refractive index
- electrochemical
- contactless conductivity

For the optical methods, the polymer protection layer that surrounds the quartz capillary has to be removed. The sensitivity of UV-VIS methods is limited by the short optical path imposed by the diameter of the fibre. Fluorescence is a more sensitive method, but often requires a 'derivatisation' of the reactant if it is not fluorescent itself.

Electrochemical detectors comprise potentiometric, conductometric and amperometric detectors. For potentiometric detectors, ion selective microelectrodes are usually used. Amperometric microelectrodes are also among the most sensitive methods for detecting redox components. However, electrochemical detection cannot be used directly in the electric field lines used for electro-osmotic pumping. To circumvent this difficulty, the detector can be placed at the end of the capillary as in a wall jet cell configuration (see §7.4.1). If the detector is placed directly in the capillary, it is important to electrically uncouple the high voltage electric circuit from the 3-electrode measuring circuit (see §7.1).

Conductive coupling can be used to measure the conductivity of the solution in the capillary by placing electrodes outside. In this case, it is important to operate at higher frequencies compared to classical conductometry with the electrodes directly in contact with the solution.

In addition to the detection methods cited above, note that coupling with mass spectrometry can be done relatively easily using an electro-spray ionisation.

6.3 ELECTROPHORETIC METHODS OF ANALYTICAL SEPARATION

There are about five types of electrophoretic separation.

6.3.1 Moving boundary

The phenomenon of the ***moving boundary*** can be simply illustrated by considering the system in Figure 6.15.

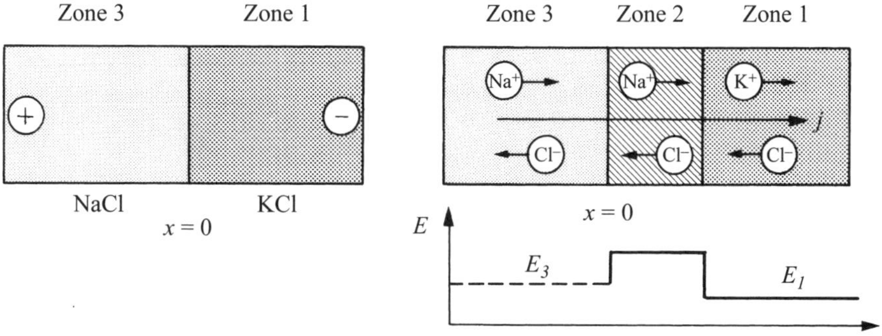

Fig. 6.15 Scheme of the moving boundary principle.

At time $t = 0$, we form an interface between a KCl solution (on the right, zone 1) and a NaCl solution (on the left, zone 3). From an experimental point of view, a 3-way tap allows the formation of such interfaces. We then apply an electric field from left to right. Given that the mobility of the Na^+ ions ($u_{Na^+} = 51.9{\cdot}10^{-9}$ m·V^{-1}·s^{-1}) is lower than the mobility of the K^+ ions ($u_{K^+} = 76.2{\cdot}10^{-9}$ m^2·V^{-1}·s^{-1}), the cations will migrate at different speeds, thus generating an intermediary zone of NaCl (zone 2) with a different concentration from zone 3. The boundary between zones 2 and 3 remains immobile while the current passes.

Given that the current density j is constant across the three zones, but that the concentrations and therefore the conductivities are different, the electric field will be different in each zone. The electric field in zone 2 may be obtained by measuring the progress velocity $\boldsymbol{v}_{12}$ of the moving boundary between zones 1 and 2. In effect, using equation (4.18), we have

$$\boldsymbol{v}_{12} \quad = \quad u_{\mathrm{Na^+},2}\boldsymbol{E}_2 \tag{6.64}$$

The difference in electric field between zones 1 and 2 has for consequence to maintain the separation boundary sharp (*self-sharpening effect*). For example, if a sodium ion went into zone 1, it would find itself in a zone where the field would be too weak with respect to its mobility and it would therefore rejoin zone 2. To calculate the concentrations of the different ions in the different zones, we have to consider the variation in volume of zone 2 per unit charge, W_{12}, defined by

$$W_{12} \quad = \quad \frac{\Delta V_2}{\Delta Q} \tag{6.65}$$

where ΔQ represents the charge that passes during the time Δt and ΔV_2 the variation in volume of the zone 2 formed. The quantity W_{12} is a function of the progress velocity of the interface 1 | 2,

$$W_{12} \quad = \quad \frac{\Delta V_2}{\Delta Q} \quad = \quad \frac{v_{12} S \Delta t}{\Delta Q} \quad = \quad \frac{v_{12}}{j} \tag{6.66}$$

where S is the cross-section of the tube and j the current density. The amount of sodium in zone 2 equals the flux of sodium going into this zone during the period Δt

$$\left(J_{\mathrm{Na^+},2}\right)\cdot \Delta t S \quad = \quad \left(c_{\mathrm{Na^+},2}\right)\cdot \Delta V_2 \tag{6.67}$$

Using equation (4.21), the sodium flux is given by

$$J_{\mathrm{Na^+},2} \quad = \quad v_{\mathrm{Na^+},2}\, c_{\mathrm{Na^+},2} \quad = \quad u_{\mathrm{Na^+},2}\, c_{\mathrm{Na^+},2} E_2 \quad = \quad \frac{u_{\mathrm{Na^+},2}\, c_{\mathrm{Na^+},2}}{\sigma_2} j \tag{6.68}$$

and we can write that

$$W_{12} \quad = \quad \frac{v_{12}}{j} \quad = \quad \frac{u_{\mathrm{Na^+},2} E_2}{j} \quad = \quad \frac{u_{\mathrm{Na^+},2}}{\sigma_2} \tag{6.69}$$

In the same way, the amount of potassium initially present in zone 2 equals the potassium flux going out of this zone during the time Δt, and thus we have

$$\frac{u_{K^+,1}}{\sigma_1} = W_{12} \tag{6.70}$$

Relations (6.60) & (6.70) allow the calculation of the progress velocity of the boundary

$$v_{12} = \frac{u_{Na^+,2}\,j}{\sigma_2} = \frac{u_{K^+,1}\,j}{\sigma_1} \tag{6.71}$$

Furthermore, using equations (4.16) & (4.21), we have

$$\sigma_1 = F\left[u_{K^+,1} - u_{Cl^-,1}\right]c_{K^+,1} \tag{6.72}$$

and

$$\sigma_2 = F\left[u_{Na^+,2} - u_{Cl^-,2}\right]c_{Na^+,2} \tag{6.73}$$

We deduce from this that

$$\frac{c_{Na^+,2}}{c_{K^+,1}} = \frac{u_{Na^+,2}}{u_{Na^+,2} - u_{Cl^-,2}} \cdot \frac{u_{K^+,1} - u_{Cl^-,1}}{u_{K^+,2}} \tag{6.74}$$

These equations show that the concentration of NaCl behind the KCl lead solution is solely determined by the salt concentration in the head zone and the electric mobility of the different species.

EXAMPLE

> Consider a system where the initial concentration of KCl in zone 1 is 0.01 M, and let's calculate the concentration of NaCl in zone 2.
> With the data of Table 4.1 and by making use of equation (4.16), equation (6.74) shows that this concentration is 0.008 M.
> It is interesting to note that this value does not depend on the initial concentration of NaCl. Of course, from an experimental point of view, the moving boundary is observed if the initial concentration of NaCl is of the same order of magnitude as that in the head zone. Thus, if this initial concentration is 0.01 M, the distribution of the electric field in the three zones is that shown in Figure 6.15.

Now consider a more complicated situation where zone 1 contains a mixture of NaCl and KCl. As before, a mass balance for the K^+ ion gives

$$\left(J_{K^+,1} - J_{K^+,2}\right)\Delta t S = \left(\frac{u_{K^+,1}c_{K^+,1}}{\sigma_1} - \frac{u_{K^+,2}c_{K^+,2}}{\sigma_2}\right)\Delta Q = \left(c_{K^+,1} - c_{K^+,2}\right)V_2 \tag{6.75}$$

which is

$$\frac{u_{K^+,1}c_{K^+,1}}{\sigma_1} - \frac{u_{K^+,2}c_{K^+,2}}{\sigma_2} = W_{12}\left(c_{K^+,1} - c_{K^+,2}\right) \tag{6.76}$$

In the same way for the chloride ion, we have

$$\frac{u_{\mathrm{Cl^-},1}c_{\mathrm{Cl^-},1}}{\sigma_1} - \frac{u_{\mathrm{Cl^-},2}c_{\mathrm{Cl^-},2}}{\sigma_2} = W_{12}\left(c_{\mathrm{Cl^-},1} - c_{\mathrm{Cl^-},2}\right) \tag{6.77}$$

For the sodium ion, equation (6.69) remains applicable, and the conservation of the flux at the boundary between zones 2 and 3 allows the calculation of the concentration of Na^+ ions in the intermediary zone

$$\frac{u_{\mathrm{Na^+},2}c_{\mathrm{Na^+},2}}{u_{\mathrm{Na^+},3}c_{\mathrm{Na^+},3}} = \frac{\sigma_2}{\sigma_3} \tag{6.78}$$

The concentrations in the different zones can be calculated by the solution of the system of equations given above. The analytical resolution can prove difficult and numerical methods are often required.

6.3.2 Isotachophoresis

This technique, whose name tells us that all the species migrate at the same velocity, is used for separating species with identical charge. Let's look, by way of an example at a separation of cations from a mixture of salts with common anions (e.g. Cl^-), placed between a solution with high electric mobility (e.g. HCl) and a solution with very low mobility (e.g. an organic salt such as tetrabutylammonium chloride TBACl). The passing of a current not only causes a separation of the cations, but also leads to the formation of distinct bands.

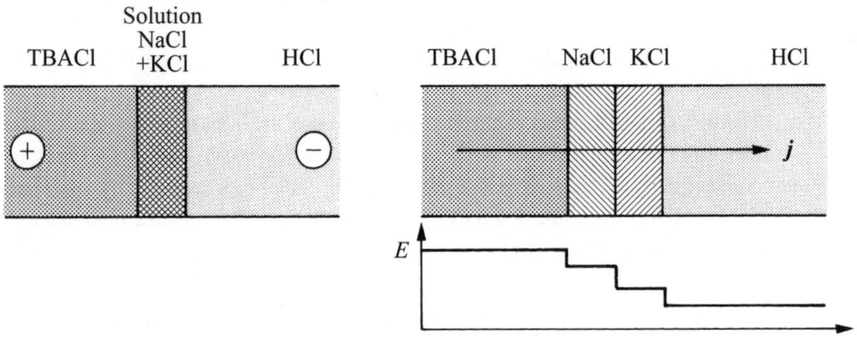

Fig. 6.16 Scheme of the principle of isotachophoresis.

In effect, when the potential is applied, potential gradients form in such a way that all the ions move at the same velocity; the electric field is weak in the zones where the ions are highly mobile, and strong in the zones where the ions are not very mobile; the electric field is thus constant in each band. So the separation between the bands is very clear, as shown in Figure 6.17.

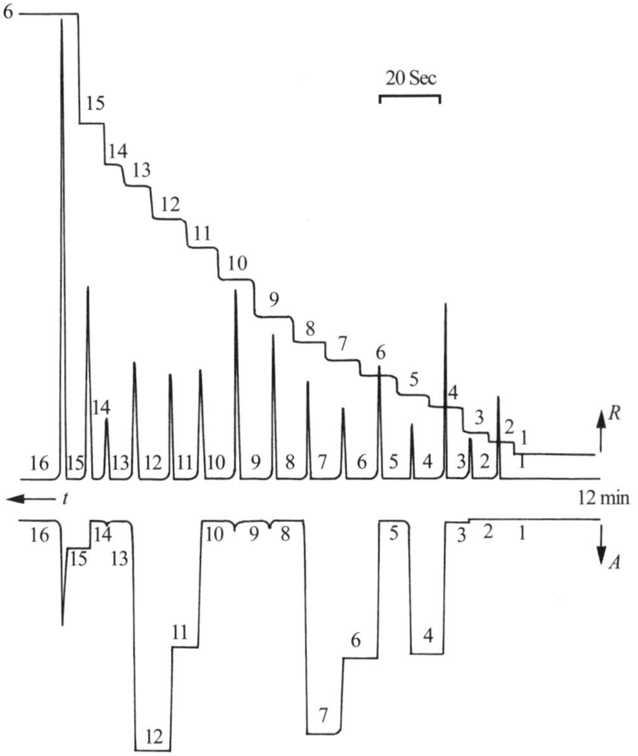

Fig. 6.17 Isotachophoretic separation of a series of anions. Current density: 60mA·cm^{-2}. Conductometric (top) and UV detection (bottom). 1=chloride, 2=sulfate, 3=oxalate, 4=naphthalene trisulfate, 5=malonate, 6=pyrazole-3.5-dicarboxylate, 7=naphthalene disulfonate, 8=adipate, 9=acetate, 10=β-chloropropionate, 11=benzolate, 12=naphthalene monosulfonate, 13=glutamate, 14=enanthate, 15=benzyldiaspartate, 16=morpholinoethane sulfonate (Electrophoresis, a survey of techniques and applications, Z. Deyl ed., Elsevier, Amsterdam, 1979).

6.3.3 Zone electrophoresis

It is possible to separate small samples by using electrophoresis in an electrolyte solution (often a pH buffer) supported in a rigid matrix such as filter paper or in a capillary tube. The solid matrix prevents convection and ensures that the only mass transports possible are diffusion and migration.

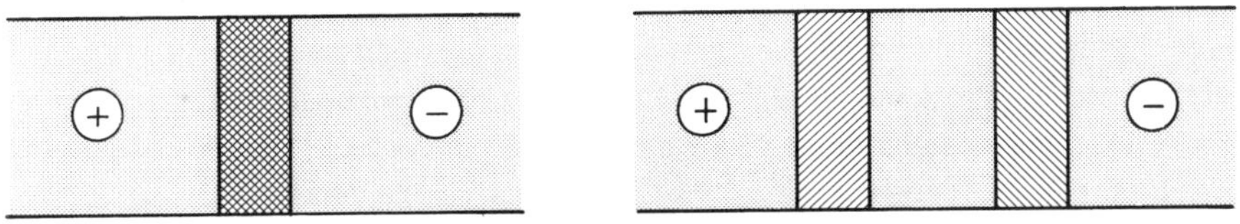

Fig. 6.18 Scheme of the principle of zone electrophoresis.

Zone electrophoresis is the technique generally employed in capillary electrophoresis for separating the charged ions or molecules and even certain proteins. However, this method cannot be used for the separation of DNA or proteins denatured by charged surfactants (*vide infra*).

6.4 ELECTROPHORETIC SEPARATION OF BIOPOLYMERS

6.4.1 PAGE Electrophoresis

Electrophoresis on paper is not much used nowadays. However, electrophoresis in polyacrylamide gels (***PAGE*** = *PolyAcrylamide Gel Electrophoresis*) remains the classic technique for the separation of proteins. This separation method based on zone electrophoresis has allowed the spectacular advances in molecular biology in the second half of the twentieth century. Polyacrylamide is a three-dimensional network obtained by polymerisation of acrylamide (CH_2=CH–$CONH_2$) using methylenebisacrylamide (CH_2=CH–CO–NH–CH_2–NH-CO–CH=CH_2) for the reticulation, in the presence of a catalyst such as tetramethylenediamine (TEMED).

Protein separation

Electrophoresis on a gel allows the separation of proteins according to their size, by using the porosity of the gel. The proteins are first denatured with detergents such as **SDS** (Sodium Dodecylsulfate) (1.4 g of SDS per g of proteins) and by reducing agents such as 2-mercaptoethanol to break the disulfur bridges (–S–S–). The proteins, denatured in this way can be assimilated into polyelectrolyte chains 'wrapped up' in ionic surfactant molecules. One of the goals of the denaturing process is to fix the mass to charge ratio that implies that the denatured proteins have nearly all the same electrophoretic mobility.

The presence of the gel in fact allows the separation of the proteins principally by their size. The distance covered in the gel is then directly proportional to the logarithm of their molar mass. Models such as the reptation model presented in §6.4.2 can be used to describe the separation mechanism.

One of the major drawbacks of this technique is the slowness of the operations of making the gel, running the electrophoretic separation and staining the proteins for visual identification, either with dyes (Coomassie Blue) or with silver salts.

For molecules whose mass exceeds 1 MDa, the pores of PAGE are too fine to allow migration. In this case, we can use agarose gels that have a greater porosity. Electrophoresis in an agarose gel – a negatively charged polysaccharide – is also affected by electro-osmosis that causes a flow of the solvent towards the cathode. This flow is quite large, since the majority of proteins, even negatively charged, arrive at the cathode and only the immunoglobulins go towards the anode.

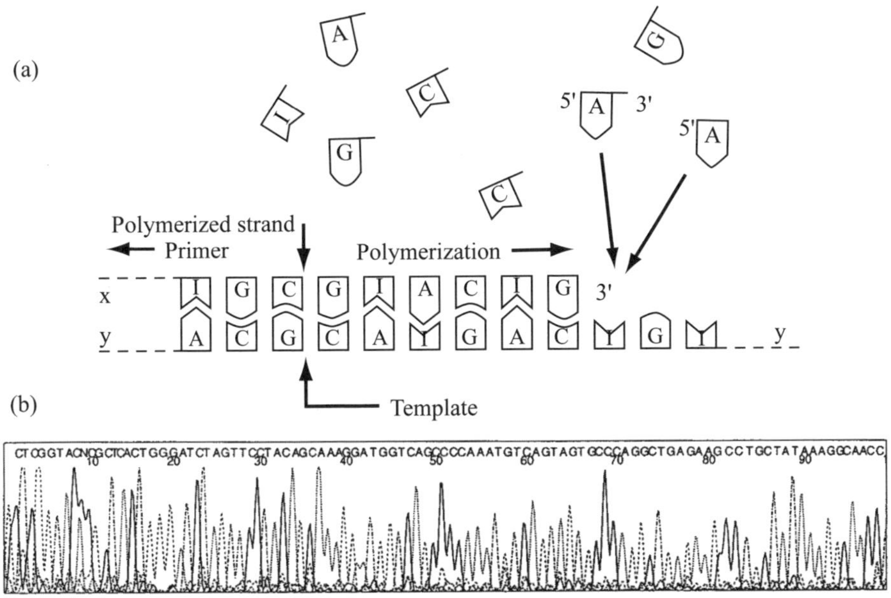

Fig. 6.19 Schematic representation of a sequencing process ("four-color Sanger") starting from many copies of the single-stranded DNA to be sequenced, bearing a known "marker" at the beginning of the unknown sequence, a short oligonucleotide "primer" complementary to this marker is hybridized (i.e., paired) to the marker, in the presence of DNA polymerase and free nucleotides. This hybridization initiates reconstruction by the polymerase of a single strand complementary to the unknown sequence (a). Including in the nucleotide bath in which the polymerization takes place a small fraction of fluorescently labelled dideoxynucleotides (one different dye for each nucleotide type), which miss the OH group necessary for further extension of the strand, one is able to synthesize at random complementary strands with all possible arrest points (i.e., all possible lengths with an integer number of nucleotides). These newly synthesized single-stranded DNA's are then separated electrophoretically by size [see electrophoregram in (b)]: consecutive peaks correspond to DNA fragments differing by one base, and each line corresponds to one given nucleotide. Automated analysis of the data allows the determination of the sequence (symbols above the peaks). The symbol *N* indicates ambiguous determination. In the present case, the sequence determination was faultless up to 435 bases. (Adapted from J.-L. Viovy, *Rev. Mode. Phys.*, 72 (2000) 813-872).

Separation of nucleic acids and the sequencing of DNA

Electrophoresis in a gel is one of the methods for separating nucleic acids. Gel electrophoresis in a capillary has recently been used on a large scale for the sequencing of the human genome. Fragments of single stranded DNA labelled with fluorescent molecules are generated by enzyme reactions, and analysed in parallel to allow rapid reading of the sequences as shown in Figure 6.19.

As with the denatured proteins, the separation of double stranded ds-DNA by electrophoresis in solution is not possible because the ratio of mass to charge is constant. Thus, the electrophoretic mobility of a ds-DNA chain is practically independent of its composition in elementary bases. Nevertheless, the use of agarose gel, hydroxyethycellulose or entangled polymers which play the role of transitory gels allow electrophoretic separation. In the case of planar gels, the constant electric field is sometimes replaced by orthogonal pulsed fields applied alternately.

6.4.2 Reptation theory

Movement of an ideal chain

One of the ways of treating the movement of a polymer in solution is to consider that the position of the chain corresponds to random steps like those described in chapter 4. To simplify, let's consider a chain in a flat geometry.

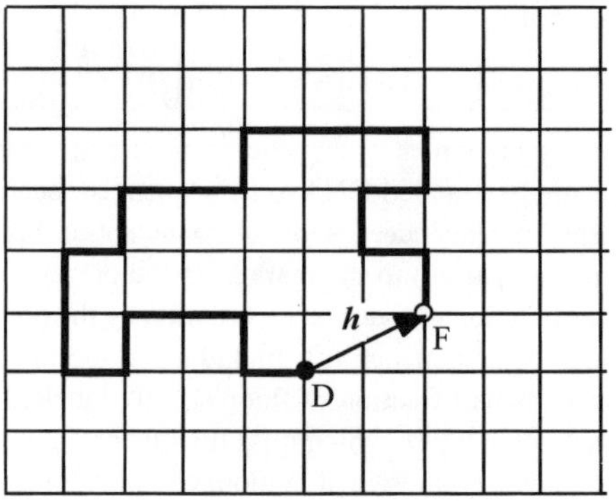

Fig. 6.20 Flat chain, or random steps in a plane.

The vector $\boldsymbol{h}$ that links the beginning and end of the chain is the sum of N step vectors $\boldsymbol{a}_i$, i.e.

$$\boldsymbol{h} = \sum_i \boldsymbol{a}_i \tag{6.79}$$

Because of the random character of the chain, the step vectors have independent orientations, having the consequence that in the calculation of the average root mean square, the cross terms cancel out

$$<\boldsymbol{h}^2> = \sum_i \sum_j <\boldsymbol{a}_i \cdot \boldsymbol{a}_j> = \sum_i <\boldsymbol{a}_n^2> = Na^2 \tag{6.80}$$

This relationship is valid for long chains that we shall call 'the ideal chain'. So we define the average 'end to end' length of an ideal chain as

$$R_N = a\sqrt{N} \tag{6.81}$$

Reptation

In a gel, a biopolymer does not go forward randomly. In fact, we can consider that the biopolymer moves like a 'snake in tall grass'. The theory of the reptation of biopolymers in a gel makes a first hypothesis that the sequence of pores via which the molecule travels forms a tube whose diameter can, in a first approximation, be considered as uniform.

Thus, we shall separate the sliding of the polymer chain in its tube and the random movement of the tube.

If the polymer is flexible on the scale of the size of the pores, we name a 'blob' an element of length a characteristic of the porosity such that the linear length of the chain is

$$L = N a \tag{6.82}$$

The time required for the chain to come completely out of its tube can be calculated by defining a sliding mobility u_s of the chain in the tube. This mobility can be defined by considering a force $\boldsymbol{f}$ due to an electrochemical potential gradient and the linear sliding velocity of the chain in the tube $\boldsymbol{v}_s$ such that we have

$$\boldsymbol{v}_s = u_s \boldsymbol{f} \tag{6.83}$$

By definition, the sliding mobility must be proportional to the length of the chain and therefore inversely proportional to the number of blobs

$$u_s = 1 / N\zeta_s \tag{6.84}$$

where ζ_s is the sliding friction coefficient for one blob.

By analogy with the Stokes-Einstein law (see §4.1.2), we can also define a sliding diffusion coefficient D_s by

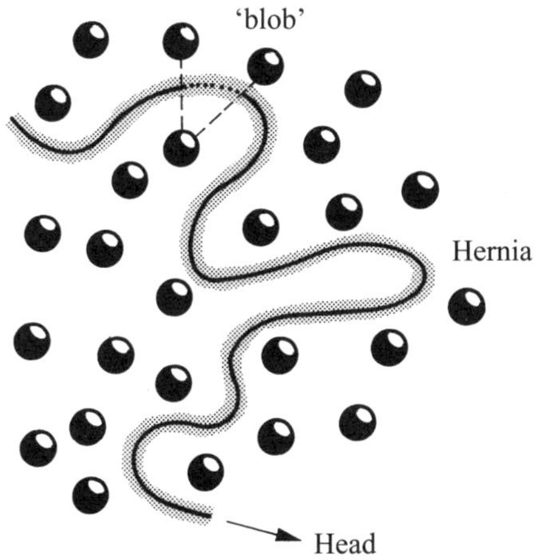

Fig. 6.21 Reptation of a biopolymer in a porous matrix.

$$D_s = RTu_s = \frac{kT}{\zeta_s} \tag{6.85}$$

Thus, we can calculate the time τ_s for the chain to slide out of its tube by using equations (4.164) or (4.173).

$$\tau_s = \frac{L^2}{2D_s} = \tau_1 N^3 \tag{6.86}$$

where L is the linear length of the chain. This equation is a scaling law that shows that the time for a polymer to slide within its tube is proportional to the cube of the number of elements in the chain. τ_1 is the time to slide through an element of the tube.

If, over the time τ_s the whole length L has slid into its tube, the displacement in space corresponds to that of the 'end-to-end' length R_N. Thus, the diffusion coefficient by reptation is given by

$$D_{rep} = \frac{R_N^2}{2\tau_s} = \frac{D_1}{N^2} = \frac{kT}{\zeta_{rep}} \tag{6.87}$$

D_1 is the diffusion coefficient of an element in the chain. Equation (6.87) shows that the diffusion coefficient by reptation of a polymer is inversely proportional to the square of its length. The friction constant ζ_{rep} of an unstructured chain is proportional to the square of its mass. A random diffusion movement in a porous matrix will therefore be faster for smaller molecules than for longer molecules.

Reptation in an electric field

For a polyelectrolyte in an electric field $\boldsymbol{E}$, each element of the chain will be subject to an electric migration force. The total force exerted on the chain in the direction of the field is

$$f = \sum_i q_i \boldsymbol{E}\cdot\hat{\boldsymbol{t}}_i = \frac{q}{a}\sum_i \boldsymbol{E}\cdot\boldsymbol{a}_i = \frac{qE}{a}\sum_i \hat{\boldsymbol{i}}\cdot\boldsymbol{a}_i = \frac{qE}{a}h_x \tag{6.88}$$

where q is the charge of a segment, and $\hat{\boldsymbol{t}}_i$ is the unit tangent vector of each element of the chain and $\hat{\boldsymbol{i}}$ the unit vector in the direction of the electric field and h_x the projection of the 'end-to-end' vector illustrated in Figure 6.20. Using equations (6.83) & (6.88), the sliding velocity is then given by

$$v_s = \frac{qEu_s}{a}h_x = \frac{qE}{Na\zeta_s}h_x = \frac{qE}{L\zeta_s}h_x \tag{6.89}$$

To find the velocity of displacement of the centre of mass of the polymer in the direction of the field, we can write

$$M\boldsymbol{R}_{cm} = \sum_i m_i \boldsymbol{r}_i \tag{6.90}$$

where M is the total mass, $\boldsymbol{R}_{cm}$ the displacement to the centre of mass, m_i the mass of a segment ($= M a_i / L$) and $\boldsymbol{r}_i$ its position vector. Differentiating with respect to time, we have

$$M\boldsymbol{V}_{\mathrm{cm}} = \sum_i M\frac{a_i}{L}\boldsymbol{v}_{\mathrm{s}i} \tag{6.91}$$

Noting that the sliding velocity of an element of the chain is a constant

$$\boldsymbol{v}_{\mathrm{s}i} = v_{\mathrm{s}}\,\hat{\boldsymbol{t}}_i \tag{6.92}$$

we get

$$\boldsymbol{V}_{\mathrm{cm}} = \frac{v_{\mathrm{s}}}{L}\sum_i a_i\,\hat{\boldsymbol{t}}_i = \frac{v_{\mathrm{s}}}{L}\sum_i \boldsymbol{a}_i = \frac{v_{\mathrm{s}}}{L}\boldsymbol{h} = \tau_{\mathrm{s}}^{-1}\boldsymbol{h} \tag{6.93}$$

Thus, the velocity of the centre of mass on the x- axis, $V_{x\mathrm{cm}}$, is written as

$$V_{x\mathrm{cm}} = \frac{v_{\mathrm{s}}}{L}h_x \tag{6.94}$$

By combining with equation (6.89), we obtain a relation for the average velocity of a polyelectrolyte chain in an electric field

$$< V_{x\mathrm{cm}} > = \frac{qE}{\zeta_{\mathrm{s}}L^2} < h_x^2 > \tag{6.95}$$

Equation (6.95) is applicable as a first approximation for the electrophoresis of polyelectrolytes in porous gels. It shows that chains aligned with the electric field move faster than those placed perpendicular to it, and that the speed is inversely proportional to the square of the length.

Biased reptation in an electric field

An essential aspect of reptation theory is the role played by the 'head' of the chain, which has to choose which pore to enter into.

In an electric field, the movement of a polyelectrolyte will also be influenced by the electrostatic force that is applied at the head of the chain when going out of the sliding tube. In the absence of a field, this direction is completely random whilst in the presence of a field this direction is 'biased' by the field. Thus, the influence of the field on the 'charged head' of the polyelectrolyte and the Brownian motion will be responsible for the direction taken by the chain when the head goes forward.

To quantify these effects, we need to calculate $<h_x^2>$. From equation (6.79), we can calculate

$$h_x = \sum_i a_{xi} \tag{6.96}$$

and

$$h_x^2 = \sum_i\sum_j a_{xi}a_{xj} = \sum_i a_{xi}^2 + \sum_i\sum_{j\neq i} a_{xi}a_{xj} \tag{6.97}$$

If the chain is flexible, and the effect of the electric field on the orientation of the head is weak, there is little correlation between the orientations of the different segments. So, taking averages, we have

$$< h_x^2 > \;=\; N < a_x^2 > \;+\; N(N-1) < a_x >^2 \tag{6.98}$$

To take into account the influence of the field, we make the hypothesis that a_x is governed by a Boltzmann distribution of the electrostatic energy for the head segment which can move freely. The other segments inside cannot move in a lateral manner because of the gel and can only slide along the tube. By defining for each element an angle of orientation θ with respect to the direction of the field, such that

$$a_x \;=\; a\cos\theta \tag{6.99}$$

equation (6.98) becomes

$$< h_x^2 > \;=\; Na^2 < \cos^2\theta > \;+\; N(N-1)a^2 < \cos\theta >^2 \tag{6.100}$$

Now the problem is reduced to the calculation of $< \cos\theta >$ and $< \cos^2\theta >$. To do this, we can calculate the average electrostatic energy due to the interaction of the electric field with the charged head segment

$$w_e \;=\; -\tfrac{1}{2} q\boldsymbol{E}\cdot\boldsymbol{a} \;=\; -\tfrac{1}{2} qaE\cos\theta \tag{6.101}$$

Using the same reasoning as in §3.3.5 where we calculated the average ion-dipole energy of interaction, we can say that the electrostatic energy w_e is given by a Boltzmann distribution

$$< w_e > \;=\; \frac{\int_0^\pi w_e \exp^{-w_e/kT} \sin\theta d\theta}{\int_0^\pi \exp^{-w_e/kT} \sin\theta d\theta} \tag{6.102}$$

To calculate these two integrals, we put

$$x \;=\; aqE\cos\theta / 2 \tag{6.103}$$

and

$$u \;=\; aqE / 2 \tag{6.104}$$

We have then

$$\begin{aligned} < w_e > \;&=\; -\frac{\int_{-u}^{u} x\exp^{x/kT} dx}{\int_{-u}^{u} \exp^{x/kT} dx} \;=\; -\frac{\left[xkT\exp^{x/kT} - k^2T^2\exp^{x/kT}\right]_{-u}^{u}}{\left[kT\exp^{x/kT}\right]_{-u}^{u}} \\ &=\; -u\coth\left(\frac{u}{kT}\right) + kT \;=\; -u\, L\left(\frac{u}{kT}\right) \end{aligned} \tag{6.105}$$

where $L(x)$ is the Langevin function described in Figure 3.15. For small values of u, the average electrostatic energy tends towards

$$\lim_{u\to 0} < w_e > \; = \; \lim_{u\to 0}\left[u\, L\left(\frac{u}{kT}\right)\right] \; = \; \frac{u^2}{3kT} \tag{6.106}$$

The average orientation angle is then defined by

$$< \cos\theta > \; = \; \frac{< w_e >}{-u} \; = \; L\left(\frac{u}{kT}\right) \tag{6.107}$$

In the same way,

$$< w_e^2 > \; = \; \frac{\int_0^\pi w_e^2 \exp^{-w_e/kT} \sin\theta \mathrm{d}\theta}{\int_0^\pi \exp^{-w_e/kT} \sin\theta \mathrm{d}\theta} \tag{6.108}$$

which is

$$\begin{aligned} < w_e^2 > \; &= \; \frac{\int_{-u}^{u} x^2 \exp^{x/kT} \mathrm{d}x}{\int_{-u}^{u} \exp^{x/kT} \mathrm{d}x} \; = \; \frac{\left[x^2 kT \exp^{x/kT}\right]_{-u}^{u} - 2kT\int_{-u}^{u} x\exp^{x/kT}\mathrm{d}x}{\left[kT\exp^{x/kT}\right]_{-u}^{u}} \\ &= \; u^2 - 2kTu \coth\left(\frac{u}{kT}\right) + 2k^2T^2 \; = \; u^2 - 2kTu\, L\left(\frac{u}{kT}\right) \end{aligned} \tag{6.109}$$

For small values of u, the average electrostatic energy tends to

$$\lim_{u\to 0} < w_e^2 > \; = \; \lim_{u\to 0}\left[u^2 - 2kTu\, L\left(\frac{u}{kT}\right)\right] \; = \; \frac{u^2}{3} \tag{6.110}$$

and $< \cos^2\theta >$ tends to 1/3 since

$$< \cos^2\theta > \; = \; \frac{< w_e^2 >}{u^2} \; = \; 1 - \frac{2kT}{u} L\left(\frac{u}{kT}\right) \tag{6.111}$$

By substituting in equation (6.100), we have

$$\begin{aligned} \frac{< h_x^2 >}{a^2} \; &= \; N\left[< \cos^2\theta > - < \cos\theta >^2\right] + N^2 < \cos\theta >^2 \\ &= \; N\left[1 - \frac{2kT}{u} L\left(\frac{u}{kT}\right) - L^2\left(\frac{u}{kT}\right)\right] + N^2 L^2\left(\frac{u}{kT}\right) \end{aligned} \tag{6.112}$$

Finally, by doing a limited series expansion and replacing in equation (6.95) we have

$$< V_{x\mathrm{cm}} > \; = \; \frac{qE}{\zeta_s 3}\left[\frac{1}{N} + \frac{1}{3}\left(\frac{aqE}{2kT}\right)^2\right] \tag{6.113}$$

The mobility $<V_{x\mathrm{cm}}> / E$ consists then of two terms: one independent of the field and inversely proportional to N and the other independent of the length that becomes negligible for weak fields. For small values of N, the first term of equation (6.113) dominates accounting for the experimental dependence of the mobility with N^{-1}. For large values of N and for strong fields, the mobility of long polyelectrolytes becomes independent of size and dependent on the field.

The biased reptation model presented above can be amended to consider also the fluctuations of the end of the tube itself and not only the fluctuations of the first element of the chain, shown as the head in Figure 6.21. This means that we have to consider the internal modes of the tube itself. The mathematical considerations involved are too long to be presented here, but this model called the 'biased reptation model with fluctuations' corroborates rather well the experimental data.

6.4.3 Isoelectric focusing

This technique is used for separating ampholytes, such as proteins. Separation is effected in a pH gradient, from an arbitrary position. The negative species migrate towards the anode, whilst the positive species migrate to the cathode. These species become immobilised when they arrive in a pH zone where they become neutralised. The pH for which an ampholyte no longer migrates in an electric field is called the ***isoelectric point*** (pI) of the ampholyte. The pH gradient is usually created by distributing ampholytes of different pI or even polyampholytes (polymers containing for example at the same time amino and carboxylate groups). The buffer effect of these different molecules generates a pH gradient whose resolution reaches 0.001 pH units, which allows the separation of proteins with pI differences of the order of 0.02 pH units. It is important to note that isoelectric focusing is a steady state method that does not depend on the mode of application of the sample, nor on the total quantity of proteins as long as the buffer effect of the pH gradient is not disturbed.

The pH gradient can be immobilised in a gel structure by the polymerisation of acid or basic derivatives of acrylamide previously distributed in such a way as to form a pH gradient. After the polymerisation, the gradient is immobilised. (***IPG*** = *Immobilised* pH *gradient*).

The concept of focusing can be explained from transport equations. If $\boldsymbol{J}_i$ is the flux of a species i given by equation (4.10),

$$\boldsymbol{J}_i = -c_i\tilde{u}_i\,\mathbf{grad}\tilde{\mu}_i = -c_i\tilde{u}_i\,\mathbf{grad}\mu_i - c_i\tilde{u}_i z_i F\,\mathbf{grad}\phi \tag{6.114}$$

the law of the conservation of mass in the steady state regime is written as

$$\frac{\partial c_i}{\partial t} = -\operatorname{div}\boldsymbol{J}_i = 0 \tag{6.115}$$

By only considering unidirectional transport, this expression becomes

$$-\frac{\partial}{\partial x}\left[-\frac{c_i\tilde{u}_i RT}{c_i}\frac{\partial c_i}{\partial x}_i - c_i\tilde{u}_i z_i F\frac{\partial\phi}{\partial x}\right] = 0 \tag{6.116}$$

If we make the hypothesis that the applied electric field is not perturbed by the presence of a migrating protein or ampholyte (excess of supporting electrolyte) and

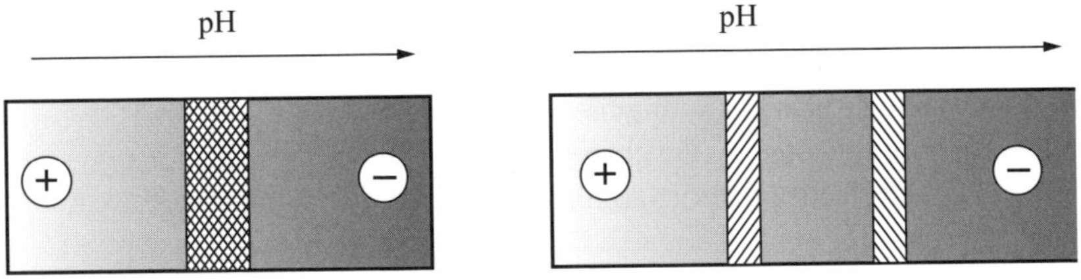

Fig. 6.22 Scheme of the principle of isoelectric focusing. The ampholytes (the cross-hatched zones) migrate when an electric field is applied parallel to the pH gradient.

that it is therefore constant across the gel, the only terms in equation (6.116) depending on the distance x are the concentration and the charge of the proteins. The term in square brackets in equation (6.116) is therefore constant and equal to zero at the isoelectric point, where the concentration is at its maximum, and the charge is zero. Thus, it appears that the concentration gradient is directly proportional to the electric field E

$$RT\frac{\partial c_i}{\partial x} = c_i z_i FE \tag{6.117}$$

The maximum concentration corresponds to the centre of the band, since as soon as a species leaves the band by diffusion, the electric field makes it 'get back in line'.

Titration of a protein in a pH gradient gel

We take a pH gradient gel, and apply along its length a solution containing one protein only. Then, we apply an electric field perpendicular to the pH gradient. The proteins situated in a pH zone higher than their pI become negatively charged and

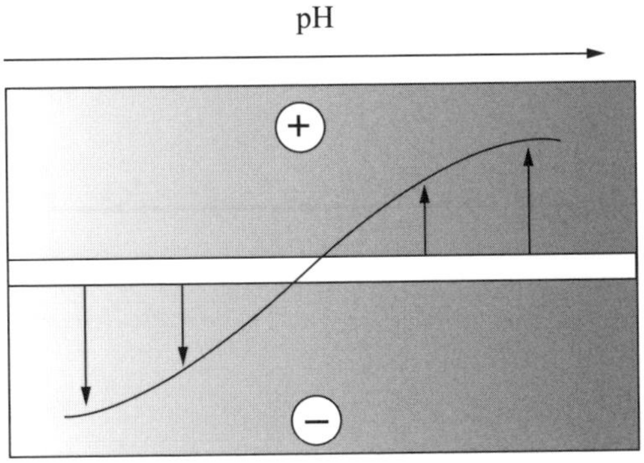

Fig. 6.23 Scheme of the principle of a pH titration of a protein deposited in a fine band all along the length of a pH gradient gel. The proteins migrate when an electric field is applied perpendicular to the pH gradient.

migrate to the anode, whilst those situated in a pH zone lower than their pI become positively charged and migrate to the cathode. Only proteins in a pH zone that corresponds to their pI will not migrate.

The distance travelled by the charged proteins depends on their charge at a given pH. In effect, the electrophoretic mobility is proportional to the charge.

Resolution of isoelectric focusing

In order to calculate the width of an isoelectric band such as shown in Figure 6.22, we need to solve the differential equation (6.117) all the while taking into account that the charge varies as a function of the pH and therefore of the distance. By linearising, the variation in the charge as a function of the pH around the corresponding pI value,

$$z_i = -p_i x \tag{6.118}$$

the negative sign being due to the charge becoming more negative when x increases (increasing pH values), we can write equation (6.117) in the form

$$RT\frac{\partial c_i}{\partial x} = -c_i p_i FE\, x \tag{6.119}$$

Integrating equation (6.119) then gives us

$$c_i = c_i^{\max} \exp^{-\frac{p_i FEx^2}{2RT}} \tag{6.120}$$

The steady state distribution in isoelectric focusing is therefore a Gaussian one, whose variance depends not only on the electric field applied and the temperature, but also on the factor p_i

$$\sigma = \frac{RT}{p_i FE} \tag{6.121}$$

The parameter p_i can be defined as being

$$p_i = -\left(\frac{\mathrm{d}z_i}{\mathrm{dpH}}\right)\left(\frac{\mathrm{dpH}}{\mathrm{d}x}\right) \tag{6.122}$$

The resolution of two Gaussian distributions being defined as

$$R_\sigma = \frac{\Delta \mathrm{pI}}{\overline{\sigma}} = \frac{2\Delta \mathrm{pI}}{\sigma_1 + \sigma_2} \tag{6.123}$$

we see that we can thus optimise the resolution by choosing a gel with an appropriate gradient.

'Off-gel' isoelectric separation

A method recently developed for the isoelectric purification of proteins in solution consists of streaming a solution of proteins under a pH gradient gel. Thus, by applying a p.d. parallel to the pH gradient, certain electric field lines pass through the channel

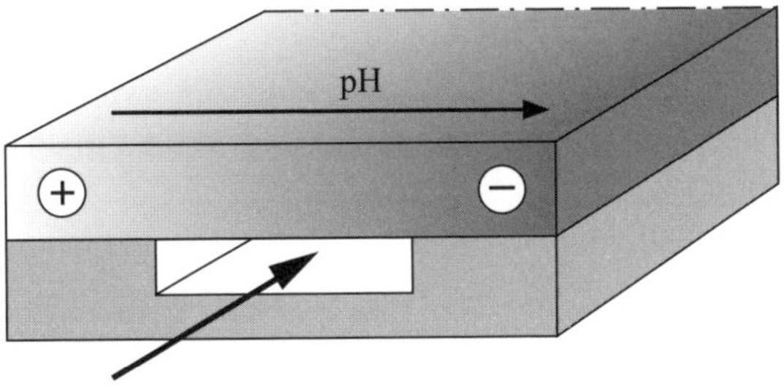

Fig. 6.24 Scheme of the principle of 'off-gel' isoelectric separation.

situated under the gel and take with them certain proteins, according to their charge. At the solution | gel junction, the penetration of the proteins into the gel will depend on the local pH.

Only the proteins whose pI is greater than the pH above the channel will be able to migrate towards the cathode, and conversely, only the proteins whose pI is less than this pH will be able to migrate towards the anode.

Thus, at the exit of the channel, the solution contains essentially proteins whose pI corresponds to the pH range in the gel covering the channel.

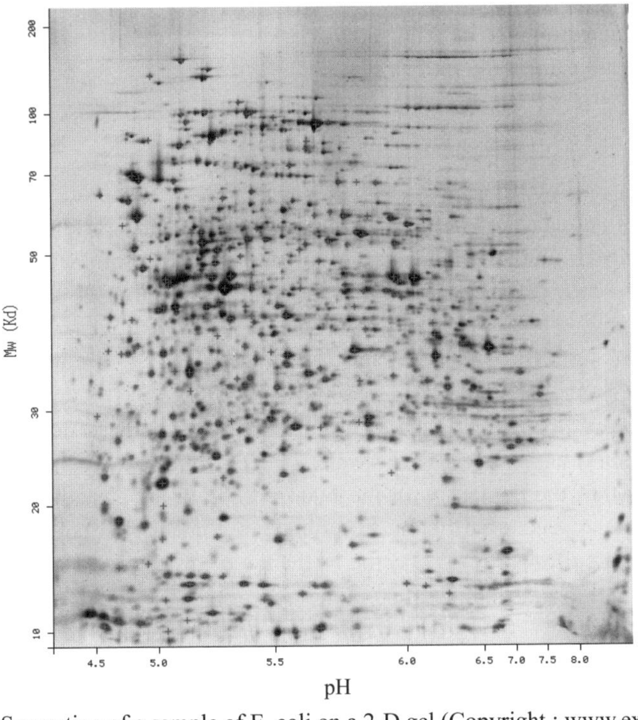

Fig. 6.25 Separation of a sample of E. coli on a 2-D gel (Copyright : www.expasy.ch).

6.4.4 Electrophoresis on a 2D gel

2D electrophoresis comprises, as its name indicates, two methods of separation. The first consists of separating the proteins according to their pI on an pH gradient gel (separation by electric focusing).

The second consists of doing a SDS-PAGE separation to differentiate the proteins by their molecular mass.

The gels obtained such as that shown in Figure 6.25 can then be compared to databanks to identify the proteins.

6.5 ION CHROMATOGRAPHY

6.5.1 Ion exchange chromatography

This method consists of passing electrolytes over columns containing ion exchange resins and can be used equally well for organic or mineral species.

For all chromatographic techniques, the capacity factor is defined as a function of a distribution coefficient $K_{D,i}$

$$k_i' = \phi K_{D,i} = \frac{V_s}{V_m}\frac{c_{i,s}}{c_{i,m}} \tag{6.124}$$

where ϕ is called the ***phase ratio,*** defined as the relation between the volumes of the stationary phase (V_S) and the mobile phase (V_M), c representing the concentrations.

The key question is to know how to evaluate the distribution coefficient between the stationary phase and the mobile phase.

$$K_{D,i} = \exp^{-\frac{\Delta G_{tr}^{\ominus,m\to s}}{RT}} = \frac{a_i^s}{a_i^m} \tag{6.125}$$

Even if it is easy in partition chromatography, such as HPLC, to define the stationary phase and the physico-chemical retention mode (e.g. distribution between two phases), the definition of the stationary phase is not so easy in ion chromatography.

In order to simplify, we can say that the stationary phase is a fine layer of solvent containing the hydration molecules of the fixed charges of the polymer which in a first approximation can be taken as the diffuse layer. It is difficult to define the thickness (few nanometers) and especially the water structure in this layer. Nevertheless, it is highly likely that the contact with the hydrophobic polymer and the fixed charges induces a solvent structure different from that in the bulk of the solution.

Consider a sulfonate resin and the exchange of a proton for a sodium ion. In a basic medium, this exchange is irreversible

$$R—SO_3^- \text{ --- } H^+ + NaOH \Rightarrow R—SO_3^- \text{ --- } Na^+ + H_2O$$

whereas in a neutral medium, this reaction becomes reversible

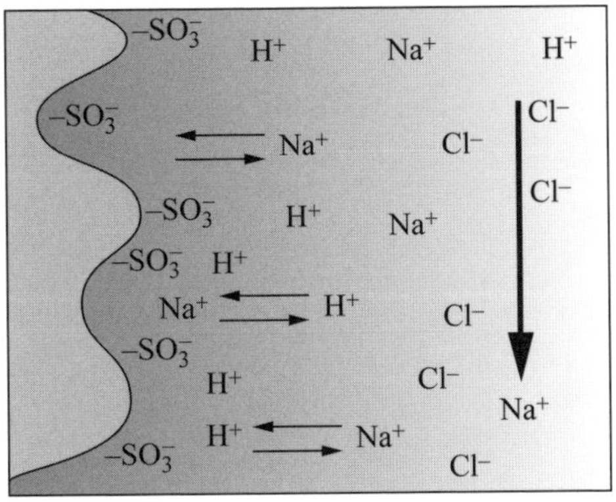

Fig. 6.26 Scheme of the principle of ion chromatography by ion exchange.

$$R - SO_3^- \text{ --- } H^+ \quad + \text{ NaCl} \quad \Leftrightarrow \quad R - SO_3^- \text{ --- } Na^+ \quad + \text{ HCl}$$

Because of the Donnan exclusion principle, the chloride does not penetrate into the exchanger if it is a porous resin. Given that the counter-ion does not intervene, we can write a law of mass action

$$K_{H^+}^{Na^+} = \frac{[Na^+]_s[H^+]_m}{[Na^+]_m[H^+]_s} \tag{6.126}$$

Table 6.1 Equilibrium constants for the exchange of cations on polystyrene sulfonate at 4% reticulation (Dowex 50, Dionex, USA).

Cation	$K_{H^+}^{M^{z+}}$	Cation	$K_{H^+}^{M^{z+}}$
H^+	1	Mg^{2+}	2.23
Li^+	0.76	Ca^{2+}	3.14
Na^+	1.2	Sr^{2+}	3.56
NH_4^+	1.44	Ba^{2+}	5.66
K^+	1.72	Co^{2+}	2.45
Rb^+	1.86	Ni^{2+}	2.61
Cs^+	2.02	Cu^{2+}	2.49
Ag^+	3.58	Zn^{2+}	2.37
Tl^+	5.08	Pb^{2+}	4.97
		UO_2^{2+}	1.79

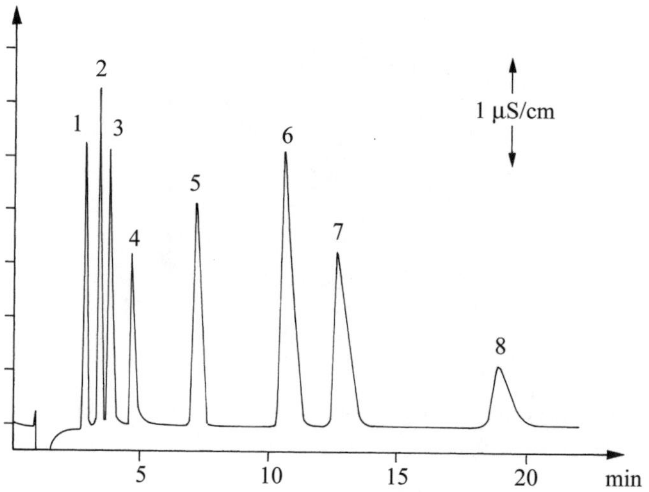

Fig. 6.27 Chromatography by cation exchange. Metrosep column. Eluent 4mM tartaric acid +1mM picolinic acid. Flow rate 1ml·mn^{-1}: 1. Lithium, 2. Sodium, 3. Ammonium, 4. Potassium, 5. Calcium, 6. Magnesium, 7. Strontium, 8. Barium. Picolinic acid is added to complex possible traces of heavy metals (Metrohm, CH).

The constant $K_{H^+}^{Na^+}$ is called the ***selectivity constant*** between the stationary and mobile phases. This can be measured, and certain tendencies deduced from it. For alkali metals, we have

$Cs^+ > Rb^+ > K^+ > Na^+ > Li^+$

whilst for alkaline earths metals, the series is :

$Ba^{2+} > Sr^{2+} > Ca^{2+} > Mg^{2+}$

It is possible to compare the selectivity coefficient to the distribution coefficient, and we have

$$K_D = \frac{[Na^+]_s}{[Na^+]_m} = K_{H^+}^{Na^+}\frac{[H^+]_s}{[H^+]_m} \qquad (6.127)$$

6.5.2 Chromatography by ionic exclusion

The separation is based on the Donnan exclusion principle, steric exclusion and adsorption phenomena. This technique is used for the separation of weak acids from acids that are completely dissociated at the pH of the eluent

In the example shown in Figure 6.28, the basic form of the acetic acid that is anionic is electrostatically repelled from the diffuse layer around the fixed charges of the polymer cation exchanger where the solution velocity is less than in the bulk. This phenomenon is the same as that described in §2.6.2. On the other hand, the neutral

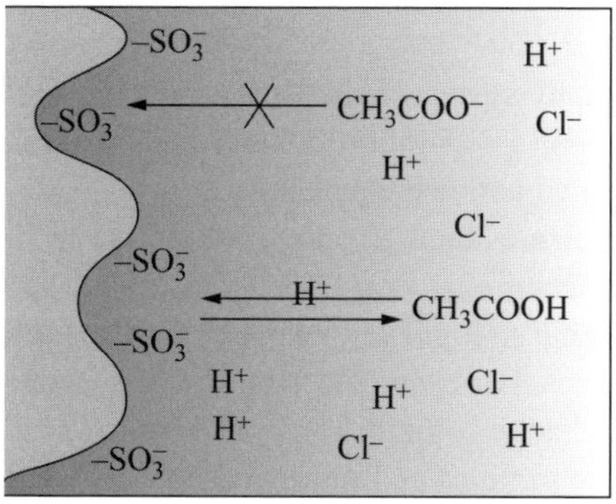

Fig. 6.28 Scheme of the principle of chromatography by ionic exclusion.

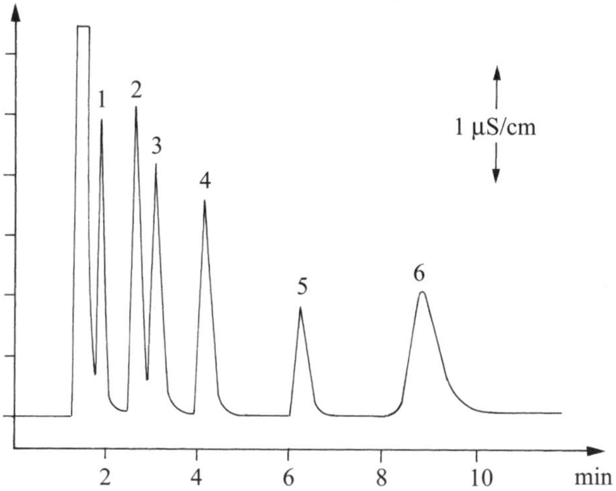

Fig. 6.29 Separation of acids. Column: Metrohm Hamilton PRP-X300. Eluent 1.5 mM perchloric acid. Flow rate 1ml·min^{-1}. 1. Tartrate, 2. Malate, 3. Citrate, 4. Lactate, 5. Acetate, 6. Succinate (Metrohm, CH).

form of the acid can reside in this stationary layer, and consequently its progress in the column will be retarded.

6.5.3 Ion pair chromatography

Here, the stationary phase is an apolar medium capable of extracting salts in the form of ion pairs. This method is very useful for separating ionic surfactants and also organometallic complexes.

6.5.4 Experimental methods

An ion chromatography system consists of a pump to make the eluent circulate, an injector (10 to 100 μl), a separation column and if necessary an ionic strength suppressor column, situated above the detector.

The limits of detection are about 10 ppb for injections of 50 μl, but can be lowered using pre-concentration techniques.

The reason for the ionic strength suppression column is to facilitate the conductometric detection of samples. For example, consider the separation of a mixture of KCl and NaCl on a cation exchange column. The eluent being an acid, its conductivity will be higher than that of the samples and the signal obtained will be negative on a high background signal. If, after the separation column, we pass the eluent and the sample through an anion exchange column, we neutralise the eluent to form water with a low conductivity, and we exchange in the same way the chloride ions of the sample with hydroxides. Thus, the passage of the sample in front of the conductometric detector now gives a positive signal, namely that of a NaOH sample on a water background. If the conductometric detectors have a very good signal-to-noise ratio, the measurements can be made without a suppression column.

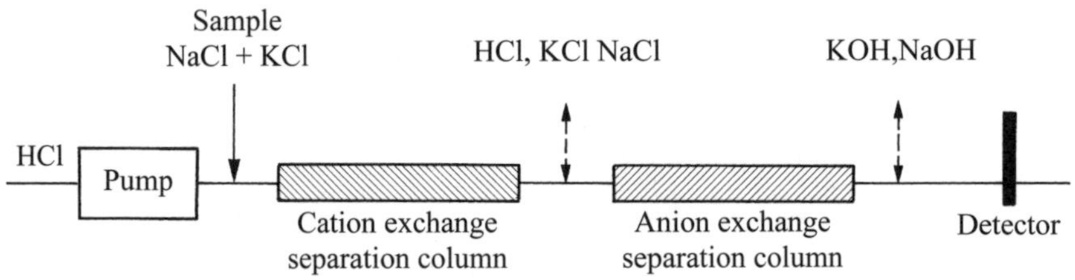

Fig. 6.30 Scheme of the principle of a cation ionic chromatography system with an anion exchange suppression column.

These columns require frequent servicing, and other systems have appeared based on using ion exchange membranes such as shown in Figure 6.31 for the neutralisation of bicarbonate for anion exchange chromatography. In this approach, the neutralisation is effected by a counter-current passage of sulfuric acid. To reduce the consumption of acid, new systems based on electrodialysis (see §6.6.1) have recently been developed.

The detectors used in ion chromatography are mainly electrochemical, although UV VIS detectors can be used in certain special cases.

The most common method is either direct (sample with higher conductivity than the eluent) or indirect conductometry (sample with lower conductivity than the eluent).

For certain redox components, amperometric detectors are often used, e.g. for the analysis of sugars in alkaline media.

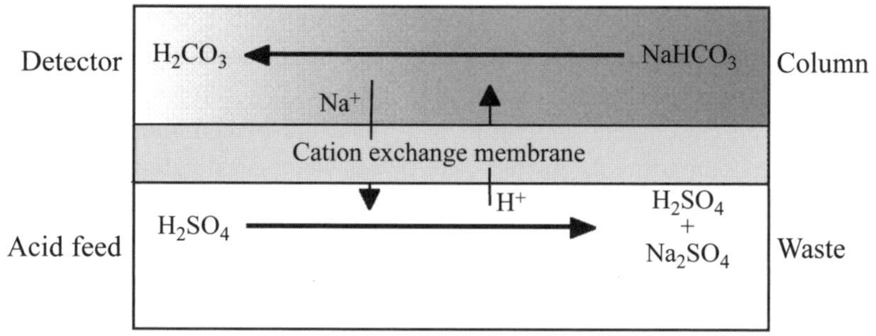

Fig. 6.31 Membrane suppressor.

6.6 INDUSTRIAL METHODS OF ELECTROCHEMICAL SEPARATION

Ion exchange membranes have allowed the development of several industrial or analytical processes.

6.6.1 Electrodialysis

The purpose of electrodialysis is mainly to desalinate solutions. The most important application of this is undoubtedly the desalination of seawater according to the principle shown below, but we can also note the simultaneous manufacture of soda and sulfuric acid from sodium sulfate. The separation of amino acids and proteins is a more recent application.

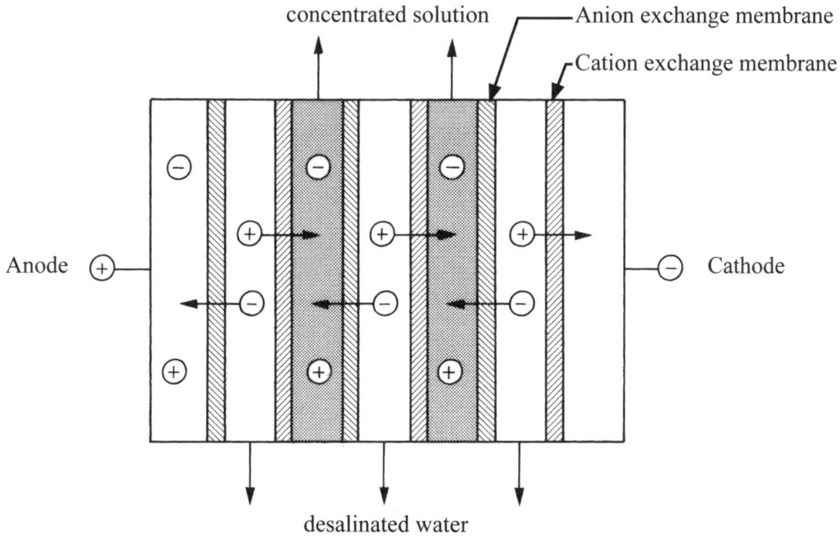

Fig. 6.32 Working principle of desalination by electrodialysis.

The principle of electrodialysis is shown in Figure 6.32. Under the effect of an electric field from the anode to the cathode, the ions in the solution to be desalinated migrate towards the respective ion exchange membranes according to their charge and concentrate in the enrichment compartments. These are placed alternately between the desalination compartments. In the case of seawater desalination, the electrode reactions can be used to make soda at the cathode (reduction of the protons and enrichment of the sodium ions) and hydrochloric acid at the anode (oxidation of water and enrichment of the chloride ions).

6.6.2 Donnan dialysis

Sometimes, the separation of ions is more economical using a pH gradient rather than an electrical potential gradient. This is the case for the recuperation of metal ions from dilute solutions, e.g. that of Cu^{2+} from microelectronic plant effluents. In order to understand how we can concentrate metallic ions using a cation exchange membrane, consider the example of Figure 6.33. A concentrated acid solution is used to impose a Galvani potential difference between the two aqueous phases. Neglecting mass transport in solution, we can write the equality of the electrochemical potential of H^+ between the two aqueous phases (1) & (2) separated by the membrane as

$$\mu^{\ominus}_{H^+} + RT\ln a^1_{H^+} + F\phi^1 = \mu^{\ominus}_{H^+} + RT\ln a^2_{H^+} + F\phi^2 \qquad (6.128)$$

Thus, the Galvani potential between the two phases is written as

$$\Delta^2_1\phi = \phi^2 - \phi^1 = \frac{RT}{F}\ln\left(\frac{a^1_{H^+}}{a^2_{H^+}}\right) \qquad (6.129)$$

We can also write the equality of the electrochemical potentials for Cu^{2+} as

$$\mu^{\ominus}_{Cu^{2+}} + RT\ln a^1_{Cu^{2+}} + 2F\phi^1 = \mu^{\ominus}_{Cu^{2+}} + RT\ln a^2_{Cu^{2+}} + 2F\phi^2 \qquad (6.130)$$

and assuming that the excess of acid controls entirely the Galvani potential difference, we can calculate the ratio of concentrations of Cu^{2+} between the two phases

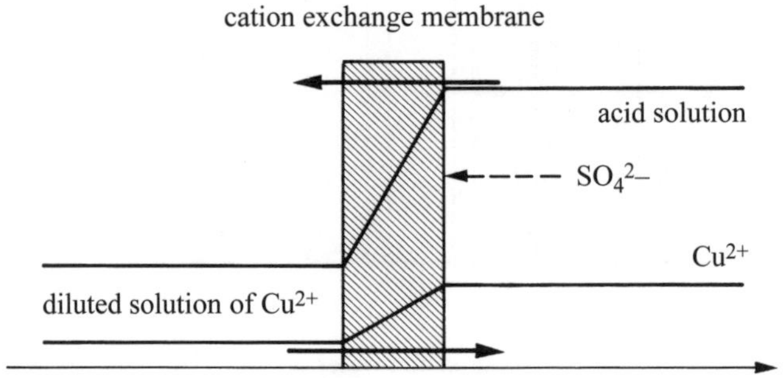

Fig. 6.33 Schematic illustration of the Donnan dialysis principle.

$$\left(\frac{a^1_{Cu^{2+}}}{a^2_{Cu^{2+}}}\right) = \left(\frac{a^1_{H^+}}{a^2_{H^+}}\right)^2 \tag{6.131}$$

We can see therefore that we can concentrate the copper ions against their concentration gradient from 30 ppm to 3 000 ppm. In this case, the distribution of the protons between the two phases fixes the Galvani potential, which in turn induces the transfer of the copper ions. We have therefore a chemical method, rather than an electrical method, for establishing a potential.

6.6.3 Ion exchange dialysis.

We have seen that anion exchange membranes have a rather poor selectivity for the transfer of protons. This weakness can be used as advantage, for example in de-acidifying electrodeposition baths.

As shown in Figure 6.34, an anion exchange membrane allows the dialysis of acids.

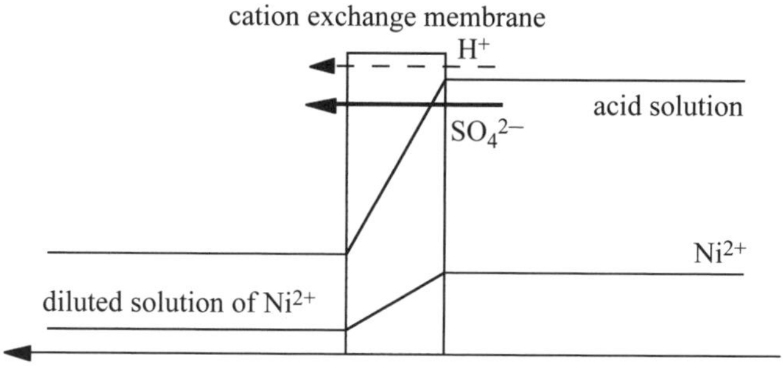

Fig. 6.34 Schematic illustration of the ion exchange dialysis principle.

CHAPTER 7

STEADY STATE AMPEROMETRY

7.1 ELECTROCHEMICAL KINETICS

We saw in chapter 2 how we could measure the electrode potential E by using a reference electrode and a voltmeter. We also showed that if the reaction

$$\text{Ox} + ne^- \rightleftarrows \text{Red}$$

takes place at a working electrode, the potential of this electrode at equilibrium obeys the Nernst law (2.8), which in a general manner can be written as

$$E_{\text{eq}} = E^{\ominus}_{\text{O/R}} + \frac{RT}{nF}\ln\left(\frac{a_{\text{O}}}{a_{\text{R}}}\right) = E^{\ominus\prime}_{\text{O/R}} + \frac{RT}{nF}\ln\left(\frac{c_{\text{O}}}{c_{\text{R}}}\right) \tag{7.1}$$

where $E^{\ominus}_{\text{O/R}}$ is the standard redox potential and $E^{\ominus\prime}_{\text{O/R}}$ is the formal redox potential or the apparent standard redox potential as defined by equation (2.25).

(N.B. From here on in this book, values linked to oxidised and reduced species will have indices of just O and R, in order to lighten the notation)

Now, instead of measuring the electrode potential, let's impose this value with the help of a 3-electrode setup [Working electrode (WE), Reference electrode (RE), Counter-electrode (CE)] and an instrument called a ***potentiostat***. A potentiostat controls the electrode potential, i.e. the potential difference between the working electrode and the reference electrode, by passing a current between the working electrode and the counter-electrode. In effect, it is not recommended to send an electric current through the reference electrode, since the equilibrium defining the p.d. between the reference electrode and the solution (see for example Figure 2.4) would be destroyed, and the potential of the reference electrode would no longer be constant.

Therefore, the role of a potentiostat is to send a current through the circuit:

Potentiostat – Counter electrode – solution – working electrode – Potentiostat

As shown in the circuit of Figure 7.1, the nature of the current is of course: electronic conduction in the counter electrode – electrochemical reaction at the counter-electrode – ionic conduction in solution – electrochemical reaction at the working electrode – electronic conduction in the working electrode.

The potential of the working electrode is measured with respect to the reference electrode as shown in Figure 7.1. The measuring circuit comprises the part of the solution situated between the reference electrode and the working electrode (see

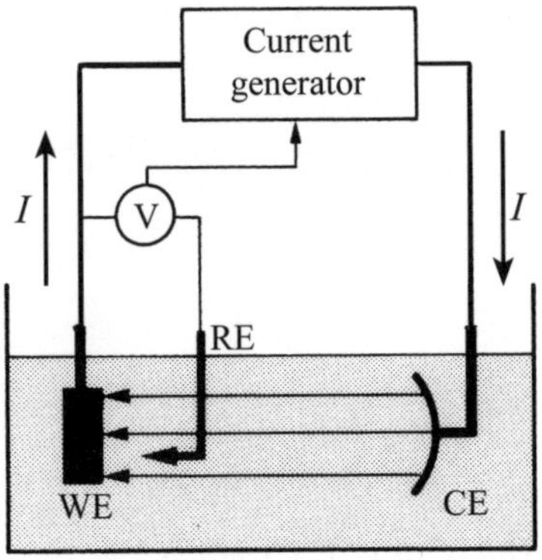

Fig. 7.1 A 3-electrode electrochemical cell with potentiostatic control of the working electrode potential.

§7.8). Only a very small current, a few pA flows in this voltage measuring circuit, thus preserving the electrochemical equilibrium at the reference electrode. The low value of the current in this measuring circuit is controlled by the high input impedance of the voltmeter.

Thus, so to speak, to 'impose' the potential of the working electrode with respect to the reference electrode, the potentiostat in fact applies a current between the counter-electrode and the working electrode. The voltmeter circuit measures the working electrode potential with respect to the reference electrode potential and the value measured is then compared to the electrode potential value that we wish to apply, and via a feedback system, the potentiostat applies the required current so that the measured potential is equal to the one desired. The response time of a standard potentiostat is of the order of a microsecond.

What happens if we impose an electrode potential E that is different from the equilibrium potential E_{eq} given by Nernst's law (7.1)?

When the applied electrode potential is more positive than the equilibrium value ($E > E_{eq}$), we displace the interfacial redox equilibrium at the working electrode towards the oxidation of the reduced species, and the current resulting from the electrons being transferred to the electrode is called the ***anodic current***, considered by convention to be positive (Careful when reading the literature and some other textbooks, some do not respect international conventions!).

When the applied electrode potential is more negative than the equilibrium value ($E < E_{eq}$), we displace the interfacial redox equilibrium towards the reduction of the oxidized species, and the resulting current is called the ***cathodic current***, considered by convention to be negative.

The current associated to these displacements of redox equilibria is supplied by the potentiostat. The range of current usually employed varies from picoamperes for microelectrodes (see §7.4.2) to amperes in the case of large electrodes.

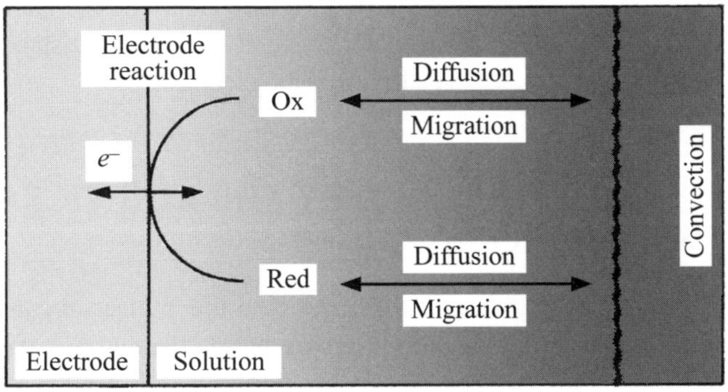

Fig. 7.2 General scheme for a redox reaction at an electrode.

These considerations between current and electrode potential were obtained from thermodynamic principles. In fact, the amount of current crossing the working electrode | solution interface is a measure of the kinetics of the redox reaction, which can be broken down into three steps, as shown in Figure 7.2:

- flux of the solution reactants towards the electrode (diffusion-migration)
- interfacial electron transfer reaction
- flux of the products from the surface towards the solution (diffusion-migration)

It is customary to recognise three cases:

- The mass transfer, diffusion/migration flux of the reactants and the products of the electrode reaction, is rapid compared with the electron transfer reaction. The reaction is said to be '***irreversible***'. This expression has nothing to do with the chemical reversibility of the reaction, which indicates whether the reaction can happen in both directions, but indicates simply in electrochemical jargon that the limiting slow step for the reaction is the electron transfer reaction at the electrode. Note also that the notion of electrochemical reversibility also has no relation to thermodynamic reversibility.
- The electron transfer is rapid compared to the arrival rate of the reactants at the electrode and the departure rate of the products. The reaction is then said to be '***reversible***'. In the same way, this expression signifies that the limiting slow step for the reaction is the arrival flux of reactants equal to the departure flux of the products of the reaction.
- The two phenomena take place in comparable time-scales. The reaction is then known as '***quasi-reversible***'.

In this chapter, we shall consider electrochemical reactions in a steady state regime (i.e. time is not a variable). The volume of solution containing the redox species is considered big enough that the quantities of species oxidised or reduced at the electrode remain negligible, and that the bulk reactant concentrations are considered as constant. Also, we will consider mostly electrodes of planar geometry. Therefore, the concentrations in solution $c(x)$ depend only on the distance x from the electrode, $x = 0$ corresponding to the surface of the electrode.

7.2 CURRENT CONTROLLED BY THE KINETICS OF THE REDOX REACTIONS

Oxidation is a first order reaction with respect to the interfacial concentration of the reduced species, and the anodic current, representing the number of electrons per second transferred from the reduced species in solution is thus written as

$$I_a = nFA\, k_a\, c_R(0) \tag{7.2}$$

Reduction can be considered as a pseudo-first order reaction with respect to the concentration of the oxidised species, if we consider that there is an excess of electrons in the metal with respect to the interfacial concentration of the oxidised species in solution, and the cathodic current is then written as

$$I_c = -nFA\, k_c\, c_O(0) \tag{7.3}$$

where F is the Faraday constant, A the area of the electrode, and k_a and k_c are the electrochemical rate constants for oxidation and reduction respectively (units = m·s^{-1} or more commonly cm·s^{-1}). I is the current in amperes (C·s^{-1}) and j the current density ($j = I/A$) . It is important to note that the product kc represents a mass flux (mol·m^{-2}·s^{-1}).

Equations (7.2) & (7.3) show that the current measured using a potentiostat is proportional to the rate of the electrochemical reaction. In classical chemical kinetics, the variations in concentration of the reactants and products are measured over time to determine the rate of the reactions, whereas in electrochemical kinetics we measure directly the rate of the redox reaction.

Unlike the rate constants of chemical reactions that depend essentially on the temperature, electrochemical rate constants depend also on the applied electrode potential, that is to say the Gibbs energy used to displace the equilibrium. The question is now to find out how electrochemical rate constants depend on the electrode potential.

7.2.1 Standard case of an ideal solution

In the standard case, the bulk concentrations of the oxidised and reduced species are taken equal, $[c_O(\infty) = c_R(\infty) = c]$ such that the equilibrium electrode potential E_{eq} is equal to the standard redox potential $E^{\ominus}_{O/R}$ if the solutions are ideally dilute, i.e. if we can consider the activity coefficients equal to unity, or otherwise the equilibrium electrode potential E_{eq} is equal to the formal redox potential $E^{\ominus\prime}_{O/R}$.

At equilibrium, the rate of oxidation is equal to the rate of reduction $[k_a c_R(0) = k_c c_O(0)]$. Given that we have made the hypothesis that the mass transfer is rapid with respect to the kinetics of the electron transfer reaction, we can make the additional hypothesis that the concentrations at the surface of the electrode are equal to those inside the bulk of the solution [$c_O(0) = c_O(\infty) = c_R(\infty) = c$], as long as the conversion rates of the reactants (oxidation or reduction) remain negligible.

Thus, the activation energy barrier shown in Figure 7.3 is symmetrical, since in the standard case we have *de facto* $k_a = k_c$. This rate constant value is called the ***standard rate constant*** $k^{\ominus}$. According to the transition state theory, $k^{\ominus}$ can be written as

$$k^{\ominus} = \delta\left(\frac{kT}{h}\right)e^{-\Delta G^{\ominus}_{act}/RT} \tag{7.4}$$

where δ is a distance and $\Delta G^{\ominus}_{act}$ the standard Gibbs activation energy of the electron transfer reaction. δ can be considered in a first approximation as the minimum distance separating the reactants from the electrode.

The current measured at equilibrium is of course zero, since it is the sum of the anodic and cathodic currents which themselves are not zero, but equal in absolute values.

Now, with the aid of the potentiostat, let's impose an electrode potential increased by the quantity $E - E^{\ominus\prime}_{O/R}$, written shortly $E - E^{\ominus\prime}$.

This can be done experimentally, for example by maintaining the inner potential ϕ^M of the working electrode constant and lowering the inner potential ϕ^S of the solution by the value $E - E^{\ominus\prime}$ [$\phi^S = \phi^S_{eq} - (E - E^{\ominus\prime})$].

Thus, the standard electrochemical potential of O defined by equation (1.63), $\mu^{\ominus}_O + z_O F\phi^S$, decreases by an amount $z_O F(E - E^{\ominus\prime})$, whilst the standard electrochemical potential of R, $\mu^{\ominus}_R + z_R F\phi^S$, is reduced by an amount $z_R F(E - E^{\ominus\prime})$. Given that $z_O - z_R = n$, the number of electrons exchanged, the barrier is no longer symmetrical since its two sides do not decrease by the same amount.

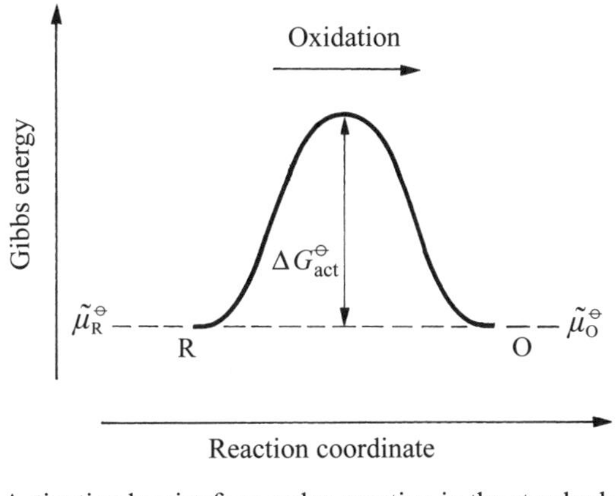

Fig. 7.3 Activation barrier for a redox reaction in the standard case.

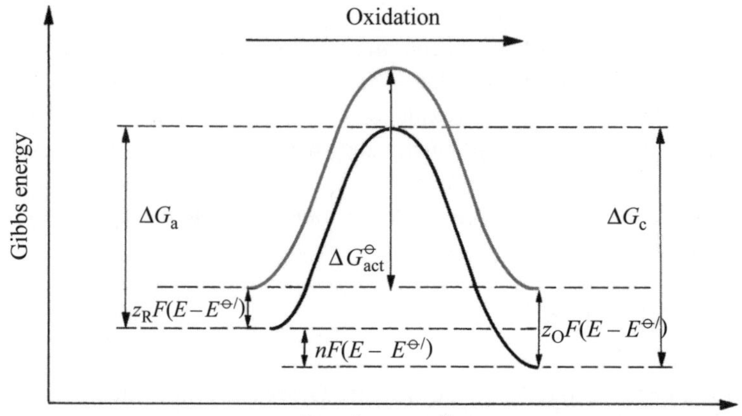

Fig. 7.4 Influence of the working electrode potential on the activation energy barrier for an oxidation, in the standard case.

The activation energy for oxidation, symbolically represented by the height of the barrier only decreases by a fraction α of the energy difference $nF\,(E - E^{\ominus\prime})$

$$\Delta G_a = \Delta G^{\ominus}_{act} - \alpha nF(E - E^{\ominus\prime}) \tag{7.5}$$

as shown in Figure 7.4. α represents the fraction of the Gibbs energy operating on the transition state and is called the ***charge transfer coefficient***.

In a similar fashion, the activation energy for the reduction is increased by the complement to $\alpha nF\,(E - E^{\ominus\prime})$ which is

$$\Delta G_c = \Delta G^{\ominus}_{act} + (1-\alpha)nF(E - E^{\ominus\prime}) \tag{7.6}$$

The total current, which is the sum of the anodic and cathodic currents, is obtained by combining equations (7.2) to (7.6)

$$\begin{aligned} I &= nF\,A\,k^{\ominus}\left[c_R\, e^{\alpha nF(E-E^{\ominus\prime})/RT} - c_O\, e^{-(1-\alpha)\,nF(E-E^{\ominus\prime})/RT}\right] \\ &= nF\,A\,k^{\ominus}c\left[e^{\alpha nF(E-E^{\ominus\prime})/RT} - e^{-(1-\alpha)\,nF(E-E^{\ominus\prime})/RT}\right] \end{aligned} \tag{7.7}$$

Equation (7.7) shows that the the more positive the applied electode potential compared to the formal redox potential, the larger will be the anodic current and the more the cathodic current will become negligible. The current measured will then be the anodic current.

Conversely, the more negative the applied electrode potential with respect to the formal redox potential, the larger will be the absolute value of the cathodic current, and the more the anodic current will become negligible. The measured current will then be the cathodic current.

7.2.2 General case

In the general case where the concentration of the oxidised and reduced species in solution are not equal ($c_R(\infty) \neq c_O(\infty)$), the activation energy barrier is no longer symmetrical at equilibrium, and consequently, the anodic and cathodic activation energies are no longer equal. Nevertheless, at equilibrium, we still have equality between the rates of oxidation and reduction [$k_a c_R(0) = k_c c_O(0)$], and the inequality of the surface concentrations means that the anodic and cathodic rate constants cannot be equal.

From the transition state theory, the reaction rate is written as :

$$v = \delta\left(\frac{kT}{h}\right)\frac{\gamma_R(0)}{\gamma^{\#}}e^{-\Delta G_a^{eq}/RT}c_R(0) - \delta\left(\frac{kT}{h}\right)\frac{\gamma_O(0)}{\gamma^{\#}}e^{-\Delta G_c^{eq}/RT}c_O(0) \quad (7.8)$$

taking into account the activity coefficients of the reactants and that of the activated complex (#), and where $\Delta G_a^{eq}$and ΔG_c^{eq}represent the anodic and cathodic Gibbs energies of activation at equilibrium, shown in Figure 7.5.

At equilibrium, the global reaction rate is zero, and we have

$$\Delta G_c^{eq} = \Delta G_a^{eq} + RT\ln\left(\frac{\gamma_O(0)c_O(0)}{\gamma_R(0)c_R(0)}\right) \quad (7.9)$$

As before, if we displace the equilibrium and impose an electrode potential increase by the amount $E - E_{eq}$, the anodic and cathodic activation energies become, by the same reasoning

$$\Delta G_a = \Delta G_a^{eq} - \alpha\, nF(E - E_{eq}) \quad (7.10)$$

and

$$\Delta G_c = \Delta G_c^{eq} + (1-\alpha)\, nF(E - E_{eq}) \quad (7.11)$$

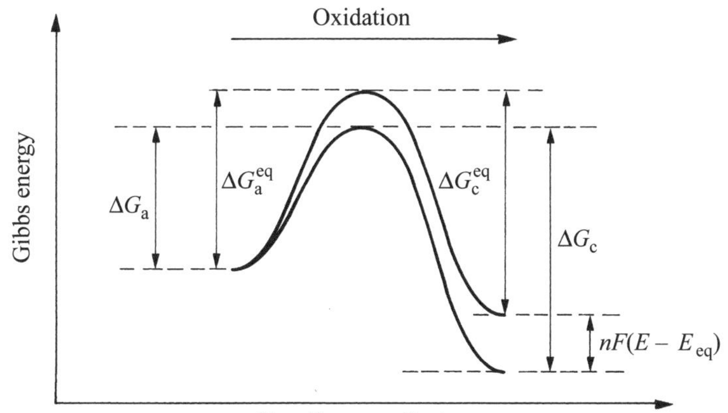

Fig. 7.5 Influence of the working electrode polarisation on the activation barrier for an oxidation in the general case.

From these two equations, we can express the variation of the anodic and cathodic rate constants as a function of the electrode potential by regrouping all the terms that are independent of the electrode potential in the pre-exponential term

$$k_a = k_a^o\, e^{\alpha nF E/RT} \tag{7.12}$$

and

$$k_c = k_c^o\, e^{-(1-\alpha) nF E/RT} \tag{7.13}$$

At the formal redox potential, the activation barrier is symmetrical, and the two rate constants k_a^o and k_c^o are linked to the standard rate constant $k^\ominus$ by

$$k^\ominus = k_a^o\, e^{\alpha nF E^{\ominus\prime}/RT} = k_c^o\, e^{-(1-\alpha) nF E^{\ominus\prime}/RT} \tag{7.14}$$

The total current is then written as

$$I = nF A\left[k_a^o\, c_R(0)\, e^{\alpha nF E/RT} - k_c^o\, c_O(0)\, e^{-(1-\alpha) nF E/RT}\right] \tag{7.15}$$

or as

$$I = nFA\, k^\ominus \left[c_R(0)\, e^{\alpha nF(E-E^{\ominus\prime})/RT} - c_O(0)\, e^{-(1-\alpha) nF(E-E^{\ominus\prime})/RT}\right] \tag{7.16}$$

In general, it is desirable to express the current as a function of the applied potential perturbation $E - E_{eq}$, which is called the ***overpotential*** and has the symbol η

$$\eta = E - E_{eq} \tag{7.17}$$

Since the equilibrium potential is given by the Nernst equation (7.1),

$$E_{eq} = E^{\ominus\prime} + \frac{RT}{nF}\ln\left(\frac{c_O(\infty)}{c_R(\infty)}\right) \tag{7.18}$$

we get

$$E - E^{\ominus\prime} = \eta + \frac{RT}{nF}\ln\left(\frac{c_O(\infty)}{c_R(\infty)}\right) \tag{7.19}$$

Thus, we can express the current as a function of the difference in electrode potential imposed on the system to displace the equilibrium

$$I = I_o\left[\left(\frac{c_R(0)}{c_R(\infty)}\right) e^{\alpha nF \eta/RT} - \left(\frac{c_O(0)}{c_O(\infty)}\right) e^{-(1-\alpha) nF \eta/RT}\right] \tag{7.20}$$

I_o is the ***exchange current***, which represents the anodic current and the absolute value of the cathodic current crossing the interface at equilibrium, given by

$$I_o = n\, FA\, k^\ominus \left[c_R(\infty)\right]^{1-\alpha} \left[c_O(\infty)\right]^{\alpha} \tag{7.21}$$

If the current at the electrode is sufficiently small that the concentrations at the surface of the electrode remain equal to the concentrations in the solutions (i.e. if the

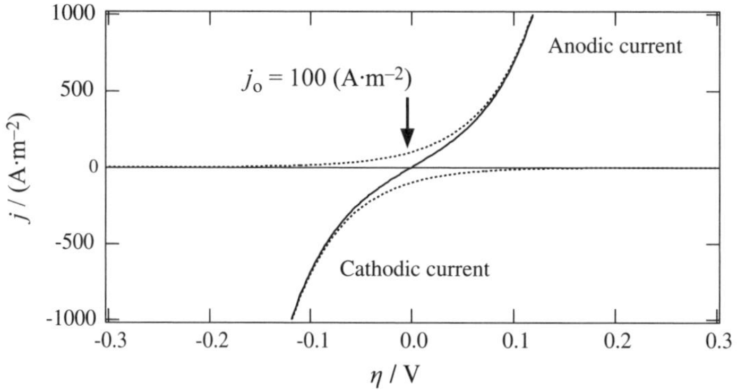

Fig. 7.6 Polarisation curves for a kinetically controlled electrochemical reaction (n=1, α = 0.5, T=298K).

mass transfer can be considered as infinitely rapid) and if the volume of the solution is large enough that the concentrations in solution remain constant, we obtain the ***Butler-Volmer equation***, which is the principal equation of electrochemical kinetics

$$I = I_o \left[e^{\alpha nF\eta/RT} - e^{-(1-\alpha)nF\eta/RT} \right] \tag{7.22}$$

Figure 7.6 shows the variation of the current density as a function of the overpotential. The exchange current density j_o is the value of the anodic current density for $\eta = 0$ V.

The shape of the current-potential curve depends on the value of the exchange current. For large exchange current values, a small deviation of the electrode potential with respect to the equilibrium potential causes an electrochemical reaction. For small exchange current values, it is necessary to impose a large overpotential for the reaction to take place, as shown in Figure 7.7.

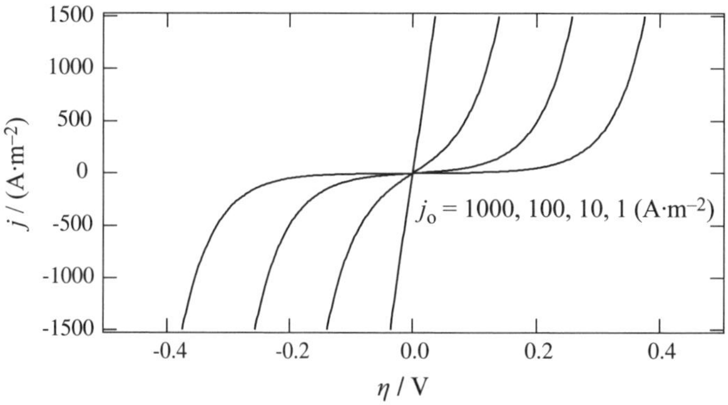

Fig. 7.7 Influence of the exchange current on current-potential curves (n=1, α = 0.5, T=298K).

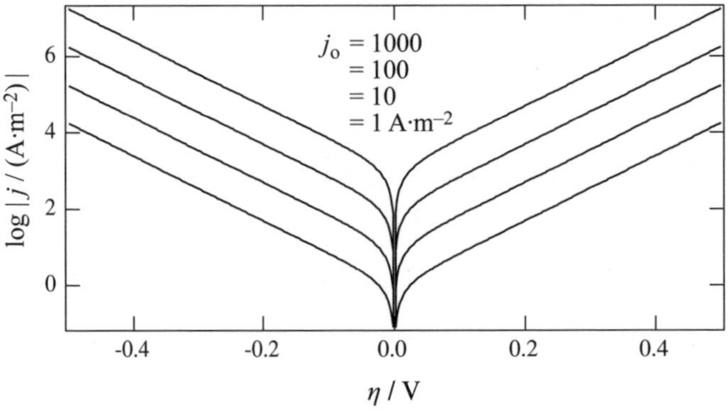

Fig. 7.8 Tafel plots (n=1, α= 0.5, T=298K). The extrapolated values for log|j_o| are respectively 3, 2, 1 and 0.

The Butler-Volmer equation is often shown in the form of log|I| or log|j| graphs as a function of the overpotential η, known as ***Tafel plots*** (Figure 7.8). These graphs allow us to measure graphically the exchange current by extrapolation of the two lines at η = 0 V, and the slope allows us to calculate the value of the charge transfer coefficient.

When $nF\eta/RT$ is well below unity, we can linearise the Butler-Volmer equation (7.22) to obtain

$$I = I_o nF\eta / RT \tag{7.23}$$

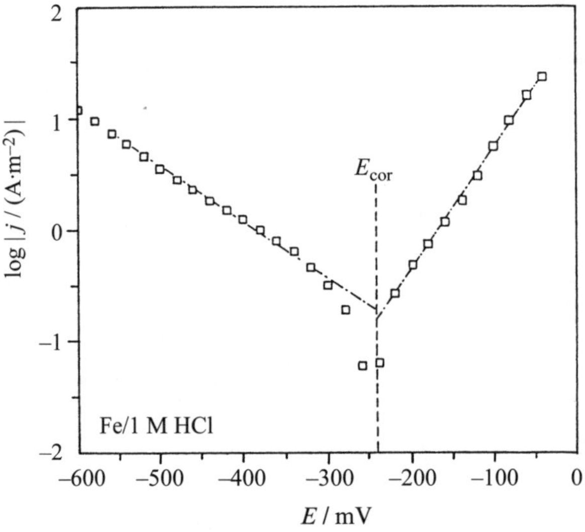

Fig. 7.9 Tafel curves for the corrosion of iron [D. Landolt, Traité des Matériaux, Vol.12, *Corrosion et Chimie de surface des métaux* PPUR, 3rd ed. 2003].

By analogy with Ohm's law, the term RT/nFI_0 is called the ***charge transfer resistance***.

The Butler-Volmer equation underlines the great difference between a classical chemical reaction and an electrochemical reaction. For a classical chemical reaction, the only way to vary the reaction rate is to vary the temperature (Arrhenius' law). On the other hand, for an electrochemical reaction, we can see that the electrode potential controls the reaction rate represented by the current. If we take, for example, an average value for the charge transfer coefficient ($\alpha = 0.5$), we can see that an increase of 1 V leads to an increase in the ratio I/I_0 by a factor of $3 \cdot 10^8$ ($= \exp^{F/2RT}$).

In certain cases, oxidation and reduction reactions may be different. This is the case, for example, for an iron electrode in an acid solution. Oxidation corresponds to the corrosion and dissolution of the iron electrode, and reduction corresponds to the reduction of the protons. The Tafel curves obtained then have different slopes, as shown in Figure 7.9. We define the corrosion potential by the intercept of the two Tafel lines and the corrosion current at this potential corresponds to the exchange current defined above.

7.3 REVERSIBLE SYSTEMS: CURRENT LIMITED BY DIFFUSION

7.3.1 Diffusion layer

If the electrochemical reaction

$$\mathrm{Ox} + ne^- \rightleftarrows \mathrm{Red}$$

is infinitely rapid with respect to the rate of arrival of the reactants and departure of the products, the system is said to be reversible. For planar electrodes where the arrival of the reactants happens by semi-infinite linear diffusion, a system is reversible if $k^{\ominus} > 0.01$ to $0.1\ \mathrm{cm \cdot s^{-1}}$, to give an order of magnitude (see equation (7.57) for more detail). For reversible systems, the Nernst equation is always valid for the interfacial concentrations

$$E \quad = \quad E^{\ominus\prime} + \frac{RT}{nF} \ln\left(\frac{c_{\mathrm{O}}(0)}{c_{\mathrm{R}}(0)}\right) \tag{7.24}$$

thus creating concentration gradients between the surface of the electrode and the bulk of the solution. The current, which is the flux of electrons crossing the interface, is thus limited by the flux of species in solution reaching the electrode. We have seen that the flux of species in solution could be either a convection flux, a migration flux, a diffusion flux, or even an osmotic one. Close to any solid wall in solution, the convection flux tends to zero and the layer adjacent to the wall, where the convection is negligible, is called the ***diffusion layer*** or ***Nernst layer***. If we have also added to the solution an electro-inactive salt (the supporting electrolyte) such that the transport number of the electroactive species is negligible, the flux that limits the current will be a diffusion flux.

Consider an oxidation limited by the diffusion of the reduced species in solution to the electrode. We can write that

$$I_a = n\,FA\,D_R \left(\frac{\partial c_R(x)}{\partial x}\right)_{x=0} \tag{7.25}$$

where the signs are determined by the convention that anodic currents are taken as positive and cathodic currents as negative. Of course, the arrival flux of the reduced species at the electrode is equal to the departure flux of the oxidised species from the electrode.

$$I_a = n\,FA\,D_R \left(\frac{\partial c_R(x)}{\partial x}\right)_{x=0} = -n\,FA\,D_O \left(\frac{\partial c_O(x)}{\partial x}\right)_{x=0} \tag{7.26}$$

We can obtain a steady state current if the thickness of the diffusion layer can be fixed at a value δ either hydrodynamically (e.g. using rotating electrodes), or by the actual geometry of the electrodes (e.g. using microelectrodes). In effect, the concentrations on both sides of the diffusion layer are fixed by the Nernst equation, the equality of the fluxes at the electrode surface given by equation (7.26), and by the convection in the bulk solution that maintains the bulk concentration constant. The steady state diffusion current can then be described by the Nernst approximation

$$I_a = nFA\,D_R \left(c_R(\infty) - c_R(0)\right)/\,\delta_R = -nFA\,D_O \left(c_O(\infty) - c_O(0)\right)/\,\delta_O \tag{7.27}$$

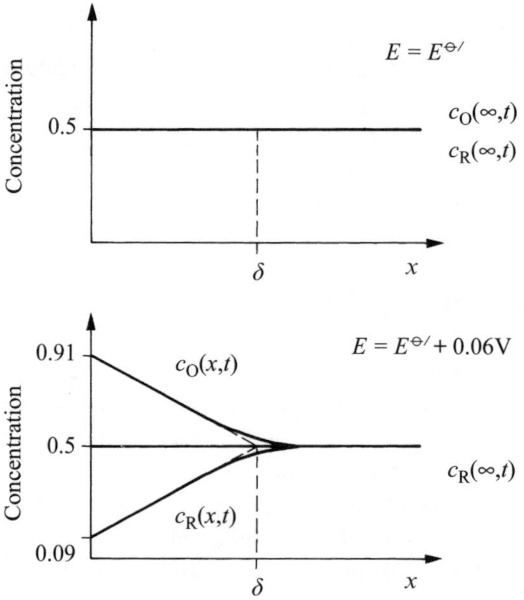

Fig. 7.10 Concentration profiles for a solution containing an equimolar mixture of oxidised and reduced species at the standard potential and at an electrode potential value of $E^{\ominus\prime} + 0.06\text{V}$ corresponding to an oxidation (n=1). The ratio c_O/c_R is then 10. $\delta=\delta_O=\delta_R$ and $D = D_O = D_R$.

The notion of a diffusion layer thickness provides an expression for the concentration gradient simply as a function of the difference between the interfacial and the bulk concentrations.

Consider the example of Figure 7.10 of a solution containing an equimolar mixture of oxidised and reduced species. At equilibrium, the electode potential is equal to the formal redox potential and the concentrations of the two species are taken equal to 0.5 in solution.

Let's now apply an electrode potential value 0.06 V more positive than the formal redox potential to oxidise the reduced species. Instantly, the interfacial concentrations are fixed by the Nernst equation ($n = 1$) which imposes the ratio $c_O(0)/c_R(0)$ to be equal to 10. On the other hand, the equality of the fluxes imposes the concentration gradients to be equal in absolute value, if the diffusion coefficients D_O and D_R are equal. The interfacial concentration of the oxidised species is thus 0.909 whilst that of the reduced species is 0.0909. As we shall see below, the thickness of the diffusion layer is not an intrinsic value, but depends on the nature of the diffusing species and their diffusion coefficient.

A statistical analysis (see §4.6) shows that the time τ to establish a diffusion layer is

$$\tau = \frac{\delta^2}{2D} \tag{7.28}$$

Thus, for an average value of the diffusion coefficient D of $5 \cdot 10^{-6}$ $cm^2 \cdot s^{-1}$, and for thicknesses of the diffusion layer of the order of micrometres, the characteristic times vary as in Table 7.1.

Table 7.1 Characteristic diffusion times across the Nernst layer.

δ / μm	τ / s
1	10^{-3}
10	0.1
100	10

7.3.2 Limiting diffusion current

For an oxidation, if the electrode potential value imposed is sufficiently large, the ratio $c_O(0)/c_R(0)$ satisfying the Nernst equation also becomes very large. Given that the interfacial concentration $c_O(0)$ cannot reach very large values, this means that the surface concentration of the reduced species $c_R(0)$ tends to zero. In equation (7.27), the surface concentration can therefore be neglected in comparison with the concentration in the solution [$c_R(0) << c_R(\infty)$]. The anodic current thus attains a limiting value

$$I_{da} = \frac{nFA\, D_R c_R(\infty)}{\delta_R} = nFA\, m_R c_R(\infty) \tag{7.29}$$

where m_R is a mass transfer coefficient used above all in electrochemical engineering.

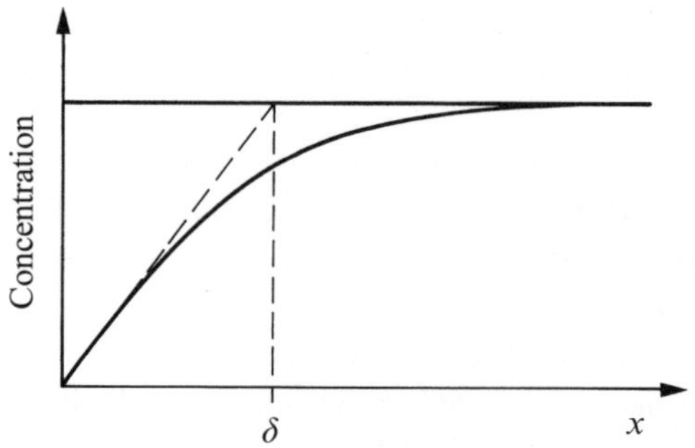

Fig. 7.11 Concentration profile of a reacting species, corresponding to the limiting current.

The notion of a limiting diffusion current allows us to make a graphic definition of the thickness of the diffusion layer, as shown in Figure 7.11. It is important to note that the notion of a limiting diffusion current requires that the concentration gradients at the electrode reach a maximum slope, but this does not mean that the whole profile of the concentrations should not evolve further once the limiting diffusion current has been reached.

7.3.3 Current-potential curve

To obtain a relation between the current and the electrode potential, we have to write the Nernst law at the electrode and substitute in it the interfacial concentrations expressed as a function of the steady state limiting diffusion currents

$$I_{da} = nFA\,D_R\,c_R(\infty)/\delta_R \tag{7.30}$$

and

$$I_{dc} = -nFA\,D_O\,c_O(\infty)/\delta_O \tag{7.31}$$

By combining these equations with equation (7.27), we can express the interfacial concentrations as a function of the limiting diffusion currents

$$c_R(0) = c_R(\infty) - \frac{\delta_R I}{nFAD_R} = \frac{\delta_R}{nFAD_R}(I_{da} - I) \tag{7.32}$$

and

$$c_O(0) = c_O(\infty) + \frac{\delta_O I}{nFAD_O} = \frac{\delta_O}{nFAD_O}(I - I_{dc}) \tag{7.33}$$

By substituting these values into the Nernst equation (7.1), we then obtain

$$E = E^{\ominus\prime}_{O/R} + \frac{RT}{nF}\ln\left[\frac{D_R\,\delta_O}{D_O\,\delta_R}\right] + \frac{RT}{nF}\ln\left[\frac{I_{dc} - I}{I - I_{da}}\right] \tag{7.34}$$

The shape of the curve given by equation (7.34) is shown in Figure 7.12.

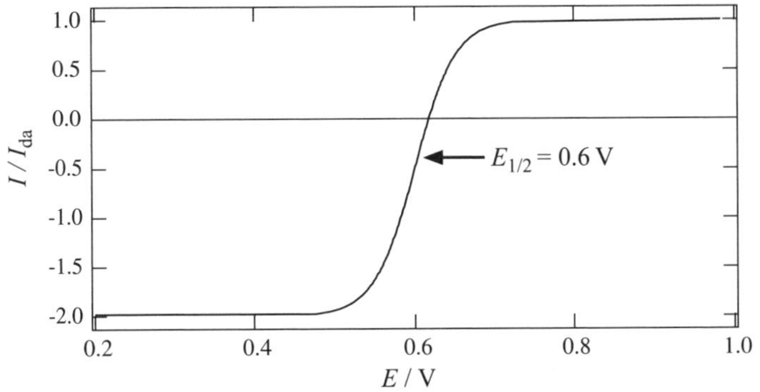

Fig. 7.12 Current-potential curve for a reaction limited by the diffusion of species in solution. $n = 1$, $c_O(\infty) = 2\, c_R(\infty)$, $\delta = \delta_O = \delta_R$ and $D = D_O = D_R$.

When $I = (I_{da} + I_{dc})/2$, the last term of equation (7.34) is zero and the potential is then called the ***half-wave potential***

$$E_{1/2} = E^{\ominus\prime}_{O/R} + \frac{RT}{nF} \ln\left[\frac{D_R\, \delta_O}{D_O\, \delta_R}\right] \tag{7.35}$$

Often $D_O \approx D_R$ and therefore $\delta_O = \delta_R$. The measurement of the half-wave potential is then a method for experimentally determining the formal redox potential, or even the standard redox potential.

In the case where only one species (oxidised or reduced) is present in solution, the current-potential curve varies from a zero current up to the limiting diffusion current (see Figure 7.13). Thus for an oxidation, equation (7.34) reduces to

$$E = E_{1/2} + \frac{RT}{nF} \ln\left[\frac{I}{I_{da} - I}\right] \tag{7.36}$$

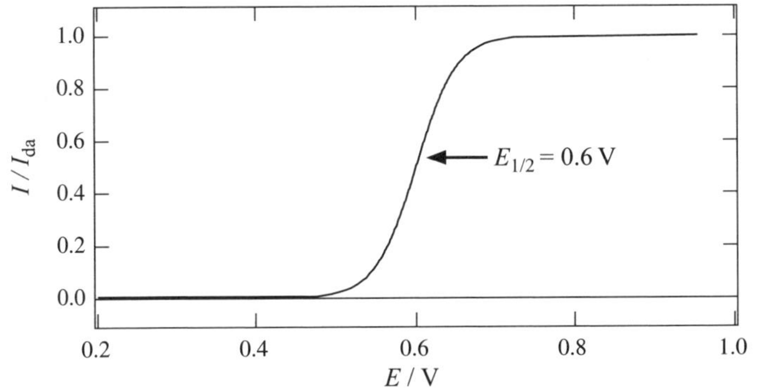

Fig. 7.13 Current-potential curve for a reaction limited by the diffusion of species in solution. $n = 1$, $c_O(\infty) = 0$, $\delta = \delta_O = \delta_R$ and $D = D_O = D_R$.

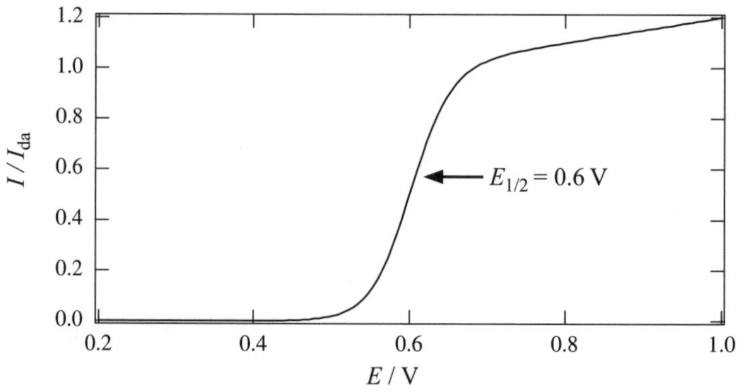

Fig. 7.14 Illustration of equation (7.37) with $n = 1$ and $B = 0.5$.

In this case, we trace a graph ln $[I/(I_{da}-I)]$ as a function of the potential to get a straight line with the slope nF/RT. Such a graph is often drawn to verify the reversibility of the electrode reaction (see §7.5.1) and to determine the half-wave potential.

From an experimental point of view as shown below in Figures 7.16 and 7.19, the limiting diffusion current does not necessarily appear clearly as a horizontal line, but rather as a slope. There can be several reasons for this, linked either to the capacitative current or to 'parasite' reactions. An empirical equation can be used to take this problem into account (Figure 7.14).

$$I = I_{da}\left(1+B\cdot\left(E-E_{1/2}\right)\right)\left[\frac{\exp^{nF\left(E-E_{1/2}\right)/RT}}{1+\exp^{nF\left(E-E_{1/2}\right)/RT}}\right] \qquad (7.37)$$

The case $B = 0$ corresponds to equation (7.36)

7.4 ELECTRODES WITH A DIFFUSION LAYER OF CONTROLLED THICKNESS

7.4.1 Rotating disc electrode

The rotating disc electrode, developed in the 1950s, was one of the first techniques capable of generating mass transfer controlled steady state diffusion currents. The principle is based on the fact that the rotation of a cylinder on its axis of symmetry causes a pumping of the liquid in which it is immersed towards the disc electrode at the end of the cylinder as shown in Figure 7.15.

This hydrodynamic movement controls the thickness of the diffusion layer δ as a function of the angular velocity ω (s^{-1} or Hz):

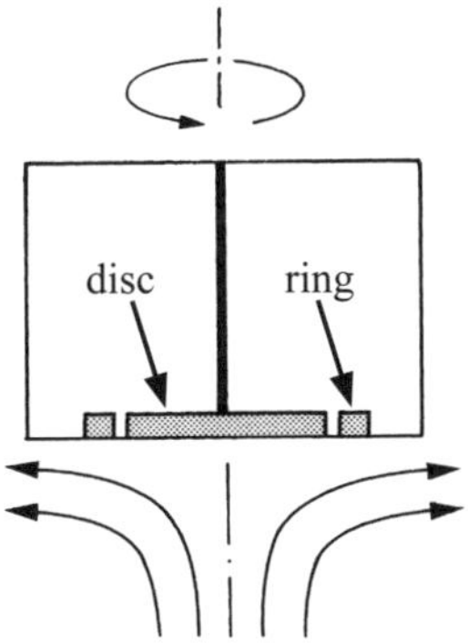

Fig. 7.15 Hydrodynamic flux under a disc electrode and a rotating ring electrode.

$$\delta = 1.61\, D^{1/3}\, \nu^{1/6}\, \omega^{-1/2} \tag{7.38}$$

δ is given in centimeters if the diffusion coefficient D and the kinematic viscosity ν (equal to the ratio of the viscosity to the density) are both given in $cm^2{\cdot}s^{-1}$ ($\nu = 0.01$ $cm^2{\cdot}s^{-1}$ for water).

In the laboratory, rotation frequencies ($f = \omega/2\pi$ = rotations$\cdot s^{-1}$) vary approximately between

$$2 < \frac{\omega}{2\pi} < 50\ \text{Hz}$$

since at higher velocites, there is a risk of turbulence. In practice, current-potential curves are measured at different rotation velocities as shown in Figure 7.16.

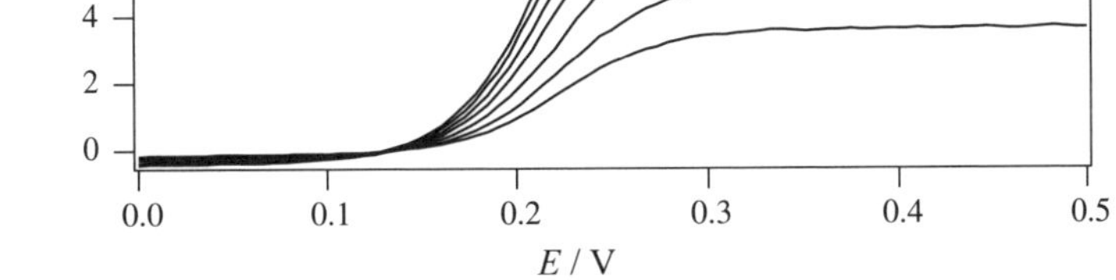

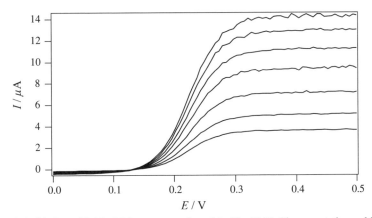

Fig. 7. 16 Oxidation of 0.25mM ferrocenemethanol in 50mM NaCl on a rotating gold electrode (Diameter=3mm). Angular velocities : 200, 400, 800, 1400, 2000, 2600 & 3000 rpm. Potential scan rate =10mV$\cdot s^{-1}$ forward and reverse. At higher rotation speeds, current instabilities are noticeable (Olivier Bagel, EPFL).

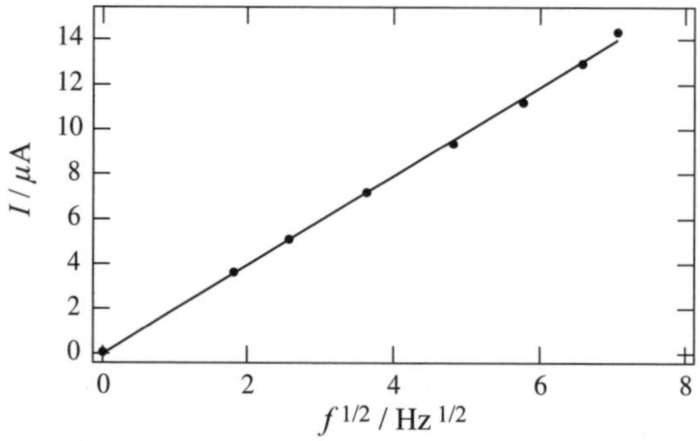

Fig. 7. 17 Variation of the anodic limiting diffusion current as a function of the square root of the rotation velocity calculated from the data in Figure 7.16. Slope = 2 μA·s$^{1/2}$. This gives a value for the diffusion coefficient of D=6.9 10^{-6} cm^2·s^{-1}.

Next, we plot the limiting diffusion current as a function of the inverse of the square root of the angular velocity (see Figure 7.17) to calculate the diffusion coefficient of the reactant, if we know its concentration, or to calculate the concentration if we know its diffusion coefficient.

For a rotating disc electrode, the half-wave potential is given by equations (7.35) and (7.38) i.e.

$$E_{1/2} = E^{\ominus\prime} + \frac{RT}{nF} \ln \left[\frac{D_R^{2/3}}{D_O^{2/3}} \right] \tag{7.39}$$

Rotating electrodes can have, not only a disc electrode as described above, but also a concentric ring electrode outside the disc electrode. This arrangement with both the disc and the concentric ring allows electrochemical study on the ring electrode of the species generated on the disc electrode that, under the effect of the pumping shown in Figure 7.15, pass under the ring. Obviously, only a fraction of the species generated can be oxidised or reduced on the ring electrode. This fraction depends on the geometric characteristics of the disc and the ring, and we define the ***collection factor*** as the ratio of ring/disc limiting currents.

The 'wall-jet' electrode has a jet of solution projected against a static electrode like a watering hose against a wall, and is another technique for which the hydrodynamics is similar in nature to the rotating electrode.

7.4.2 Microelectrodes

One of the major achievements in electroanalytical chemistry in the 1980s was the introduction of microelectrodes, i.e. electrodes of which one of the characteristic dimensions is of the order of a few micrometres (the radius in the case of microdiscs and microhemispheres, band width in the case of microbands, etc.). The characteristics

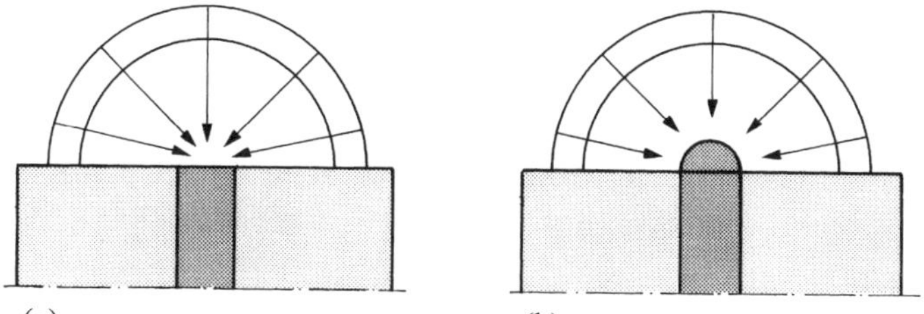

Fig. 7.18 Cross-section of microelectrodes – microdiscs and microhemispheres. Spherical diffusion: the equiconcentration curves are hemispheres.

of microelectrodes is the establishment of a cylindrical or hemispherical diffusion flux leading to a large increase in mass transfer. The diffusion layers thus created are either hemispherical in the case of microdiscs and microhemispheres (see Figure 7.18), or hemicylindrical in the case of microbands.

The thickness of the diffusion layer is then entirely governed by the geometry of the electrode and hardly influenced by the convection in solution. Microdisc and microhemisphere electrodes have the characteristics to generate steady state limiting currents controlled by diffusion.

Microhemispheres

For a hemispherical electrode with a radius of r_{hs}, the expression for the diffusion layer δ is

$$\delta = r_{hs} \tag{7.40}$$

and the steady state diffusion current is then directly proportional to the radius r_{hs}, to the diffusion coefficient D and the bulk concentration c of the reactant in solution.

$$I_d = 2\pi\, nFD\, c\, r_{hs} \tag{7.41}$$

To demonstrate this relationship, consider the molar flux J_m, which is the number of moles of reactants per unit time going into the concentric hemispheres. It is equal to the product of the diffusion flux due to the concentration gradient, here given in spherical coordinates, and the area of the concentric hemisphere of radius r

$$J_m = 2\pi r^2 D\, \frac{dc}{dr} \tag{7.42}$$

whatever the radius r. Integration at constant flux allows us to obtain the concentration at the electrode surface

$$\left[c\right]_{r=r_{hs}}^{r=\infty} = \left[\frac{-J}{2\pi D r}\right]_{r=r_{hs}}^{r=\infty} \tag{7.43}$$

which is

$$c(r_{hs},t) \;=\; c(\infty,t) - \frac{J}{2\pi D r_{hs}} \tag{7.44}$$

For the limiting diffusion current, the concentration at the surface of the hemispherical electrode is zero, and consequently, we come back to equation (7.41).

Microdiscs

For a microdisc of radius r_d, it is possible to show by a rather long calculation that the thickness of the diffusion layer δ is

$$\delta \;=\; \frac{\pi r_d}{4} \tag{7.45}$$

and the steady state diffusion current is then proportional to the concentration in the bulk of the solution $c = c(\infty)$, to the diffusion coefficient D of the reactant and to the radius r_d.

$$I_d = 4\, nFD\, c\, r_d \tag{7.46}$$

Figure 7.19 shows an oxidation reaction on a microdisc electrode.

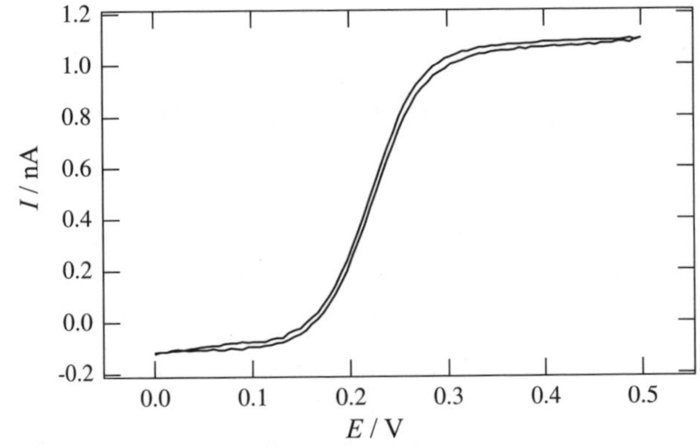

Fig. 7. 19 Oxidation of ferrocenemethanol on a platinum microdisc electrode. Radius: 10 μm. Same conditions as in Figure 7.16. Potential scan rate =1mV·s^{-1} (Olivier Bagel, EPFL).

Recessed microdiscs

For recessed microdiscs as illustrated in Figure 7.20, we have two diffusion geometries: linear diffusion in the microcylinder above the electrode, and hemispherical diffusion such as that obtained on a microdisc electrode.

The equation for the continuity of the flux of the reactants towards the electrode gives us that

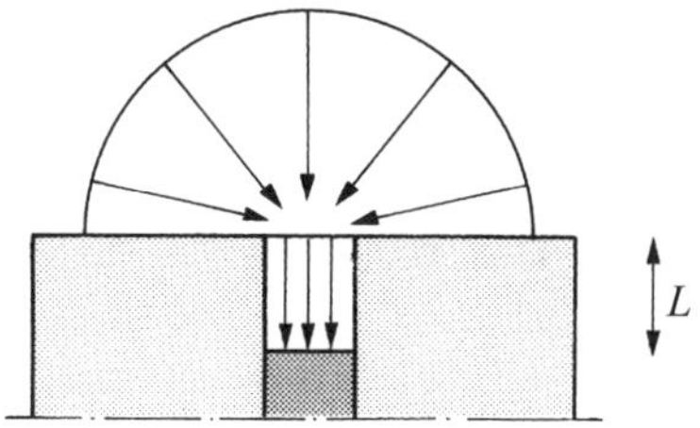

Fig. 7.20 Recessed microdisc electrode.

$$I_d = nFADc_L / L = nFAD(c - c_L) / \delta \tag{7.47}$$

where c_L is the reactant concentration in the steady state regime at the top of the cylinder where linear diffusion takes place and where δ is given by equation (7.45). By elimination of the concentration c_L from equation (7.47), we obtain by substitution

$$I_d = \frac{nFADc}{L + \delta} \tag{7.48}$$

This recessed microelectrode behaves like a micoelectrode for which the apparent thickness of the diffusion layer is $L + \delta$.

Half-wave potential

For all these geometries of microelectrodes, the thickness of the diffusion layer is entirely determined by the geometry of the electrode and independent of the diffusion coefficient. Thus, the half-wave potential is given by

$$E_{1/2} = E^{\ominus\prime} + \frac{RT}{nF} \ln\left[\frac{D_R}{D_O}\right] \tag{7.49}$$

Microelectrodes have many advantages compared to electrodes of millimeter (or more) dimensions. In particular, they allow voltammetry in highly resistant media such as frozen solvents, organic solvents without supporting electrolyte, or even in supercritical fluids. Effectively, the ohmic drop between the working electrode and the reference electrode is smaller for microelectrodes. This point will be taken up in more detail in §7.8.

7.4.3 Band electrode in a laminar flow

Microband and microhemicylindrical electrodes do not provide limiting diffusion currents in stagnant solutions. The hemicylindrical diffusion observed for these systems is not as efficient as the spherical diffusion for the transport of reactants towards the electrode. However, if band electrodes are placed in flow channels of a thickness such that the flow is laminar (see Figure 7.21), then it is possible to measure limiting diffusion currents.

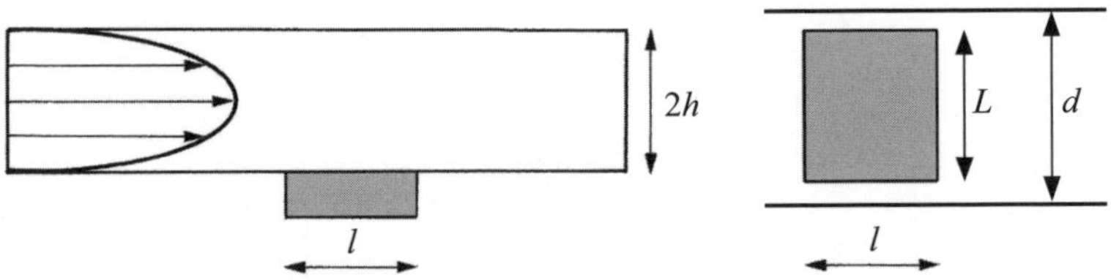

Fig 7.21 Band electrode in a microchannel with laminar flow. Side view and top view.

$$I = 0.925nFcL(lD)^{2/3}\left(\frac{F_V}{h^2 d}\right)^{1/3} \quad (7.50)$$

where c and D are the bulk concentration and the diffusion coefficient of the reacting species, l and L are the width and length of the band, F_V is the volumic flow rate and $2h$ and d are the height and width of the channel.

7.4.4 Membrane-covered electrode

If an electrode is covered with a porous membrane as illustrated in Figure 7.22, e.g. a polymer membrane, where the diffusion coefficient is smaller than that in solution, the thickness of the membrane can be considered *de facto* as the thickness of the diffusion layer.
The limiting diffusion current is the

$$I = nFAcD_m / \delta_m \quad (7.51)$$

where D_m and δ_m are the diffusion coefficient in the membrane and the membrane thickness respectively. This will be demonstrated more rigorously at the end of chapter 8. This approach to control the diffusion layer thickness is often used in the design of single-use (not to say disposable) electrodes in biosensing applications.

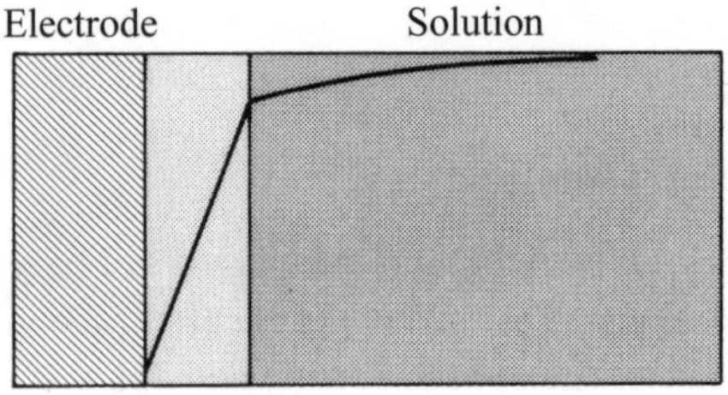

Fig. 7.22 Concentration profile of the reactant for an electrode covered with a porous membrane. The concentration gradient takes place mainly in the membrane.

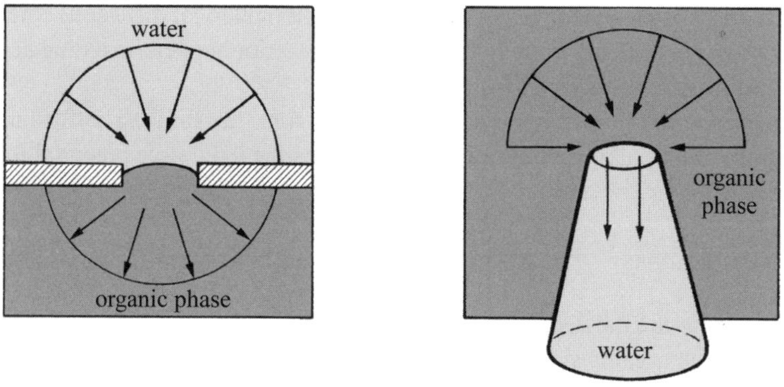

Fig. 7.23 Liquid | liquid microinterfaces supported on a microhole in a polymer film and on the tip of a glass micropipette.

7.4.5 Liquid | liquid micro-interfaces

There are two principal methods for constructing liquid | liquid micro-interfaces. The first is to micro-manufacture microholes e.g. by laser photo-ablation in polymer films (such as PolyEthylene Terephthalate (PET)). By placing a liquid phase either side of the film, we thus make liquid | liquid micro-interfaces (see Figure 7.23). The passage of ions across these micro-interfaces is characterised by limiting diffusion currents such as those obtained for metallic microelectrodes, the major difference being that the diffusion coefficients of the ion in the two adjacent phases are not the same.

The second method is to use glass micropipettes obtained by pulling glass capillaries. These micropipettes have tips whose diameter can be less than a micrometre. The microinterfaces on micropipettes have the peculiarity that the movement of ions (ingress) from the exterior to the interior of the micropipettes

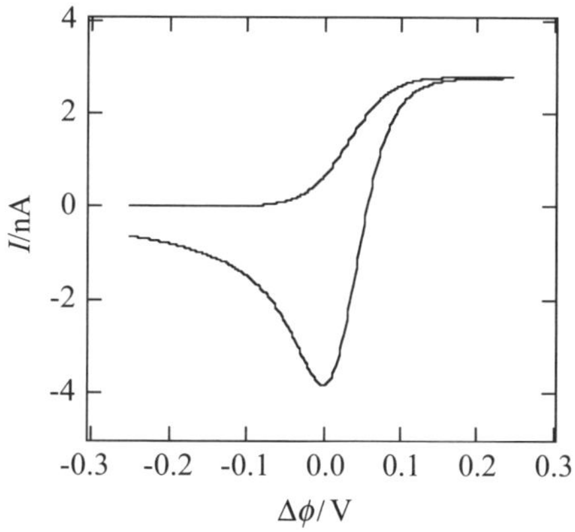

Fig. 7.24 Simulation of an ion transfer reaction across a liquid | liquid interface on the tip of a pipette with a 20 μm diameter, $\Delta\phi^{\ominus}_{tr} = 0$ V (R. Ferrigno, EPFL).

gives rise to a steady state limiting diffusion current (spherical diffusion) whereas the return movement (egress) from the interior to the exterior happens in a time dependent fashion (semi-infinite linear diffusion).

For micro-interfaces supported by a polymer film, the limiting diffusion current corresponds to that of an embedded electrode. Thus the half-wave potential for an ion transfer reaction is given by

$$\Delta_{\rm o}^{\rm w}\phi_{1/2} \;=\; \Delta_{\rm o}^{\rm w}\phi^{\ominus\prime} + \frac{RT}{zF}\ln\left[\frac{D^{\rm w}}{D^{\rm o}}\frac{L^{\rm o}+\pi r_{\rm d}/4}{L^{\rm w}+\pi r_{\rm d}/4}\right] \tag{7.52}$$

where $L^{\rm o}$ and $L^{\rm w}$ are the depths of the interface compared to the surface of the film on, respectively, the organic side and the aqueous side.

7.5 QUASI-REVERSIBLE SYSTEMS: CURRENT LIMITED BY KINETICS AND DIFFUSION

7.5.1 Current-potential curve

In electrochemical jargon, a reaction is quasi-reversible if the current obtained is controlled at the same time by the kinetics of the electrochemical reaction at the electrode and by the mass transfer. Thus, the current can be written as a diffusion current

$$I \;=\; nFA\,D_{\rm O}\left(c_{\rm O}(0)-c_{\rm O}(\infty)\right)/\,\delta_{\rm O} \;=\; -nFA\,D_{\rm R}\left(c_{\rm R}(0)-c_{\rm R}(\infty)\right)/\,\delta_{\rm R} \tag{7.53}$$

and as a kinetically controlled current

$$I \;=\; nFA\left(k_{\rm a}\,c_{\rm R}(0)-k_{\rm c}\,c_{\rm O}(0)\right) \tag{7.54}$$

By eliminating from this equation the concentrations at the electrode surface given by equations (7.32) and (7.33), we get an expression for the current as a function of the electrochemical rate constants

$$I \;=\; \frac{nFA\left(k_{\rm a}\,c_{\rm R}(\infty)-k_{\rm c}\,c_{\rm O}(\infty)\right)}{1+\dfrac{k_{\rm a}\,\delta_{\rm R}}{D_{\rm R}}+\dfrac{k_{\rm c}\,\delta_{\rm O}}{D_{\rm O}}} \tag{7.55}$$

If the reverse reaction can be ignored (high overpotential), this equation reduces, e.g. for an oxidation, to

$$I \;=\; \frac{nFA\,k_{\rm a}\,c_{\rm R}(\infty)}{1+\dfrac{k_{\rm a}\,\delta_{\rm R}}{D_{\rm R}}} \tag{7.56}$$

Equation (7.56) is useful for determining whether an electrode reaction is controlled by diffusion or by kinetics:

$$\text{If}\quad \frac{k_{\rm a}\,\delta_{\rm R}}{D_{\rm R}} \gg 1 \quad\text{then}\quad I \;=\; \frac{nFA\,D_{\rm R}\,c_{\rm R}(\infty)}{\delta_{\rm R}} \quad\text{diffusion controlled} \tag{7.57}$$

$$\text{If} \quad \frac{k_a\,\delta_R}{D_R} << 1 \quad \text{then} \quad I = nFA\,k_a\,c_R(\infty) \quad \text{kinetics controlled} \tag{7.58}$$

For a rotating electrode, equation (7.55) can be written as

$$\frac{1}{I} = \frac{1}{nFA\left(k_a\,c_R(\infty) - k_c\,c_O(\infty)\right)}\left[1 + \frac{k_a\,D_R^{-2/3} + k_c\,D_O^{-2/3}}{0.62\,\nu^{-1/6}\,\omega^{1/2}}\right] \tag{7.59}$$

In the potential zone where $k_a >> k_c$, this equation reduces to the ***Koutecký-Levich*** equation

$$\frac{1}{I} = \frac{1}{nFA\,k_a\,c_R(\infty)}\left[1 + \frac{k_a\,D_R^{-2/3}}{0.62\,\nu^{-1/6}\,\omega^{1/2}}\right] \tag{7.60}$$

Thus, by plotting I^{-1} as a function of $\omega^{-1/2}$ for different electrode potential values, we obtain the rate constant k_a as a function of the potential and we can then verify whether the system obeys the Butler-Volmer law given by equation (7.22).

Figure 7.25 shows that the results from Figure 7.16 correspond more or less to a reversible reaction, since all the lines converge towards the origin ($k_a \to \infty$). Nevertheless, a closer study of the data in Figure 7.25 allows us to plot $\log k_a$ as a function of the potential when $\omega \to \infty$, as shown in Figure 7.26. Thus, the value obtained at the half-wave potential $E_{1/2} = 0.225$ V gives us the standard rate constant i.e. $k_a^{\ominus} = 4 \cdot 10^{-2}$ cm·s^{-1}. This value, obtained for one concentration only of ferrocenemethanol and one concentration only of the supporting electrolyte is just

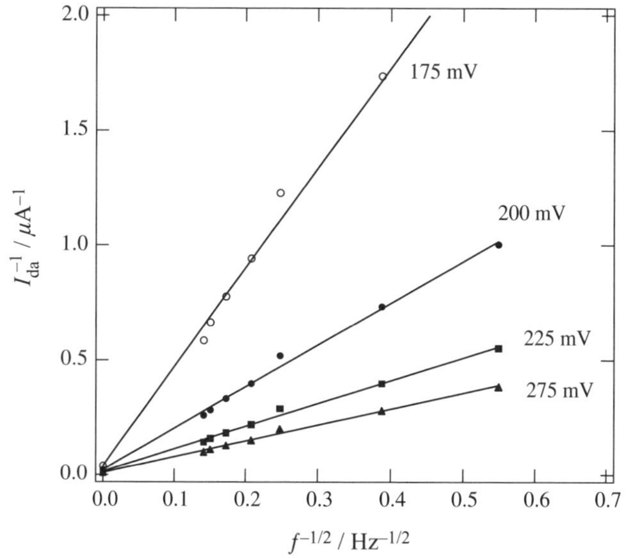

Fig 7.25 Application of equation (7.60) to the data from Figure 7.16.

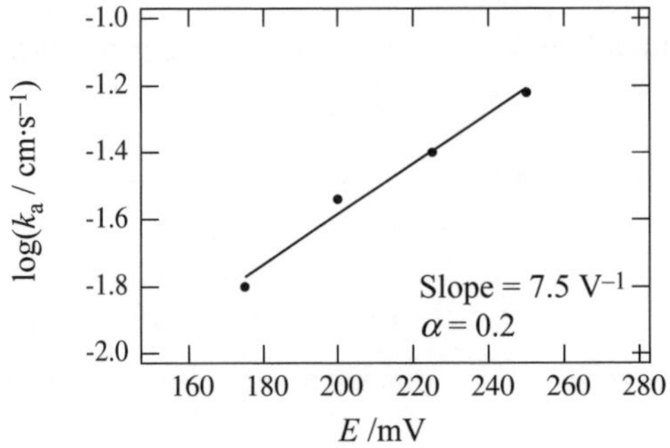

Fig. 7.26 Potential dependence of the standard rate constants obtained from equation (7.60) from the data of Figure 7.25.

given to illustrate that the methodology reaches its limits when the points converge on the origin ($k_a \cong 10^{-2}$ cm·s^{-1}). This value obtained can be considered as the upper limit; the methodology not allowing access to higher values.

In the same way, for hemispherical microelectrodes of different sizes, we have

$$\frac{1}{I} = \frac{1+r\left[\dfrac{k_a}{D_R}+\dfrac{k_c}{D_O}\right]}{nFA\left(k_a\, c_R(\infty)-k_c\, c_O(\infty)\right)} \tag{7.61}$$

For an oxidation where $k_a >> k_c$, this equation simplifies to

$$\frac{1}{I} = \frac{1}{nFA\, k_a\, c_R(\infty)} + \frac{r}{nFA\, D_R\, c_R(\infty)} \tag{7.62}$$

A graph of the reciprocal of the current as a function of the radius of the microelectrode used allows the measurement of the anodic rate constant.

If, for practical reasons, it is not possible to vary the mass transport to the electrode, it is nevertheless possible to extract kinetic information by measuring just one steady state current/potential curve.

As equations (7.12) & (7.14) show, the ratio of the oxidation and reduction rate constants is a function of the applied potential

$$\frac{k_a}{k_c} = \exp^{nF\left(E-E^{\ominus\prime}\right)/RT} \tag{7.63}$$

Thus, by combining this equation and the interfacial concentration values expressed as a function of their limiting currents (7.32) and (7.33), we get

$$I = nFA\, k_a\left[c_R(\infty)\left(1-\frac{I}{I_{da}}\right) - c_O(\infty)\left(1-\frac{I}{I_{dc}}\right)e^{-nF\left(E-E^{\ominus\prime}\right)/RT}\right] \tag{7.64}$$

In the case where $c_O(\infty)=0$, the current equation becomes

$$I = nFA\,k_a \left[c_R(\infty)\left(1 - \frac{I}{I_{da}}\right) - \frac{\delta_O\, I\, e^{-nF\left(E-E^{\ominus\prime}\right)/RT}}{nFA\,D_O} \right] \tag{7.65}$$

or again, substituting in equations (7.12) and (7.14)

$$\frac{I}{I_{da}} = \frac{nFA\,c_R(\infty)k^{\ominus}e^{\alpha nF(E-E^{\ominus\prime})/RT}}{\left[I_{da} + nFA\,c_R(\infty)\,k^{\ominus}e^{\alpha nF(E-E^{\ominus\prime})/RT} + \frac{\delta_O}{D_O} I_{da} e^{-(1-\alpha)nF\left(E-E^{\ominus\prime}\right)/RT}\right]} \tag{7.66}$$

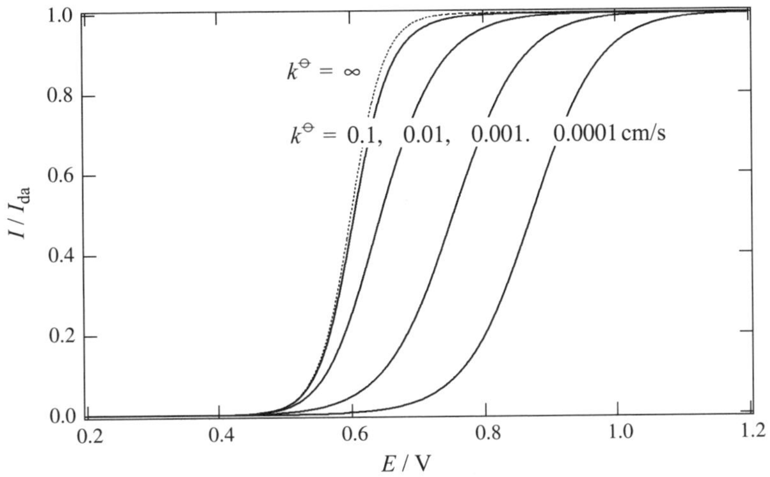

Fig. 7.27 Kinetic influence on the reaction for quasi-reversible systems, $n = 1$, $D = 10^{-5}$ cm^2·s^{-1}, δ=5.10^{-4} cm, $\alpha = 0.5$, $E^{\ominus\prime} = 0.6$ V.

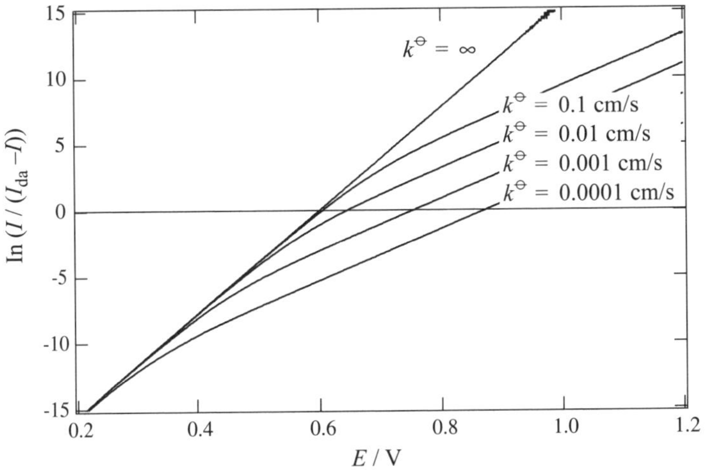

Fig. 7.28 Graph of $\ln(I/(I_{da} - I))$ as a function of potential for the values of Figure 7.27.

Figure 7.27 shows the kinetic effect of the electrode reaction on the steady state current-potential curve: the smaller the diffusion coefficient, the more apparent the kinetic effect.

The kinetic effect is very visible on graphs of ln $[I/(I_{da}-I)]$ as a function of the potential. In effect, the linear behaviour obtained for reversible systems is no longer respected, as shown in Figure 7.28.

7.5.2 Influence of the ohmic drop

If, for experimental reasons, the electrode potential cannot be precisely controlled (e.g. reference electrode far away, a very resistive solution, electrode materials not very conductive...), the potential applied with the potentiostat is no longer equal to the electrode potential, but also contains an additional factor associated to the ohmic drop. Thus we have:

$$E_{\text{electrode}} = E_{\text{applied}} - R_S I \tag{7.67}$$

where R_S is the equivalent resistor in series. The potential current curve is plotted of course as a function of the applied potential and not of the real potential. The curves produced in this way are deformed as shown in Figure 7.29.

7.5.3 Comments on the experimental measurement of rate constants

It can never be repeated often enough that to obtain reliable rate constant data for a redox reaction, we have not only to repeat the measurements for different reactant concentrations, but also for different concentrations of the supporting electrolytes.

Enemy number one of electrode kinetic measurements is the un-compensated ohmic drop (see §7.8). The better the geometry of the cell is, and the higher the concentration of the supporting electrolyte is, the smaller is the ohmic drop and the more reliable are the results. Therefore, by increasing the concentration of the

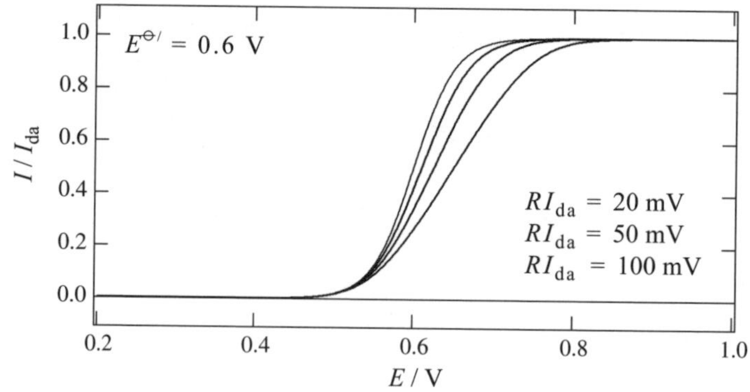

Fig. 7.29 Influence of the ohmic drop on a current-potential curve in a steady state regime ($n = 1$).

supporting electrolyte, we decrease the ohmic drop and the rate constant values measured should not vary.

Enemy number two in the case of solid electrodes is without any doubt the state of the electrode surface. Polishing electrodes is a process often considered as tiresome. Nevertheless, good polishing is a *sine qua non* condition for reliable measurements. This is illustrated in Figure 7.30. The experimental data are similar to those used in Figure 7.25 with the one difference being the polishing of the electrode, for which Figure 7.30 can be considered as 'botched'. Thus, on a badly polished electrode a reversible reaction can appear as quasi-reversible: the straight lines in Figure 7.30 do not converge on the origin. The standard rate constant obtained from this figure is one order of magnitude smaller than the one obtained from the data in Figure 7.25. It is useful to remember that the rate constants of electrochemical reactions measured by amperometry are limited in the upper range by the mass transfer.

A badly-polished electrode is not electro-active in a uniform manner across its whole surface in contact with the solution, and it can be shown that current-potential curves measured on partially blocked electrodes have the characteristics of slow electrochemical reactions.

Given the practical difficulties in obtaining reliable experimental rate constant values, it is not surprising that some values published in the literature are rather due to experimental factors than to the reaction rate itself.

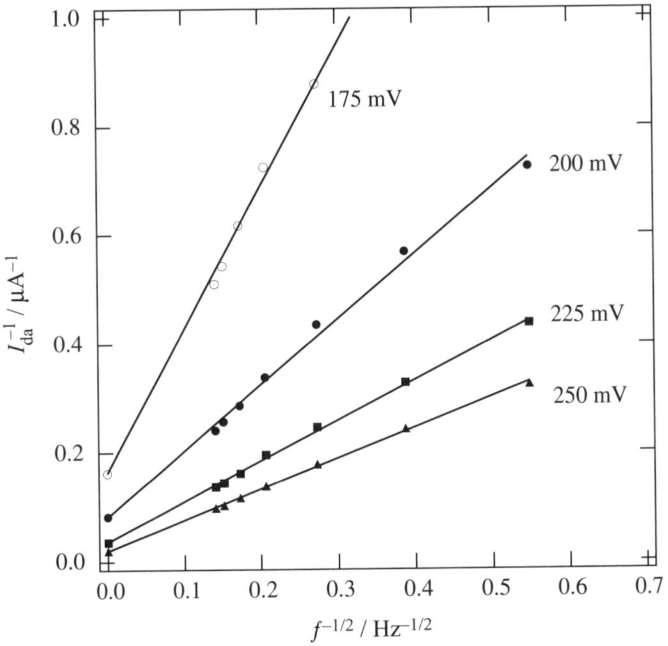

Fig. 7.30 Results similar to those in Figure 7.25 but on a 'badly polished' electrode. These measurements give a value for the standard rate constant of $2.5 \cdot 10^{-3}$ cm·s^{-1} (Olivier Bagel, EPFL).

7.6 IRREVERSIBLE SYSTEMS: CURRENT LIMITED BY KINETICS AND DIFFUSION

In electrochemical jargon, a system is said to be irreversible if the kinetics of the electron transfer reactions is slow. To give an order of magnitude, the standard rate constant $k^{\ominus}$ would be less than 10^{-3} cm·s^{-1}(see equation (7.58) for more detail). Thus for an oxidation, we can ignore the reverse reaction, i.e. the cathodic current. And so, by combining the Butler-Volmer equation

$$\begin{aligned} I &= nFA\,k^{\ominus}\left[c_{\mathrm{R}}(0)\,e^{\alpha nF(E-E^{\ominus\prime})/RT} - c_{\mathrm{O}}(0)\,e^{-(1-\alpha)nF(E-E^{\ominus\prime})/RT}\right] \\ &\cong nFA\,k^{\ominus}\,c_{\mathrm{R}}(0)\,e^{\alpha nF(E-E^{\ominus\prime})/RT} \end{aligned} \tag{7.68}$$

with the interfacial concentration given by equation (7.32), we get

$$I = \frac{k^{\ominus}\,\delta_{\mathrm{R}}}{D_{\mathrm{R}}}\left(I_{\mathrm{da}} - I\right)\;e^{\alpha nF(E-E^{\ominus\prime})/RT} \tag{7.69}$$

By expanding the above equation, we can express the current-potential curve in the same way as we did for reversible systems.

$$E = E^{\ominus\prime} + \frac{RT}{\alpha nF}\ln\left(\frac{D_{\mathrm{R}}}{k^{\ominus}\,\delta_{\mathrm{R}}}\right) + \frac{RT}{\alpha nF}\ln\left(\frac{I}{I_{\mathrm{da}} - I}\right) \tag{7.70}$$

The following graph shows that for slow reactions, we find the same current-potential curves as those shown in Figure 7.27.

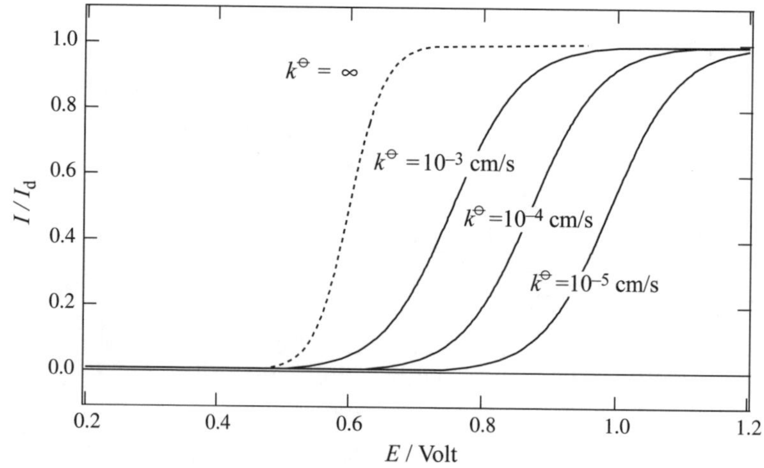

Fig. 7.31 Evolution of the current-potential curves for irreversible systems, $n = 1$, $D = 10^{-5}$ cm^2·s^{-1}, $\delta = 5$ µm, $\alpha = 0.5$, $E^{\ominus\prime} = 0.6$ Volt.

7.7 QUASI-REVERSIBLE SYSTEMS: CURRENT LIMITED BY DIFFUSION, MIGRATION AND KINETICS

When the concentration of a supporting electrolyte is not sufficient, it is no longer possible to neglect the migration. The migration diffusion flux is written as

$$\boldsymbol{J} = -\left[D\,\mathbf{grad}c(x) + \frac{zFDc(x)}{RT}\mathbf{grad}\phi(x)\right] = -De^{-zF\phi(x)/RT}\,\mathbf{grad}\left[c(x)\cdot e^{zF\phi(x)/RT}\right] \tag{7.71}$$

where $\phi(x)$ is the potential in solution. In a steady state regime, the flux is constant and we deduce from this that

$$J\int_0^\delta e^{zF\phi(x)/RT}\mathrm{d}x = -\frac{D}{\delta}\left[c(\infty)\cdot e^{zF\phi(\infty)/RT} - c(0)\cdot e^{zF\phi(0)/RT}\right] \tag{7.72}$$

Thus, the current for reversible systems is given by

$$\begin{aligned} I &= nFA\,D_{\mathrm{O}}\left(c_{\mathrm{O}}(0)\cdot e^{z_{\mathrm{O}}F\phi(0)/RT} - c_{\mathrm{O}}(\infty)\right)/\,\delta_{\mathrm{O}}\int_0^{\delta_{\mathrm{O}}} e^{z_{\mathrm{O}}F\phi(x)/RT}\mathrm{d}x \\ &= -nFA\,D_{\mathrm{R}}\left(c_{\mathrm{R}}(0)\cdot e^{z_{\mathrm{R}}F\phi(0)/RT} - c_{\mathrm{R}}(\infty)\right)/\,\delta_{\mathrm{R}}\int_0^{\delta_{\mathrm{R}}} e^{z_{\mathrm{R}}F\phi(x)/RT}\mathrm{d}x \end{aligned} \tag{7.73}$$

or, for a kinetically controlled current, for quasi-reversible systems by

$$I = nFA\left(k_{\mathrm{a}}\,c_{\mathrm{R}}(0) - k_{\mathrm{c}}\,c_{\mathrm{O}}(0)\right) \tag{7.74}$$

By eliminating from this equation the concentrations at the electrode given for equations (7.32) and (7.33), we get an expression for the current as a function of the electrochemical rate constants

$$I = \frac{nFA\left(k_{\mathrm{a}}\,c_{\mathrm{R}}(\infty)e^{-z_{\mathrm{R}}F\phi_2/RT} - k_{\mathrm{c}}\,c_{\mathrm{O}}(\infty)e^{-z_{\mathrm{O}}F\phi_2)/RT}\right)}{1+\dfrac{k_{\mathrm{a}}\,\delta_{\mathrm{R}}e^{-z_{\mathrm{R}}F\phi_2/RT}\int_0^{\delta_{\mathrm{R}}} e^{z_{\mathrm{R}}F\phi(x)/RT}\mathrm{d}x}{D_{\mathrm{R}}} + \dfrac{k_{\mathrm{c}}\,\delta_{\mathrm{O}}e^{-z_{\mathrm{O}}F\phi_2/RT}\int_0^{\delta_{\mathrm{O}}} e^{z_{\mathrm{O}}F\phi(x)/RT}\mathrm{d}x}{D_{\mathrm{O}}}} \tag{7.75}$$

by putting $\phi(0) = \phi_2$. If the reverse reaction can be ignored (high overpotential), this equation reduces, e.g. for an oxidation, to

$$I = \frac{nFA\,k_{\mathrm{a}}\,c_{\mathrm{R}}(\infty)e^{-z_{\mathrm{R}}F\phi_2/RT}}{1+\dfrac{k_{\mathrm{a}}\,\delta_{\mathrm{R}}e^{-z_{\mathrm{R}}F\phi_2/RT}\int_0^{\delta_{\mathrm{R}}} e^{z_{\mathrm{R}}F\phi(x)/RT}\mathrm{d}x}{D_{\mathrm{R}}}} \tag{7.76}$$

Thus, taking the reciprocal of equation (7.76) we get

$$\frac{1}{I} = \frac{1}{nFA\,k_{\mathrm{a}}\,c_{\mathrm{R}}(\infty)}\left[1+\frac{k_{\mathrm{a}}\,\delta_{\mathrm{R}}\int_0^{\delta_{\mathrm{R}}} e^{z_{\mathrm{R}}F\phi(x)/RT}\mathrm{d}x}{D_{\mathrm{R}}}\right] \tag{7.77}$$

Comparison with equation (7.56) shows the role of migration.

7.8 EXPERIMENTAL ASPECTS OF AMPEROMETRY

7.8.1 Ohmic drop at a planar electrode

We saw in §7.1 that amperometry is done using three electrodes and a potentiostat. Here we shall explain how to position the electrodes with respect to each other.

The purpose of the potentiostat is to make the current flow between the working electrode and the counter-electrode. If we have two parallel electrodes of the same size, the current lines are normal to the electrodes, and the equipotentials are parallel to the surface of the electrodes as shown in Figure 7.32.

In this case, to control the potential between the working electrode and the solution, the reference electrode needs to be placed as close as possible to the surface of the working electrode. Placing a reference electrode directly next to the working electrode is practically speaking not simple, and it is customary to place a salt bridge terminating with a capillary, known as a ***Luggin capillary***.

In the example in Figure 7.32, the difference of the inner potentials of the anode and the cathode $\phi^{A} - \phi^{C}$ is the sum of the metal | solution potential difference at the anode $\phi^{A} - \phi^{SA}$, the potential drop $\phi^{SA} - \phi^{SC}$ in the solution due to the current passing through a resistive medium (Ohm's law) and the potential difference at the solution | metal interface at the cathode $\phi^{SC} - \phi^{C}$. This is shown schematically in Figure 7.33.

When measuring the electrode potential, a negligible current (a few picoamps) flows between the cathode and the reference electrode, the whole salt bridge is therefore at the same potential (Figure 7.34). Thus, by placing the tip of the Luggin capillary on an equipotential, the difference in potential between the reference electrode and the cathode $\phi^{R} - \phi^{C}$ is the sum of the potential difference at the surface of the reference electrode $\phi^{R} - \phi^{SR}$, the potential drop inside the capillary $\phi^{SR} - \phi^{LC}$ which is negligible, the ohmic drop in that part of the solution between the top of the capillary and the cathode where the current flows between the anode and the cathode

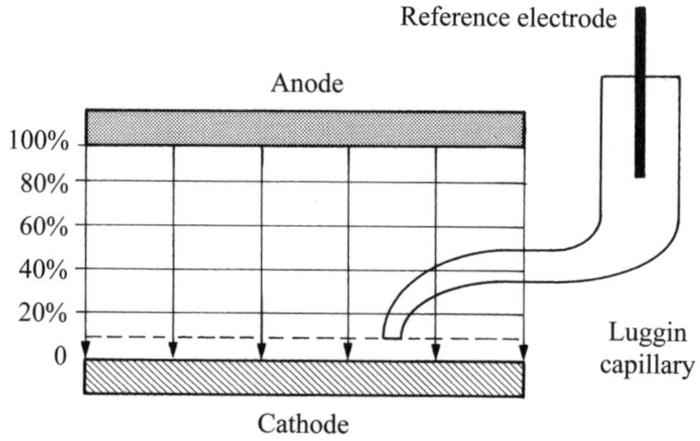

Fig. 7.32 Distribution of equipotential lines (horizontals) and current lines (vertical arrows from the anode to the cathode) in a cell with parallel plane electrodes.

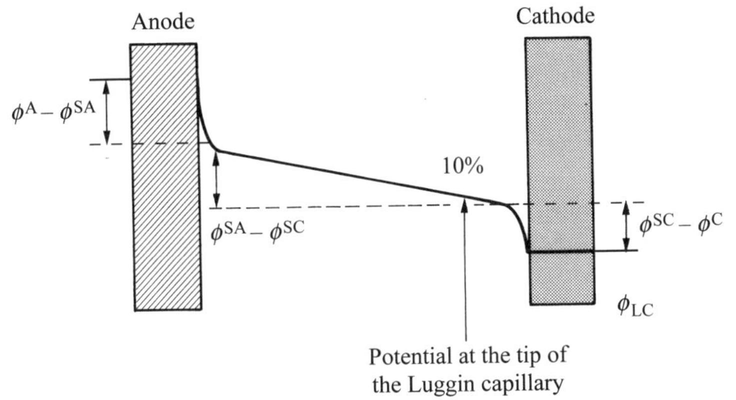

Fig. 7.33 Distribution of the potential between the anode and the cathode.

$\phi^{LC} - \phi^{SC}$ (=10% of $\phi^{SA} - \phi^{SC}$ in the example in Figure 7.32) and the potential difference at the surface of the cathode $\phi^{SC} - \phi^{C}$.

We saw in §2.1.2 that the electrode potential defined as the potential difference E between the working cathode and the reference electrode is written as

$$E \;=\; \phi^{Cu^{II}} - \phi^{Cu^{I}} \;=\; \phi^{Cu^{II}} - \phi^{C} + \phi^{C} - \phi^{R} + \phi^{R} - \phi^{Cu^{I}} \tag{7.78}$$

where Cu^{II} and Cu^{I} refer to the copper wires linking the working electrode (the cathode in the example of Figure 7.32) and the reference electrode respectively to the voltmeter. Thus, the measured electrode potential $E_{measured}$ is equal to the electrode potential E plus part of the ohmic drop in solution $\phi^{LC} - \phi^{SC}$ (=10% of $\phi^{SA} - \phi^{SC}$). This last contribution is called the ***ohmic drop***. We shall thus write for equation (7.67)

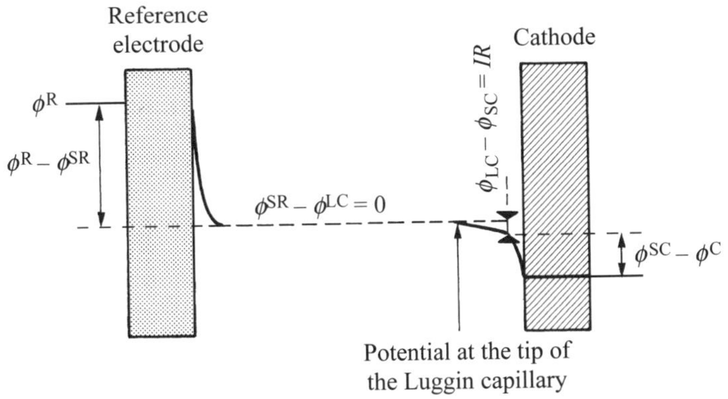

Fig. 7.34 Potential distribution between the reference electrode and the cathode.

$$E_{\text{applied}} = E_{\text{measured}} = E + IR_S \tag{7.79}$$

where I is the current flowing between the working electrode and the counter-electrode and R_S is the resistance of the solution for the part between the equipotential at the tip of the Luggin capillary and the equipotential in solution at the working electrode surface.

It is good to remember that the potential difference between the working electrode and the solution includes a thermodynamic term (see chapter 2) and a kinetic overpotential term (see §7.2).

The configuration of parallel electrodes such as shown in Figure 7.32 is sometimes used in electrosynthesis where it is often useful to separate the anode and the cathode by a membrane, either porous or ion-exchange. In this way the products generated on the two electrodes do not mix. Electrosynthesis cells are rarely operated by a potentiostat, and it is more normal to work either with the anode/cathode potential difference imposed, or using imposed currents.

7.8.2 Ohmic drop at a hemispherical electrode

In analytical electrochemistry, it is usual that the working electrode should have a smaller surface area than the counter electrode. In this way, the current lines converge towards the working electrode and the equipotential lines to the normal current lines will be closer when those converge, and more spaced out when they diverge. This has the advantage that the placement of the Luggin capillary is less critical, as shown in Figure 7.35 for a hemispherical electrode.

In the case of planar electrodes the ohmic drop depends mainly on the position of the Luggin capillary and the resistivity of the solution. Using equation (4.28), we have

$$IR_S = jS \cdot \frac{\rho l}{S} = \rho l j \tag{7.80}$$

where j is the current density, l the distance between the working electrode and the tip of the Luggin capillary, and ρ the resistivity of the solution. The ohmic drop is therefore independent of the size of the electrode.

In the case of microelectrodes, the ohmic drop decreases with the size of the electrode. This is easily demonstrated for hemispherical electrodes. Effectively, in this case the ohmic drop is given by equation (4.29) written as

$$IR_S = jS \cdot \frac{\rho}{2\pi}\left(\frac{1}{r_{\text{electrode}}} - \frac{1}{R}\right) \cong j2\pi r_{\text{electrode}}^2 \cdot \frac{\rho}{2\pi}\left(\frac{1}{r_{\text{electrode}}}\right) = \rho j r_{\text{electrode}} \tag{7.81}$$

making the hypothesis that R, the distance between the tip of the Luggin capillary and the centre of the microhemisphere is large with respect to its radius. Thus, equation (7.81) shows that the ohmic drop decreases with the size of the electrode, and that the position of the Luggin capillary is unimportant. Also, for microelectrodes where the currents are extremely small, it is possible to use the reference electrode as a counter-electrode. In this case, the potentiostat can be replaced by a simple power supply.

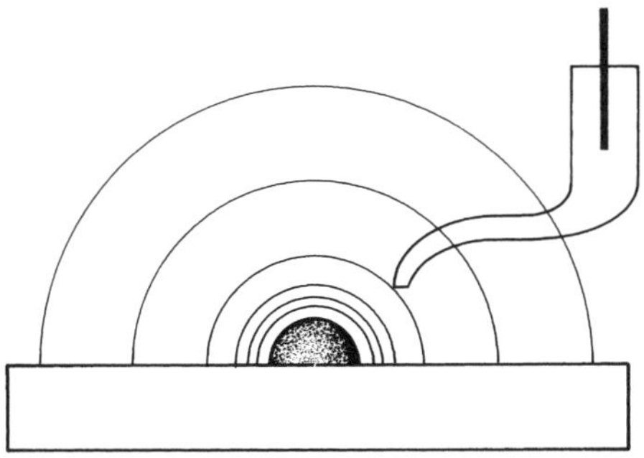

Fig. 7.35 Schematic potential distribution for a 'mini working electrode' taken here as hemispherical for example.

7.8.3 Compensating for the ohmic drop

To alleviate the problems due to the ohmic drop, it is often possible to add to the required value of the electrode potential, a potential term equal to IR_S. There are different experimental means to compensate for the ohmic drop, the most common being an analog electronic method based on the principle of 'positive feedback', the other being digital when using a digital potentiostat.

Nevertheless, it is useful to remember that the best way of combatting ohmic drop is to start with a good cell and electrode geometry, using Luggin capillaries. Electronic compensation is never better than partial, and cannot be used as a complete solution to the problem.

Concerning the measurement of reaction kinetics, the electrochemical impedance described in chapter 9 is recommended because this technique measures the ohmic drop and does not require it to be compensated for.

CHAPTER 8

PULSE VOLTAMMETRY

In this chapter, we shall study the current response as a function of time of a redox system when the electrode potential is varied by a sequence of potential steps or potential pulses. These methods should more generally be called chronoamperometry or transient amperometry, but are usually referred to as pulse voltammetry. We shall restrict the present study to reversible electrode reactions, i.e. those that are limited by the diffusion of the reactants and products. We will therefore make the hypothesis of infinitely rapid electrochemical reactions. Thus, the interfacial concentration of the oxidised and reduced species are imposed by the Nernst equation, whilst the concentrations in solution will depend on the distance from the electrode and on the time; they will thus be notated $c(x,t)$.

8.1 CHRONOAMPEROMETRY FOLLOWING A POTENTIAL STEP

8.1.1 Chronoamperometry with semi-infinite linear diffusion for a complete interfacial oxidation.

Chronoamperometry is a technique where the current is measured as a function of time. The simplest transient amperometric method consists in monitoring the current response following a stepwise variation of the electrode potential from an electrode potential value where only one redox species is stable in solution, either the oxidised form or the reduced form, to an electrode potential value where either a simple reduction or a simple oxidation occurs. In order to simplify the mathematical treatment, we shall only consider anodic oxidations, i.e we shall study the current response of a stepwise variation of electrode potential from an equilibrium value more negative than the standard redox potential of the redox couple under study ($E_{eq} < E^{\ominus}$–120mV) to an applied electrode potential value greater than the standard redox potential value ($E_{appl} > E^{\ominus}$+120mV). With these conditions, the ratio c_O/c_R is lower than 0.01 at the equilibrium electrode potential value, and it is greater than 100 at the applied step potential value.

We shall consider an electrochemically reversible oxidation,

$$\text{Red} \longrightarrow \text{Ox} + n\, e^-$$

that is one whose rate is very fast and for which the current is therefore controlled by diffusion. If the electrode is planar and of classic dimensions (a few millimetres or more), the diffusion of the reactants to the electrode and that of the products away from the electrode occurs in one dimension on the x-axis perpendicular to the surface of the electrode.

With the conditions described above, the solution initially contains only the reduced species with a concentration value $c_R(x,0)$ equal to the bulk concentration c_R. To satisfy the Nernst equation, the concentration of oxidised species is strictly speaking ($c_O = c_R \exp^{-nF(E_{eq} - E^{\ominus\prime})/RT} < 0.01 c_R$), but we shall consider it as negligible and take the following initial conditions :

$$c_R(x,0) \quad = \quad c_R \text{ and } c_O(x,0) \quad \approx \quad 0 \tag{8.1}$$

The electrolyte solution being considered as semi-infinite, we make the hypothesis that we have an infinite reservoir of reduced species, and therefore write:

$$\lim_{x\to\infty} c_R(x,t) = c_R \quad \text{and} \quad \lim_{x\to\infty} c_O(x,t) \approx 0 \tag{8.2}$$

Finally, we shall consider a step variation of the electrode potential from the equilibrium potential where the redox couple is reduced to a potential value where it is *completely oxidised* at the electrode, which implies that the interfacial concentration of R becomes negligible, and then we can write a boundary condition for the interfacial concentration of the reduced species

$$c_R(0,t) \quad \approx \quad 0 \text{ for } t > 0 \tag{8.3}$$

In this situation, where only one reduced reactant is present in solution at equilibrium, and is then completely oxidised at the surface of the electrode following the application of a potential step, we only need to consider the mass transfer of this reactant to the electrode to calculate the current response.

For electrochemically reversible reactions, the current is limited by the rate of arrival of the reactants at the electrode (see Figure 7.2). In the presence of a supporting electrolyte, we can neglect migration, and the current is controlled by a diffusion flux. Consequently, it can be written as

$$I \quad = \quad nFA\, D_R \left(\frac{\partial c_R(x,t)}{\partial x} \right)_{x=0} \tag{8.4}$$

The differential equation to be solved is the one for the conservation of mass, also called ***Fick's second equation***

$$\frac{\partial c_R(x,t)}{\partial t} \quad = \quad D_R \frac{\partial^2 c_R(x,t)}{\partial x^2} \tag{8.5}$$

This type of differential equation is reasonably simple to solve using the ***Laplace transformation*** defined for a function F of the variable t by :

$$L\{F(t)\} \quad \equiv \quad \int_0^\infty e^{-st} F(t)\, \mathrm{d}t \quad = \quad \overline{F}(s) \tag{8.6}$$

Among the properties of the transformed quantities, note the transform of the derivative of the function F

$$L\left\{\frac{\mathrm{d}F(t)}{\mathrm{d}t}\right\} = s\overline{F}(s) - F(0) \tag{8.7}$$

and more generally the transform of the nth derivative

$$L\left\{F^{(n)}\right\} = s^n\overline{F}(s) - s^{n-1}F(0) - s^{n-2}F'(0) - \ldots - F^{(n-1)}(0) \tag{8.8}$$

The Laplace transform of Fick's equation (8.5) is written

$$s\bar{c}_{\mathrm{R}}(x,s) - c_{\mathrm{R}}(x,0) = D_{\mathrm{R}}\frac{\partial^2\bar{c}_{\mathrm{R}}(x,s)}{\partial x^2} \tag{8.9}$$

which, after rearrangement, reads

$$\frac{\partial^2\bar{c}_{\mathrm{R}}(x,s)}{\partial x^2} - \frac{s}{D_{\mathrm{R}}}\bar{c}_{\mathrm{R}}(x,s) = -\frac{c_{\mathrm{R}}(x,0)}{D_{\mathrm{R}}} \tag{8.10}$$

The solution of this equation, expressed here as the sum of the solution of the homogeneous equation and of a particular solution, is

$$\bar{c}_{\mathrm{R}}(x,s) = \frac{c_{\mathrm{R}}(x,0)}{s} + A(s)\,e^{-\sqrt{\frac{s}{D_{\mathrm{R}}}}x} + B(s)\,e^{\sqrt{\frac{s}{D_{\mathrm{R}}}}x} \tag{8.11}$$

The constants A and B can be determined from the boundary conditions. For the bulk concentration of the reduced species, knowing that $F(t) = a$ then $L\{F(t)\} = a/s$, we have first of all

$$\lim_{x\to\infty} \bar{c}_{\mathrm{R}}(x,s) = \frac{c_{\mathrm{R}}}{s} \tag{8.12}$$

which leads to $B = 0$ and therefore equation (8.11) reduces to

$$\bar{c}_{\mathrm{R}}(x,s) = \frac{c_{\mathrm{R}}}{s} + A(s)\,e^{-\sqrt{\frac{s}{D_{\mathrm{R}}}}x} \tag{8.13}$$

The constant A can be calculated, knowing that the concentration of R at the electrode is zero. Thus, we have:

$$\bar{c}_{\mathrm{R}}(0,s) = 0 \tag{8.14}$$

The Laplace transform of the concentration of the reduced species is therefore written

$$\bar{c}_{\mathrm{R}}(x,s) = \frac{c_{\mathrm{R}}}{s} - \frac{c_{\mathrm{R}}}{s}\,e^{-\sqrt{\frac{s}{D_{\mathrm{R}}}}x} \tag{8.15}$$

The current being equal to the diffusion flux of the reactants towards the electrode

$$-J_{\mathrm{R}}(0,t) = \frac{I(t)}{nFA} = D_{\mathrm{R}}\left[\frac{\partial c_{\mathrm{R}}(x,t)}{\partial x}\right]_{x=0} \tag{8.16}$$

the Laplace transform of the current is given by

$$\frac{\bar{I}(s)}{nFA} = D_R\left[\frac{\partial \bar{c}_R(x,s)}{\partial x}\right]_{x=0} \tag{8.17}$$

or

$$\bar{I}(s) = \frac{nFA\, D_R^{1/2}\, c_R}{s^{1/2}} \tag{8.18}$$

The inverse transform of $1/\sqrt{s}$ being $1/\sqrt{\pi t}$, the limiting anodic current controlled by the diffusion is written as

$$I_{da}(t) = nFA\, c_R \sqrt{\frac{D_R}{\pi t}} \tag{8.19}$$

This is called the ***Cottrell equation***, and shows that, following a step in electrode potential, the current that is proportional to the slope of the concentration profile at the surface of the electrode, decreases as $1/\sqrt{\pi t}$ to tend to zero in the absence of convection in the solution. The current is called a ***limiting current*** by analogy with steady state diffusion currents treated in §7.3.2 as the interfacial concentration of the reactant R falls to zero when the potential step is applied. The difference between the present case and that illustrated in Figure 7.11 is that we here allow the diffusion layer thickness to extend to infinity (semi-infinite linear diffusion).

From an experimental point of view, the diffusion current is masked at very short times by the charging current of the double layer. At very long times, the current decay is limited by the extension of the diffusion process into the convection zone, where the solution is homogenised, fixing in an arbitrary way a diffusion layer thickness.

For the concentration profile of the reduced species, the inverse transform of $\exp - \sqrt{s/k}\,x / s$ being $\operatorname{erfc}\left[x/2\sqrt{kt}\right]$, we get from equation (8.15)

$$c_R(x,t) = c_R\left\{1 - \operatorname{erfc}\left(\frac{x}{2\sqrt{D_R t}}\right)\right\} \tag{8.20}$$

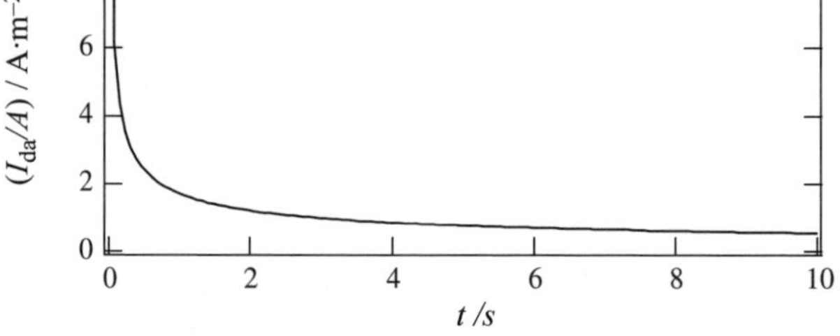

Fig. 8.1 Variation of current density according to the Cottrell equation (8.19), $D = 10^{-5}$ cm^2·s^{-1}, $c = 1$ mM.

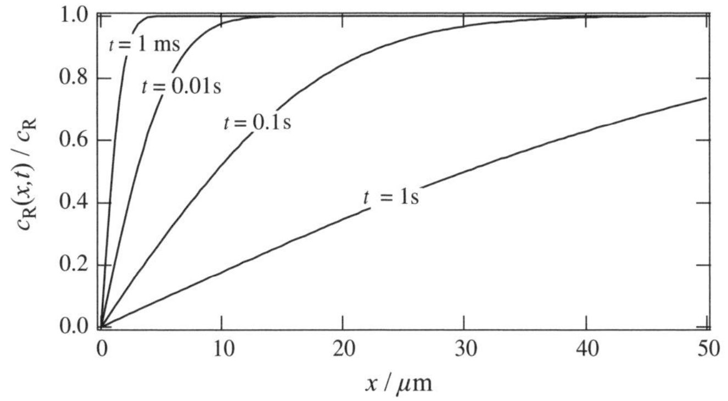

Fig. 8.2 Concentration profiles of the reduced species R after a potential step at which R is completely oxidised at the electrode. The diffusion is linear and semi-infinite.

or again

$$c_{\mathrm{R}}(x,t) = c_{\mathrm{R}} \operatorname{erf}\left[\frac{x}{2\sqrt{D_{\mathrm{R}}t}}\right] \tag{8.21}$$

The concentration profiles obtained during reactions limited by diffusion are illustrated in Figure 8.2.

8.1.2 Chronoamperometry with spherical diffusion for a complete interfacial oxidation

For a spherical electrode such as a mercury-drop electrode, the differential equation to be solved is the Fick equation in spherical coordinates

$$\frac{\partial c_{\mathrm{R}}(r,t)}{\partial t} = D_{\mathrm{R}}\left[\frac{\partial^2 c_{\mathrm{R}}(r,t)}{\partial r^2} + \frac{2}{r}\frac{\partial c_{\mathrm{R}}(r,t)}{\partial r}\right] \tag{8.22}$$

To do this, consider the following change of variable

$$v_{\mathrm{R}}(r,t) = r\, c_{\mathrm{R}}(r,t) \tag{8.23}$$

which takes us to an equation similiar to equation (8.5)

$$\frac{\partial v_{\mathrm{R}}(r,t)}{\partial t} = D_{\mathrm{R}}\frac{\partial^2 v_{\mathrm{R}}(r,t)}{\partial r^2} \tag{8.24}$$

The Laplace transform of this equation has the solution

$$\bar{v}_{\mathrm{R}}(r,s) = \frac{v_{\mathrm{R}}(r,0)}{s} + A(s)\, e^{-\sqrt{\frac{s}{D_{\mathrm{R}}}}\,r} \tag{8.25}$$

By using the Laplace transform of the interfacial boundary condition, we have

$$\bar{v}_{\mathrm{R}}(r_{\mathrm{e}},s) = 0 \tag{8.26}$$

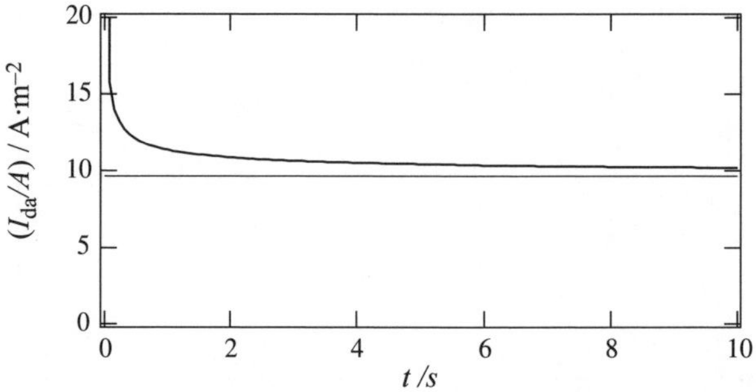

Fig. 8.3 Variation in the current density limited by spherical diffusion according to equation (8.30), $D = 10^{-5}$ cm^2·s^{-1}, $c = 1$ mM, $r_e = 10$ μm.

where r_e is the radius of the electrode, and the constant $A(s)$ is given by

$$A(s) \;=\; -\frac{v_R(r_e,0)}{s} e^{\sqrt{\frac{s}{D}}\, r_e} \tag{8.27}$$

The Laplace transform of the concentration of the reduced species is then

$$\bar{c}_R(r,s) \;=\; \frac{\bar{v}_R(r,s)}{r} \;=\; \frac{c_R}{s} \;-\; \frac{r_e c_R}{s\,r} e^{-\sqrt{\frac{s}{D_R}}(r-r_e)} \tag{8.28}$$

and thus

$$\left(\frac{\partial \bar{c}_R(r,s)}{\partial r}\right)_{r=r_e} \;=\; \frac{c_R}{s}\left[\frac{1}{r_e}+\sqrt{\frac{s}{D_R}}\right] \tag{8.29}$$

Therefore, the current defined in terms of the diffusion flux is written as

$$I_{da}(t) \;=\; nFA\, D_R\left(\frac{\partial c_R}{\partial r}\right)_{r=r_e} \;=\; nFA\, D_R c_R\left[\sqrt{\frac{1}{D_R \pi t}}+\frac{1}{r_e}\right] \tag{8.30}$$

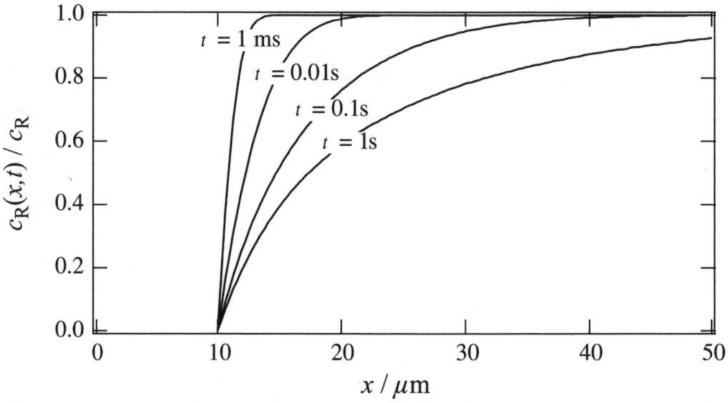

Fig. 8.4 Concentration profiles for symmetrically spherical diffusion of the reduced species R after a potential step where R is completely oxidised at the electrode.

The principal difference between a planar and a spherical electrode comes from the fact that the limiting current tends to the non-zero steady state limiting current value given by equation (7.41).

The concentration profile is determined by taking the inverse transform of equation (8.28) to obtain

$$c_R(x,t) \quad = \quad c_R\left\{1-\frac{r_e}{r}\,\mathrm{erfc}\left(\frac{r-r_e}{2\sqrt{D_R t}}\right)\right\} \tag{8.31}$$

A comparison of Figures 8.2 and 8.4 shows that spherical diffusion happens in a zone that is closer to the electrode. For longer times than those shown in Figure 8.4, the slope at the origin becomes constant when the steady state current becomes established. Nevertheless, it is important to note that even if the slope at the origin attains a limiting value, the concentration profiles continue to evolve away from the electrode over longer periods.

8.1.3 Chronocoulometry with semi-infinite linear diffusion for complete interfacial oxidation

Measuring the charge flowing, rather than the current, presents several advantages. Firstly, the integration gives a better signal-to-noise ratio. Secondly, the signal measured increases with time, and the effects of the double layer are also easily taken into account.

Integration of the relation (8.19) with time gives

$$Q(t) \quad = \quad 2nF\,A\,c_R\sqrt{\frac{D_R t}{\pi}} \tag{8.32}$$

The graph of Q as a function of $\sqrt{t}$ gives a straight line going through the origin. If the intercept at $x = 0$ is positive, this means that either the double layer charge is large, or that the species R is adsorbed on the electrode. Conversely, if the intercept at $y = 0$ is positive, this means that the reaction is limited by kinetics.

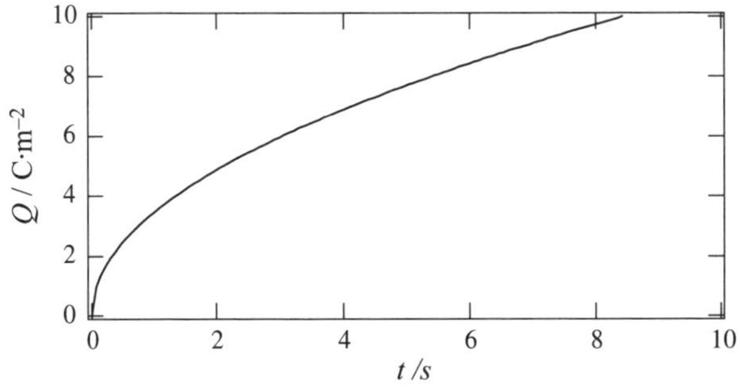

Fig. 8.5 Charge variation for a reaction controlled by linear diffusion after a potential step, according to equation (8.32), $D = 10^{-5}$ cm^2·s^{-1}, $c = 1$ mM.

8.1.4 Chronoamperometry with semi-infinite linear diffusion for a partial interfacial oxidation

Consider again a system containing only a single reduced species, i.e. a system for which the equilibrium potential is at least 120 mV more negative than the formal redox potential.

$$E_{eq} = E^{\ominus\prime} + \frac{RT}{nF}\ln\left(\frac{c_O(x,0)}{c_R(x,0)}\right) \tag{8.33}$$

An electrode potential step from the equilibrium potential to an applied potential E greater than the equilibrium potential causes a partial interfacial oxidation and consequently a variation in the interfacial concentrations. If we make the hypothesis that the electrode reactions are electrochemically reversible, then Nernst's law applies to the interfacial concentrations

$$E = E^{\ominus\prime} + \frac{RT}{nF}\ln\left(\frac{c_O(0,t)}{c_R(0,t)}\right) \tag{8.34}$$

In this general case, we must take into account the diffusion of the two species, namely that of the reduced form to the electrode and that of the oxidised form away from the electrode. We then solve the two Fick equations

$$\frac{\partial c_O(x,t)}{\partial t} = D_O\frac{\partial^2 c_O(x,t)}{\partial x^2} \tag{8.35}$$

and

$$\frac{\partial c_R(x,t)}{\partial t} = D_R\frac{\partial^2 c_R(x,t)}{\partial x^2} \tag{8.36}$$

with the initial conditions

$$c_R(x,0) = c_R \quad \text{and} \quad c_O(x,0) \approx 0 \tag{8.37}$$

and the conditions linked to the hypothesis of a semi-infinite bulk solution

$$\lim_{x\to\infty} c_R(x,t) = c_R \quad \text{and} \quad \lim_{x\to\infty} c_O(x,t) \approx 0 \tag{8.38}$$

As before, it is good to remember that the Nernst equation forbids that the interfacial concentrations become absolutely zero. One must remain aware of the fact that equations (8.37) and (8.38) are only approximations.

The third condition is the equality of the diffusion fluxes at the interface

$$D_O\left(\frac{\partial c_O(x,t)}{\partial x}\right)_{x=0} + D_R\left(\frac{\partial c_R(x,t)}{\partial x}\right)_{x=0} = 0 \tag{8.39}$$

To solve these differential equations, we shall use the Laplace transform method as before (see equation (8.13)). Thus, the transforms of the solutions of the Fick equations are

$$\overline{c}_R(x,s) = \frac{c_R}{s} + A(s)\,e^{-\sqrt{\frac{s}{D_R}}x} \tag{8.40}$$

and

$$\bar{c}_{\mathrm{O}}(x,s) \;=\; B(s)\, e^{-\sqrt{\frac{s}{D_{\mathrm{O}}}}x} \tag{8.41}$$

The transform of the boundary condition expressing the conservation of the flux is written as

$$D_{\mathrm{O}}\left[\frac{\partial \bar{c}_{\mathrm{O}}(x,s)}{\partial x}\right]_{x=0} \;+\; D_{\mathrm{R}}\left[\frac{\partial \bar{c}_{\mathrm{R}}(x,s)}{\partial x}\right]_{x=0} \;=\; 0 \tag{8.42}$$

or as

$$-A(s)\sqrt{D_{\mathrm{R}} s} \;-\; B(s)\sqrt{D_{\mathrm{O}} s} \;=\; 0 \tag{8.43}$$

From this, we deduce that

$$B(s) \;=\; -\xi\, A(s) \tag{8.44}$$

with

$$\xi \;=\; \sqrt{\frac{D_{\mathrm{R}}}{D_{\mathrm{O}}}} \tag{8.45}$$

The Laplace transforms of the concentrations then read

$$\bar{c}_{\mathrm{R}}(x,s) \;=\; \frac{c_{\mathrm{R}}}{s} \;+\; A(s)\, e^{-\sqrt{\frac{s}{D_{\mathrm{R}}}}x} \tag{8.46}$$

and

$$\bar{c}_{\mathrm{O}}(x,s) \;=\; -A(s)\,\xi\, e^{-\sqrt{\frac{s}{D_{\mathrm{O}}}}x} \tag{8.47}$$

The condition of electrochemical reversibility (fast electron transfer reactions) is written via the Nernst equation with the interfacial concentrations by introducing a dimensionless number θ such that:

$$\theta \;=\; \frac{c_{\mathrm{R}}(0,t)}{c_{\mathrm{O}}(0,t)} \;=\; \exp\left[-\frac{nF}{RT}\left(E - E^{\ominus\prime}\right)\right] \tag{8.48}$$

The transform of this equation becomes

$$\bar{c}_{\mathrm{R}}(0,s) \;=\; \theta\, \bar{c}_{\mathrm{O}}(0,s) \tag{8.49}$$

or

$$\frac{c_{\mathrm{R}}}{s} \;+\; A(s) \;=\; -\xi\,\theta\, A(s) \tag{8.50}$$

The constant A is thus given by

$$A(s) \;=\; -\frac{c_{\mathrm{R}}}{s(1+\xi\,\theta)} \tag{8.51}$$

The transforms of the concentration profiles are then

$$\bar{c}_R(x,s) \;=\; \frac{c_R}{s} \;-\; \frac{c_R\, e^{-\sqrt{\frac{s}{D_R}}x}}{s(1+\xi\,\theta)} \tag{8.52}$$

and

$$\bar{c}_O(x,s) \;=\; \frac{\xi\, c_R\, e^{-\sqrt{\frac{s}{D_O}}x}}{s(1+\xi\,\theta)} \tag{8.53}$$

The inverse transforms of these equations give us

$$c_R(x,t) \;=\; c_R \;-\; \frac{c_R}{1+\xi\,\theta}\,\mathrm{erfc}\left[\frac{x}{2\sqrt{D_R t}}\right] \tag{8.54}$$

and

$$c_O(x,t) \;=\; \frac{\xi\, c_R}{1+\xi\,\theta}\,\mathrm{erfc}\left[\frac{x}{2\sqrt{D_O t}}\right] \tag{8.55}$$

The transform of the current is then

$$\bar{I}(s) \;=\; nFAD_R\left[\frac{\partial \bar{c}_R(x,s)}{\partial x}\right]_{x=0} \;=\; \frac{nFAD_R^{1/2} c_R}{s^{1/2}(1+\xi\theta)} \tag{8.56}$$

and so the inverse transform is

$$I(t) \;=\; \frac{nFAc_R}{1+\xi\theta}\sqrt{\frac{D_R}{\pi t}} \;=\; \frac{I_{da}(t)}{1+\xi\theta} \tag{8.57}$$

where $I_{da}(t)$ is the limiting anodic diffusion current, which is the current obtained when the reduced species is completely oxidised at the surface of the electrode.

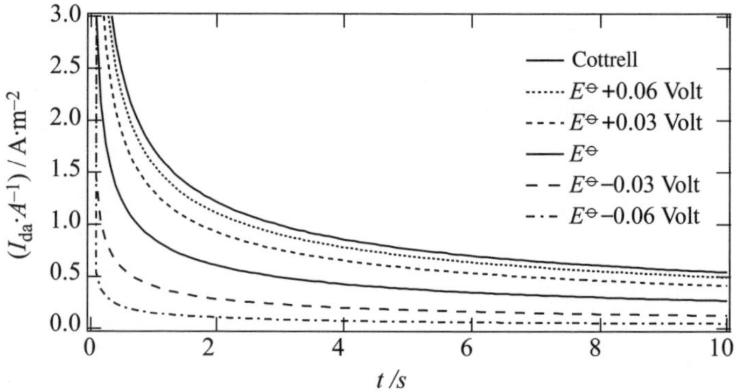

Fig. 8.6 Chronoamperograms at different potentials around the formal redox potential, according to equation (8.57), $D = 10^{-5}$ cm^2·s^{-1}, $c = 1$ mM.

The interfacial concentrations can simply be calculated from the current values

$$c_R(0,t) = c_R\left(1-\frac{1}{1+\xi\theta}\right) = c_R\left[1-\frac{I(t)}{I_{da}(t)}\right] \tag{8.58}$$

and

$$c_O(0,t) = c_R\left(\frac{\xi}{1+\xi\theta}\right) = \xi\, c_R\left[\frac{I(t)}{I_{da}(t)}\right] \tag{8.59}$$

EXERCISE

Proceed in the same way for a reduction ($c_R(x,0) = 0$) and find the expression for the current

$$I(t) = -\frac{nFA\xi\theta\, c_O}{1+\xi\theta}\sqrt{\frac{D_O}{\pi t}} = \frac{\xi\theta}{1+\xi\theta} I_{dc}(t) \tag{8.60}$$

Also show that the concentration profiles are given by

$$c_R(x,t) = \frac{\theta\, c_O}{1+\xi\theta}\operatorname{erfc}\left[\frac{x}{2\sqrt{D_R t}}\right] \tag{8.61}$$

and

$$c_O(x,t) = c_O - \frac{\xi\theta\, c_O}{1+\xi\theta}\operatorname{erfc}\left[\frac{x}{2\sqrt{D_O t}}\right] \tag{8.62}$$

8.2 POLAROGRAPHY

8.2.1 Dropping mercury electrode

Polarography is the oldest and most established of electroanalytical techniques, if perhaps not the most practical. Unfortunately, polarography is based on a dropping mercury electrode, and the restrictions, imposed in several countries on the use of mercury in the laboratory, make it rather out-of-fashion nowadays.

In polarography, the mercury drips, drop by drop from the tip of a capillary tube. Originally, the dripping was natural, but in most modern devices, the dripping is forced by mechanical means so that the speed of formation of the drops is reproducible, and not dependent on the height of mercury in the reservoir with respect to the capillary.

The main advantage of polarography is the reproducibility of the measurements due to the fact that each measurement is made on a freshly-formed electrode. The major disadvantage of this technique is due to the fact that mercury oxidises very easily.

$$\left[E^{\ominus}_{\frac{1}{2}Hg_2^{2+}/Hg}\right]_{SHE} = 0.796\ V \quad \text{and} \quad \left[E^{\ominus}_{Hg^{2+}/\frac{1}{2}Hg_2^{2+}}\right]_{SHE} = 0.991\ V$$

In the presence of certain anions such as chlorides, these standard redox potentials decrease to about 0.2 V. Therefore, the electrode is above all used for studying reductions, particularly those of metal ions whose reduced form makes an amalgam with the mercury.

8.2.2 Staircase polarography with sampling (*TAST Polarography*)

With each new drop, a new diffusion layer is freshly established by the drop passing through the solution as it falls. Thus, at the formation of each drop, the following initial conditions for a reduction are reset at every drop:

$$c_{\mathrm{O}}(x,0) = c_{\mathrm{O}} \qquad \text{and} \qquad c_{\mathrm{R}}(x,0) \approx 0 \tag{8.63}$$

and the bulk conditions for a semi-infinite solution are

$$\lim_{x\to\infty} c_{\mathrm{O}}(x,t) = c_{\mathrm{O}} \qquad \text{and} \qquad \lim_{x\to\infty} c_{\mathrm{R}}(x,t) \approx 0 \tag{8.64}$$

When the reduction is that of a metal ion that forms an amalgam in the mercury, the boundary conditions imposed on the reduced species are different if we take into account the finite size of the mercury drop. Equation (8.64) applies, strictly speaking only to reductions where the reduced species remains in the electrolyte solution.

For each drop, we apply an electrode potential value, and after a certain time τ (of the order of a second), we measure the current before making the drop fall.

Consider a reduction with a formal redox potential $E^{\ominus\prime}$. At applied potentials $E >> E^{\ominus\prime}$, the oxidised species is stable in solution and the cathodic current for each drop is zero. At potentials $E << E^{\ominus\prime}$ where the oxidised species is completely reduced at the surface of the mercury drop electrode, we have for each drop a limiting

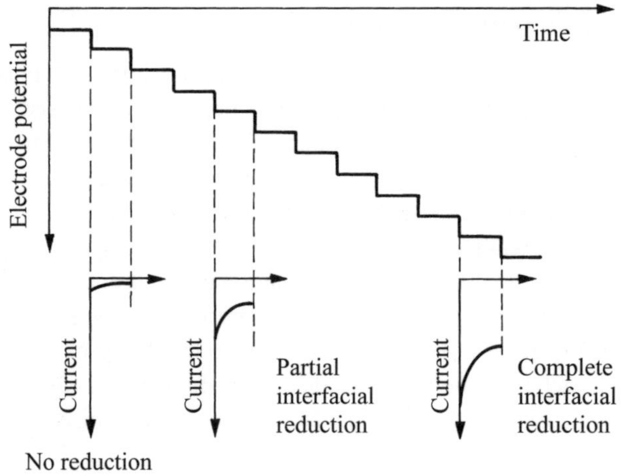

Fig. 8.7 Staircase potential curve for a reduction, with schematic evolution of the response current for each step.

cathodic current given by the Cottrell equation (8.19). Thus, the current measured at the end of each drop is the same for every drop. Between these two regions, there is an intermediary zone where the reduction current sampled at the end of each drop is given by equation (8.60)

$$I(\tau) \quad = \quad -nFAc_{\rm O}\frac{\xi\theta}{1+\xi\theta}\sqrt{\frac{D_{\rm O}}{\pi t}} \quad = \quad \frac{\xi\theta}{1+\xi\theta}I_{\rm dc}(\tau) \tag{8.65}$$

The advantage of measuring the current at the end of the lifetime of the drop is that we don't have to take into account the variation with time of the drop area as it expands. In this way, the area in equation (8.65) can be considered as constant.

Thus, if $E >> E^{\ominus\prime}$, the ratio θ defined by equation (8.48) tends to zero and the cathodic current defined by equation (8.65) is quasi-zero; if $E << E^{\ominus\prime}$, θ tends to infinity, and the current tends to the limiting current value for a complete reduction at the surface of the electrode. Looking at Figure 8.6, it is obvious that the longer the lifespan of the drop, the later is the sampling and the lower is the signal measured. By expanding the adimensional term θ, expression (8.65) can be rewritten in the form

$$E \quad = \quad E^{\ominus\prime} + \frac{RT}{nF}\ln\left(\frac{D_{\rm R}^{1/2}}{D_{\rm O}^{1/2}}\right) + \frac{RT}{nF}\ln\left(\frac{I_{\rm dc}(\tau) - I(\tau)}{I(\tau)}\right) \tag{8.66}$$

In this way, we again find a relation between the current and the electrode potential similar to that in (7.34), obtained for steady state methods.

Figure 8.8 shows a ***polarogram*** for the reduction of Pb^{2+} in solution. We can observe a 'sloping' base line that renders difficult the reading of the limiting cathodic current, and thus the half-wave potential value. Equation (7.37) can help in the analysis of this kind of polarogram. From an analytical point of view, the direct proportionality between the limiting current $I_{\rm dc}$ and the concentration represents the most important aspect of this technique, with detection limits for reactant concentrations in the micromolar range.

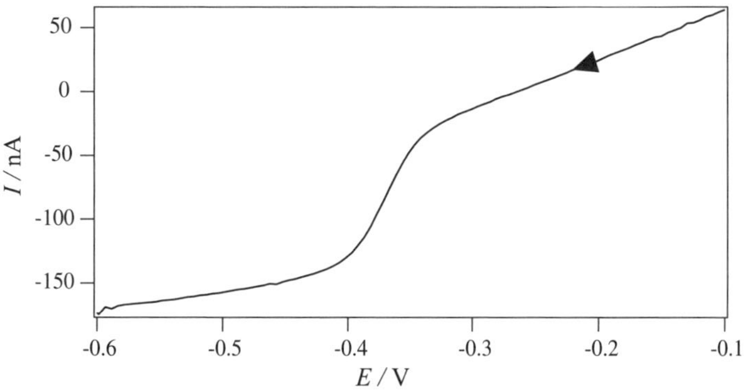

Fig. 8.8 Polarogram obtained by staircase polarography with sampling for a 10^{-5} M solution of Pb^{2+} in 0.05 M HCl. Step height: 5 mV, start potential: –0.1 V (Olivier Bagel, EPFL).

8.2.3 Normal pulse polarography

We have just seen that by applying to each drop a different potential in the form of a step function, the current measured at the end of each drop will be correspondingly smaller as the lifespan of the drop is longer. For mechanical reasons, the lifespans of the drops are of the order of a second and it is interesting to offset the application of the potential with respect to the birth of the drop as shown in Figure 8.9 in order to measure larger currents. Thus, in normal pulse polarography, for a reduction, a rest potential $E_r >> E^{\ominus\prime}$ is applied during the growth period of the drop, and before it falls a potential pulse (negative for a reduction) is applied for tens of milliseconds. The current is then sampled at the end of the pulse. For each drop, the value of the pulse potential is increased (by a negative value for a reduction).

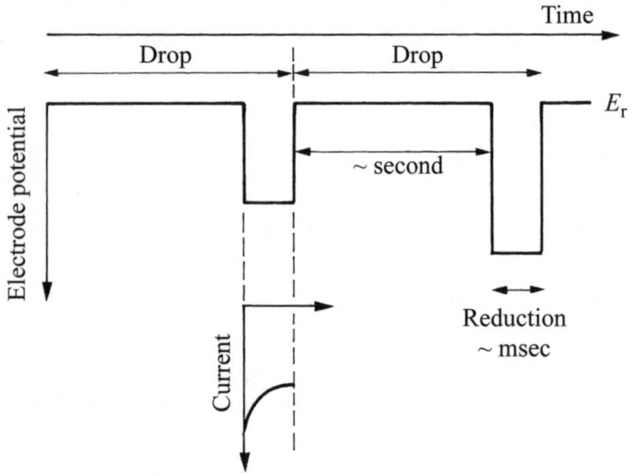

Fig. 8.9 Normal pulse potential waveform and the current response shown schematically.

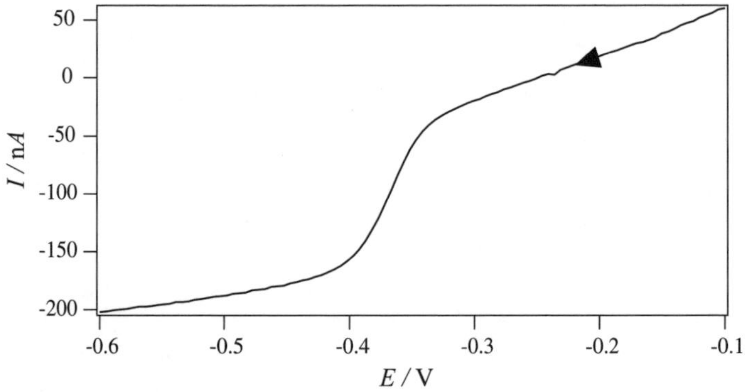

Fig. 8.10 Polarogram obtained using normal pulse polarography for a 10^{-5} M solution of Pb^{2+} in 0.05 M HCl. Pulse height increment: 5 mV, duration 40 ms. Start potential: –0.1 V (Olivier Bagel, EPFL).

This technique generally known as ***normal pulse polarography*** gives results with a better signal-to-noise ratio than Tast polarography. The pulse duration during which the reduction occurs at the surface of the mercury drop is about 50 ms and the limits of detection are of the order of 10^{-6} to 10^{-7} M.

Figure 8.10 shows a normal pulse polarogram for the same system as in Figure 8.8. Comparing the two figures shows that the results obtained using these two methods are naturally quite similar, the limiting cathodic current being slightly larger for normal pulse polarography.

8.2.4 Differential pulse polarography

An adaptation of staircase polarography and normal pulse polarography, designated as ***differential pulse polarography*** allows the detection of lower concentrations down to the nanomolar range. The principle of this technique is, as in staircase polarography, to impose during the greater part of the lifespan of a drop, a constant potential, onto which is superimposed a potential pulse of fixed height, at the end of the life of the drop, as shown in Figure 8.11.

During the application of the staircase potential (plateau potential), the interfacial concentrations follow the Nernst equation and thus, after a while, the concentrations of O and R near the electrode can be considered as being those imposed by the Nernst equation. For a reduction, these surface concentrations can be calculated as a function of the plateau potential E_p, and by applying equations (8.61) and (8.62), the apparent bulk concentrations imposed by the potential E_p are

$$c_R(0,t) = \frac{\theta_p\, c_O}{1+\xi\,\theta_p} = c_R^{app} \tag{8.67}$$

and

$$c_O(0,t) = \frac{c_O}{1+\xi\,\theta_p} = c_O^{app} \tag{8.68}$$

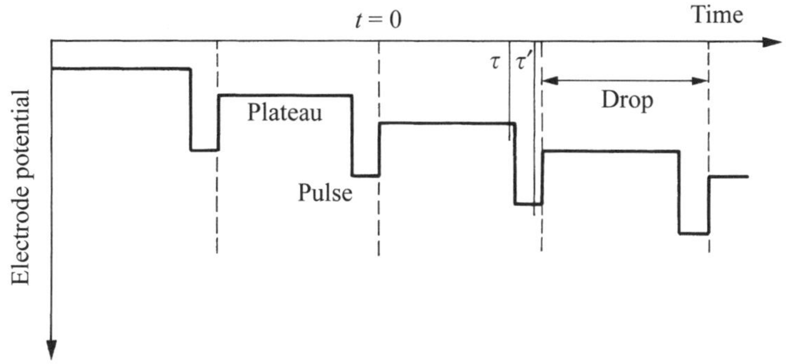

Fig. 8.11 Potential profile for differential pulse polarography. The current is measured at the end of each plateau ($t = \tau$) and at the end of each pulse ($t = \tau'$).

These concentrations now represent the boundary conditions of the Fick equations for calculating the current resulting from the potential pulse.

The differential equations are still the same Fick equations (8.35) and (8.36) and by defining the start of the pulse as the time origin ($t = 0$), the initial conditions are

$$c_R(x,0) = c_R^{app} \quad \text{and} \quad c_O(x,0) = c_O^{app} \tag{8.69}$$

The other pseudo-bulk boundary conditions are

$$\lim_{x\to\infty} c_R(x,t) = c_R^{app} \quad \text{and} \quad \lim_{x\to\infty} c_O(x,t) = c_O^{app} \tag{8.70}$$

Of course, this condition is not, strictly speaking, absolutely correct, but holds well when the length of the pulse is short with respect to the duration of the plateau.

The transforms of the solutions of the Fick equations are

$$\bar{c}_R(x,s) = \frac{c_R^{app}}{s} + A(s)\,e^{-\sqrt{\frac{s}{D_R}}x} \tag{8.71}$$

and

$$\bar{c}_O(x,s) = \frac{c_O^{app}}{s} + B(s)\,e^{-\sqrt{\frac{s}{D_O}}x} \tag{8.72}$$

The transform of the boundary condition expressing the conservation of flux (8.39) still gives

$$B(s) = -\xi\,A(s) \quad \text{with} \quad \xi = \sqrt{\frac{D_R}{D_O}} \tag{8.73}$$

Nernst's equation written adimensionally (8.48) thus gives

$$\frac{c_R^{app}}{s} + A(s) = \theta\left[\frac{c_O^{app}}{s} - \xi\,A(s)\right] \tag{8.74}$$

from which we calculate

$$A(s) = \frac{\theta c_O^{app} - c_R^{app}}{s(1+\xi\,\theta)} \tag{8.75}$$

Thus the transforms of the concentration profiles are

$$\bar{c}_R(x,s) = \frac{c_R^{app}}{s} + \frac{\theta c_O^{app} - c_R^{app}}{s(1+\xi\,\theta)}\,e^{-\sqrt{\frac{s}{D_R}}x} \tag{8.76}$$

and

$$\bar{c}_O(x,s) = \frac{c_O^{app}}{s} - \frac{\xi(\theta c_O^{app} - c_R^{app})}{s(1+\xi\,\theta)}\,e^{-\sqrt{\frac{s}{D_O}}x} \tag{8.77}$$

The transform of the current is given by

$$\bar{I}(s) \quad = \quad -nFAD_{\rm O}\left[\frac{\partial \bar{c}_{\rm O}(x,s)}{\partial x}\right]_{x=0} \quad = \quad -\frac{nFA\,\xi\left[\theta c_{\rm O}^{\rm app}-c_{\rm R}^{\rm app}\right]}{(1+\xi\theta)}\sqrt{\frac{D_{\rm O}}{s}} \qquad (8.78)$$

By inverse transformation, the current reads

$$I(t) \quad = \quad -\frac{nFA\,\xi\left(\theta c_{\rm O}^{\rm app}-c_{\rm R}^{\rm app}\right)}{1+\xi\theta}\sqrt{\frac{D_{\rm O}}{\pi t}} \qquad (8.79)$$

By further substituting in the values of the apparent concentrations, we finally get

$$I(t) \quad = \quad -\frac{nFA\,\xi\left(\theta-\theta_p\right)c_{\rm O}}{(1+\xi\theta)\left(1+\xi\theta_p\right)}\sqrt{\frac{D_{\rm O}}{\pi t}} \qquad (8.80)$$

As shown in Figure 8.11, the differential current (pulse – plateau) is obtained by subtracting from the pulse current $I(\tau')$ the residual plateau current $I(\tau)$. In order to better represent the difference $I(\tau') - I(\tau)$ as a function of the potential, it is useful to introduce the parameters Θ and σ defined by :

$$\Theta \quad = \quad \exp\left[-\frac{nF}{RT}\left(E_{\rm p}+\frac{\Delta E}{2}-E^{\ominus\prime}\right)\right] \qquad (8.81)$$

and

$$\sigma \quad = \quad \exp\left[-\frac{nF}{RT}\frac{\Delta E}{2}\right] \qquad (8.82)$$

Thus, the adimensional parameters of Nernst's equation are

$$\theta_{\rm p} \quad = \quad \Theta/\sigma \quad \text{and} \quad \theta \quad = \quad \Theta\sigma \qquad (8.83)$$

The current difference due to the potential step at $E = E_{\rm p} + \Delta E$ is then equal to :

$$I(\tau')-I(\tau) \quad = \quad -\frac{nFA\,\xi\,\Theta\left(\sigma^2-1\right)c_{\rm O}}{(\sigma+\xi\Theta)(1+\xi\Theta\sigma)}\sqrt{\frac{D_{\rm O}}{\pi(\tau'-\tau)}} \qquad (8.84)$$

Putting $u = \xi\ \Theta$, we show easily that the difference $I(\tau') - I(\tau)$ goes through a maximum when u=1 , i.e. when

$$E_{\rm max} \quad = \quad E^{\ominus\prime} \quad + \quad \frac{RT}{nF}\ln\sqrt{\frac{D_{\rm R}}{D_{\rm O}}} \quad - \quad \frac{\Delta E}{2} \qquad (8.85)$$

and

$$\left[I(\tau')-I(\tau)\right]_{\rm max} \quad = \quad -nFAc_{\rm O}\sqrt{\frac{D_{\rm O}}{\pi(\tau'-\tau)}}\,\frac{(\sigma-1)}{(\sigma+1)} \qquad (8.86)$$

The general shape of the difference $I(\tau') - I(\tau)$ is represented in Figure 8.12.

Figure 8.13 shows the polarogram for the reduction of Pb^{2+} with the same experimental conditions as for Figures 8.8 and 8.10. Comparing to the other polarograms, we can note that the base line is better and that the peak obtained can be more easily used for analytical purposes.

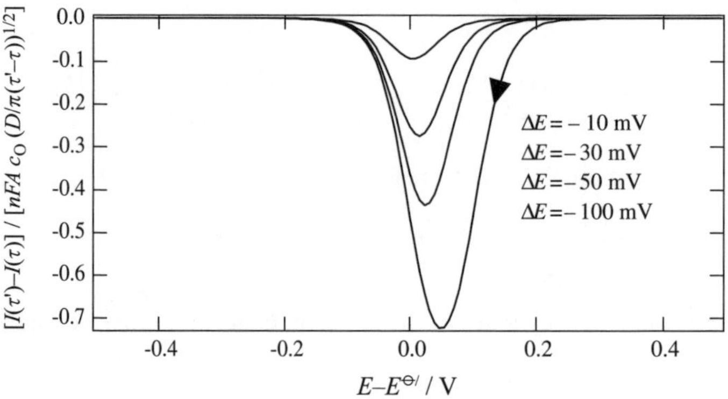

Fig. 8.12 Differential pulse polarograms for different pulse heights according to equation (8.86). Start potential = 0.5 V. $\xi = 1$.

From a purely analytical point of view, it is worth noticing that differential pulse polarography is one of the most sensitive electrochemical techniques, allowing the detection of analytes at sub-micromolar concentrations. In general, it is not advisable to use equation (8.86) directly for determining the analyte concentration, but rather to use internal calibration methods such as the standard addition method. Apart from metal ions, the technique can be applied to the reduction of a good number of organic molecules.

This technique, described here under the heading of polarography, is also applicable to redox reactions on solid electrodes – equally well for oxidations and reductions. We then speak of differential pulse voltamperometry or more simply ***differential pulse voltammetry***. For oxidations, the potential jump for each plateau and for each pulse is positive, whilst it is negative for reductions.

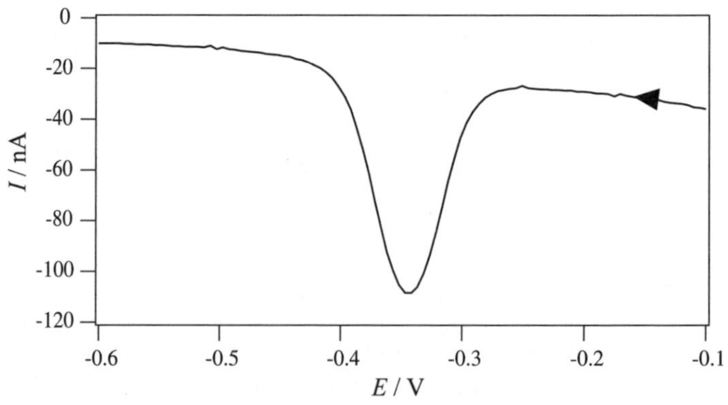

Fig. 8.13 Differential pulse polarogram for a 10^{-5} M solution of Pb^{2+} in 0.05 M HCl. Step height 5 mV, duration 400 ms. Pulse height 50 mV, duration 40 ms. Start potential –0.1 V (Olivier Bagel, EPFL).

8.3 SQUARE WAVE VOLTAMMETRY

To understand more generally pulse voltammetry with different applied potential waveforms, it is useful to look first at a mathematical methodology based on the ***superposition principle.***

8.3.1 Superposition principle

Consider a system in equilibrium at a potential E_0 with c_R^0 and c_O^0 the initial concentrations of the reduced and oxidised species. At a time $t = 0$, we impose a potential E_1 for a time τ, then a step change in potential from the potential E_1 to the potential E_2 as shown in Figure 8.14.

To calculate the concentration profiles as a function of time following a potential jump from E_1 to E_2 , we need first to calculate the concentration profiles at the time τ. To do this, we shall follow the method we used previously for chronoamperometry for a partial oxidation (see §8.1.4), but taking c_R^0 and c_O^0 for the initial concentrations values of the reduced and oxidised species.

The differential equations are still the Fick equations for the oxidised species

$$\frac{\partial c_O(x,t)}{\partial t} = D_O \frac{\partial^2 c_O(x,t)}{\partial x^2} \tag{8.87}$$

and for the reduced species

$$\frac{\partial c_R(x,t)}{\partial t} = D_R \frac{\partial^2 c_R(x,t)}{\partial x^2} \tag{8.88}$$

with now as boundary conditions

$$c_R(x,0) = c_R^0 \quad \text{and} \quad c_O(x,0) = c_O^0 \tag{8.89}$$

$$\lim_{x\to\infty} c_R(x,t) = c_R^0 \quad \text{and} \quad \lim_{x\to\infty} c_O(x,t) = c_O^0 \tag{8.90}$$

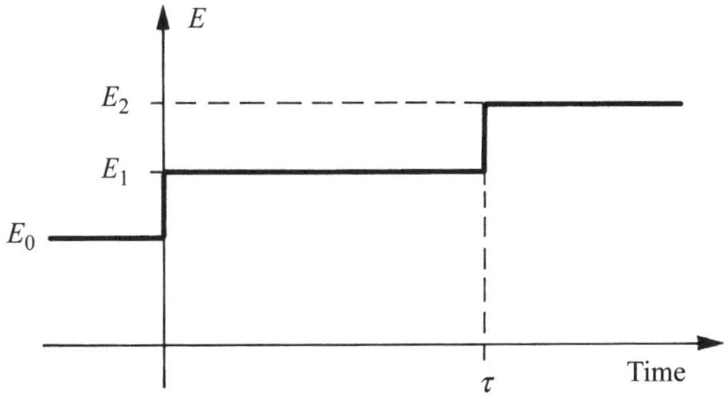

Fig. 8.14 Step potential variation.

To solve these differential equations, we shall use the Laplace transformation method as before. Thus, the transforms of the solutions of the Fick equations are (see equations (8.71) and (8.72))

$$\bar{c}_{\mathrm{R}}(x,s) = \frac{c_{\mathrm{R}}^{0}}{s} + A(s)\, e^{-\sqrt{\frac{s}{D_{\mathrm{R}}}}x} \tag{8.91}$$

and

$$\bar{c}_{\mathrm{O}}(x,s) = \frac{c_{\mathrm{O}}^{0}}{s} + B(s)\, e^{-\sqrt{\frac{s}{D_{\mathrm{O}}}}x} \tag{8.92}$$

taking into account the boundary condition expressing the conservation of the flux with

$$D_{\mathrm{O}}\left[\frac{\partial \bar{c}_{\mathrm{O}}(x,s)}{\partial x}\right]_{x=0} + D_{\mathrm{R}}\left[\frac{\partial \bar{c}_{\mathrm{R}}(x,s)}{\partial x}\right]_{x=0} = 0 \tag{8.93}$$

or

$$-A(s)\sqrt{D_{\mathrm{R}}s} - B(s)\sqrt{D_{\mathrm{O}}s} = 0 \tag{8.94}$$

Substituting equations (8.91) and (8.92) into equation (8.94), we obtain a relation between the Laplace transforms of the interfacial concentrations and the concentrations in solution

$$\sqrt{D_{\mathrm{R}}}\left[\bar{c}_{\mathrm{R}}(0,s)-\frac{c_{\mathrm{R}}^{0}}{s}\right] + \sqrt{D_{\mathrm{O}}}\left[\bar{c}_{\mathrm{O}}(0,s)-\frac{c_{\mathrm{O}}^{0}}{s}\right] = 0 \tag{8.95}$$

which is, taking the inverse transform,

$$\sqrt{D_{\mathrm{R}}}\, c_{\mathrm{R}}(0,t) + \sqrt{D_{\mathrm{O}}}\, c_{\mathrm{O}}(0,t) = \sqrt{D_{\mathrm{R}}}\, c_{\mathrm{R}}^{0} + \sqrt{D_{\mathrm{O}}}\, c_{\mathrm{O}}^{0} \tag{8.96}$$

If the coefficients D_{O} and D_{R} are equal, this equation reduces to

$$c_{\mathrm{R}}(0,t) + c_{\mathrm{O}}(0,t) = c_{\mathrm{R}}^{0} + c_{\mathrm{O}}^{0} = c_{\mathrm{Total}} \tag{8.97}$$

Equation (8.94) allows us also to calculate $B(s)$ as a function of $A(s)$. The electrochemical reversibility condition is again written in its adimensional form

$$\theta_1 = \frac{c_{\mathrm{R}}(0,t)}{c_{\mathrm{O}}(0,t)} = \exp\left[-\frac{nF}{RT}\left(E_1 - E^{\ominus\prime}\right)\right] \tag{8.98}$$

or

$$\bar{c}_{\mathrm{R}}(0,s) = \theta_1\, \bar{c}_{\mathrm{O}}(0,s) \tag{8.99}$$

We then get

$$\frac{c_{\mathrm{R}}^{0}}{s} + A(s) = \theta_1\left[\frac{c_{\mathrm{O}}^{0}}{s} - \xi A(s)\right] \tag{8.100}$$

still with $\xi = \sqrt{D_{\mathrm{R}}/D_{\mathrm{O}}}$.

This allows the calculation of the constant A

$$A(s) = \frac{\theta_1 c_{\rm O}^0 - c_{\rm R}^0}{s(1+\xi\,\theta_1)} \tag{8.101}$$

Thus, the transforms of the concentration profiles are

$$\bar{c}_{\rm R}(x,s) = \frac{c_{\rm R}^0}{s} + \left[\frac{\theta_1 c_{\rm O}^0 - c_{\rm R}^0}{s(1+\xi\,\theta_1)}\right] e^{-\sqrt{\frac{s}{D_{\rm R}}}x} \tag{8.102}$$

and

$$\bar{c}_{\rm O}(x,s) = \frac{c_{\rm O}^0}{s} - \left[\frac{\theta_1 c_{\rm O}^0 - c_{\rm R}^0}{s(1+\xi\,\theta_1)}\right] \xi\, e^{-\sqrt{\frac{s}{D_{\rm O}}}x} \tag{8.103}$$

The inverse transforms of these equations give us the concentration profiles at the time τ

$$c_{\rm R}(x,\tau) = c_{\rm R}^0 + \left[\frac{\theta_1 c_{\rm O}^0 - c_{\rm R}^0}{1+\xi\,\theta_1}\right] \mathrm{erfc}\left[\frac{x}{2\sqrt{D_{\rm R}\tau}}\right] \tag{8.104}$$

and

$$c_{\rm O}(x,\tau) = c_{\rm O}^0 - \xi\left[\frac{\theta_1 c_{\rm O}^0 - c_{\rm R}^0}{1+\xi\,\theta_1}\right] \mathrm{erfc}\left[\frac{x}{2\sqrt{D_{\rm O}\tau}}\right] \tag{8.105}$$

These expressions can be rearranged to make the interfacial concentrations imposed by the potential E_1 appear, that is $c_{\rm R}^1(0,t) = c_{\rm R}^1$ and $c_{\rm O}^1(0,t) = c_{\rm O}^1$. In order to do this, we use equation (8.96) coming from the hypothesis of the equality of the flux of products and reactants, which is now written as

$$\xi c_{\rm R}^1 + c_{\rm O}^1 = \xi c_{\rm R}^0 + c_{\rm O}^0 \tag{8.106}$$

By substituting equations (8.98) & (8.106) into equations (8.104) & (8.105), and knowing that $\mathrm{erfc}(x) = 1 - \mathrm{erf}(x)$, we get

$$c_{\rm R}(x,\tau) = c_{\rm R}^1 - \left[c_{\rm R}^1 - c_{\rm R}^0\right] \mathrm{erf}\left[\frac{x}{2\sqrt{D_{\rm R}\tau}}\right] \tag{8.107}$$

and

$$c_{\rm O}(x,\tau) = c_{\rm O}^1 - \left[c_{\rm O}^1 - c_{\rm O}^0\right] \mathrm{erf}\left[\frac{x}{2\sqrt{D_{\rm O}\tau}}\right] \tag{8.108}$$

At times $t > \tau$, we again need to solve the Fick equations in the form

$$\frac{\partial c(x,t-\tau)}{\partial(t-\tau)} = D\frac{\partial^2 c(x,t-\tau)}{\partial x^2} \tag{8.109}$$

By using the following theorem

$$\int_0^\infty e^{-st}\, F(t-\tau)\mathrm{d}t = e^{-s\tau} f(s) \tag{8.110}$$

the transform of the differential equation is

$$D\frac{\partial^2\bar{c}(x,s)}{\partial x^2}e^{-s\tau} \;=\; s\,e^{-s\tau}\,\bar{c}(x,s) \;-\; c(x,\tau) \tag{8.111}$$

or again

$$\frac{\partial^2\bar{c}(x,s)}{\partial x^2} \;-\; \frac{s}{D}\bar{c}(x,s) \;=\; -\frac{c(x,\tau)\,e^{s\tau}}{D} \tag{8.112}$$

Thus, by expanding, we obtain for the reduced species

$$\frac{\partial^2\bar{c}_{\mathrm{R}}(x,s)}{\partial x^2} \;-\; \frac{s}{D_{\mathrm{R}}}\bar{c}_{\mathrm{R}}(x,s) \;=\; -\frac{c_{\mathrm{R}}^1}{D_{\mathrm{R}}}e^{s\tau} \;+\; \left[\frac{\left(c_{\mathrm{R}}^1 - c_{\mathrm{R}}^0\right)e^{s\tau}}{D_{\mathrm{R}}}\right]\mathrm{erf}\left[\frac{x}{2\sqrt{D_{\mathrm{R}}\tau}}\right] \tag{8.113}$$

It is possible to solve this equation analytically: however, it is laborious and ends up with rather complicated expressions.

A more elegant way of treating the problem is to decompose the potential function as the sum of two potential functions such that

$$E \;=\; E_{\mathrm{I}} \;+\; E_{\mathrm{II}} \qquad t \;>\; 0 \tag{8.114}$$

with

$$E_{\mathrm{I}} \;=\; E_1 \qquad t \;>\; 0 \tag{8.115}$$

and

$$E_{\mathrm{II}} \;=\; 0 \qquad t \;\leq\; \tau \tag{8.116}$$

$$E_{\mathrm{II}} \;=\; E_2 \;-\; E_1 \qquad t \;>\; \tau \tag{8.117}$$

In the same way, the concentrations can be considered as the superposition of two virtual concentrations

$$c_{\mathrm{R}}(x,t) \;=\; c_{\mathrm{RI}}(x,t) + c_{\mathrm{RII}}(x,t) \tag{8.118}$$

with as boundary conditions

$$c_{\mathrm{RI}}(0,t) \;=\; c_{\mathrm{R}}^1 \qquad t \;>\; 0 \tag{8.119}$$

and

$$c_{\mathrm{RII}}(0,t) \;=\; 0 \qquad t \;\leq\; \tau \tag{8.120}$$

$$c_{\mathrm{RII}}(0,t) \;=\; c_{\mathrm{R}}^2 \;-\; c_{\mathrm{R}}^1 \qquad t \;>\; \tau \tag{8.121}$$

The virtual concentration profiles are then obtained by solving the respective Fick equations:

$$\frac{\partial c_{\mathrm{RI}}(x,t)}{\partial t} \;=\; D_{\mathrm{R}}\frac{\partial^2 c_{\mathrm{RI}}(x,t)}{\partial x^2} \tag{8.122}$$

and

$$\frac{\partial c_{\text{RII}}(x,t-\tau)}{\partial t} = D_{\text{R}} \frac{\partial^2 c_{\text{RII}}(x,t-\tau)}{\partial x^2} \tag{8.123}$$

For the first virtual concentration c_{RI}, the boundary conditions are :

$$c_{\text{RI}}^0(x,0) = c_{\text{R}}^0 \text{ and } \lim_{x\to\infty} c_{\text{RI}}^0(x,t) = c_{\text{R}}^0 \tag{8.124}$$

Thus, expressed as a function of the interfacial concentration imposed by the potential E_1, the concentration profile of the virtual concentration c_{RI} is (see equation (8.107)):

$$c_{\text{RI}}(x,t) = c_{\text{R}}^1 - \left[c_{\text{R}}^1 - c_{\text{R}}^0\right] \text{erf}\left[\frac{x}{2\sqrt{D_{\text{R}}t}}\right] \qquad t > 0 \tag{8.125}$$

In the same way, for the second virtual concentration, the boundary condition is :

$$c_{\text{RII}}(x,t) = 0 \qquad t \leq \tau \tag{8.126}$$

and so, by analogy with equation (8.104), we have

$$c_{\text{RII}}(x,t-\tau) = \left[c_{\text{R}}^2 - c_{\text{R}}^1\right] \text{erfc}\left[\frac{x}{2\sqrt{D_{\text{R}}(t-\tau)}}\right] \qquad t > \tau \tag{8.127}$$

The concentration profiles corresponding to the potential step are then given as the sum of the profiles $c_{\text{RI}}(x)$ and $c_{\text{RII}}(x,t)$.

Knowing that

$$\frac{\partial}{\partial x}\left(\text{erf}\left[\frac{x}{2\sqrt{D_{\text{R}}t}}\right]\right) = \frac{1}{\sqrt{\pi D_{\text{R}}t}} \exp\left[-\frac{x^2}{4D_{\text{R}}t}\right] \tag{8.128}$$

the expression for the current as a function of time is then :

$$I = nFAD_{\text{R}}\left(\frac{\partial c_{\text{R}}}{\partial x}\right)_{x=0} = nFA\left[\left(c_{\text{R}}^0 - c_{\text{R}}^1\right)\sqrt{\frac{D_{\text{R}}}{\pi t}} + \left(c_{\text{R}}^1 - c_{\text{R}}^2\right)\sqrt{\frac{D_{\text{R}}}{\pi(t-\tau)}}\right] \tag{8.129}$$

where the second term disappears when $t < \tau$.

This approach, calculating the current for a potential step is called the ***superposition principle***. It can be applied to a whole range of pulse techniques such as staircase voltammetry, differential pulse voltammetry, square wave voltammetry, etc.

8.3.2 Constant amplitude alternate pulse voltammetry

This technique resembles sine wave AC voltammetry, which is discussed later in §9.4, but here using a square wave of period τ as shown on the next page.

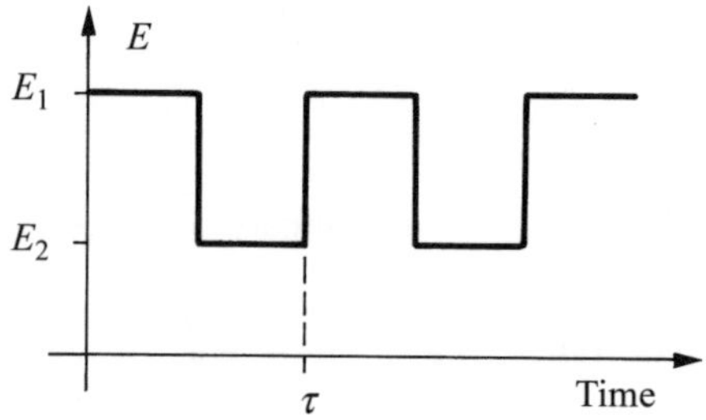

Fig. 8.15 Constant amplitude alternate pulse variation, also called square-wave variation.

The applied potential can be described by the following two equations

$$E(t) = \begin{cases} E_1 & 0 < t < \tau/2 \\ E_2 & \tau/2 < t < \tau \end{cases} \tag{8.130}$$

and

$$E(t) = E(t+\tau) \tag{8.131}$$

c_R^1 and c_R^2 are the interfacial concentrations corresponding to the potentials E_1 and E_2. By applying the superposition principle, we can see that the concentration profiles can be written in the form of a series

$$c_R(x,t) = c_{RI}(x,t) + c_{RII}(x,t) + c_{RIII}(x,t) + ... \tag{8.132}$$

The boundary conditions for the virtual concentration c_{RI} for the reduced species are as before :

$$\lim_{x\to\infty} c_{RI}(x,t) = c_R^0 \quad \text{and} \quad \lim_{x\to\infty} c_{RI}(x,t) = c_R^0 \tag{8.133}$$

and therefore the concentration profile of the virtual concentration c_{RI} is given by :

$$c_{RI}(x,t) = c_R^1 - \left[c_R^1 - c_R^0\right] \operatorname{erf}\left[\frac{x}{2\sqrt{D_R t}}\right] \qquad t > 0 \tag{8.134}$$

Furthermore, for the virtual concentration c_{RII} the interfacial conditions are also :

$$c_{RII} = 0 \qquad t \leq \tau/2 \tag{8.135}$$

$$c_{RII} = c_R^2 - c_R^1 \qquad t > \tau/2 \tag{8.136}$$

and the corresponding concentration profiles are :

$$c_{RII}(x,t) = 0 \qquad t \leq \tau/2 \tag{8.137}$$

$$c_{RII}(x,t-\tau/2) = \left[c_R^2 - c_R^1\right] \operatorname{erfc}\left[\frac{x}{2\sqrt{D_R(t-\tau/2)}}\right] \qquad t > \tau/2 \tag{8.138}$$

In the same way, the concentration profiles of the virtual concentration c_{RIII} are also:

$$c_{\mathrm{RIII}}(x,t) = 0 \qquad t \leq \tau \tag{8.139}$$

$$c_{\mathrm{RIII}}(x,t-\tau) = \left[c_{\mathrm{R}}^{1} - c_{\mathrm{R}}^{2}\right] \mathrm{erfc}\left[\frac{x}{2\sqrt{D_{\mathrm{R}}(t-\tau)}}\right] \qquad t > \tau \tag{8.140}$$

Generalising, the global concentration profile for any interval of time between $(n\tau)/2$ and $((n+1)\tau)/2$ is given by :

$$\begin{aligned} c_{\mathrm{R}}(x,t) &= c_{\mathrm{R}}^{1} - \left[c_{\mathrm{R}}^{1} - c_{\mathrm{R}}^{0}\right] \mathrm{erf}\left[\frac{x}{2\sqrt{D_{\mathrm{R}}t}}\right] \\ &+ \sum_{j=1}^{n} (-1)^{j}\left[c_{\mathrm{R}}^{1} - c_{\mathrm{R}}^{2}\right] \mathrm{erfc}\left[\frac{x}{2\sqrt{D_{\mathrm{R}}(t-j\tau/2)}}\right] \end{aligned} \tag{8.141}$$

Thus, using equation (8.128), the current is expressed in the form

$$\frac{I}{nFA} = \left[c_{\mathrm{R}}^{0} - c_{\mathrm{R}}^{1}\right]\sqrt{\frac{D_{\mathrm{R}}}{\pi t}} + \sum_{j=1}^{n}(-1)^{j+1}\left[c_{\mathrm{R}}^{1} - c_{\mathrm{R}}^{2}\right]\sqrt{\frac{D_{\mathrm{R}}}{\pi(t-j\tau/2)}} \tag{8.142}$$

If we define the alternating current ΔI as the difference $I(2j+1) - I(2j)$, then we have:

$$\Delta I = 2nFA\left(c_{\mathrm{R}}^{1} - c_{\mathrm{R}}^{2}\right)\sqrt{\frac{2D_{\mathrm{R}}}{\pi\tau}} \tag{8.143}$$

If the potentials E_1 and E_2 are far from the formal redox potential, the difference $c_{\mathrm{R}}^{1} - c_{\mathrm{R}}^{2}$ will be small. On the contrary, this difference will be the largest if the potentials are very close. Effectively, this means that the difference in interfacial concentrations imposed by the Nernst equation can be expressed as a function of the potential.

For an oxidation where the initial concentration of oxidised species is zero, by combining the two equations (8.98) and (8.106) we have:

$$c_{\mathrm{R}}^{1} - c_{\mathrm{R}}^{2} = c_{\mathrm{R}}^{0}\left[\frac{\xi\theta_1}{1+\xi\theta_1} - \frac{\xi\theta_2}{1+\xi\theta_2}\right] \tag{8.144}$$

By putting $E_1 = E + \Delta E/2$ and $E_2 = E - \Delta E/2$, the second term of equation (8.144) varies with the electrode potential E as shown in Figure 8.16. Such a curve can be obtained experimentally points by points by varying the value of the electrode potential E or by adding a slow potential ramp to the waveform illustrated in Figure 8.15. This technique is sometimes called Barker's square wave voltammetry.

By analogy with differential pulse polarography, we can define the parameter σ such that :

$$\sigma = \exp\left[-\frac{nF}{RT}\frac{\Delta E}{2}\right] \tag{8.145}$$

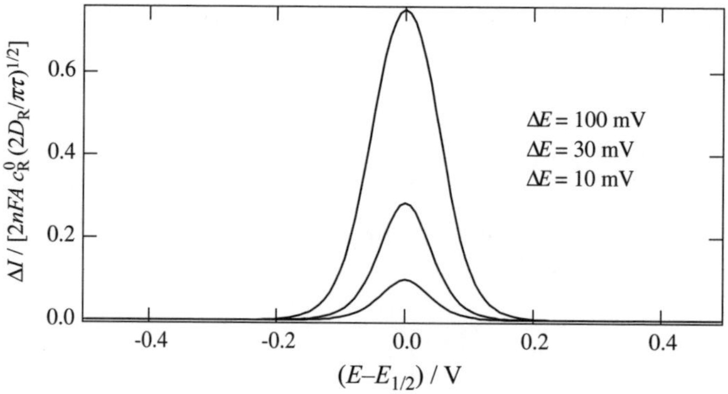

Fig. 8.16 Constant amplitude alternate pulse voltammetry as a function of the applied potential according to equation (8.143). $\xi = 1$.

Thus we have:

$$\left[\frac{\xi\theta_1}{1+\xi\theta_1} - \frac{\xi\theta_2}{1+\xi\theta_2}\right] = \left[\frac{\xi\theta\sigma}{1+\xi\theta\sigma} - \frac{\xi\theta/\sigma}{1+\xi\theta/\sigma}\right] = \frac{\xi\theta\,(\sigma^2-1)}{(1+\sigma\xi\theta)\,(\sigma+\xi\theta)} \tag{8.146}$$

By differentiating with respect to $u = \xi\theta$, we can easily show that this curve passes through a maximum when $u = 1$, i.e. when:

$$E_{\text{max}} = E_{1/2} = E^{\ominus\prime} + \frac{RT}{nF}\ln\left(\sqrt{\frac{D_\text{R}}{D_\text{O}}}\right) \tag{8.147}$$

The curve is therefore centred on the half-wave potential. It is interesting to compare this curve to the one in Figure (9.37) obtained using A.C. sine wave voltammetry.

8.3.3 Staircase voltammetry

The theory of cyclic voltammetry developed later in chapter 10 is only strictly valid when an analog ramp generator that delivers a linear potential function is used. Most computerised apparatus have, nowadays, a staircase ramp generator of the type

$$E = E_i + j\Delta E \qquad j\tau \le t < (j+1)\tau \tag{8.148}$$

where ΔE is the potential increment and τ the duration of the step.

In order to solve this problem, we can again use the superposition principle and write the concentration in the form of a series

$$c(x,t) = c_\text{I}(x,t) + c_\text{II}(x,t) + c_\text{III}(x,t) + \ldots \tag{8.149}$$

The boundary conditions for the virtual concentration c_RI for the reduced species are as before:

$$c_\text{RI}(x,0) = c_\text{R}^0 \quad \text{and} \quad \lim_{x\to\infty} c_\text{RI}(x,t) = c_\text{R}^0 \tag{8.150}$$

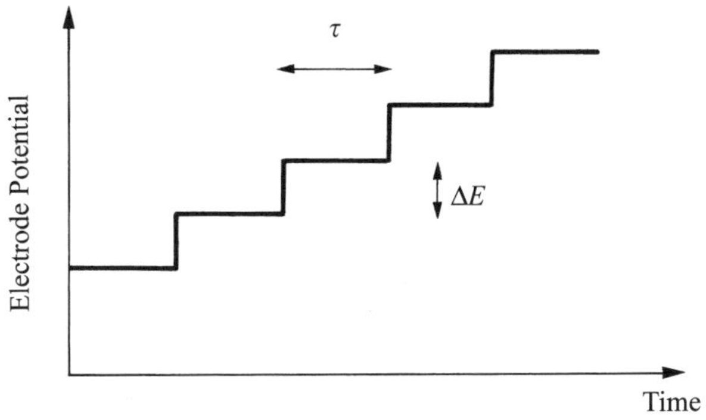

Fig. 8.17 Staircase potential ramp.

and so the concentration profile for the function c_{RI} is given by :

$$c_{\mathrm{RI}}(x,t) = c_{\mathrm{R}}^{1} - \left[c_{\mathrm{R}}^{1} - c_{\mathrm{R}}^{0}\right]\mathrm{erf}\left[\frac{x}{2\sqrt{D_{\mathrm{R}}t}}\right] \tag{8.151}$$

Furthermore, for the virtual concentration c_{RII} the interfacial conditions are also:

$$c_{\mathrm{RII}} = 0 \qquad t \leq \tau \tag{8.152}$$

$$c_{\mathrm{II}} = c_{\mathrm{R}}^{2} - c_{\mathrm{R}}^{1} \qquad t > \tau \tag{8.153}$$

and the corresponding concentration profile is:

$$c_{\mathrm{RII}}(x,t) = 0 \qquad t \leq \tau \tag{8.154}$$

$$c_{\mathrm{RII}}(x,t-\tau) = \left[c_{\mathrm{R}}^{2} - c_{\mathrm{R}}^{1}\right]\mathrm{erfc}\left[\frac{x}{2\sqrt{D_{\mathrm{R}}(t-\tau)}}\right] \qquad t > \tau \tag{8.155}$$

In the same way, the concentration profile of c_{RIII} is also:

$$c_{\mathrm{RIII}}(x,t) = 0 \qquad t \leq 2\tau \tag{8.156}$$

$$c_{\mathrm{RIII}}(x,t-2\tau) = \left[c_{\mathrm{R}}^{3} - c_{\mathrm{R}}^{2}\right]\mathrm{erfc}\left[\frac{x}{2\sqrt{D_{\mathrm{R}}(t-2\tau)}}\right] \qquad t > 2\tau \tag{8.157}$$

Generalising, the global concentration profile for any time interval between $(n\tau)/2$ and $((n+1)\tau)/2$ is given by :

$$c_{\mathrm{R}}(x,t) = c_{\mathrm{R}}^{1} - \left[c_{\mathrm{R}}^{1} - c_{\mathrm{R}}^{0}\right]\mathrm{erf}\left[\frac{x}{2\sqrt{D_{\mathrm{R}}t}}\right] + \sum_{j=1}^{n-1}\left[c_{\mathrm{R}}^{j+1} - c_{\mathrm{R}}^{j}\right]\mathrm{erfc}\left[\frac{x}{2\sqrt{D_{\mathrm{R}}(t-j\tau)}}\right] \tag{8.158}$$

Thus, the current is expressed in the form

$$\frac{I}{nFA} = \left[c_R^0 - c_R^1\right]\sqrt{\frac{D_R}{\pi t}} - \sum_{j=1}^{n-1}\left[c_R^{j+1} - c_R^j\right]\sqrt{\frac{D_R}{\pi(t-j\tau)}} \tag{8.159}$$

If we sample at the end of each potential step, the current is

$$\frac{I(n)}{nFA} = \sqrt{\frac{D_R}{\pi\,\tau}}\left[\frac{c_R^0 - c_R^1}{\sqrt{n}} - \sum_{j=1}^{n-1}\frac{c_R^{j+1} - c_R^j}{\sqrt{n-j}}\right] \tag{8.160}$$

For an oxidation where the concentration of oxidised species is initially zero, we have:

$$\frac{I(n)}{nFA} = c_R^0\sqrt{\frac{D_R}{\pi\,\tau}}\left[\frac{1}{\sqrt{n}} - \frac{\xi\theta_1}{(1+\xi\theta_1)\sqrt{n}} - \sum_{j=1}^{n-1}\frac{1}{\sqrt{n-j}}\left(\frac{\xi\theta_{j+1}}{1+\xi\theta_{j+1}} - \frac{\xi\theta_j}{1+\xi\theta_j}\right)\right] \tag{8.161}$$

which is

$$\frac{I(n)}{nFA} = c_R^0\sqrt{\frac{D_R}{\pi\,\tau}}\left[\frac{1}{\sqrt{n}} - \frac{\xi\theta_n}{(1+\xi\theta_n)} + \sum_{j=1}^{n-1}\frac{\xi\theta_j}{1+\xi\theta_j}\left(\frac{1}{\sqrt{n-j}} - \frac{1}{\sqrt{n-j+1}}\right)\right] \tag{8.162}$$

Figure 8.18 shows the current calculated point by point according to equation (8.162), the higher the potential step height, the fewer the number of points. The results are similar to classic linear sweep voltammetry as described in chapter 10 when the step height is small enough for the potential staircase to resemble a potential ramp.

These calculations show that if we wish to increase the sweep rate to do linear sweep voltammetry, it is preferable to reduce the duration τ of the potential plateau, rather than increase the height ΔE of the potential steps. Otherwise the peak potential varies like in the presence of an ohmic drop.

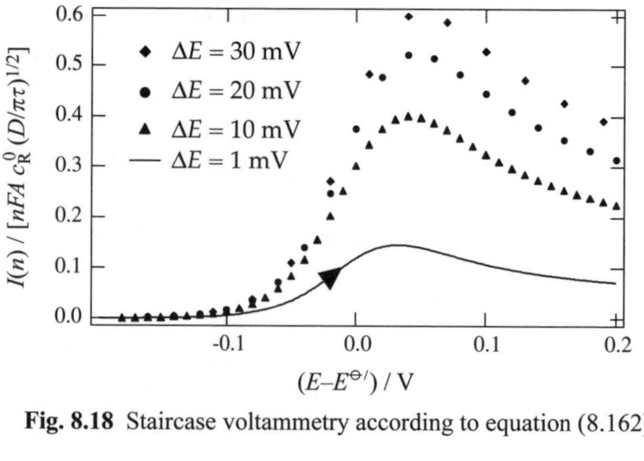

Fig. 8.18 Staircase voltammetry according to equation (8.162).

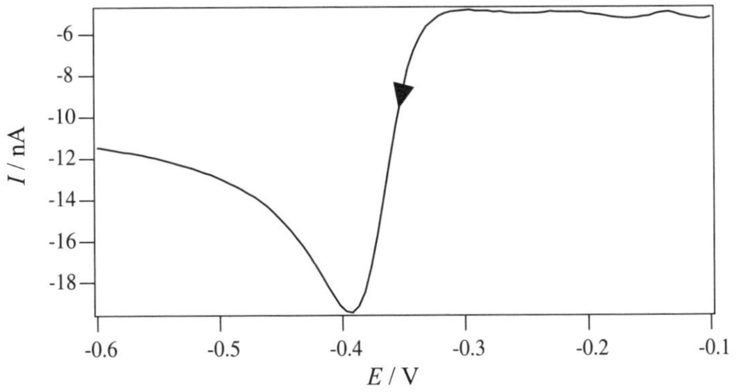

Fig. 8.19 Staircase voltammetry on a static drop of mercury for a 10^{-5} M solution of Pb^{2+} in 0.05 M HCl. Step height 5 mV, duration 100 ms, start potential: –0.1 V (Olivier Bagel, EPFL).

The graph in Figure 8.19 shows the response of staircase voltammetry for the reduction of Pb^{2+} on a drop of mercury. Notice that the peak current is less than that obtained using differential pulse polarography (see Figure 8.13) or the limiting cathodic current in Figure 8.10 for normal pulse polarography.

It is also interesting to compare staircase polarography, that results in a wave similar to that we have on a rotating electrode or a microelectrode, and staircase voltammetry. The presence of a peak with the second method is due to the fact that the electrode is not renewed as in polarography, and therefore the diffusion layer between each measurement remains the same.

8.3.4 Square Wave Voltammetry

This type of voltammetry is a combination of constant amplitude alternative pulse voltammetry and staircase voltammetry. Therefore, the resulting potential waveform is as shown in Figure 8.20.

The corresponding electrode potential for this scheme is

$$E = E_i + \mathrm{int}\left(\frac{j+1}{2}\right)\Delta E_s + (-1)^j E_{sw} \tag{8.163}$$

where E_i is the start potential, ΔE_s is the height of the potential step between two plateaux, int( ) represents the integer truncation function and E_{sw} is the amplitude of the alternate square signal. τ is the period of this signal and $1/\tau$ the frequency.

Applying the superposition principle as before, we get:

$$\begin{aligned} c_R(x,t) &= c_R^1 - \left[c_R^1 - c_R^0\right]\mathrm{erf}\left[\frac{x}{2\sqrt{D_R t}}\right] \\ &+ \sum_{j=1}^{n-1}\left[c_R^{j+1} - c_R^j\right]\mathrm{erfc}\left[\frac{x}{2\sqrt{D_R(t - j(\tau/2))}}\right] \end{aligned} \tag{8.164}$$

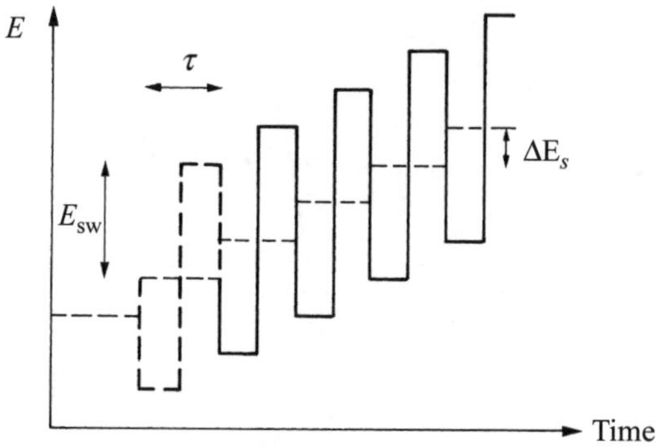

Fig. 8.20 Square wave voltammetry for an oxidation.

Thus the current is again expressed in the form

$$\frac{I}{nFA} = \left[c_R^0 - c_R^1\right]\sqrt{\frac{D_R}{\pi t}} - \sum_{j=1}^{n-1}\left[c_R^{j+1} - c_R^j\right]\sqrt{\frac{D_R}{\pi(t - j(\tau/2))}} \tag{8.165}$$

Furthermore, if we sample at the end of each potential pulse, the current is

$$\frac{I(n)}{nFA} = \sqrt{\frac{2D_R}{\pi\,\tau}}\sum_{j=0}^{n-1}\frac{c_R^j - c_R^{j+1}}{\sqrt{n-j}} \tag{8.166}$$

For an oxidation where the initial concentration of oxidised species is zero, we have:

$$\begin{aligned}\frac{I(n)}{nFA} &= c_R^0\sqrt{\frac{2D_R}{\pi\,\tau}}\sum_{j=0}^{n-1}\frac{1}{\sqrt{n-j}}\left(\frac{\xi\theta_j}{1+\xi\theta_j} - \frac{\xi\theta_{j+1}}{1+\xi\theta_{j+1}}\right)\\ &= c_R^0\sqrt{\frac{2D_R}{\pi\,\tau}}\sum_{j=0}^{n-1}\frac{1}{\sqrt{n-j}}\left(Q_j - Q_{j+1}\right)\end{aligned} \tag{8.167}$$

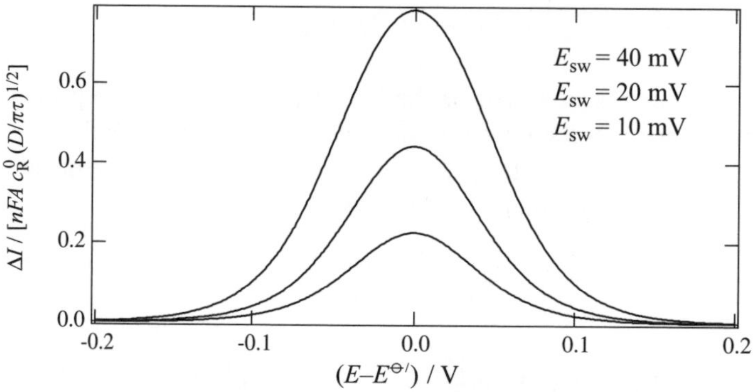

Fig. 8.21 Square-wave voltammetry according to equation (8.169).

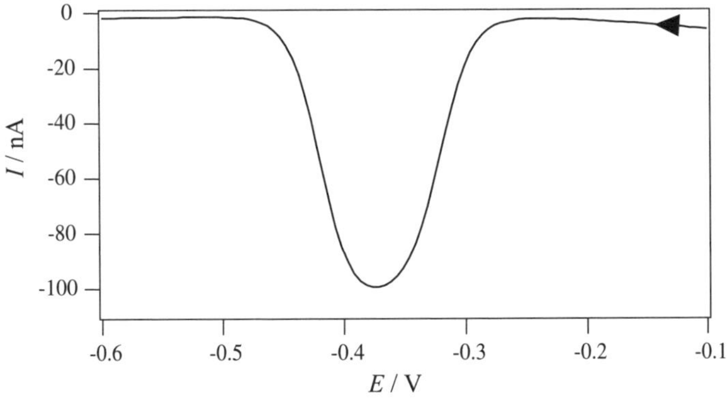

Fig. 8.22 Square wave voltammetry on a static drop of mercury for a 10^{-5} M solution of Pb^{2+} in 0.05 M HCl. Step height 5 mV, amplitude: 50 mV, frequency 20 Hz. Start potential: –0.1 V (Olivier Bagel, EPFL)

For each period, we calculate the difference $\Delta I(k) = I(2k+1) - I(2k)$

$$\Delta I(k) \quad = \quad nFAc_{\mathrm{R}}^{0}\sqrt{\frac{2D_{\mathrm{R}}}{\pi\,\tau}}\left[\frac{Q_0 - Q_1}{\sqrt{2k+1}} \quad - \quad \sum_{j=1}^{2k}\frac{1}{\sqrt{2k+1-j}}\left(Q_{j-1} - 2Q_j + Q_{j+1}\right)\right] \tag{8.168}$$

Neglecting the first term of the expression in brackets, this reduces to:

$$\Delta I(k) \quad = \quad nFAc_{\mathrm{R}}^{0}\sqrt{\frac{2D_{\mathrm{R}}}{\pi\,\tau}}\left[\sum_{j=0}^{2k}\frac{1}{\sqrt{2k-j}}\left(\frac{\xi\theta_{j-1}}{1+\xi\theta_{j-1}} - 2\frac{\xi\theta_j}{1+\xi\theta_j} + \frac{\xi\theta_{j+1}}{1+\xi\theta_{j+1}}\right)\right] \tag{8.169}$$

The curve obtained is then a peak centred on the formal redox potential.

The results in Figure 8.22 still for the reduction of Pb^{2+} show that continuous square wave voltammetry gives peaks similar to those obtained using differential pulse voltammetry shown in Figure 8.13. This is not surprising given the similarity of the applied signals; the only difference being in the duration of the potential plateau before the pulse that in the first case is equal to the pulse duration, but is much larger than the pulse duration in the second case.

8.4 STRIPPING VOLTAMMETRY

8.4.1 Anodic stripping voltammetry

One of the most sensitive methods for the detection of heavy metals is ***anodic stripping voltammetry***. The principle is to concentrate heavy metals existing in the form of ion traces in solution into a mercury electrode by reduction. This accumulation is generally done using reduction of the trace metal ions at a constant

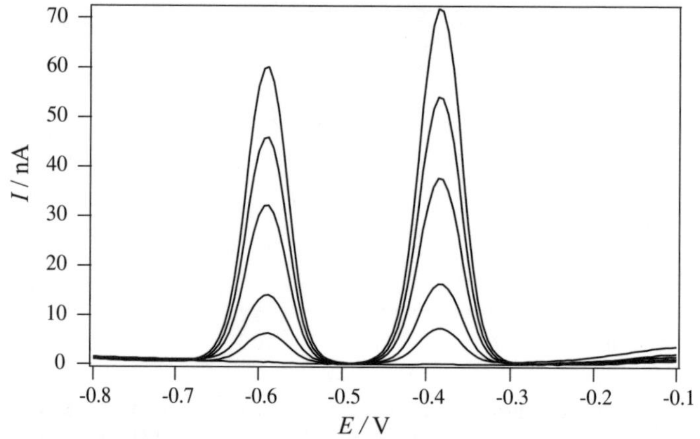

Fig 8.23 Stripping voltammetry of Cd and Pb. Experimental conditions Electrolyte 0.05 M HCl, 30 s of degassing under agitation, pre-concentration in the drop 10s at –1.1 V. Differential pulse stripping: staircase (height 5 mV, duration 0.4 ps), pulse (height 50 mV, duration 0.04 s). [Cd]& [Pb] = 10^{-7} M, $2.5\cdot10^{-7}$ M, $5\cdot10^{-7}$ M, $7.5\cdot10^{-7}$ M & 10^{-6} M (Olivier Bagel, EPFL).

electrode potential. After a certain time, of the order of a few minutes, a potential ramp either continuous or pulsed is applied from the accumulation potential to a potential that is higher, but just below that of the dissolution of mercury. This potential ramp can be produced in various ways, the most common being that of differential pulse voltammetry. During this ramp, the heavy metals are sequentially re-oxidised and the corresponding anodic current is measured.

The mathematical description of this technique depends mainly on the type of mercury electrodes used. These can be classified in two main families:

- Hanging mercury drop electrodes (HDME)
- Mercury film electrodes electrodeposited on iridium or vitreous carbon electrodes.

The mathematical aspects of these techniques hold little interest from an analytical point of view, since they are mainly used with the aid of calibration procedures, such as for example the standard addition method.

8.4.2 Cathodic stripping voltammetry

This category covers various variations on the same theme. The main idea is to allow the adsorption of metal-ligand complexes (preferably hydrophobic) at the surface of the electrode and to apply a potential ramp towards negative potentials and then to measure the cathodic current corresponding to the reduction of the adsorbed metal complexes.

8.5 THIN LAYER VOLTAMMETRY

8.5.1 Chronoamperometry in finite linear diffusion for complete interfacial oxidation

With the miniaturisation of electrochemical detectors, the size of the measuring cell can have dimensions smaller than the thickness of the diffusion layers. In this case, the size and the geometry of the cell play a major role in the mass transfer. For simplicity, we shall treat here only systems known as 'thin layer systems' that is to say a planar electrode close to a planar wall parallel to the electrode and situated at a distance δ from it. Still aiming for simplicity, we shall consider a complete interfacial oxidation under finite linear diffusion control, assuming that only the reduced species is initially present in solution and where the initial conditions (8.1) are therefore still valid. The difference between this and the case of semi-infinite linear diffusion is that we consider here the impoverishment of the reduced species in the volume of the cell and the condition imposed by the wall is a condition of zero flux

$$\left[\frac{\partial c_{\mathrm{R}}(x,t)}{\partial x}\right]_{x=\delta} = 0 \tag{8.170}$$

The solution of Fick's equation (8.5) remains

$$\bar{c}_{\mathrm{R}}(x,s) = \frac{c_{\mathrm{R}}(x,0)}{s} + A(s)\,e^{-\sqrt{\frac{s}{D_{\mathrm{R}}}}x} + B(s)\,e^{\sqrt{\frac{s}{D_{\mathrm{R}}}}x} \tag{8.171}$$

where the constant $B(s)$ can no longer be considered as zero. To determine the constants $A(s)$ and $B(s)$, we take the derivative of equation (8.171) at the wall where the flux is zero

$$-A(s)\sqrt{\frac{s}{D_{\mathrm{R}}}}\,e^{-\sqrt{\frac{s}{D_{\mathrm{R}}}}\delta} + B(s)\sqrt{\frac{s}{D_{\mathrm{R}}}}\,e^{\sqrt{\frac{s}{D_{\mathrm{R}}}}\delta} = 0 \tag{8.172}$$

By substitution, we therefore get

$$A(s) = -\frac{c_{\mathrm{R}}}{s}\,\frac{e^{\sqrt{\frac{s}{D_{\mathrm{R}}}}\delta}}{e^{\sqrt{\frac{s}{D_{\mathrm{R}}}}\delta} + e^{-\sqrt{\frac{s}{D_{\mathrm{R}}}}\delta}} \tag{8.173}$$

and

$$B(s) = -\frac{c_{\mathrm{R}}}{s}\,\frac{e^{-\sqrt{\frac{s}{D_{\mathrm{R}}}}\delta}}{e^{\sqrt{\frac{s}{D_{\mathrm{R}}}}\delta} + e^{-\sqrt{\frac{s}{D_{\mathrm{R}}}}\delta}} \tag{8.174}$$

where c_R is the initial concentration of the reduced species. So the Laplace transform of the current is given by:

$$\bar{I}(s) = nFAD_R\left[\frac{\partial\bar{c}_R(x,s)}{\partial x}\right]_{x=0} = nFAD_R\sqrt{\frac{s}{D_R}}[-A(s)+B(s)] \qquad (8.175)$$

which is, by expanding

$$\bar{I}(s) = nFAc_R\sqrt{\frac{D_R}{s}}\left[\frac{e^{\sqrt{\frac{s}{D_R}}\delta}-e^{-\sqrt{\frac{s}{D_R}}\delta}}{e^{\sqrt{\frac{s}{D_R}}\delta}+e^{-\sqrt{\frac{s}{D_R}}\delta}}\right] = nFAc_R\sqrt{\frac{D_R}{s}}\tanh\left[\sqrt{\frac{s}{D_R}}\delta\right] \qquad (8.176)$$

When δ tends to infinity, the hyperbolic tangent tends to unity and we come back to equation (8.18), The current response to the potential step is the Cottrel equation. Conversely, when δ tends to zero, equation (8.176) reduces to:

$$\bar{I}(s) = nFAc_R\delta \qquad (8.177)$$

since $\lim_{x\to 0}\tanh x = x$, and the inverse transform of equation (8.177) is then

$$I(t) = nFAc_R\delta\cdot\text{Dirac}(t) \qquad (8.178)$$

This equation means that the reduced species is instantaneously consumed.

Between these two limits, the current response to a potential step has no easy analytical solution and the best way to circumvent this difficulty is to recur to computer simulations such as finite difference or finite element. Results obtained by simulation are shown in Figure 8.24 for different thin-layer thicknesses.

The concentrations of the reactant in a thin layer cell decrease first close to the electrode and then through the whole cell as shown in Figure 8.25.

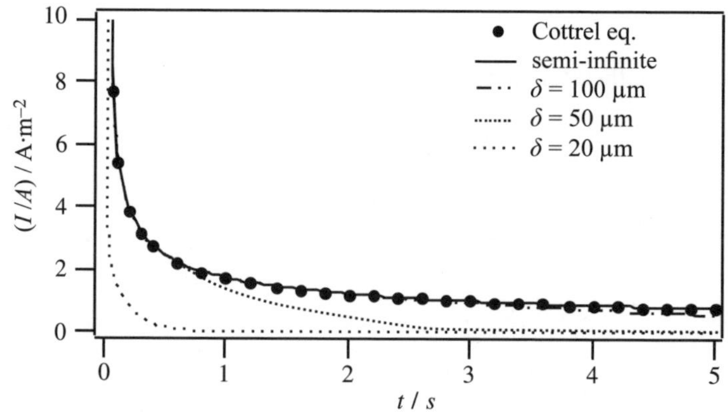

Fig. 8.24 Current response for a potential jump in a thin-layer cell of thickness δ. Same conditions as in Figure 8.1 (Jacques Josserand, EPFL).

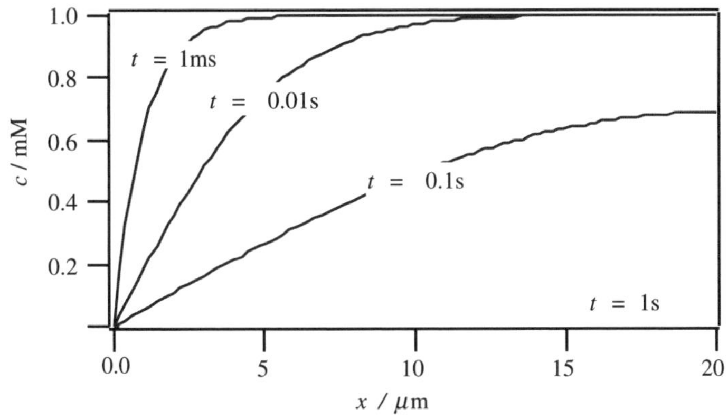

Fig. 8.25 Concentrations profiles for a 20μm thin layer cell. Same conditions as in Figure 8.1 (Jacques Josserand, EPFL).

8.5.2 Electrode covered with a thin membrane

Consider an electrode covered with a fine membrane and placed in a solution as shown Figure 7.22. The solutions of the Fick equations for the membrane and the solution are

$$\bar{c}_{\mathrm{Rm}}(x,s) = \frac{c_{\mathrm{R}}(x,0)}{s} + A(s)\, e^{-\sqrt{\frac{s}{D_{\mathrm{Rm}}}}x} + B(s)\, e^{\sqrt{\frac{s}{D_{\mathrm{Rm}}}}x} \tag{8.179}$$

and

$$\bar{c}_{\mathrm{Rs}}(x,s) = \frac{c_{\mathrm{R}}(x,0)}{s} + C(s)\, e^{-\sqrt{\frac{s}{D_{\mathrm{Rm}}}}x} + D(s)\, e^{\sqrt{\frac{s}{D_{\mathrm{Rm}}}}x} \tag{8.180}$$

As in the case for a non-covered electrode, we can immediately conclude that the constant $D(s)$ is zero, since the concentration cannot tend to infinity.

At the interface between the membrane and the solution, there is a condition of equality of the fluxes and one for the concentrations, which are respectively

$$D_{\mathrm{Rm}}\left(\frac{\partial c_{\mathrm{Rm}}(x,t)}{\partial x}\right)_{x=\delta} = D_{\mathrm{Rs}}\left(\frac{\partial c_{\mathrm{Rs}}(x,t)}{\partial x}\right)_{x=\delta} \tag{8.181}$$

and

$$c_{\mathrm{Rm}}(\delta,t) = c_{\mathrm{Rs}}(\delta,t) \tag{8.182}$$

Using the Laplace transforms of these equations, by substitution we have

$$\sqrt{D_{\mathrm{Rm}}}\left[-A(s)\, e^{-\sqrt{\frac{s}{D_{\mathrm{Rm}}}}\delta} + B(s)\, e^{\sqrt{\frac{s}{D_{\mathrm{Rm}}}}\delta}\right] = -\sqrt{D_{\mathrm{Rs}}}\,C(s)\, e^{-\sqrt{\frac{s}{D_{\mathrm{Rm}}}}\delta} \tag{8.183}$$

and

$$\frac{c_{\mathrm{Rm}}(x,0)}{s}+A(s)\,e^{-\sqrt{\frac{s}{D_{\mathrm{Rm}}}}\delta}+B(s)\,e^{\sqrt{\frac{s}{D_{\mathrm{Rm}}}}\delta} \;=\; \frac{c_{\mathrm{Rs}}(x,0)}{s}+C(s)\,e^{-\sqrt{\frac{s}{D_{\mathrm{Rm}}}}\delta} \tag{8.184}$$

For a complete oxidation, the interfacial concentration is quasi-zero, and also we have

$$\frac{c_{\mathrm{Rm}}(x,0)}{s}+A(s)+B(s) \;=\; 0 \tag{8.185}$$

It is possible to solve this system of three equations with three unknowns. By substitution, we get

$$A(s)=-\frac{c_{\mathrm{Rm}}(x,0)}{s}\left[1+\frac{\left(\sqrt{D_{\mathrm{Rm}}}-\sqrt{D_{\mathrm{Rs}}}\right)e^{-\sqrt{\frac{s}{D_{\mathrm{Rm}}}}\delta}}{\left(\sqrt{D_{\mathrm{Rm}}}+\sqrt{D_{\mathrm{Rs}}}\right)e^{\sqrt{\frac{s}{D_{\mathrm{Rm}}}}\delta}-\left(\sqrt{D_{\mathrm{Rm}}}-\sqrt{D_{\mathrm{Rs}}}\right)e^{-\sqrt{\frac{s}{D_{\mathrm{Rm}}}}\delta}}\right] \tag{8.186}$$

and

$$B(s)=\frac{c_{\mathrm{Rm}}(x,0)}{s}\left[\frac{\left(\sqrt{D_{\mathrm{Rm}}}-\sqrt{D_{\mathrm{Rs}}}\right)e^{-\sqrt{\frac{s}{D_{\mathrm{Rm}}}}\delta}}{\left(\sqrt{D_{\mathrm{Rm}}}+\sqrt{D_{\mathrm{Rs}}}\right)e^{\sqrt{\frac{s}{D_{\mathrm{Rm}}}}\delta}-\left(\sqrt{D_{\mathrm{Rm}}}-\sqrt{D_{\mathrm{Rs}}}\right)e^{-\sqrt{\frac{s}{D_{\mathrm{Rm}}}}\delta}}\right] \tag{8.187}$$

As before, we define the Laplace transform of the current as

$$\bar{I}(s) \;=\; nFAD_{\mathrm{Rm}}\left(\frac{\partial c_{\mathrm{Rm}}(x,t)}{\partial x}\right)_{x=0} \tag{8.188}$$

which, by substitution becomes

$$\bar{I}(s) \;=\; nFA\sqrt{sD_{\mathrm{Rm}}}\left[-A(s)+B(s)\right] \tag{8.189}$$

$$\bar{I}(s) \;=\; nFA\sqrt{D_{\mathrm{Rm}}}\left[c_{\mathrm{Rm}}(x,0)\sqrt{s}+\frac{2c_{\mathrm{Rm}}(x,0)\sqrt{s}\left(\sqrt{D_{\mathrm{Rm}}}-\sqrt{D_{\mathrm{Rs}}}\right)e^{-\sqrt{\frac{s}{D_{\mathrm{Rm}}}}\delta}}{\left(\sqrt{D_{\mathrm{Rm}}}+\sqrt{D_{\mathrm{Rs}}}\right)e^{\sqrt{\frac{s}{D_{\mathrm{Rm}}}}\delta}-\left(\sqrt{D_{\mathrm{Rm}}}-\sqrt{D_{\mathrm{Rs}}}\right)e^{-\sqrt{\frac{s}{D_{\mathrm{Rm}}}}\delta}}\right] \tag{8.190}$$

This equation does not have an easily-expressed inverse transform. However, we can study the boundary conditions.

- $D_{Rm} = D_{Rs}$. The second term in the square brackets of equation (8.190) cancels out and we find ourselves with Cottrell's equation. The membrane plays no part.
- $D_{Rm} << D_{Rs}$. Equation (8.190) reduces to

$$\bar{I}(s) \quad = \quad nFAc_{Rm}(x,0)\sqrt{sD_{Rm}}\coth\sqrt{\frac{s}{D_{Rm}}}\delta \tag{8.191}$$

Furthermore, if $\delta\sqrt{s/D_{Rm}} \rightarrow 0$ then $\coth(\delta\sqrt{s/D_{Rm}} \rightarrow \delta^{-1}\sqrt{D_{Rm}/s}$ and the inverse transformation of the current is

$$I(t) \quad = \quad nFAc_{Rm}(x,0)D_{Rm} / \delta \tag{8.192}$$

In this way, we demonstrate that we have a steady state current when an electrode is covered with a membrane, where the diffusion coefficient of the reacting species is smaller than that in solution.

8.6 AMPEROMETRIC DETECTORS FOR CHROMATOGRAPHY

Amperometric detection is sometimes used in HPLC chromatography for analyses of organic components such as sugar for example. The most common methods are constant potential amperometry and pulse voltammetry. The types of measuring cell include flow cell or wall jet cell as shown schematically in Figure 8.26.

In the in-line version, the reference electrode is placed 'upstream' of the working electrode and the counter-electrode is placed 'downstream' so that the reaction products on the counter-electrode do not perturb the measurements. For the same reasons, the reference electrode and the counter-electrode are well-spaced from the working electrode in the 'wall-jet' version (see §7.4.1).

For the in-line version, the thickness of the diffusion layer is fixed by the hydrodynamics of the flow and we again find conditions similar to those described by Figure 7.21. If the flow is slow, we approach the conditions described in §8.5.1. Thus, as a function of the geometric characteristics of the cell, the proportion of oxidised or reduced analyte varies approximately from 10 to 90%.

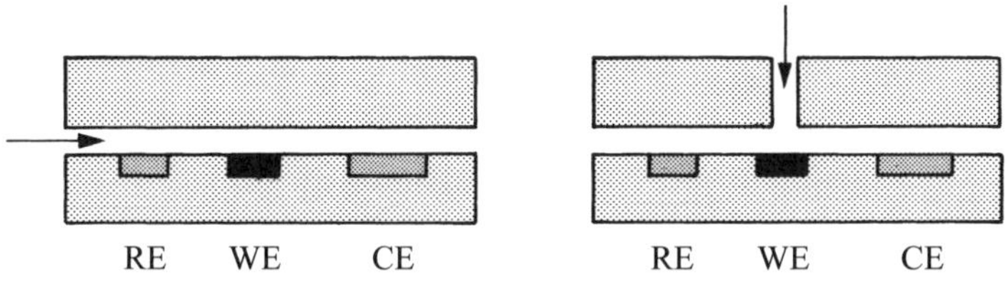

Fig. 8.26 Amperometric detector: in-line version and 'wall-jet' version. RE = Reference electrode, WE = Working electrode, CE = Counter-electrode.

In the 'wall-jet' version, the thickness of the diffusion layer is controlled by the arrival of the solution, as in the case of the rotating electrode.

The major problem with amperometric detectors is the fouling of the electrodes. To alleviate this, it is usual to clean the electrodes by submitting the working electrode to a series of strong anodic and cathodic pulses.

In certain cases, it is preferable to oxidise or reduce 100% of the sample passing in front of the detector. For this, porous electrodes are used to optimise the surface-to-volume ratio, and often the charge passed is measured rather than the current in order to increase the signal-to-noise ratio.

CHAPTER 9

ELECTROCHEMICAL IMPEDANCE

In the last chapter, we studied the current response of an electrochemical system for different electrode potential excitations (potential step, square wave, etc.). We considered only the influence of the mass transport on the response as a function of time, neglecting the kinetics of the electrode reaction. In other words, we consistently made the hypothesis that the Nernst equation applied to the interfacial concentrations. It is, of course, possible for the techniques studied previously, such as potential step amperometry, cyclic voltammetry, square-wave voltammetry, etc. to take into account kinetic effects by introducing Butler-Volmer type equations as boundary conditions of the diffusion equations. However, even though those techniques just mentioned can be used to study the kinetics of an electrode reaction, the result is often corrupted by side-effects such as the charging currents of the double layer observed on a time-scale of the order of a millisecond, or by the ohmic drop associated to the experimental setup. We have already seen in chapters 7 and 8 that the response of reversible electrochemical systems studied in the presence of an ohmic drop unfortunately resembled the response of kinetically slow systems. The best way of differentiating the kinetics of an electrode reaction from experimental side-effects is to use an excitation function covering a large time domain. The most common of these techniques is electrochemical impedance where the electrode potential excitation function is a sine wave of variable frequencies.

9.1 TRANSFER FUNCTION

9.1.1 AC response

As a first approximation, we can consider an electrochemical system as linear, i.e. that the current response for small potential perturbations is linear, and the potential response for small imposed current perturbations is also linear.

To illustrate this, consider a steady state current-potential curve such as that in Figure 9.1, and let's examine the system response if we vary the electrode potential sinusoidally at low frequencies around a constant value E_c with a small amplitude ΔE. The current response follow the steady state curve at the same frequency around the constant current value I_c with an amplitude ΔI which reflects the slope of the steady state curve. The two functions have the same frequency, but the current can be dephased with respect to the potential. Thus, basically, we can write

$$E = E_c + \Delta E \sin \omega t \tag{9.1}$$

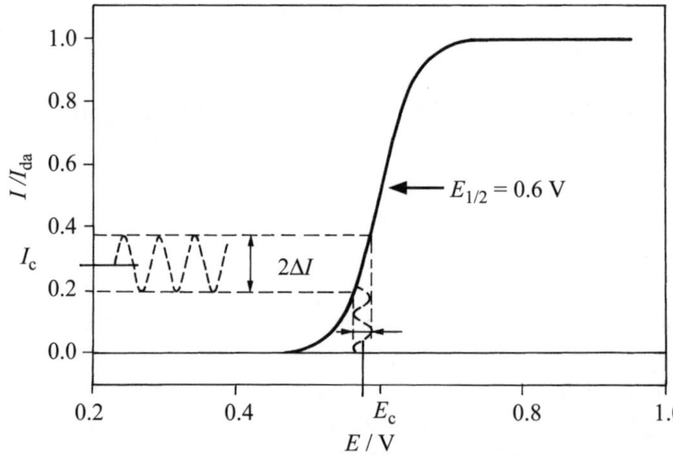

Fig. 9.1 Linear current response to a sinusoidal potential excitation of small amplitude around a constant value E_c.

and

$$I = I_c + \Delta I \sin(\omega t + \phi) \tag{9.2}$$

where ω is the angular frequency, also called the pulsation, and ϕ is the phase angle between the two signals.

9.1.2 Linear Systems

In a more general way, we can say that for any linear system, there is a ratio between the input function $x(t)$ and the output function $y(t)$.

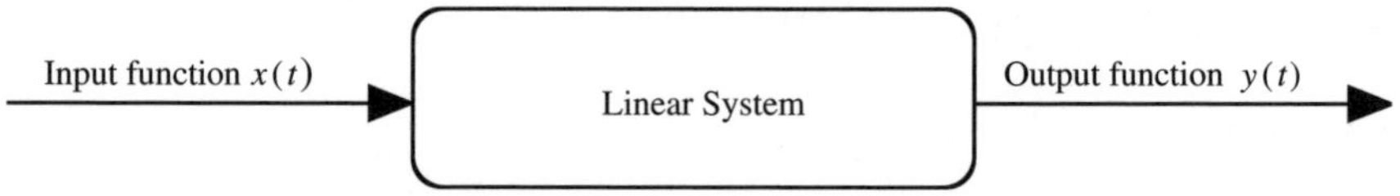

Fig. 9.2 Linear response of a system.

A ***linear system*** with localised parameters is defined by a differential equation of order n

$$\sum_{i=1}^{n} a_i \frac{\partial^i y(t)}{\partial t^i} = \sum_{i=1}^{m} b_i \frac{\partial^i x(t)}{\partial t^i} \tag{9.3}$$

or by a system of n first-order differential equations. A linear system obeys the superposition principle that states that the output function of a linear combination of input functions is equal to the linear combination of the respective output functions.

Therefore, if the input functions x_1 and x_2 have the output functions y_1 and y_2 respectively, then the output function of $x_3 = \alpha x_1 + \beta x_2$ is $y_3 = \alpha y_1 + \beta y_2$. In equation (9.3) the coefficients a_i and b_i are independent of time.

In a time-independent regime, equation (9.3) reduces to

$$a_0 y = b_0 x \tag{9.4}$$

and the output function is directly proportional to the input function (linear system). When the input function consists of small variations (pulses, sine-wave, noise, etc.) around a fixed value x_o

$$x(t) = x_o + \Delta x(t) \tag{9.5}$$

the output function varies around the corresponding static value y_o

$$y(t) = y_o + \Delta y(t) \tag{9.6}$$

In this case, equation (9.3) is written as

$$\sum_{i=1}^{n} a_i \frac{\partial^i \Delta y(t)}{\partial t^i} = \sum_{i=1}^{m} b_i \frac{\partial^i \Delta x(t)}{\partial t^i} \tag{9.7}$$

The Laplace transform of a function $F(t)$ previously defined by equation (8.6) can also be defined by considering the variable in the Laplace plane, s, as a complex number.

$$s = \sigma + \mathrm{j}\omega \tag{9.8}$$

The Laplace transform is then

$$\overline{F}(s) = L\{F(t)\} = \int_0^\infty F(t)e^{-st}\mathrm{d}t = \int_0^\infty F(t)e^{-(\sigma+\mathrm{j}\omega)t}\mathrm{d}t \tag{9.9}$$

By using the property of derivatives already used in the last chapter (see equation (8.8)), the Laplace transform of equation (9.7) becomes

$$L\left\{\sum_{i=1}^{n} a_i \frac{\partial^i \Delta y(t)}{\partial t^i}\right\} = \sum_{i=1}^{n} a_i s^i \Delta\overline{y}(s) = L\left\{\sum_{i=1}^{m} b_i \frac{\partial^i \Delta x(t)}{\partial t^i}\right\} = \sum_{i=1}^{m} b_i s^i \Delta\overline{x}(s) \tag{9.10}$$

making the hypothesis that all the partial derivatives are zero at the origin.

The ***transfer function*** of a linear system is defined as the ratio of the Laplace transforms of the variations of the output to the input functions.

$$H(s) = \frac{\Delta\overline{y}(s)}{\Delta\overline{x}(s)} = \frac{\sum_{i=1}^{m} b_i s^i}{\sum_{i=1}^{n} a_i s^i} \tag{9.11}$$

9.1.3 Impedance and admittance

For a sinusoidal excitation, we have

$$\Delta x(t) = \Delta x \sin \omega t \tag{9.12}$$

the Laplace transform of which is

$$\Delta\bar{x}(s) = \Delta x \frac{\omega}{s^2+\omega^2} \tag{9.13}$$

The output function is

$$\Delta\bar{y}(s) = H(s)\cdot\Delta\bar{x}(s) = \Delta x \frac{H(s)\cdot\omega}{s^2+\omega^2} \tag{9.14}$$

In view of the complex nature of s, it is useful to distinguish what we can call the ***frequency domain*** which depends on the angular frequency ω, and which consists of taking $s = \mathrm{j}\omega$ in the Laplace transform.

Thus, in the inverse transform of equation (9.14) the frequency domain corresponds to the sinusoidal part of the output function

$$\Delta y(t) = \Delta x\,|H(\omega)|\sin(\omega t+\phi) \tag{9.15}$$

where $|H(\omega)|$ is the modulus of $H(\omega)$ defined as a function of the real part $\mathrm{Re}H(\omega)$ and the imaginary part $\mathrm{Im}H(\omega)$

$$|H(\omega)| = \sqrt{[\mathrm{Re}\,H(\omega)]^2+[\mathrm{Im}\,H(\omega)]^2} \tag{9.16}$$

and ϕ is the argument defined by

$$\phi = \mathrm{Arg}[H(\omega)] = \arctan\left[\frac{\mathrm{Im}\,H(\omega)}{\mathrm{Re}\,H(\omega)}\right] \tag{9.17}$$

For an electric circuit or an electrochemical system, the transfer function from the potential (input function) to the current (output function) is called the ***admittance*** of the system and has the symbol Y. In the frequency domain, it is defined by

$$\Delta\bar{I}(\omega) = Y(\omega)\Delta\bar{E}(\omega) \tag{9.18}$$

In the same way, the transfer function from the current (input function) to the potential (output function) is called the ***impedance*** of the system and has the symbol Z. In the frequency domain, it is defined by

$$\Delta\bar{E}(\omega) = Z(\omega)\Delta\bar{I}(\omega) \tag{9.19}$$

Clearly, the admittance is the inverse of the impedance.

9.1.4 The Nyquist diagram

The transfer function in the frequency domain being a complex number, it is useful to represent it by plotting the imaginary part as a function of the real part.

$$H(\omega) = \mathrm{Re}H(\omega) + \mathrm{j}\,\mathrm{Im}H(\omega) \tag{9.20}$$

In electrochemistry in fact, it is more customary to trace $-\mathrm{Im}H(\omega)$ as a function of $\mathrm{Re}H(\omega)$. The graph, obtained by doing this, is called a ***Nyquist diagram***.

EXAMPLE

Consider a system with a 'first order' transfer function (one pole in the denominator)

$$H(\omega) = \frac{K}{1+\mathrm{j}\omega\tau}$$

where τ is the time constant of the system. Let's plot its Nyquist diagram. By multiplying both the numerator and the denominator by the conjugated complex, we have

$$H(\omega) = \frac{K}{1+\omega^2\tau^2}[1-\mathrm{j}\omega\tau]$$

The Nyquist diagram then gives a semi-circle going from the point at coordinates (K,0) when the angular frequency tends to zero, and finishing at the origin when the angular frequency tends to infinity.

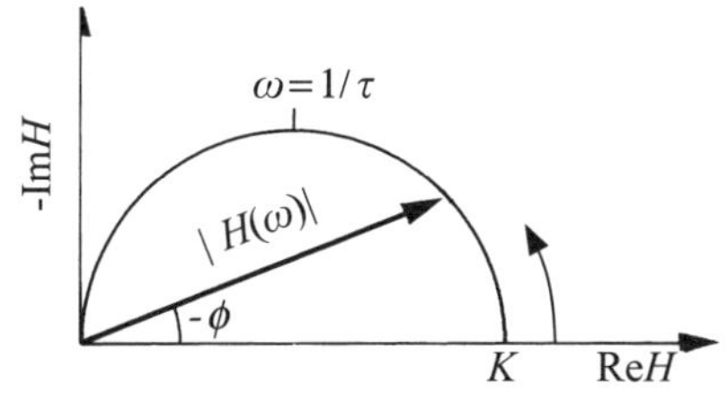

9.1.5 The Bode Diagram

Another way of representing a complex number is to plot the modulus as a function of the angular frequency. In electrochemistry, it is also the custom to plot $\log|H(\omega)|$ as a function of $\log(\omega)$ and the graph obtained by doing this is called a ***Bode diagram***.

9.2 ELEMENTARY CIRCUITS

Before studying the transfer function of an electrochemical system, it is interesting to look at the response of some electric circuits containing simple elements such as resistors and capacitors

9.2.1 Resistor

The potential difference at the terminals of a resistor is given by Ohm's law

$$\Delta E(t) = R\,\Delta I(t) \qquad (9.21)$$

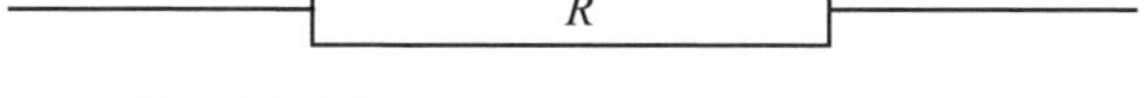

Fig. 9.3 Schematic representation of a resistor.

and therefore the Laplace transform is simply

$$\Delta\bar{I}(\omega) = R^{-1}\Delta\bar{E}(\omega) \tag{9.22}$$

The admittance of a resistor is therefore

$$Y(\omega) = \frac{\Delta\bar{I}(\omega)}{\Delta\bar{E}(\omega)} = R^{-1} \tag{9.23}$$

If the alternating potential is written as

$$\Delta E(t) = \Delta E \sin\omega t \tag{9.24}$$

the inverse transform of equation (9.22) in the frequency domain is given by equations (9.16) and (9.17). Thus, the alternating current I_{ac} is

$$I_{ac} = \Delta I(t) = \Delta E|Y(\omega)|\sin(\omega t + \arctan(0)) = \frac{\Delta E}{R}\sin\omega t \tag{9.25}$$

Thus, we arrive again at equation (9.21). A resistor does not introduce dephasing, and Ohm's law applies equally to alternating currents and potentials.

9.2.2 Capacitor

The potential difference at the terminals of a capacitor is proportional to its charge (see page 11)

$$Q(t) = C\,E(t) \tag{9.26}$$

The current is defined as the variation with the charge with time

$$I(t) = \frac{\mathrm{d}Q(t)}{\mathrm{d}t} = \frac{C\mathrm{d}E(t)}{\mathrm{d}t} \tag{9.27}$$

and therefore the Laplace transform, using equation (8.8), is given by

$$\Delta\bar{I}(s) = Cs\Delta\bar{E}(s) \tag{9.28}$$

The admittance in the frequency domain is obtained, by taking $s = \mathrm{j}\omega$

$$Y(\omega) = \frac{\Delta\bar{I}(\omega)}{\Delta\bar{E}(\omega)} = \mathrm{j}C\omega \tag{9.29}$$

If the potential is given by equation (9.24), then the inverse transform of equation (9.29) from equations (9.16) and (9.17) gives

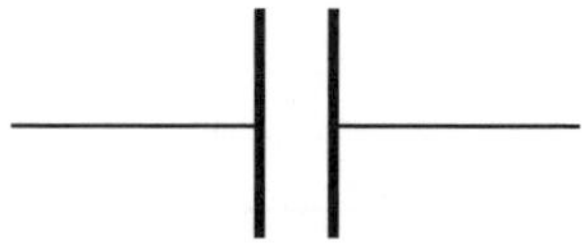

Fig. 9.4 Schematic representation of a capacitor.

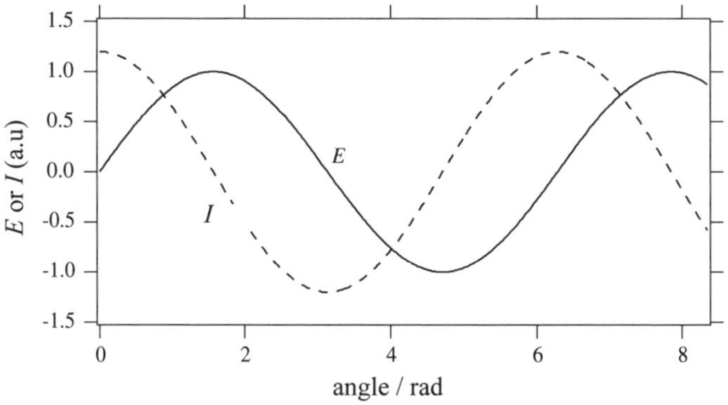

Fig. 9.5 Alternating current and potential at the terminals of a capacitor. The current is advanced by 90°.

$$
\begin{aligned}
I_{ac} &= \Delta I(t) = \Delta E\,|Y(\omega)|\sin(\omega t + \arctan(\infty)) \\
&= \Delta E\, C\omega \sin(\omega t + \pi/2) = \Delta E\, C\omega\cos(\omega t)
\end{aligned}
\tag{9.30}
$$

The alternating current at the terminals of a capacitor is therefore dephased with respect to the applied potential by 90°. This is shown in Figure 9.5 that demonstrates the variations of potential and current as a function of the angle ωt.

9.2.3 Resistor – capacitor in series

The impedance of two circuits in series is the sum of the impedances of the elements of the circuit. In effect, the potential at the terminals of the circuit is equal to the sum of the potentials at the terminals of each element, and the current is the same through all the elements of a series circuit.

$$
Z = \frac{\Delta \bar{E}}{\Delta \bar{I}} = \frac{\Delta \bar{E}_1 + \Delta \bar{E}_2}{\Delta \bar{I}} = Z_1 + Z_2 \tag{9.31}
$$

Taking a resistor and a capacitor in series, we then have

$$
Z(\omega) = Z_R(\omega) + Z_C(\omega) = R - \frac{j}{C\omega} \tag{9.32}
$$

The Nyquist diagram of the impedance gives a straight vertical line, which at high frequencies tends to the point $Z_R = R$. On the contrary, the diagram representing the

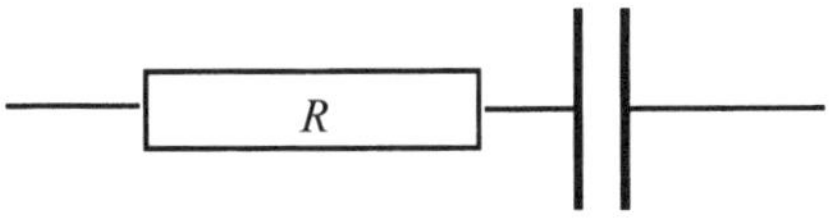

Fig. 9.6 Circuit with a resistor and a capacitor in series.

imaginary part of the admittance as a function of its real part forms a semi-circle tending to the origin when the frequency tends to zero, and tending to the point $Y_R = R^{-1}$ at high frequencies. The angular frequency at the apex of the semi-circle corresponds to the product RC. RC^{-1} represents the time constant of the circuit.

This equivalent circuit corresponds to that of an ideally polarisable working electrode such as the one shown in Figure 5.5. The resistor represents the resistance of the solution between the working electrode and the reference electrode (see Figures (7.32) and (7.33)), and the capacitor corresponds to the capacity of the double layer (see equation (5.97)). Therefore, if we are trying to measure the capacity of a polarisable electrode, it is advisable to make measurements at different frequencies, and to plot $-Z_I$ as a function of ω^{-1}, because the slope of this graph is the inverse of the capacity, as shown in equation (9.32).

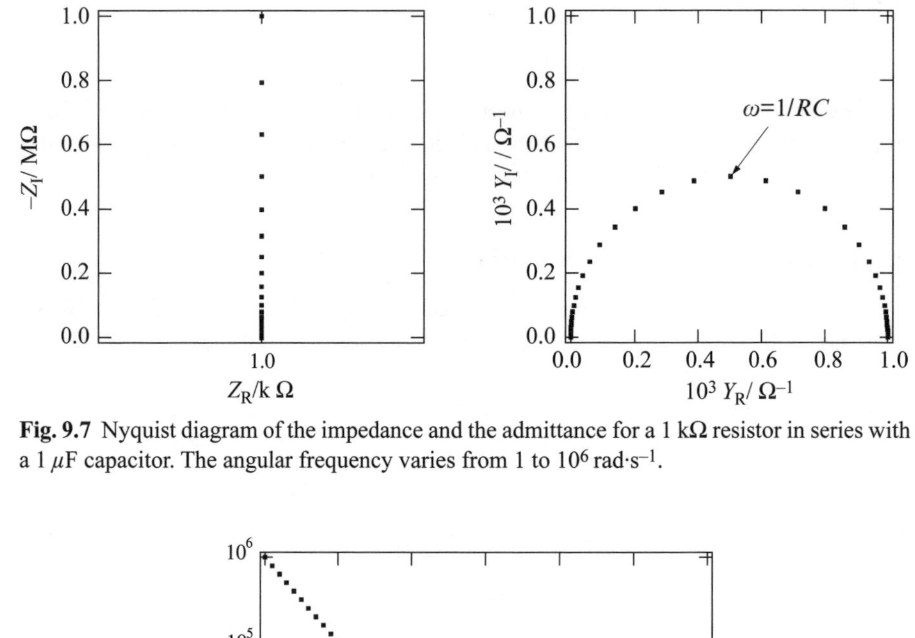

Fig. 9.7 Nyquist diagram of the impedance and the admittance for a 1 kΩ resistor in series with a 1 μF capacitor. The angular frequency varies from 1 to 10^6 rad·s^{-1}.

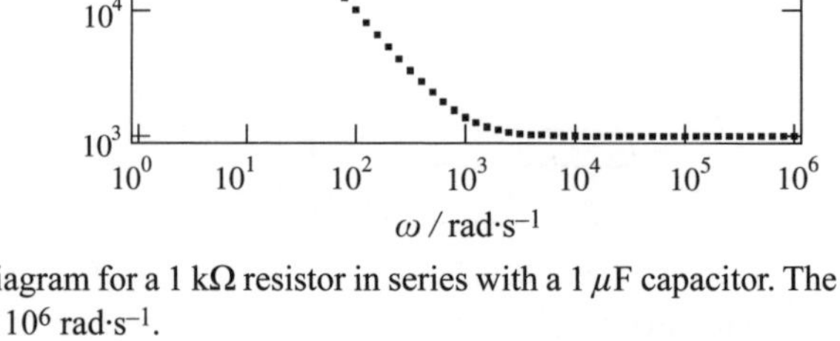

Fig. 9.8 Bode diagram for a 1 kΩ resistor in series with a 1 μF capacitor. The angular frequency varies from 1 to 10^6 rad·s^{-1}.

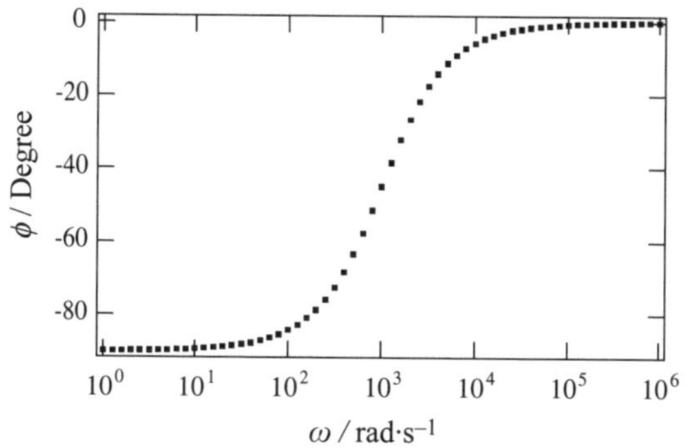

Fig. 9.9 Phase angle variation for a 1 kΩ resistor in series with a 1 μF capacitor. The angular frequency varies from 1 to 10^6 rad·s^{-1}

The Bode diagram can be illustrated schematically by two straight lines that intercept at the angular frequency $1/RC$. At high frequencies, the impedance of the circuit tends to that of a pure resistance as shown in Figure 9.8.

The phase angle diagram of the impedance in the frequency domain shows that the phase varies from −90° at low frequencies, where the influence of the capacitor dominates, to 0° at high frequencies, where the influence of the resistor dominates.

9.2.4 Resistor – capacitor in parallel

The admittance of a circuit of elements in parallel is the sum of the admittances of the individual elements. The current through the circuit is the sum of the currents through each element, and the potential is the same at the terminals of all the elements.

$$Y \quad = \quad \frac{\Delta \bar{I}}{\Delta \bar{E}} \quad = \quad \frac{\Delta \bar{I}_1 + \Delta \bar{I}_2}{\Delta \bar{E}} \quad = \quad Y_1 + Y_2 \tag{9.33}$$

For a resistor and a capacitor in parallel, we have

$$Y(\omega) \quad = \quad Y_R(\omega) + Y_C(\omega) \quad = \quad \frac{1}{R} + \mathrm{j}\omega C \tag{9.34}$$

By taking the inverse, the impedance is

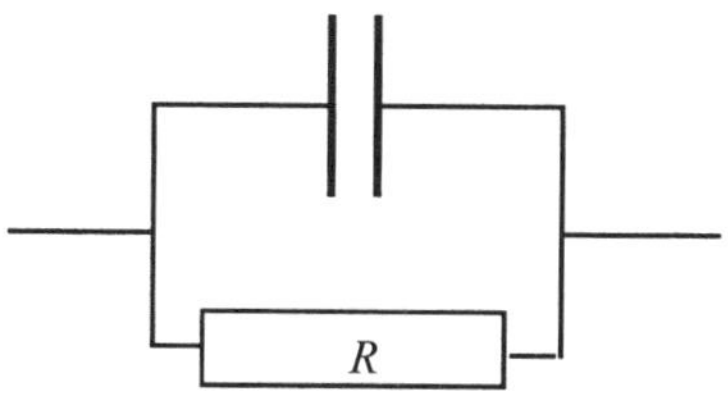

Fig. 9.10 Circuit with a resistor and a capacitor in parallel.

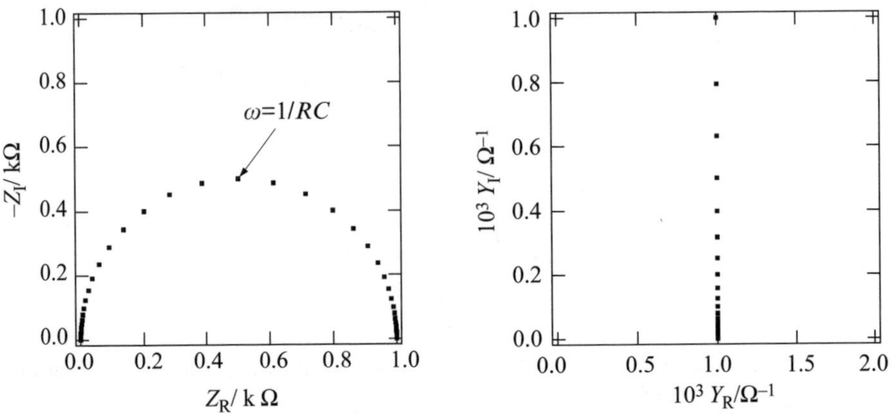

Fig. 9.11 Nyquist diagram of the impedance and admittance for a 1 kΩ resistor in parallel with a 1 μF capacitor. The angular frequency varies from 1 to 10^6 rad·s^{-1}.

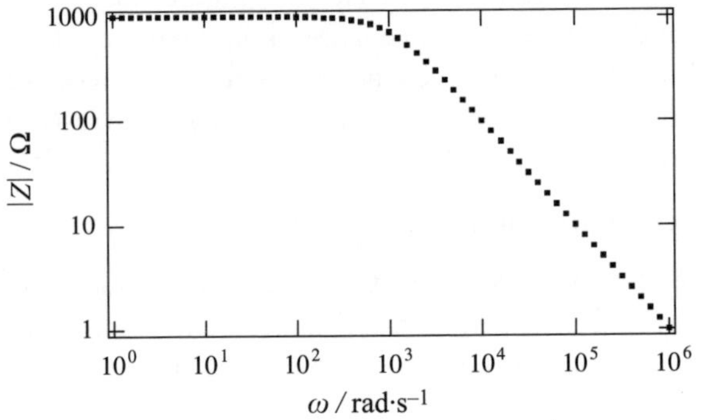

Fig. 9.12 Bode diagram for a 1 kΩ resistor in parallel with a 1 μF capacitor. The angular frequency varies from 1 to 10^6 rad·s^{-1}.

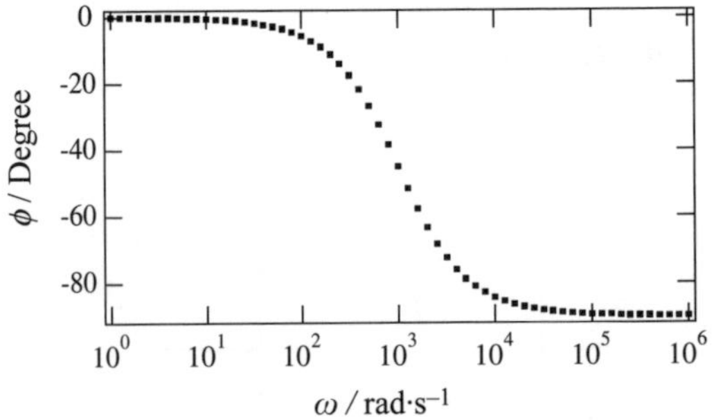

Fig. 9.13 Phase angle variation for a 1 kΩ resistor in parallel with a 1 μF capacitor. The angular frequency varies from 1 to 10^6 rad·s^{-1}.

$$Z(\omega) = \frac{R}{1 + j\omega RC} \tag{9.35}$$

We can see that the impedance of this circuit is a first order transfer function.

The Nyquist diagram of the impedance is therefore a semi-circle whose angular frequency at the apex corresponds to the inverse of the time constant of the circuit, RC. At low frequencies, the impedance tends to the point $Z_R = R$, and at high frequencies it tends to the origin.

Conversely, the Nyquist diagram of the admittance is now a straight line cutting the real axis at $1/R$ when the frequency tends to zero.

The Bode diagram can be drawn with two straight lines that intercept at the angular frequency $1/RC$ as shown in Figure 9.12.

The diagram representing the phase angle of the impedance in the frequency domain shows that the phase varies from 0° at low frequencies where the influence of the resistance dominates to –90° at high frequencies where the influence of the capacitance dominates.

From an electrochemical point of view, this circuit corresponds to a faradaic charge transfer reaction in parallel with the capacitance of the double layer.

9.2.5 Resistor in series with a parallel resistor-capacitor circuit

The impedance of this circuit is the sum of the impedances of the resistance R_1 and the circuit comprising the resistor R_2 and the capacitor in parallel. From equation (9.35), we have

$$Z(\omega) = R_1 + \frac{R_2}{1 + j\omega R_2 C} \tag{9.36}$$

The Nyquist diagram of the impedance is a semi-circle translated on the real axis by R_1. At high frequencies, the impedance tends to the point $Z_R = R_1$ and at low frequencies to the point $Z_R = R_1 + R_2$. The angular frequency at the top of the semi-circle corresponds, as before, to the time constant R_2C.

The Nyquist diagram of the admittance (see Figure 9.16) is now no longer a straight line, but is also a semi-circle with a maximum for an angular frequency $(R_1 + R_2) / R_1R_2C$. At high frequencies, the admittance tends to the point $Y_R = 1/R_1$ and at low frequencies to point $Y_R = 1/(R_1 + R_2)$.

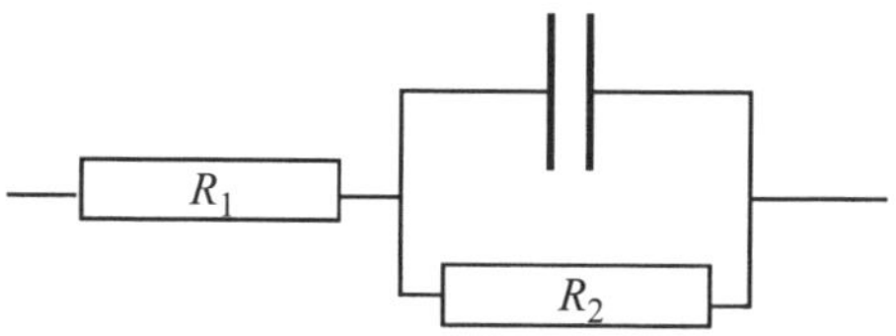

Fig. 9.14 Circuit with a resistance in series with a circuit comprising a resistance in parallel with a capacitance.

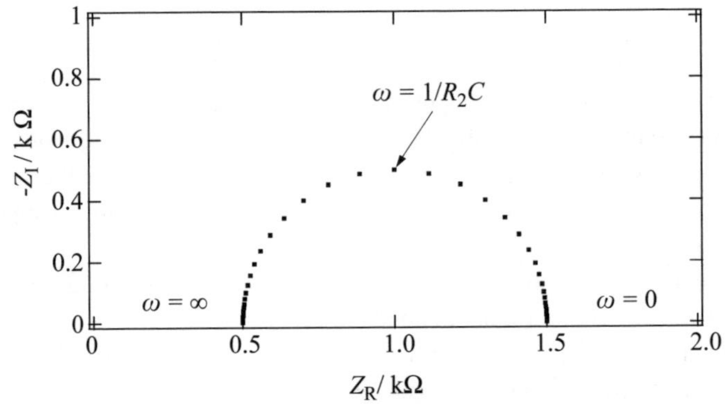

Fig. 9.15 Nyquist diagram of the impedance for a 500 Ω resistor in series with a group comprising a 1 kΩ resistor in parallel with a 1 μF capacitor. The angular frequency varies from 1 to 10^6 rad·s^{-1}.

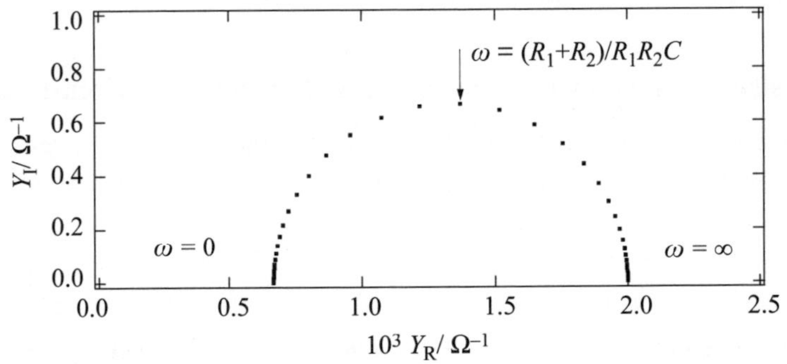

Fig. 9.16 Nyquist diagram of the admittance for a 500 Ω resistance in series with a 1 kΩ resistance in parallel with a 1 μF capacitance. The angular frequency varies from 1 to 10^6 rad·s^{-1}.

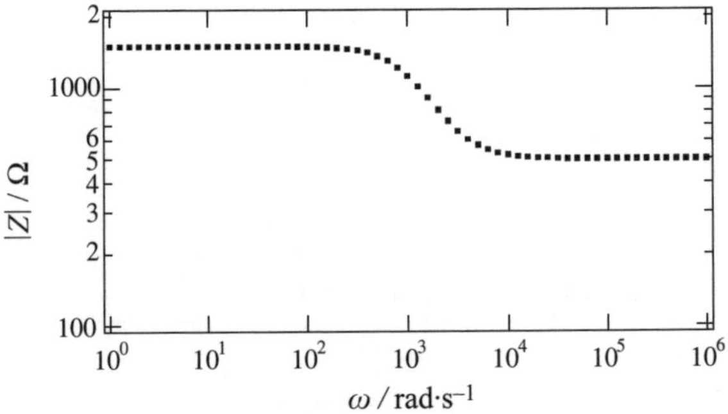

Fig. 9.17 Bode diagram for a 500 Ω resistor in series with a 1 kΩ resistor in parallel with a 1 μF capacitor. The angular frequency varies from 1 to 10^6 rad·s^{-1}.

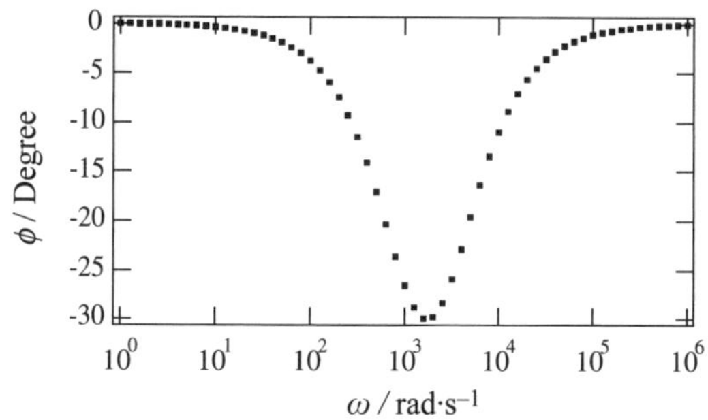

Fig. 9.18 Phase variation for a 500 Ω resistor in series with a 1 kΩ resistor in parallel with a 1 μF capacitor. The angular frequency varies from 1 to 10^6 rad·s^{-1}.

The Bode diagram can be drawn with three straight lines as shown in Figure 9.17. At low frequencies, the system behaves as two resistor in series, whereas at high frequencies it tends towards the behaviour as in Figure 9.8.

The phase angle presents a minimum at an angular frequency between the two characteristic angular velocities $1/R_2C$ and $(R_1 + R_2)/R_1R_2C$ as shown in Figure 9.18.

9.3 IMPEDANCE OF AN ELECTROCHEMICAL SYSTEM

9.3.1 Impedance of a redox reaction

To calculate the transfer function of an electrochemical system, we need to linearise the equation expressing the current as a function of the potential and the interfacial concentrations. To do this, we shall make a limited first order series expansion, thereby making the partial derivatives of the current with respect to these variables to appear

$$\Delta I = \left(\frac{\partial I}{\partial E}\right)\Delta E + \left(\frac{\partial I}{\partial c_{\rm R}(0,t)}\right)\Delta c_{\rm R}(0,t) + \left(\frac{\partial I}{\partial c_{\rm O}(0,t)}\right)\Delta c_{\rm O}(0,t) \tag{9.37}$$

The Laplace transform of this equation gives

$$\Delta \bar{I} = \left(\frac{\partial I}{\partial E}\right)\Delta \bar{E} + \left(\frac{\partial I}{\partial c_{\rm R}(0,t)}\right)\Delta \bar{c}_{\rm R}(0,s) + \left(\frac{\partial I}{\partial c_{\rm O}(0,t)}\right)\Delta \bar{c}_{\rm O}(0,s) \tag{9.38}$$

since the partial derivatives refer to the static situation and do not vary at the frequency $\omega/2\pi$ like the potential, the interfacial concentrations and the current.

The definition (9.19) of the faradaic impedance gives

$$Z_{\rm f} = \frac{\Delta \bar{E}}{\Delta \bar{I}} = \left(\frac{\partial E}{\partial I}\right) - \left(\frac{\partial I}{\partial c_{\rm R}(0,t)}\right)\left(\frac{\partial E}{\partial I}\right)\frac{\Delta \bar{c}_{\rm R}(0,s)}{\Delta \bar{I}} - \left(\frac{\partial I}{\partial c_{\rm O}(0,t)}\right)\left(\frac{\partial E}{\partial I}\right)\frac{\Delta \bar{c}_{\rm O}(0,s)}{\Delta \bar{I}} \tag{9.39}$$

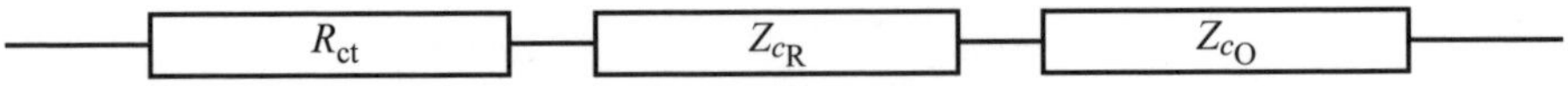

Fig. 9.19 Faradaic impedance of a redox reaction, the sum of the charge transfer resistance and the concentration impedances of the reactants and the reaction products.

We can see that the faradaic impedance is the sum of the three impedances. The first is the resistance of the charge tranfer linked to the kinetics of the charge transfer reaction

$$Z_{R_{tc}} = \left(\frac{\partial E}{\partial I}\right) \tag{9.40}$$

The two other terms of equation (9.39) are the impedances of concentrations linked to the mass transport and the kinetics of the charge transfer reaction

$$Z_{c_R} = -\left(\frac{\partial I}{\partial c_R(0,t)}\right)\left(\frac{\partial E}{\partial I}\right)\frac{\Delta\bar{c}_R(0,s)}{\Delta\bar{I}} \tag{9.41}$$

$$Z_{c_O} = -\left(\frac{\partial I}{\partial c_O(0,t)}\right)\left(\frac{\partial E}{\partial I}\right)\frac{\Delta\bar{c}_O(0,s)}{\Delta\bar{I}} \tag{9.42}$$

9.3.2 Charge transfer impedance

If the kinetics of the redox reaction obeys the Butler-Volmer law (see equation (7.15)), then we have

$$I(t) = nFA\left[k_a^o\, c_R(0,t)\, e^{\alpha\, nF\, E(t)/RT} - k_c^o\, c_O(0,t)\, e^{-(1-\alpha)\, nF\, E(t)/RT}\right] \tag{9.43}$$

By differentiating this equation with respect to the potential, for a steady state value of the potential E, we get

$$R_{ct}^{-1} = \frac{n^2F^2}{RT} A\left[\alpha\, k_a^o\, c_R(0)\, e^{\alpha\, nF\, E/RT} + (1-\alpha)\, k_c^o\, c_O(0)\, e^{-(1-\alpha)\, nF\, E/RT}\right] \tag{9.44}$$

where $c_R(0)$ and $c_O(0)$ are the steady state values of the interfacial concentrations. At the equilibrium potential, this expression reduces to

$$R_{ct}^{-1} = \frac{nFI_o}{RT} \tag{9.45}$$

Thus, the charge transfer impedance is

$$Z_{R_{ct}} = R_{ct} = \frac{RT}{nFI_o} \tag{9.46}$$

We arrive again at the definition of the charge transfer resistance that we have already obtained by linearising the Butler-Volmer equation at the equilibrium potential (see equation (7.23)).

9.3.3 Concentration impedance

The concentration impedances defined by equations (9.41) and (9.42) are the products of three terms; the first being the partial derivative of the current with respect to the interfacial concentration, the second being the charge transfer resistance, and the ratio of the Laplace transform of the interfacial concentrations to the Laplace transform of the current.

The first term is simply obtained by differentiating equation (9.43) with respect to the interfacial concentration of the reduced species

$$\frac{\partial I}{\partial c_{\mathrm{R}}(0,t)} = nF\,A\,k_{\mathrm{a}}^{\mathrm{o}}\,e^{\alpha\,nF\,E/RT} \tag{9.47}$$

and with respect to that of the oxidised species

$$\frac{\partial I}{\partial c_{\mathrm{O}}(0,t)} = -nF\,A\,k_{\mathrm{c}}^{\mathrm{o}}\,e^{-(1-\alpha)\,nF\,E/RT} \tag{9.48}$$

In order to calculate the ratios in the Laplace plane, we need to solve the Fick equations

$$\frac{\partial c_{\mathrm{R}}(x,t)}{\partial t} = D_{\mathrm{R}}\,\frac{\partial^2 c_{\mathrm{R}}(x,t)}{\partial x^2} \tag{9.49}$$

and

$$\frac{\partial c_{\mathrm{O}}(x,t)}{\partial t} = D_{\mathrm{O}}\,\frac{\partial^2 c_{\mathrm{O}}(x,t)}{\partial x^2} \tag{9.50}$$

with the boundary conditions for a semi-infinite linear diffusion

$$\begin{aligned} c_{\mathrm{R}}(x,0) &= c_{\mathrm{R}} \quad \text{and} \quad c_{\mathrm{O}}(x,0) = c_{\mathrm{O}} \\ \lim_{x\to\infty} c_{\mathrm{R}}(x,t) &= c_{\mathrm{R}} \quad \text{and} \quad \lim_{x\to\infty} c_{\mathrm{O}}(x,t) = c_{\mathrm{O}} \end{aligned} \tag{9.51}$$

As we saw several times in Chapter 8, the transforms of the Fick equations are

$$\bar{c}_{\mathrm{R}}(x,s) = \frac{c_{\mathrm{R}}}{s} + A(s)\,e^{-\sqrt{\frac{s}{D_{\mathrm{R}}}}x} \tag{9.52}$$

and

$$\bar{c}_{\mathrm{O}}(x,s) = \frac{c_{\mathrm{O}}}{s} - \xi\,A(s)\,e^{-\sqrt{\frac{s}{D_{\mathrm{O}}}}x} \tag{9.53}$$

taking into account the boundary condition for the conservation of the flux with

$$\xi = \sqrt{\frac{D_{\rm R}}{D_{\rm O}}} \tag{9.54}$$

The current being equal to the diffusion flux of the reactants towards the electrode, the Laplace transform of the current is given by

$$\frac{\bar{I}(s)}{nFA} = D_{\rm R}\left[\frac{\partial \bar{c}_{\rm R}(x,s)}{\partial x}\right]_{x=0} = -A(s)\sqrt{sD_{\rm R}} \tag{9.55}$$

We deduce from this a ratio between the Laplace transforms of the interfacial concentrations and the current

$$\bar{c}_{\rm R}(x,s) = \frac{c_{\rm R}}{s} - \frac{\bar{I}(s)}{nFA\sqrt{sD_{\rm R}}}\, e^{-\sqrt{\frac{s}{D_{\rm R}}}x} \tag{9.56}$$

and

$$\bar{c}_{\rm O}(x,s) = \frac{c_{\rm O}}{s} + \frac{\bar{I}(s)}{nFA\sqrt{sD_{\rm O}}}\, e^{-\sqrt{\frac{s}{D_{\rm O}}}x} \tag{9.57}$$

Thus, the terms we are looking for to calculate the concentration impedances are

$$\frac{\Delta\bar{c}_{\rm R}(0,s)}{\Delta\bar{I}} \approx \frac{{\rm d}\bar{c}_{\rm R}(0,s)}{{\rm d}\bar{I}} = -\frac{1}{nFA\sqrt{sD_{\rm R}}} \tag{9.58}$$

and

$$\frac{\Delta\bar{c}_{\rm O}(0,s)}{\Delta\bar{I}} \approx \frac{{\rm d}\bar{c}_{\rm O}(0,s)}{{\rm d}\bar{I}} = \frac{1}{nFA\sqrt{sD_{\rm O}}} \tag{9.59}$$

By replacing the Laplace variable by jω to access the frequency domain, we have

$$Z_{c_{\rm R}} = \frac{k_{\rm a}^{\rm o}\, e^{\alpha\, nF\, E/RT}}{\sqrt{{\rm j}\omega D_{\rm R}}} R_{\rm ct} \tag{9.60}$$

and

$$Z_{c_{\rm O}} = \frac{k_{\rm c}^{\rm o}\, e^{-(1-\alpha)\, nF\, E/RT}}{\sqrt{{\rm j}\omega D_{\rm O}}} R_{\rm ct} \tag{9.61}$$

Equations (9.60) and (9.61) tell us that the concentration impedances are proportional to the inverse of the square root of the angular frequency.

It is interesting to analyse the concentration profiles as a function of time using digital simulation methods, expressed for example in terms of dephasing (e.g. ωt, $\omega t + \pi/4$, $\omega t + \pi/2$, etc.). At each point of the solution close to the electrode, the concentration varies at the frequency of the potential perturbation and therefore at the same frequency as the interfacial concentrations.

However, the amplitude varies with the distance to the electrode. It is interesting to notice that in Figure 9.20 the thickness of the diffusion layer varies with the frequency. The higher the frequency, the more compact is the diffusion layer, of the

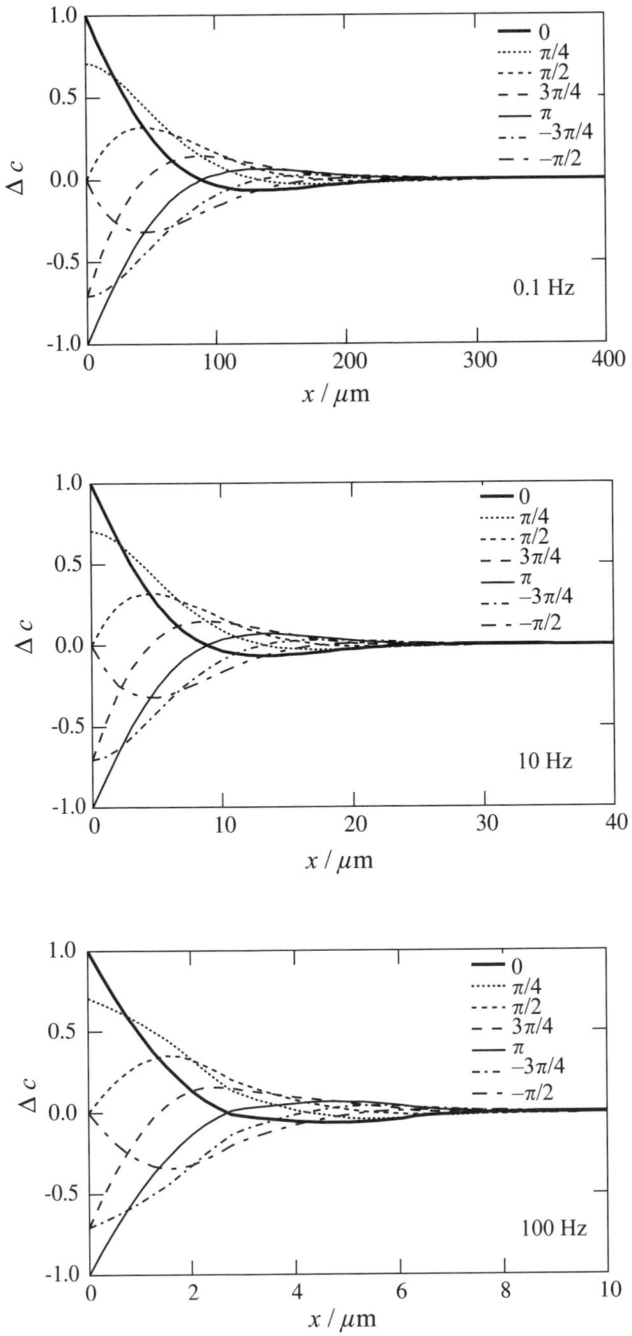

Fig. 9.20 Normalised profiles of the concentration perturbation of the reacting species for different phase angles, $D = 10^{-9}$ $m^2 \cdot s^{-1}$ [Rosaria Ferrigno,EPFL].

order of a few micrometers for a frequency of 100Hz. When the frequency decreases, the diffusion layer penetrates deeper into the solution and convection then risks perturbing the sinusoidal variation of the concentration profiles.

9.3.4 Warburg impedance

It is usual in electrochemistry to group the two concentration impedances into a single impedance, called the ***Warburg impedance***, symbolised by the letter W.

Fig. 9.21 Schematic representation of the Warburg impedance.

$$Z_W = \frac{R_{ct}}{\sqrt{j\omega}}\left[\frac{k_a^o\, e^{\alpha nF E/RT}}{\sqrt{D_R}} + \frac{k_c^o\, e^{-(1-\alpha) nF E/RT}}{\sqrt{D_O}}\right] \quad (9.62)$$

At the equilibrium potential, by introducing the exchange current I_o defined by equation (7.21), this expression reduces to

$$Z_W = \frac{R_{ct} I_o}{nFA\sqrt{j\omega}}\left[\frac{1}{c_R\sqrt{D_R}} + \frac{1}{c_O\sqrt{D_O}}\right] \quad (9.63)$$

By substituting the expression for the charge transfer (9.46), we have

$$Z_W = \frac{RT}{n^2F^2A\sqrt{j\omega}}\left[\frac{1}{c_R\sqrt{D_R}} + \frac{1}{c_O\sqrt{D_O}}\right] \quad (9.64)$$

Also, knowing that

$$\sqrt{j} = \frac{1}{\sqrt{2}} + \frac{j}{\sqrt{2}} \quad (9.65)$$

equation (9.64) becomes

$$Z_W = \frac{\sigma(1-j)}{\sqrt{\omega}} \quad (9.66)$$

with

$$\sigma = \frac{RT}{n^2F^2A\sqrt{2}}\left[\frac{1}{c_R\sqrt{D_R}} + \frac{1}{c_O\sqrt{D_O}}\right] \quad (9.67)$$

The Nyquist diagram of the impedance is a straight line with a slope of unity as shown in Figure 9.22 and its Bode diagram is a straight line with a slope of –1/2 as shown in Figure 9.23.

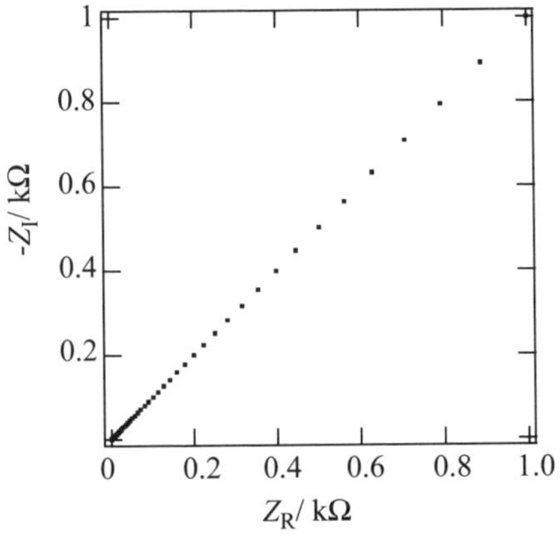

Fig. 9.22 Nyquist diagram for a Warburg impedance with $\sigma = 1000\ \Omega\cdot\mathrm{rad}^{1/2}\cdot\mathrm{s}^{-1/2}$. The angular frequency varies from 1 to 10^6 rad·s^{-1}.

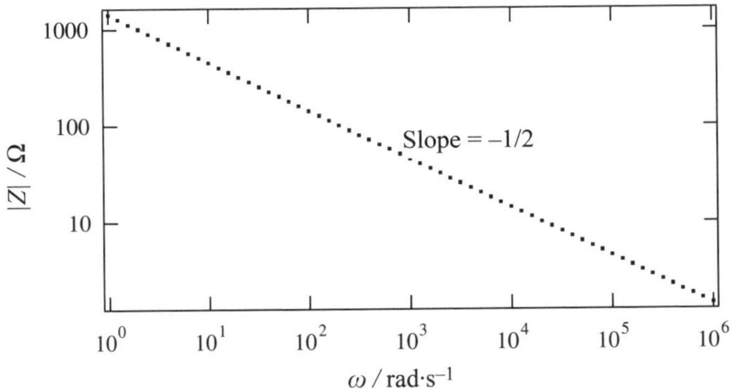

Fig. 9.23 Bode diagram for a Warburg impedance with $\sigma = 1000\ \Omega\cdot\mathrm{rad}^{1/2}\cdot\mathrm{s}^{-1/2}$. The angular frequency varies from 1 to 10^6 rad·s^{-1}.

The phase angle of the Warburg impedance is constant at –45° as shown in Figure 9.24. We get this by applying equation (9.17) since $\arctan(-1) = -\pi/4$.

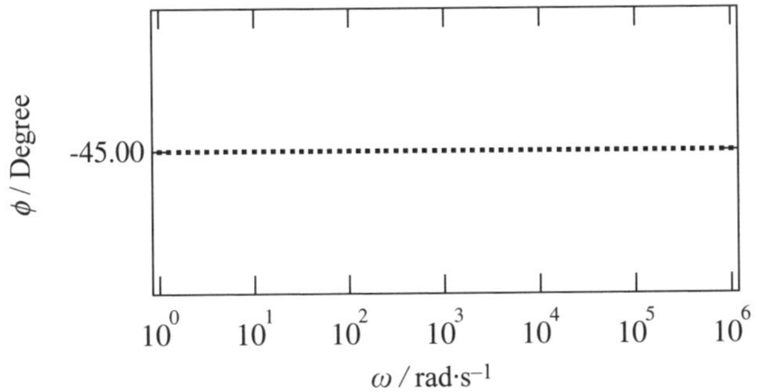

Fig. 9.24 Phase representation for a Warburg impedance with $\sigma = 1000\ \Omega\cdot\mathrm{rad}^{1/2}\cdot\mathrm{s}^{-1/2}$. The angular frequency varies from 1 to 10^6 rad·s^{-1}.

9.3.5 The Randles-Ershler circuit

During a redox reaction at an electrode, the current is in fact the sum of the faradaic current linked to the electron transfer and a capacitive current linked to the double layer charge

$$I = I_f + I_c \tag{9.68}$$

Thus, we have two elements in parallel forming what is commonly known as the ***Randles-Ershler circuit***.

The admittance of this circuit is the sum of the admittances of the elements in parallel

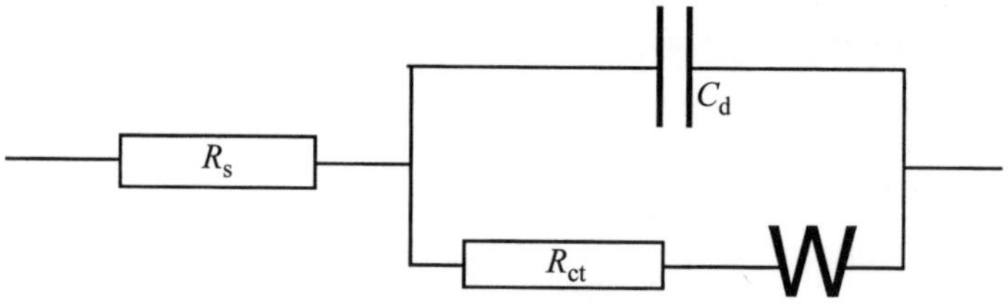

Fig. 9.25 The Randles-Ershler circuit in series with the solution resistance.

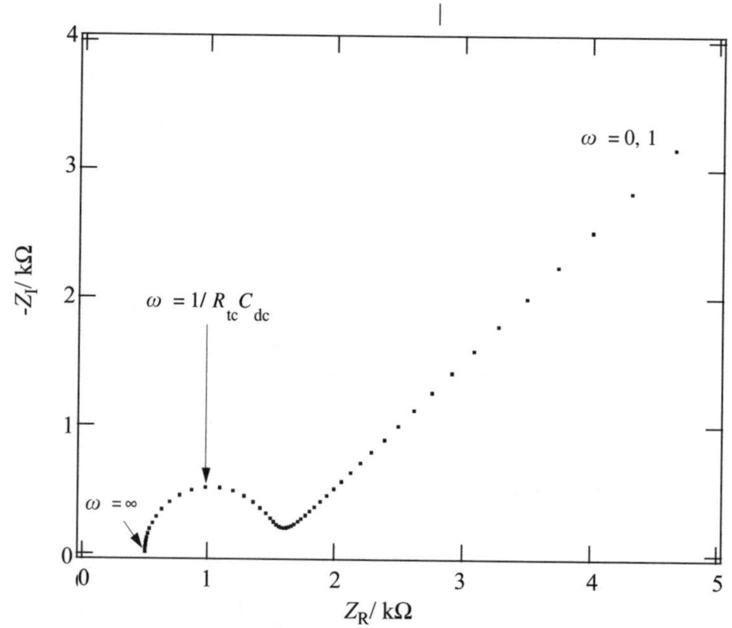

Fig. 9.26 Nyquist diagram of the impedance for a Randles-Ershler circuit with the following parameters: solution resistance: 500 Ω, charge transfer resistance: 1000 Ω, double layer capacitance: 1 μF, Warburg factor: $\sigma = 1000\ \Omega\cdot\text{rad}^{1/2}\cdot\text{s}^{-1/2}$, the angular frequency varies from 0.1 to 10^5 rad·s^{-1}.

$$Y = Y_{C_d} + Y_W = j\omega C_d + \frac{1}{R_{ct} + \frac{\sigma}{\sqrt{\omega}}(1-j)} \tag{9.69}$$

and, therefore, the impedance is the inverse of this expression.

In addition to this impedance, it is often useful to add a resistance R_s that represents the ohmic drop between the outside of the double layer and the reference electrode.

The Nyquist diagram of the Randles-Ershler circuit is a semi-circle due to the charge transfer resistance R_{ct} and the capacitance of the double layer. The top of the semi-circle corresponds to the inverse of the time constant $R_{ct}C_d$ when this constant is sufficiently small. At low frequencies, the Warburg impedance is predominant, and the Nyquist diagram becomes linear with a slope of unity as shown in Figure 9.26.

The Bode diagram, shown in Figure 9.27, comprises two linear parts with slopes of –1/2 and –1, corresponding to low frequencies where the Warburg impedance dominates, and to high frequencies where the circuit R_{ct} and C_d in parallel dominates. At very high frequencies, the slope tends to zero, when the resistance of the solution is the principal element.

These zones of predominance of the different elements, of course, resurface in the phase diagram shown in Figure 9.28. At low frequencies, the phase angle tends to –45°, and at higher frequencies we find the characteristic form of an *RC* circuit.

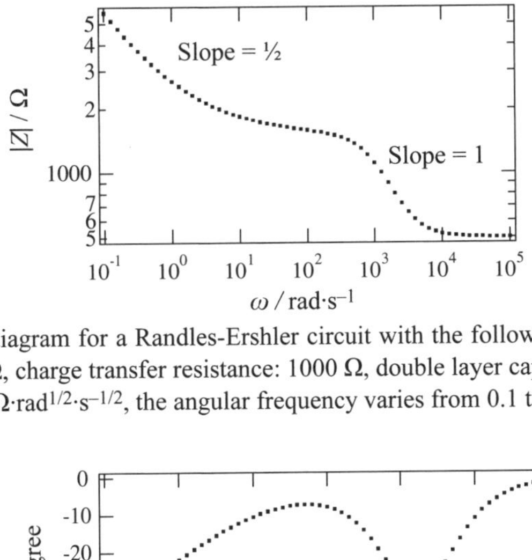

Fig. 9.27 Bode diagram for a Randles-Ershler circuit with the following parameters: solution resistance: 500 Ω, charge transfer resistance: 1000 Ω, double layer capacitance: 1 μF, Warburg factor σ = 1000 Ω·rad$^{1/2}$·s$^{-1/2}$, the angular frequency varies from 0.1 to 10^5 rad·s^{-1}.

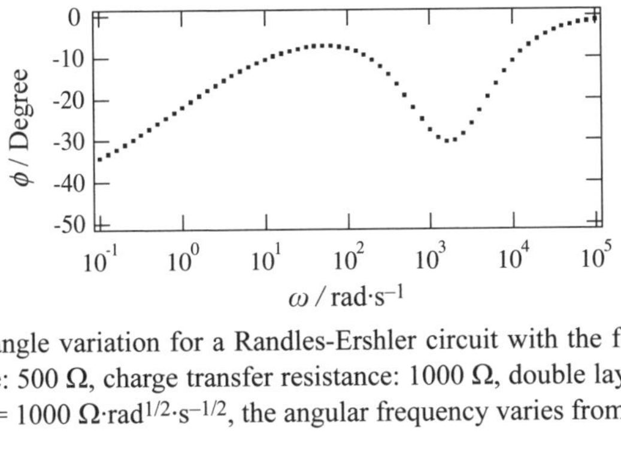

Fig. 9.28 Phase angle variation for a Randles-Ershler circuit with the following parameters: solution resistance: 500 Ω, charge transfer resistance: 1000 Ω, double layer capacitance: 1 μF, Warburg factor σ = 1000 Ω·rad$^{1/2}$·s$^{-1/2}$, the angular frequency varies from 0.1 to 10^5 rad·s^{-1}.

From an experimental point of view, the range of frequencies available is limited at high frequencies by the potentiostat, and in a more general way by the experimental setup; and at low frequencies by the duration of the acquisition time. At the level of determining an equivalent circuit, it is important to be circumspect with multi-parameter optimisation softwares. It is often useful to elucidate an equivalent circuit by successive eliminations. Firstly, it is possible in general to determine the ohmic drop and the capacitance of the double layer reasonably accurately. Thus, these two elements can be conveniently subtracted from the experimental values of the impedance, and then the optimisation software can be used directly on the remaining electrode response.

Electrochemical impedance measurements are the most reliable way of measuring the kinetics of redox reactions. The effects of the ohmic drop due to the resistance of the solution can be distinguished from the measurement of the charge transfer kinetics. However, for fast reactions, it remains sometimes difficult to separate the Warburg impedance, due to the semi-infinite diffusion, from the charge transfer resistance.

9.3.6 Concentration impedance for a diffusion layer of a fixed thickness

When the thickness δ of the diffusion layer is fixed, either hydrodynamically (e.g. using a rotating electrode), or statically (e.g. using a membrane-covered electrode), the concentration impedances can be calculated by resolving the Fick equations. Here, we shall look at systems with the diffusion layer fixed, using equations (9.49) and (9.50).

The boundary conditions for the concentration are

$$c_{\mathrm{R}}(x,0) = c_{\mathrm{R}} \quad \text{and} \quad c_{\mathrm{O}}(x,0) = c_{\mathrm{O}} \tag{9.70}$$

$$c_{\mathrm{R}}(\delta,t) = c_{\mathrm{R}} \quad \text{and} \quad c_{\mathrm{O}}(\delta,t) = c_{\mathrm{O}} \tag{9.71}$$

For the reduced species, the transform of the solution of the Fick equation is given by equation (8.11)

$$\bar{c}_{\mathrm{R}}(x,s) = \frac{c_{\mathrm{R}}}{s} + A(s)\,e^{-\sqrt{\frac{s}{D_{\mathrm{R}}}}x} + B(s)\,e^{\sqrt{\frac{s}{D_{\mathrm{R}}}}x} \tag{9.72}$$

On either side of the diffusion layer, we then have

$$\bar{c}_{\mathrm{R}}(0,s) = \frac{c_{\mathrm{R}}}{s} + A(s) + B(s) \tag{9.73}$$

and

$$\bar{c}_{\mathrm{R}}(\delta,s) = \frac{c_{\mathrm{R}}}{s} = \frac{c_{\mathrm{R}}}{s} + A(s)\,e^{-\sqrt{\frac{s}{D_{\mathrm{R}}}}\delta} + B(s)\,e^{\sqrt{\frac{s}{D_{\mathrm{R}}}}\delta} \tag{9.74}$$

The constants of equation (9.72) are therefore

$$A(s) = \frac{\bar{c}_R(0,s) - \frac{c_R}{s}}{e^{\sqrt{\frac{s}{D_R}}\delta} - e^{-\sqrt{\frac{s}{D_R}}\delta}} e^{\sqrt{\frac{s}{D_R}}\delta} \tag{9.75}$$

and

$$B(s) = -\frac{\bar{c}_R(0,s) - \frac{c_R}{s}}{e^{\sqrt{\frac{s}{D_R}}\delta} - e^{-\sqrt{\frac{s}{D_R}}\delta}} e^{-\sqrt{\frac{s}{D_R}}\delta} \tag{9.76}$$

The current is then given by

$$\frac{\bar{I}(s)}{nFA} = D_R\left[\frac{\partial \bar{c}_R(x,s)}{\partial x}\right]_{x=0} = [B(s) - A(s)]\sqrt{sD_R} \tag{9.77}$$

or again

$$\begin{aligned} \frac{\bar{I}(s)}{nFA} &= -\sqrt{sD_R}\left[\bar{c}_R(0,s) - \frac{c_R}{s}\right]\frac{e^{-\sqrt{\frac{s}{D_R}}\delta} + e^{\sqrt{\frac{s}{D_R}}\delta}}{e^{\sqrt{\frac{s}{D_R}}\delta} - e^{-\sqrt{\frac{s}{D_R}}\delta}} \\ &= -\sqrt{sD_R}\left[\bar{c}_R(0,s) - \frac{c_R}{s}\right]\coth\sqrt{\frac{s}{D_R}}\delta \end{aligned} \tag{9.78}$$

From this, we deduce an expression between the Laplace transforms of the interfacial concentration of the reduced species and that of the current

$$\bar{c}_R(x,s) = \frac{c_R}{s} + \frac{\bar{I}(s)}{nFA\sqrt{sD_R}}\left[\frac{e^{-\sqrt{\frac{s}{D_R}}\delta} e^{\sqrt{\frac{s}{D_R}}x} - e^{\sqrt{\frac{s}{D_R}}\delta} e^{-\sqrt{\frac{s}{D_R}}x}}{e^{\sqrt{\frac{s}{D_R}}\delta} + e^{-\sqrt{\frac{s}{D_R}}\delta}}\right] \tag{9.79}$$

So, the terms, we are looking for to calculate the concentration impedances, are

$$\frac{\Delta\bar{c}_R(0,s)}{\Delta\bar{I}} \approx \frac{d\bar{c}_R(0,s)}{d\bar{I}} = \frac{1}{nFA\sqrt{sD_R}}\left[\frac{e^{-\sqrt{\frac{s}{D_R}}\delta} - e^{\sqrt{\frac{s}{D_R}}\delta}}{e^{\sqrt{\frac{s}{D_R}}\delta} + e^{-\sqrt{\frac{s}{D_R}}\delta}}\right] = -\frac{\tanh\sqrt{\frac{s}{D_R}}\delta}{nFA\sqrt{sD_R}} \tag{9.80}$$

and in the same way

$$\frac{\Delta \bar{c}_{\mathrm{O}}(0,s)}{\Delta \bar{I}} \approx \frac{\mathrm{d}\bar{c}_{\mathrm{O}}(0,s)}{\mathrm{d}\bar{I}} = \frac{\tanh\sqrt{\dfrac{s}{D_{\mathrm{O}}}}\delta}{nFA\sqrt{sD_{\mathrm{O}}}} \tag{9.81}$$

By replacing the Laplace variable by $\mathrm{j}\omega$ to access the frequency domain, we have

$$Z_{c_{\mathrm{R}}} = \frac{k_{\mathrm{a}}^{\mathrm{o}}\, e^{\alpha\, nF\, E/RT} \tanh\sqrt{\dfrac{\mathrm{j}\omega}{D_{\mathrm{R}}}}\delta_{\mathrm{R}}}{\sqrt{\mathrm{j}\omega D_{\mathrm{R}}}} R_{\mathrm{ct}} \tag{9.82}$$

and

$$Z_{c_{\mathrm{O}}} = \frac{k_{\mathrm{c}}^{\mathrm{o}}\, e^{-(1-\alpha)\, nF\, E/RT} \tanh\sqrt{\dfrac{\mathrm{j}\omega}{D_{\mathrm{O}}}}\delta_{\mathrm{O}}}{\sqrt{\mathrm{j}\omega D_{\mathrm{O}}}} R_{\mathrm{ct}} \tag{9.83}$$

In these two equations, we put an index (O or R) on the thickness δ of the diffusion layer, since it can depend on the species itself, e.g. for a rotating electrode (see equation (7.38)). When ω or δ tend to infinity, equations (9.82) and (9.83) tend to the values of equations (9.60) and (9.61) respectively.

As before, we can group these two expressions together and obtain the equivalent of equation (9.64), i.e.

$$Z_{\delta} = \frac{RT}{n^2F^2A\sqrt{\mathrm{j}\omega}}\left[\frac{\tanh\sqrt{\dfrac{\mathrm{j}\omega}{D_{\mathrm{R}}}}\delta_{\mathrm{R}}}{c_{\mathrm{R}}\sqrt{D_{\mathrm{R}}}} + \frac{\tanh\sqrt{\dfrac{\mathrm{j}\omega}{D_{\mathrm{O}}}}\delta_{\mathrm{O}}}{c_{\mathrm{O}}\sqrt{D_{\mathrm{O}}}}\right] \tag{9.84}$$

Knowing that for a complex number $z = a + \mathrm{j}b$

$$\tanh z = \frac{\sinh a \cosh a + \mathrm{j} \sin b \cos b}{\sinh^2 a + \cos^2 b} \tag{9.85}$$

and using equation (9.65), we get

$$\tanh\sqrt{\frac{\mathrm{j}\omega}{D}}\delta = \tanh(1+\mathrm{j})\sqrt{\frac{\omega}{2D}}\delta = \frac{\sinh\sqrt{\dfrac{\omega}{2D}}\delta \cosh\sqrt{\dfrac{\omega}{2D}}\delta + \mathrm{j}\sin\sqrt{\dfrac{\omega}{2D}}\delta \cos\sqrt{\dfrac{\omega}{2D}}\delta}{\sinh^2\sqrt{\dfrac{\omega}{2D}}\delta + \cos^2\sqrt{\dfrac{\omega}{2D}}\delta} \tag{9.86}$$

Taking the diffusion coefficients of the two species as equal, equation (9.84) reduces to

$$Z_{\delta} = \frac{\sigma(1-\mathrm{j})}{\sqrt{\omega}}\left[\frac{\sinh\sqrt{\dfrac{\omega}{2D}}\delta \cosh\sqrt{\dfrac{\omega}{2D}}\delta + \mathrm{j}\sin\sqrt{\dfrac{\omega}{2D}}\delta \cos\sqrt{\dfrac{\omega}{2D}}\delta}{\sinh^2\sqrt{\dfrac{\omega}{2D}}\delta + \cos^2\sqrt{\dfrac{\omega}{2D}}\delta}\right] \tag{9.87}$$

where σ is the Warburg factor defined by equation (9.67).

Figure 9.29 shows the Nyquist diagram of the impedances for a fixed diffusion layer thickness, the form of this curve being a quarter lemniscate. Its main characteristic is that contrary to semi-infinite linear diffusion, the imaginary part becomes zero at low frequencies, whilst the behaviour at high frequencies is not perturbed by the limitation of the diffusion layer thickness.

Figure 9.30 shows the variation of the modulus of the impedance as a function of the frequency. Again, the resistive behaviour at low frequencies is easily observable by the horizontal straight line and the value of the resistance can be calulated from the boundary condition of equation (9.87). A simple calculation shows that this resistance is $\sigma\delta\sqrt{2/D}$, and it is clear that this resistance is directly proportional to δ.

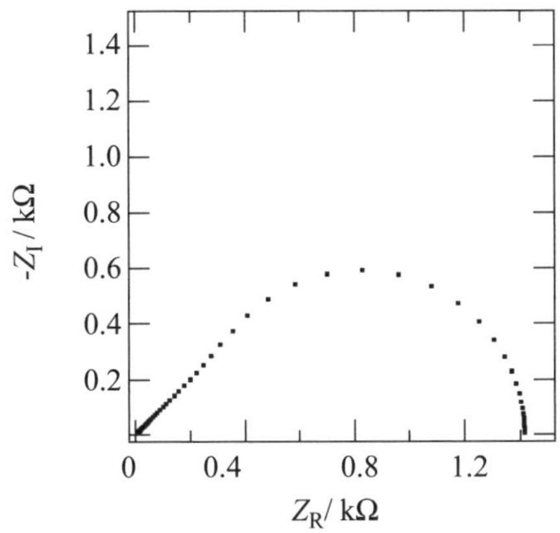

Fig. 9.29 Nyquist diagram for a concentration impedance for a fixed diffusion layer thickness with parameters $\sigma = 1000\ \Omega\cdot\text{rad}^{1/2}\cdot\text{s}^{-1/2}$ and $\delta^2/D = 1\text{s}$. The angular frequency varies from 0.01 to 10^4 rad·s^{-1}.

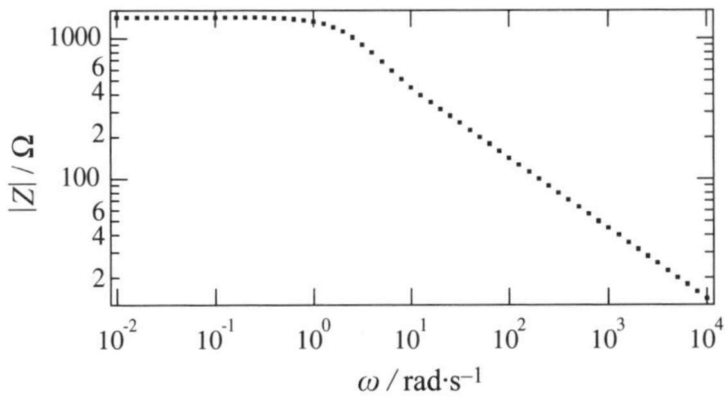

Fig. 9.30 Bode diagram for a concentration impedance for a fixed diffusion thickness with parameters $\sigma = 1000\ \Omega\cdot\text{rad}^{1/2}\cdot\text{s}^{-1/2}$ and $\delta^2/D = 1\text{s}$. The angular frequency varies from 0.01 to 10^4 rad·s^{-1}.

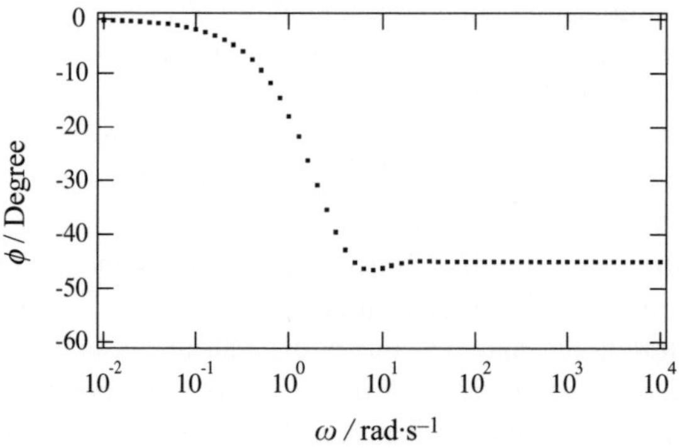

Fig. 9.31 Phase angle variation for a concentration impedance for a fixed diffusion thickness with parameters $\sigma = 1000\ \Omega\cdot\text{rad}^{1/2}\cdot\text{s}^{-1/2}$ and $\delta^2/D = 1\text{s}$. The angular frequency varies from 0.01 to 10^4 rad·s^{-1}.

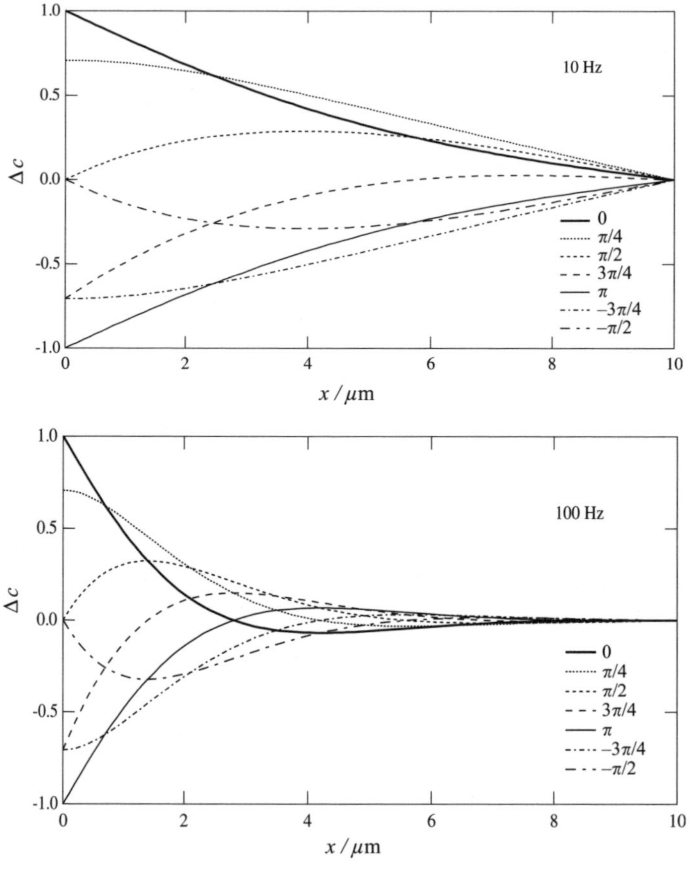

Fig. 9.32 Normalised profiles of the concentration perturbation of the reacting species for different phase angles for a fixed thickness diffusion layer of $\delta = 10\ \mu$m and $D = 10^{-9}$ m^2·s^{-1} [Rosaria Ferrigno, EPFL].

Figure 9.31 shows the phase angle variation as a function of the frequency. The resistive character is clearly recognisable at low frequencies by the zero dephasing.

As with semi-infinite linear diffusion, it is interesting to observe the shape of the concentration profiles for a system with a fixed thickness diffusion layer as shown in Figure 9.32.

At high frequencies, the limit of the thickness of the diffusion layer, which can be fixed experimentally e.g. using a rotating electrode, has no effect on the sinusoidal variations of the concentration profiles, whilst at lower frequencies, the imposition of a diffusion layer fixes the concentration at a given distance from the electrode. Figure 9.32 shows this effect very clearly for a diffusion layer thickness of 10 micrometers at a frequency of 10Hz.

9.3.7 Concentration impedance for diffusion in a thin layer cell

For an electrode covered with a thin layer of solution and an inert wall, the boundary conditions for the Fick equation are

$$c_{\mathrm{R}}(x,0) \;=\; c_{\mathrm{R}} \quad \text{and} \quad c_{\mathrm{O}}(x,0) \;=\; c_{\mathrm{O}} \tag{9.88}$$

$$\frac{\partial c_{\mathrm{R}}(\delta,t)}{\partial x} \;=\; 0 \quad \text{and} \quad \frac{\partial c_{\mathrm{O}}(\delta,t)}{\partial x} \;=\; 0 \tag{9.89}$$

For the reduced species, the transform of the solution of the Fick equation is again given by equation (9.72). The transforms of the conditions at the terminals are

$$\bar{c}_{\mathrm{R}}(0,s) \;=\; \frac{c_{\mathrm{R}}}{s} \;+\; A(s) \;+\; B(s) \tag{9.90}$$

and

$$0 \;=\; -\; A(s)\sqrt{\frac{s}{D_{\mathrm{R}}}}\,e^{-\sqrt{\frac{s}{D_{\mathrm{R}}}}\delta} \;+\; B(s)\sqrt{\frac{s}{D_{\mathrm{R}}}}\,e^{\sqrt{\frac{s}{D_{\mathrm{R}}}}\delta} \tag{9.91}$$

The constants of equation (9.72) are therefore

$$A(s) \;=\; \frac{\bar{c}_{\mathrm{R}}(0,s)-\dfrac{c_{\mathrm{R}}}{s}}{e^{\sqrt{\frac{s}{D_{\mathrm{R}}}}\delta}+e^{-\sqrt{\frac{s}{D_{\mathrm{R}}}}\delta}}\,e^{\sqrt{\frac{s}{D_{\mathrm{R}}}}\delta} \tag{9.92}$$

and

$$B(s) \;=\; \frac{\bar{c}_{\mathrm{R}}(0,s)-\dfrac{c_{\mathrm{R}}}{s}}{e^{\sqrt{\frac{s}{D_{\mathrm{R}}}}\delta}+e^{-\sqrt{\frac{s}{D_{\mathrm{R}}}}\delta}}\,e^{-\sqrt{\frac{s}{D_{\mathrm{R}}}}\delta} \tag{9.93}$$

Thus, the current is again given by equation (9.77), i.e.

$$\frac{\bar{I}(s)}{nFA} = -\sqrt{sD_R}\left[\bar{c}_R(0,s)-\frac{c_R}{s}\right]\frac{e^{\sqrt{\frac{s}{D_R}}\delta}-e^{-\sqrt{\frac{s}{D_R}}\delta}}{e^{\sqrt{\frac{s}{D_R}}\delta}+e^{-\sqrt{\frac{s}{D_R}}\delta}}$$
$$= -\sqrt{sD_R}\left[\bar{c}_R(0,s)-\frac{c_R}{s}\right]\tanh\sqrt{\frac{s}{D_R}}\delta \tag{9.94}$$

As before, we deduce from this an expression between the Laplace transforms of the interfacial concentration of the reduced species and of the current

$$\bar{c}_R(x,s) = \frac{c_R}{s} - \frac{\bar{I}(s)}{nFA\sqrt{sD_R}}\left[\frac{e^{\sqrt{\frac{s}{D_R}}\delta}\, e^{-\sqrt{\frac{s}{D_R}}x}+e^{-\sqrt{\frac{s}{D_R}}\delta}\, e^{\sqrt{\frac{s}{D_R}}x}}{e^{\sqrt{\frac{s}{D_R}}\delta}-e^{-\sqrt{\frac{s}{D_R}}\delta}}\right] \tag{9.95}$$

So the terms we are looking for to calculate the concentration impedances are

$$\frac{\Delta\bar{c}_R(0,t)}{\Delta\bar{I}} \approx \frac{\mathrm{d}\bar{c}_R(0,t)}{\mathrm{d}\bar{I}} = -\frac{1}{nFA\sqrt{sD_R}}\left[\frac{e^{\sqrt{\frac{s}{D_R}}\delta}+e^{-\sqrt{\frac{s}{D_R}}\delta}}{e^{\sqrt{\frac{s}{D_R}}\delta}-e^{-\sqrt{\frac{s}{D_R}}\delta}}\right] = -\frac{\coth\sqrt{\frac{s}{D_R}}\delta}{nFA\sqrt{sD_R}} \tag{9.96}$$

and in the same way

$$\frac{\Delta\bar{c}_O(0,t)}{\Delta\bar{I}} \approx \frac{\mathrm{d}\bar{c}_O(0,t)}{\mathrm{d}\bar{I}} = \frac{\coth\sqrt{\frac{s}{D_O}}\delta}{nFA\sqrt{sD_O}} \tag{9.97}$$

By replacing the Laplace variable by $\mathrm{j}\omega$ to access the frequency domain, we have

$$Z_{c_R} = \frac{k_a^o\, e^{\alpha nF\, E/RT}\coth\sqrt{\frac{\mathrm{j}\omega}{D_R}}\delta}{\sqrt{\mathrm{j}\omega D_R}}R_{ct} \tag{9.98}$$

and

$$Z_{c_O} = \frac{k_c^o\, e^{-(1-\alpha)\, nF\, E/RT}\coth\sqrt{\frac{\mathrm{j}\omega}{D_O}}\delta}{\sqrt{\mathrm{j}\omega D_O}}R_{ct} \tag{9.99}$$

As before, we can group these two expressions together to obtain an expression equivalent to equation (9.64). So, the impedance of the thin layer Z_{thin} is

$$Z_{\text{thin}} = \frac{RT}{n^2F^2A\sqrt{j\omega}}\left[\frac{\coth\sqrt{\frac{j\omega}{D_R}}\delta}{c_R\sqrt{D_R}} + \frac{\coth\sqrt{\frac{j\omega}{D_O}}\delta}{c_O\sqrt{D_O}}\right] \tag{9.100}$$

Knowing that for a complex number $z = a + j\,b$

$$\tanh z = \frac{\sinh a \cosh a - j \sin b \cos b}{\sinh^2 a + \sin^2 b} \tag{9.101}$$

and by using equation (9.65), we get

$$\coth\sqrt{\frac{j\omega}{D}}\delta = \coth(1+j)\sqrt{\frac{\omega}{2D}}\delta = \frac{\sinh\sqrt{\frac{\omega}{2D}}\delta \cosh\sqrt{\frac{\omega}{2D}}\delta - j\sin\sqrt{\frac{\omega}{2D}}\delta \cos\sqrt{\frac{\omega}{2D}}\delta}{\sinh^2\sqrt{\frac{\omega}{2D}}\delta + \sin^2\sqrt{\frac{\omega}{2D}}\delta} \tag{9.102}$$

Taking the diffusion coefficients of the species to be equal, equation (9.84) reduces to

$$Z_{\text{thin}} = \frac{\sigma(1-j)}{\sqrt{\omega}}\left[\frac{\sinh\sqrt{\frac{\omega}{2D}}\delta \cosh\sqrt{\frac{\omega}{2D}}\delta - j\sin\sqrt{\frac{\omega}{2D}}\delta \cos\sqrt{\frac{\omega}{2D}}\delta}{\sinh^2\sqrt{\frac{\omega}{2D}}\delta + \sin^2\sqrt{\frac{\omega}{2D}}\delta}\right] \tag{9.103}$$

where σ is the Warburg factor defined by equation (9.67).

The Nyquist diagram of the impedance shows a characteristic capacitive behaviour at low frequencies as shown in Figure 9.33.

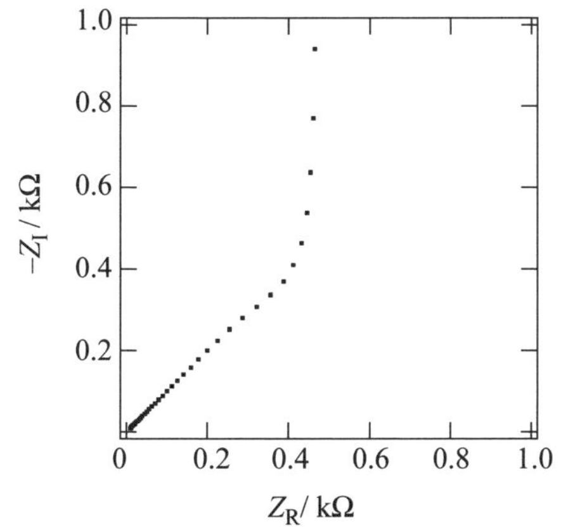

Fig. 9.33 Nyquist diagram of the concentration impedance for diffusion in a thin layer cell with $\sigma = 1000\ \Omega\cdot\text{rad}^{1/2}\cdot\text{s}^{-1/2}$ and $\delta^2/D = 1$s. The angular frequency varies from 0.01 to 10^4 rad·s^{-1}.

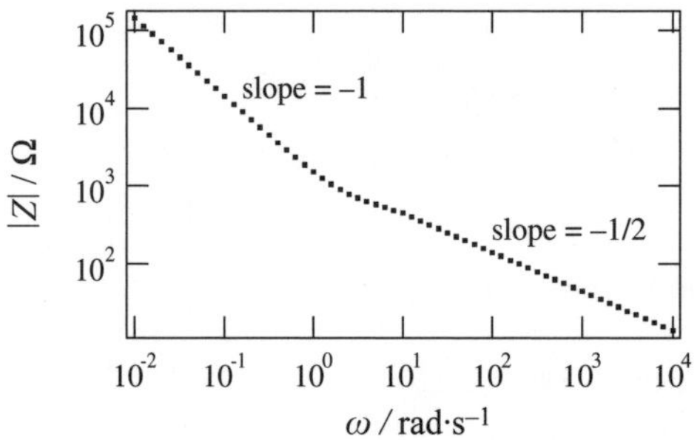

Fig. 9.34 Bode diagram of the concentration impedance for diffusion in a thin layer cell with σ = 1000 $\Omega \cdot \text{rad}^{1/2} \cdot \text{s}^{-1/2}$ and δ^2/D = 1s. The angular frequency varies from 0.01 to 10^4 $\text{rad} \cdot \text{s}^{-1}$.

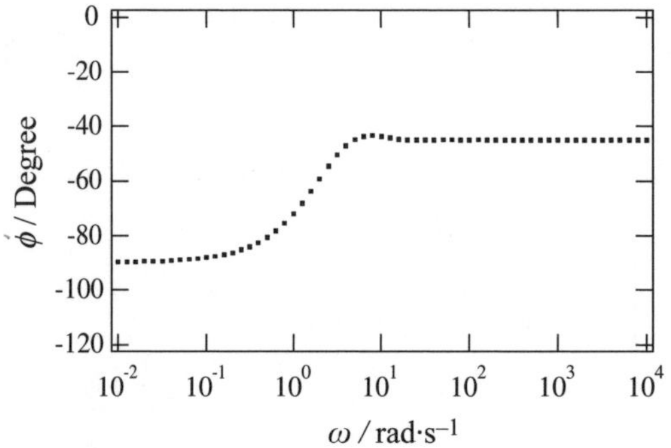

Fig. 9.35 Phase angle variation of a concentration impedance for diffusion in a thin layer cell with σ = 1000 $\Omega \cdot \text{rad}^{1/2} \cdot \text{s}^{-1/2}$ and δ^2/D = 1s. The angular frequency varies from 0.01 to 10^4 $\text{rad} \cdot \text{s}^{-1}$.

The Bode diagram and the phase angle variation as a function of the respective frequencies are shown in Figures 9.34 and 9.35. The capacitative character at low frequencies is clearly identifiable by a slope of –1 on the Bode diagram and a phase angle of –90° in Figure 9.35.

Examination of the concentration profiles in Figure 9.36 shows that at low frequencies, the whole of the concentration in the thin layer cell follows the variations of the interfacial concentrations.

9.4 AC VOLTAMMETRY

9.4.1 Voltammetry using a superimposed sinusoidal potential difference

The principle of superimposed sinusoidal potential voltamperometry, more commonly called AC voltammetry, consists of adding a small-amplitude sinusoidal potential onto a linear potential ramp, and measuring the alternating current. The average values of the interfacial concentrations correspond to the DC polarisation

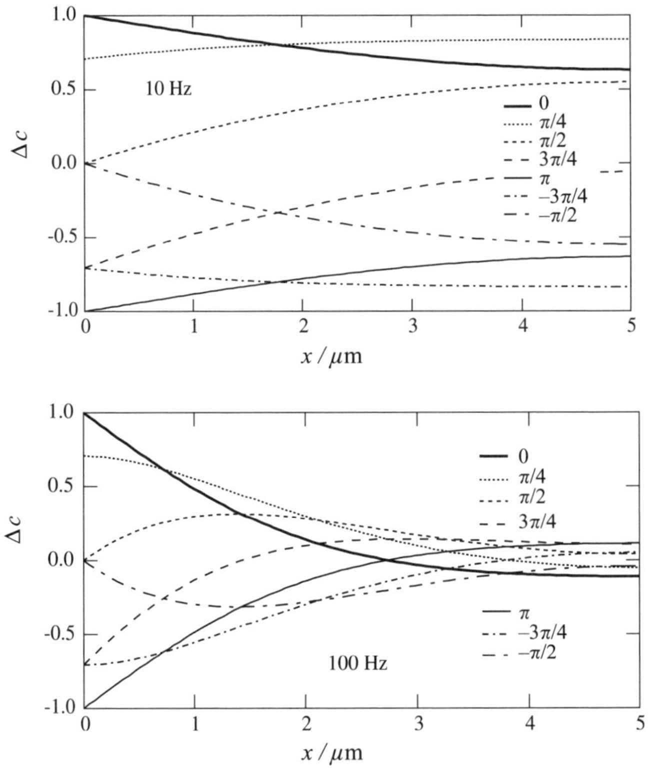

Fig. 9.36 Normalised profiles of the concentration perturbation of the reacting species for different phase angles in a thin layer of solution.

and are therefore obtained by resolving the Fick equations as we saw in §8.1.4. (see equations (8.58) and (8.59))

$$c_R(0,t) = c_R\left(\frac{\xi\theta}{1+\xi\theta}\right) = c_R^{app} \tag{9.104}$$

and

$$c_O(0,t) = c_R\left(\frac{\xi}{1+\xi\theta}\right) = c_O^{app} \tag{9.105}$$

with

$$\xi = \sqrt{\frac{D_R}{D_O}} \tag{9.106}$$

and

$$\theta = \frac{c_R(0,t)}{c_O(0,t)} = \exp\left[-\frac{nF}{RT}\left(E-E^{\ominus\prime}\right)\right] \tag{9.107}$$

To calculate the Warburg impedance at different electrode potentials along the applied potential ramp, we introduce the average values of the interfacial concentrations into equation (9.67) instead of the bulk concentrations c_R and c_O. We have already used this 'subterfuge' in §8.2.4, which consists of saying that the system behaves as if, vis a vis the small sinusoidal variations of the interfacial concentrations, the average values represent the bulk concentrations of a homogeneous solution.

$$\sigma = \frac{RT}{n^2F^2A\sqrt{2}}\left[\frac{1}{c_R^{app}\sqrt{D_R}}+\frac{1}{c_O^{app}\sqrt{D_O}}\right] \tag{9.108}$$

The faradaic diffusion impedance, which is the module of the Warburg impedance is therefore

$$Z_f = \frac{\sigma(1-j)}{\sqrt{\omega}} = \frac{RT}{n^2F^2Ac_R\sqrt{2\omega D_R}}\left[\frac{1}{\xi\theta}+2+\xi\theta\right] \tag{9.109}$$

To simplify this calculation, we put

$$\xi\theta = e^a \tag{9.110}$$

or again

$$a = \frac{c_R(0,t)\sqrt{D_R}}{c_O(0,t)\sqrt{D_O}} = -\frac{nF}{RT}\left(E-E_{1/2}\right) \tag{9.111}$$

Thus, by using the following equality,

$$\xi\theta+2+\frac{1}{\xi\theta} = e^a+e^{-a}+2 = 4\cosh^2(a/2) \tag{9.112}$$

the faradaic impedance is

$$Z_f = \frac{4RT\cosh^2(a/2)}{n^2F^2Ac_R\sqrt{2\omega D_R}}(1-j) \tag{9.113}$$

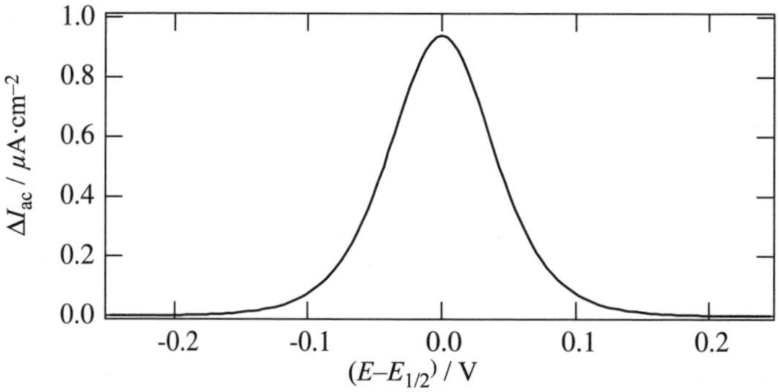

Fig. 9.37 AC Voltammogram for $\Delta E = 10$mV, $\omega = 0.01$ rad·s^{-1} and $D = 10^{-10}$ m^2·s^{-1}, $c = 1$ mM.

Thus, if we apply a potential ramp E_{dc} with a superimposed sinusoidal potential of amplitude ΔE_{ac}

$$E = E_{dc} + \Delta E_{ac}\sin(\omega t) \tag{9.114}$$

The resulting alternating current can be obtained using equations (9.15) and (9.16), i.e.

$$I_{ac} = \frac{n^2F^2Ac_R\sqrt{\omega D_R}}{4RT\cosh^2(a/2)}\Delta E_{ac}\sin\left(\omega t+\frac{\pi}{4}\right) = \Delta I_{ac}\sin\left(\omega t+\frac{\pi}{4}\right) \tag{9.115}$$

This 'bell-shaped' curve has a maximum given by

$$\Delta I_{max} = \frac{n^2F^2Ac_R\sqrt{\omega D_R}}{4RT}\Delta E_{ac} \tag{9.116}$$

as shown in Figure 9.37.

The peak current increases with the square root of the angular frequency, but is directly proportional to the amplitude of the applied alternating potential provided that its amplitude is small enough, so that the electrochemical system can be considered as linear. So for fairly small amplitudes, the width of the peak at mid-height is $90.4/n$ mV. If experimentally the width of the peak at mid-height is larger, that means that the reaction is not electrochemically reversible but is limited by the kinetics of the charge transfer reaction.

It is straightforward to show that equations (9.116) and (9.117) can be combined to give

$$E_{dc} = E_{1/2} + \frac{2RT}{nF}\ln\left[\left(\frac{\Delta I_{max}}{\Delta I_{ac}}\right)^{1/2} - \left(\frac{\Delta I_{max}-\Delta I_{ac}}{\Delta I_{ac}}\right)^{1/2}\right] \tag{9.117}$$

9.4.2 Potential-modulated reflectance (fluorescence) spectroscopy

The principle of this spectro-electrochemical technique is, for reflectance, to shine a light beam onto an electrode and measure the intensity of the reflected signal as it passes through the solution, and more particularly through the diffusion layer, as shown in Figure 9.38. To distinguish the response of the diffusion layer from the rest of the solution, it is worthwhile to modulate the electrode potential by a sinusoidal potential and to measure the relative response ($\Delta R/R$) of the reflectance. If the nature of the diffusing species permits, it is interesting to measure the fluorescence coming from the diffusion layer.

In total reflection from a metal electrode or from a liquid | liquid interface, the absorbance is given as a first approximation by the Lambert-Beer law, which, by integrating over the whole optical path gives

$$A = 2\varepsilon\int_0^\infty \frac{c(x,t)}{\cos\theta}\,dx \tag{9.118}$$

where ε is the molar absorption coefficient, θ the angle of reflection and x the normal

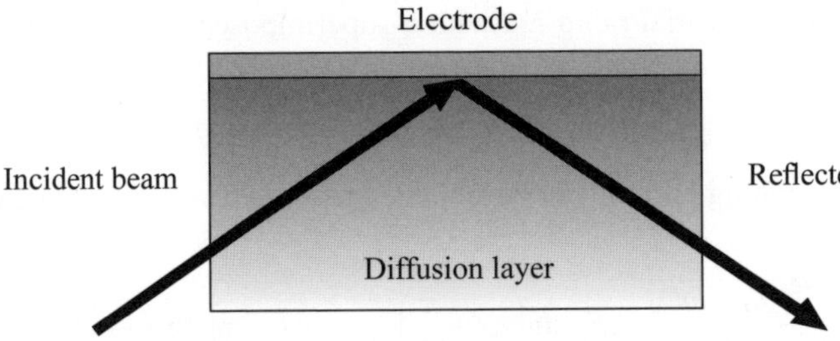

Fig. 9.38 Schematic diagram of a system for measuring the reflectance from the diffusion layer.

distance to the electrode. The quantity of absorbing species present in the diffusion layer corresponds to the charge passed, and so we have

$$A = 2\varepsilon\int_0^\infty \frac{c(x,t)}{\cos\theta}\,\mathrm{d}x = \frac{2\varepsilon}{nFA\cos\theta}\int_0^t I(t)\,\mathrm{d}t \tag{9.119}$$

The Laplace transform of the absorbance is therefore

$$\overline{A}(s) = \frac{2\varepsilon}{nFA\cos\theta}L\left\{\int_0^t I(t)\,\mathrm{d}t\right\} = \frac{2\varepsilon}{nFA\cos\theta}\frac{\overline{I}(s)}{s} = \frac{2\varepsilon}{nFA\cos\theta}\frac{\overline{E}(s)}{sZ_f(s)} \tag{9.120}$$

By using equation (9.114) as before, we have

$$\overline{A}(s) = \frac{\varepsilon}{\cos\theta}\frac{nFc_{\mathrm{R}}\sqrt{2\omega D_{\mathrm{R}}}}{2RT\cosh^2(a/2)}\frac{\overline{E}(s)}{\mathrm{j}(1-\mathrm{j})\omega} \tag{9.121}$$

thus by making the hypothesis that

$$E = E_{\mathrm{dc}} + \Delta E_{\mathrm{ac}}\sin(\omega t) \tag{9.122}$$

then taking the inverse transform, we optain

$$A(\omega) = \frac{\varepsilon}{\cos\theta}\frac{nFc_{\mathrm{R}}\sqrt{D_{\mathrm{R}}/\omega}}{2RT\cosh^2(a/2)}\Delta E_{\mathrm{ac}}\sin\left(\omega t - \frac{\pi}{4}\right) \tag{9.123}$$

From the definition of the absorbance

$$A = -\log_{10}T = -\log_{10}\left(\frac{I_\lambda}{I_\lambda^{\mathrm{o}}}\right) \tag{9.124}$$

where I_λ is the light intensity of the reflected beam and the intensity of the incident beam, we have

$$\mathrm{d}A = -\frac{\mathrm{d}I_\lambda}{I_\lambda} = -\frac{\mathrm{d}R}{R} \tag{9.125}$$

where R is the reflectance, the equivalent of the transmittance for a reflected beam. And so, by substitution, we get

$$\frac{\Delta R}{R} = -\frac{\varepsilon}{\cos\theta}\frac{nFc_{\mathrm{R}}\sqrt{D_{\mathrm{R}}/\omega}}{2RT\cosh^2(a/2)}\Delta E_{\mathrm{ac}} \tag{9.126}$$

Equation (9.123) shows that this spectro-electrochemical method allows the indirect measuring of the alternating current with a phase shift of 90°. The advantage of this approach is that only the faradaic component of the electrochemical reaction is measured, whereas the amperometric approach described in §9.4.1 can be perturbed by capacitive currents associated with the presence of the double layer.

CHAPTER 10

CYCLIC VOLTAMMETRY

10.1 ELECTROCHEMICALLY REVERSIBLE REACTIONS WITH SEMI-INFINITE LINEAR DIFFUSION

10.1.1 Integration method

Cyclic voltammetry is without any doubt the most universal electrochemical technique, used either to elucidate reaction mechanisms or to carry out quantitative analysis. In fact, it should be more accurately called cyclic voltamperometry, but cyclic voltammetry has become the accepted term.

The technique consists of varying the electrode potential in a linear fashion between two limits: the initial electrode potential E_i and the final electrode potential E_f so as to probe the reactivity of the electrochemical system over a large range of potentials in a single sweep. By varying the sweep rate, we can also probe the kinetics of the reactions and/or the mass transfer process.

For an oxidation, we usually start from an electrode potential value where no oxidation takes place and sweep, on the forward scan, the electrode potential to more positive values

$$E(t) = E_i + vt \tag{10.1}$$

After reaching the final value usually set at electrode potential values just before the oxidation of the solvent or that of the supporting electrolyte, the electrode potential is scanned back to the initial value

$$E(t) = E_f - vt \tag{10.2}$$

On the reverse scan, we reduce part of the species oxidised on the forward scan. v is called the ***sweep rate*** (or scan rate) and can vary from a few millivolts per second to a few million volts per second, according to the application and the size of the electrode.

The differential equations for cyclic voltammetry are the differential diffusion equations for the oxidised or reduced species

$$\frac{\partial c_{\mathrm{O}}(x,t)}{\partial t} = D_{\mathrm{O}} \frac{\partial^2 c_{\mathrm{O}}(x,t)}{\partial x^2} \tag{10.3}$$

and

$$\frac{\partial c_{\mathrm{R}}(x,t)}{\partial t} = D_{\mathrm{R}} \frac{\partial^2 c_{\mathrm{R}}(x,t)}{\partial x^2} \tag{10.4}$$

The boundary conditions for an oxidation on the forward scan usually assume that we start with only reduced species present in solution, and the initial conditions read

$$c_{\mathrm{R}}(x,0)=c_{\mathrm{R}} \quad \text{and} \quad c_{\mathrm{O}}(x,0)=c_{\mathrm{O}}\approx 0 \tag{10.5}$$

In effect, by fixing the initial electrode potential value more than 120 mV more negative than the standard redox potential of the oxidation process under investigation, we ensure that the surface concentration of oxidised species at the start of the scan is negligible.

If we assume that the volume of the solution is large enough, the bulk boundary conditions are

$$\lim_{x\to\infty} c_{\mathrm{R}}(x,t)=c_{\mathrm{R}} \quad \text{and} \quad \lim_{x\to\infty} c_{\mathrm{O}}(x,t)=c_{\mathrm{O}}\approx 0 \tag{10.6}$$

The equality of the diffusion fluxes at the interface creates an additional boundary condition

$$D_{\mathrm{O}}\left(\frac{\partial c_{\mathrm{O}}(x,t)}{\partial x}\right)_{x=0}+D_{\mathrm{R}}\left(\frac{\partial c_{\mathrm{R}}(x,t)}{\partial x}\right)_{x=0} = 0 \tag{10.7}$$

To solve the Fick differential equations with the Nernst equation as a boundary condition, it is preferable to define the following dimensionless parameters:

$$\frac{c_{\mathrm{R}}(0,t)}{c_{\mathrm{O}}(0,t)} = \theta\, e^{-\sigma t} = \theta S(t) \tag{10.8}$$

with

$$\theta = \exp\left[-\frac{nF}{RT}\left(E_i - E^{\ominus\prime}\right)\right] = \frac{c_{\mathrm{R}}(x,0)}{c_{\mathrm{O}}(x,0)} = \frac{c_{\mathrm{R}}}{c_{\mathrm{O}}} \tag{10.9}$$

and

$$\sigma = \frac{nF}{RT}v \tag{10.10}$$

To calculate the current-potential response, we shall first calculate the interfacial concentrations of O and R to substitute in the Nernst equation (10.8). As seen in Chapter 8, the Laplace transform for the concentration of the reduced species is obtained by solving the Laplace transform of equation (10.4)

$$\bar{c}_{\mathrm{R}}(x,s) = \frac{c_{\mathrm{R}}}{s} + A(s)\, e^{-\sqrt{\frac{s}{D_{\mathrm{R}}}}x} \tag{10.11}$$

In turn, the Laplace transform for the current is

$$\bar{I}(s) = nFA\, D_{\mathrm{R}}\left[\frac{\partial \bar{c}_{\mathrm{R}}(x,s)}{\partial x}\right]_{x=0} = nFA\, D_{\mathrm{R}}\left[-A(s)\sqrt{\frac{s}{D_{\mathrm{R}}}}\right] \tag{10.12}$$

Thus, the constant $A(s)$ can be determined as :

$$A(s) = -\frac{\bar{I}(s)}{nFA\, D_{\mathrm{R}}}\sqrt{\frac{D_{\mathrm{R}}}{s}} \tag{10.13}$$

The Laplace transform of the interfacial concentration of the reduced species is then

$$\bar{c}_{\mathrm{R}}(0,s) \;=\; \frac{c_{\mathrm{R}}}{s} \;-\; \frac{\bar{I}(s)}{nFA\,D_{\mathrm{R}}}\sqrt{\frac{D_{\mathrm{R}}}{s}} \tag{10.14}$$

In order to calculate the inverse Laplace tranform of equation (10.14), we use the convolution theorem that provides the inverse transform of a product

$$L^{-1}\left\{\bar{F}(s)\cdot\bar{G}(s)\right\} \;=\; \int_0^t F(\tau)\cdot G(t-\tau)\,\mathrm{d}\tau \tag{10.15}$$

The inverse transform of $1/\sqrt{s}$ being $1/\sqrt{\pi t}$, the interfacial concentration of the reduced species is

$$c_{\mathrm{R}}(0,t) \;=\; c_{\mathrm{R}} - \frac{1}{nFA\sqrt{\pi\,D_{\mathrm{R}}}}\int_0^t I(\tau)\,(t-\tau)^{-1/2}\,\mathrm{d}\tau \tag{10.16}$$

By introducing $f(\tau)$, the interfacial mass flux defined by

$$f(\tau) \;=\; \frac{I(\tau)}{nFA} \tag{10.17}$$

equation (10.16) becomes

$$c_{\mathrm{R}}(0,t) \;=\; c_{\mathrm{R}} - \frac{1}{\sqrt{\pi\,D_{\mathrm{R}}}}\int_0^t f(\tau)\,(t-\tau)^{-1/2}\,\mathrm{d}\tau \tag{10.18}$$

The integral of equation (10.18) is what is commonly called the ***convoluted current,***

$$\hat{I}(t) \;=\; \int_0^t f(\tau)\,(t-\tau)^{-1/2}\,\mathrm{d}\tau \tag{10.19}$$

In the same way, it is easily shown that the interfacial concentration of the oxidised species is given by:

$$c_{\mathrm{O}}(0,t) \;=\; c_{\mathrm{O}} + \frac{1}{\sqrt{\pi D_{\mathrm{O}}}}\int_0^t f(\tau)\,(t-\tau)^{-1/2}\,\mathrm{d}\tau \;\approx\; \frac{1}{\sqrt{\pi D_{\mathrm{O}}}}\int_0^t f(\tau)\,(t-\tau)^{-1/2}\,\mathrm{d}\tau \tag{10.20}$$

From the two expressions for the interfacial concentrations of O and R, and the Nernst equation, we obtain a relationship between the convoluted current and the potential

$$\theta S(t) \;=\; \frac{c_{\mathrm{R}} - \hat{I}(t)\,/\,\sqrt{\pi D_{\mathrm{R}}}}{\hat{I}(t)\,/\,\sqrt{\pi D_{\mathrm{O}}}} \tag{10.21}$$

from which we can express the convoluted current as

$$\hat{I}(t) \;=\; \frac{\sqrt{\pi D_{\mathrm{R}}}\;c_{\mathrm{R}}}{1+\xi\,\theta S(t)} \tag{10.22}$$

with ξ being a dimensionless number, previously defined by

$$\xi \;=\; \sqrt{\frac{D_{\mathrm{R}}}{D_{\mathrm{O}}}} \tag{10.23}$$

To be able to integrate equation (10.22), it is handy to make the following change of variable

$$z = \sigma\tau \quad \text{and} \quad f(\tau) = g(z) \tag{10.24}$$

so that equation (10.22) becomes

$$\int_0^{\sigma t} \frac{g(z)}{c_R\sqrt{\pi D_R \sigma}} \frac{\mathrm{d}z}{\sqrt{\sigma t - z}} = \int_0^{\sigma t} \frac{\chi(z)\,\mathrm{d}z}{\sqrt{\sigma t - z}} = \frac{1}{1+\xi\,\theta\,S(\sigma t)} \tag{10.25}$$

the function χ is dimensionless being defined as

$$\chi(z) = \frac{g(z)}{c_R\sqrt{\pi D_R\sigma}} = \frac{I(\sigma t)}{nFA\,c_R\sqrt{\pi D_R\sigma}} \tag{10.26}$$

This function χ can be calculated numerically step by step after segmenting the variable z in n intervals.

$$\int_0^{\sigma t} \frac{\chi(z)\,\mathrm{d}z}{\sqrt{\sigma t - z}} = \int_0^{\delta n} \frac{\chi(v\delta)\,\mathrm{d}(v\delta)}{\sqrt{n\delta - v\delta}} \tag{10.27}$$

with

$$z = \mathrm{v}\delta \quad \text{and} \quad \sigma t = n\delta \tag{10.28}$$

An integration by parts of equation (10.27) then gives

$$\begin{aligned}\int_0^{n\delta} \frac{\chi(v\delta)\,\mathrm{d}\,(v\delta)}{\sqrt{n\delta - v\delta}} &= \left[-2\,\chi(v\delta)\sqrt{n\delta - v\delta}\right]_0^{n\delta} + 2\int_0^{n\delta} \chi'(v\delta)\sqrt{n\delta - v\delta}\,\mathrm{d}\,(v\delta) \\ &= 2\,\chi(0)\sqrt{n\delta} + 2\int_0^{n\delta}\sqrt{n\delta - v\delta}\,\mathrm{d}\chi(v\delta)\end{aligned} \tag{10.29}$$

This last integral can be approximated in the form of a sum such that

$$\begin{aligned}\int_0^{n\delta} \frac{\chi(v\delta)\,\mathrm{d}\,(v\delta)}{\sqrt{n\delta - v\delta}} &= 2\,\chi\,(0)\sqrt{n\delta} + 2\sum_{i=0}^{n-1}\sqrt{n\delta - i\delta}\ \Big| \\ &= 2\sqrt{\delta}\sum_{i=1}^{n-1}\left[\sqrt{n-i}\ \left[\chi((i+1)\delta) - \chi(i\delta)\right]\right] + 2\sqrt{\delta}\,\chi(\delta)\sqrt{n}\end{aligned} \tag{10.30}$$

Thus, we can calculate the values of χ as a function of the electrode potential

$$2\sqrt{\delta}\left[\chi(\delta)\sqrt{n} + \sum_{i=1}^{n-1}\sqrt{n-i}\ \left[\chi((i+1)\delta) - \chi(i\delta)\right]\right] = \frac{1}{1+\xi\theta S(\delta n)} \tag{10.31}$$

This calculation is done by successive iterations. For $n = 1$, we have simply

$$2\sqrt{\delta}\,\chi(\delta) = \frac{1}{1+\xi\theta S(\delta)} \tag{10.32}$$

then for $n = 2$, we have

$$2\sqrt{\delta}\left[\chi(2\delta)\sqrt{1} + \chi(\delta)\left[\sqrt{2}-\sqrt{1}\right]\right] = \frac{1}{1+\xi\theta S(2\delta)} \tag{10.33}$$

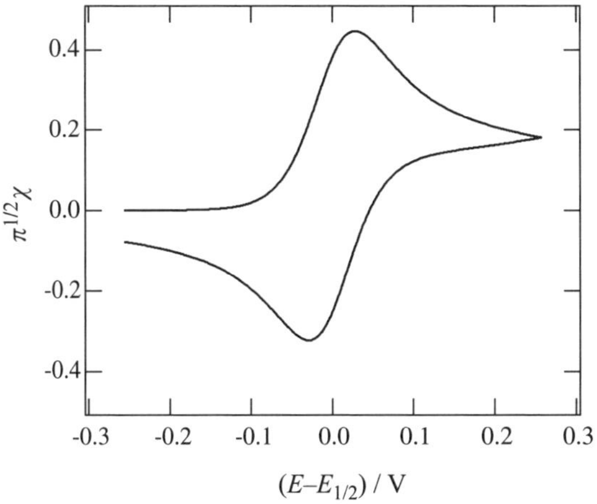

Fig. 10.1 Dimensionless cyclic voltammogram.

then for $n = 3$

$$2\sqrt{\delta}\left[\chi(3\delta)\sqrt{1} \;+\; \chi(2\delta)\left[\sqrt{2}-\sqrt{1}\right] \;+\; \chi(\delta)\left[\sqrt{3}-\sqrt{2}\right]\right] \;=\; \frac{1}{1+\xi\theta S(3\delta)} \tag{10.34}$$

We now see the iteration that permits the calculation of $\chi(n\delta)$ as a function of the previous values. A simple numerical calculation then allows us to determine the values of χ. The results of this numerical approach for the calculation of the dimensionless current χ are shown in Figure 10.1. The main results of this numerical integration are the following:

- Equation (10.26) can be re-written in the form

$$I(\sigma t) \;=\; nFA\,c_R\sqrt{\pi D_R\left(\frac{nF}{RT}\right)v}\;\;\chi(\sigma t) \tag{10.35}$$

  that shows that the current is proportional to the square root of the scan rate and to that of the diffusion coefficient.
- The maximum value of $\sqrt{\pi}\chi$ is 0.4463 as shown in Figure 10.1, and the maximum current is then simply

$$I_p \;=\; 0.4463\,nF\,A\,c_R\sqrt{\frac{nFvD_R}{RT}} \tag{10.36}$$

  This equation is known as the ***Randles-Sevcik*** equation.
- The separation between the maximum current (at $E_{1/2} + 28.5/n$ mV) and the minimum current (at $E_{1/2} - 28.5/n$ mV) is therefore $59/n$ mV .
- The potential values for the maxima and minima are independent of the scan rate.
- If the reaction kinetics is not fast enough with respect to the mass transfer, an increase in the separation of the peak potentials as the scan rate increases is observed.

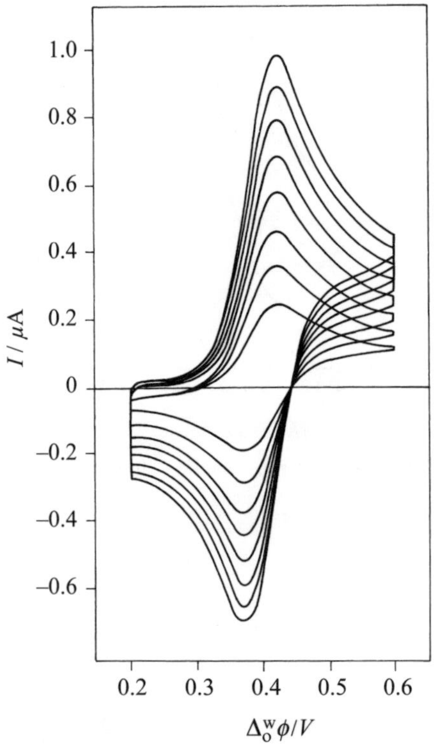

Fig. 10.2 Cyclic voltammogram of the transfer of tetramethylammonium ions across a water |1,2-dichloroethane interface [Murray Osborne, EPFL].

The methodology developed above applies to all charge transfer reactions limited by the diffusion of the reactants towards a plane surface and products away from this surface. Therefore, it applies just as well to redox reactions at an electrode as to ion transfer across a liquid|liquid interface.

From an experimental point of view, it is always interesting to record voltammograms at different scan rates as shown in Figure 10.2. A good voltammogram for a reversible system is characterised by the following points:

- The position of the peaks is independent of the scan rate. Displacement is due to either a badly compensated ohmic drop or to slow electrochemical reaction kinetics.
- The peak current is proportional to the square root of the scan rate. The ordinate at the origin of the straight line obtained by plotting the peak current as a function of the square root of the scan rate should pass through the origin. A positive displacement is in general due to a high value of the capacitive current or an offset of the potentiostat.
- An isosbestic point should be observed for $I = 0$. This can be used for working out the half-wave potential value. For $n = 1$, this point is at $E_{1/2} + 0.046$ V.

10.1.2 Convoluted current

It is interesting to come back to the notion of convoluted current, also called the ***semi-integrated current*** and to express the interfacial concentrations given by equations (10.18) & (10.20) in the form

$$c_{\mathrm{R}}(0,t) = c_{\mathrm{R}} - \frac{\hat{I}(t)}{\sqrt{\pi D_{\mathrm{R}}}} \tag{10.37}$$

and

$$c_{\mathrm{O}}(0,t) = \frac{\hat{I}(t)}{\sqrt{\pi D_{\mathrm{O}}}} \tag{10.38}$$

In the same way that we have defined the limiting diffusion currents in steady state amperometry (see equations (7.30) & (7.31)), here we can define the limiting anodic convoluted current

$$\hat{I}_{\mathrm{la}} = c_{\mathrm{R}}\sqrt{\pi D_{\mathrm{R}}} \tag{10.39}$$

so that equation (10.18) becomes

$$c_{\mathrm{R}}(0,t) = \frac{\left[\hat{I}_{\mathrm{la}} - \hat{I}(t)\right]}{\sqrt{\pi D_{\mathrm{R}}}} \tag{10.40}$$

By substituting equations (10.39) and (10.40) in the Nernst equation, we find an equation similar to equation (7.34) i.e.

$$E = E^{\ominus\prime} + \frac{RT}{nF}\ln\left[\sqrt{\frac{D_{\mathrm{R}}}{D_{\mathrm{O}}}}\right] + \frac{RT}{nF}\ln\left[\frac{\hat{I}(t)}{\hat{I}_{\mathrm{la}} - \hat{I}(t)}\right] \tag{10.41}$$

Figure 10.3 shows the variation of the convoluted current as a function of the potential

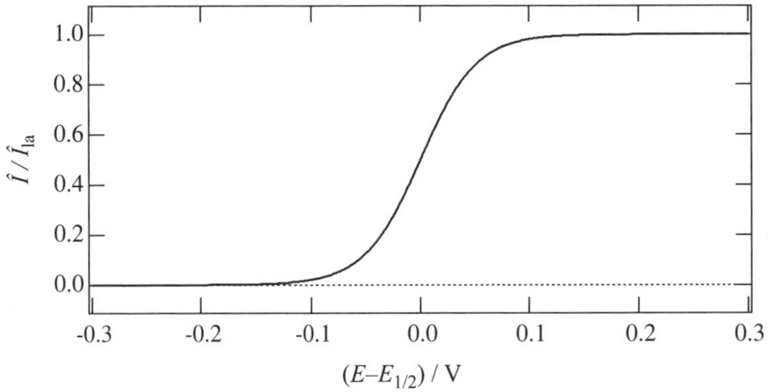

Fig. 10.3 Variation of the convoluted current (normalised by the limiting anodic convoluted current) as a function of the applied potential.

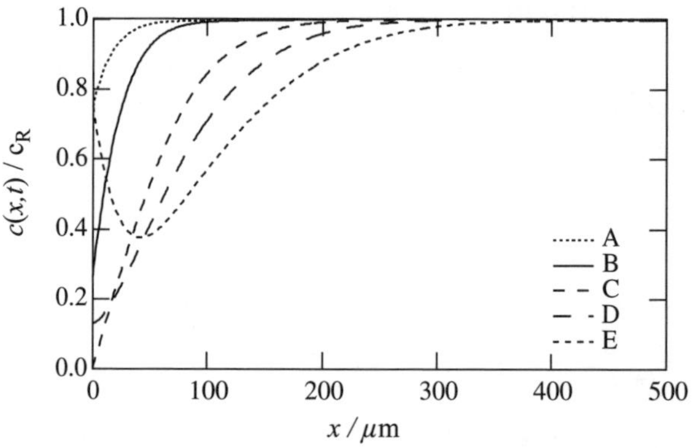

Fig. 10.4 Concentration profiles at different potentials: A: $E_{1/2} - 28$ mV ($I = I_p/2$), B: $E_{1/2} + 28$ mV (Forward peak, $I = I_p$), C: $E = E_{1/2} + 250$ mV (final potential), D: $E = E_{1/2} + 46$ mV (Return sweep, $I_p = 0$), E: $E_{1/2} - 28$ mV (Return peak). Sweep rate : 100 mV·s^{-1}, Diffusion coefficient : 10^{-9} m^2·s^{-1} [Rosaria Ferrigno, EPFL].

The interfacial concentrations can be easily calculated if we know the convoluted current as a function of the potential. In order to know the full concentration profiles at different electrode potentials, it is necessary to do a numerical simulation. The method consists of solving the Fick equations as a function of the distance from the electrode. Various methods using different algorithms have been developed. Figure 10.4 shows a few concentration profiles of the reduced species at different electrode potentials with the conditions as in Figure 10.1, the simulation being obtained by the finite element method.

It is important to note that the surface concentration of the reacting species, normalised by the bulk concentration is 0.26 when the current reaches its maximum value. A widespread error is to think that the current reaches a maximum because the interfacial concentration becomes negligible. The data in Figure 10.4 show clearly that the concentration only becomes negligible when the applied potential is much greater than the formal redox potential.

10.2 INFLUENCE OF THE KINETICS

Cyclic voltammetry is quite a convenient qualitative method for determining whether a simple electrochemical reaction is limited by diffusion (so-called reversible systems) or completely limited (irreversible systems) or partially limited by kinetics (quasi-reversible systems).

10.2.1 Electrochemically quasi-reversible reactions

For electrochemically quasi-reversible reactions, the surface concentrations are controlled both by kinetics and diffusion. As a consequence, the Nernst law (10.8) cannot longer be used as a boundary condition. The interfacial concentrations under kinetic control are given by the Butler-Volmer equation (7.16)

$$I = nF\,A\,k^{\ominus}\left[c_{\mathrm{R}}(0)\,e^{\alpha nF(E-E^{\ominus\prime})/RT} - c_{\mathrm{O}}(0)\,e^{-(1-\alpha)\,nF(E-E^{\ominus\prime})/RT}\right] \tag{10.42}$$

By using the notations of (10.8) and (10.9), this equation becomes

$$I = nF\,A\,k^{\ominus}[\theta S(t)]^{-\alpha}\left[c_{\mathrm{R}}(0) - c_{\mathrm{O}}(0)\theta S(t)\right] \tag{10.43}$$

By replacing the interfacial concentrations by equations (10.18) and (10.20), we have

$$f(\tau) = k^{\ominus}[\theta S(t)]^{-\alpha}\left[\left(c_{\mathrm{R}} - \frac{\hat{I}(\tau)}{\sqrt{\pi\,D_{\mathrm{R}}}}\right) - \theta S(t)\left(c_{\mathrm{O}} + \frac{\hat{I}(\tau)}{\sqrt{\pi\,D_{\mathrm{O}}}}\right)\right] \tag{10.44}$$

To simplify the calculations, we need to make the hypothesis that the diffusion coefficients D_{R} and D_{O} are equal. Thus equation (10.44) is rearranged as

$$\frac{f(\tau)}{k^{\ominus}c_{\mathrm{R}}[\theta S(t)]^{-\alpha}} = 1 - S(t) - \frac{1+\theta S(t)}{c_{\mathrm{R}}\sqrt{\pi\,D}}\hat{I}(\tau) \tag{10.45}$$

Again using the definitions (10.24 - 10.26), we have

$$\frac{\chi(\sigma\tau)}{\psi[\theta S(t)]^{-\alpha}} = 1 - S(t) - (1+\theta S(t))\int_0^{\sigma t}\frac{\chi(z)\,\mathrm{d}z}{\sqrt{\sigma t - z}} \tag{10.46}$$

with

$$\psi = \frac{k^{\ominus}}{\sqrt{\pi D\sigma}} = k^{\ominus}\sqrt{\frac{RT}{\pi DnFv}} \tag{10.47}$$

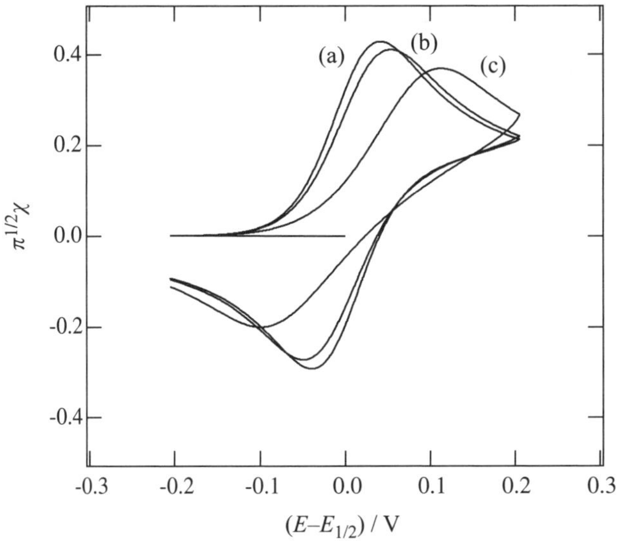

Fig. 10.5 Calculated cyclic voltammograms for quasi-reversible reactions. (a) $\psi = 1$, (b) $\psi = 0.5$ and (c) $\psi = 0.1$.

As described in §10.1, we can calculate by numerical integration the dimensionless current χ.

As the results in Figure 10.5 show, the separation of the peaks increases when the rate constant decreases or when the sweep rate increases.

Taking into account equation (10.46), it is important to note that the peak current measured experimentally will not be proportional to the square root of the sweep velocity. Effectively, the dimensionless current decreases as the sweep rate increases.

10.2.2 Influence of the ohmic drop

One of the major flaws in cyclic voltammetry is that a badly compensated potential drop for an electrochemically reversible system as studied in §10.1.1 appears as the signature of a quasi-reversible reaction kinetic as described above. This is why cyclic voltammetry is not recommended as a quantitative method to measure electrochemical reactions rates. (Certain outspoken critics go so far as to claim that some of the rate constants published in the literature are nothing more than measurements of ohmic drops). Figure 10.6 shows this effect, already described in §7.4.2 for steady state amperometry.

10.2.3 Electrochemically irreversible reactions

For electrochemically irreversible systems, the reverse reaction is so slow at electrode potentials larger than the standard redox potential that it can be ignored and so equation (10.42) reduces to

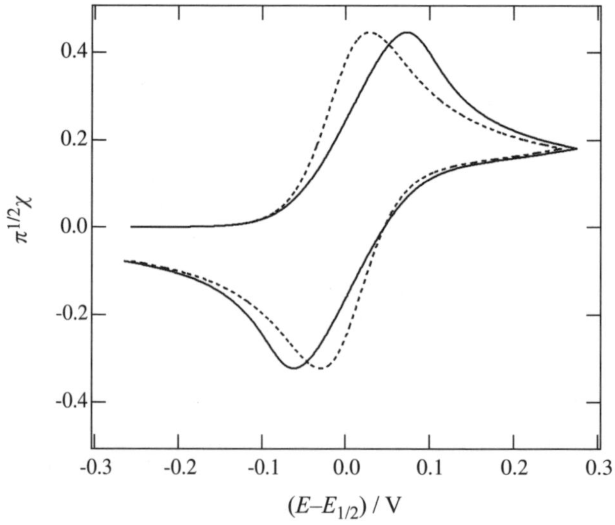

Fig. 10.6 Influence of the ohmic drop on cyclic voltammetry. Dotted line $R = 0$, continuous line $R = \Delta E/\pi^{1/2}\chi = 0.1\ \mathrm{V}^{-1}$.

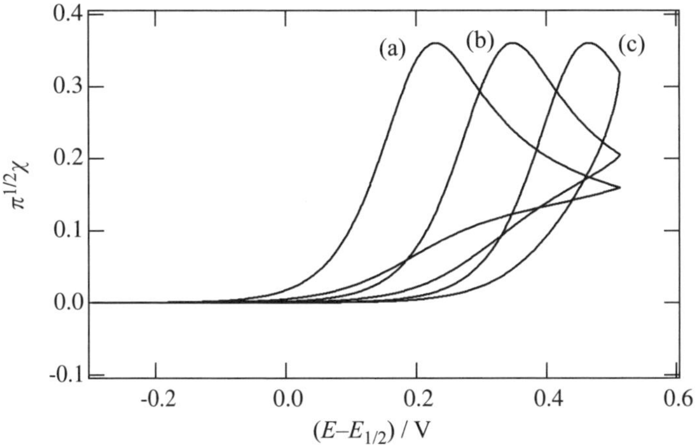

Fig. 10.7 Cyclic voltammogram of irreversible reactions: (a) $\psi = 0.01$, (b) $\psi = 0.001$, (c) $\psi = 0.0001$.

$$I = nF\,A\,k^{\ominus}\left[c_{\mathrm{R}}(0)\,e^{\alpha nF(E-E^{\ominus\prime})/RT}\right] \tag{10.48}$$

and consequently equation (10.46) becomes

$$\frac{\chi(\sigma\tau)}{\psi[\theta S(t)]^{-\alpha}} = 1-\int_0^{\sigma t}\frac{\chi(z)\,\mathrm{d}z}{\sqrt{\sigma t - z}} \tag{10.49}$$

The results obtained by numerical integration are shown Figure 10.7. As with steady state amperometry (see Figure 7.23), we see a displacement of the same curve, away from the standard redox potential when the rate constant decreases.

Since it can be observed in Figure 10.7 that the peak current does not depend on the kinetic parameter ψ, it is be, as for reversible reactions, proportional to the square root of the sweep rate.

10.3 EC REACTIONS

10.3.1 EC$_r$ reactions

One of the major advantages of cyclic voltammetry is to be able to study the mechanisms of complex reactions. To illustrate this, let's look at what we shall call an EC$_r$ mechanism (**E**lectrochemically reversible reaction at the electrode followed by a **C**hemically **r**eversible reaction).

$$\mathrm{R} \rightleftarrows \mathrm{O} + ne^- \underset{k_\mathrm{b}}{\overset{k_\mathrm{f}}{\rightleftarrows}} \mathrm{P}$$

The differential equations must now take into account the rates of the chemical reactions

$$\frac{\partial c_{\mathrm{O}}(x,t)}{\partial t} = D_{\mathrm{O}}\frac{\partial^2 c_{\mathrm{O}}(x,t)}{\partial x^2} - k_{\mathrm{f}}c_{\mathrm{O}}(x,t) + k_{\mathrm{b}}c_{\mathrm{P}}(x,t) \tag{10.50}$$

$$\frac{\partial c_{\mathrm{R}}(x,t)}{\partial t} = D_{\mathrm{R}}\frac{\partial^2 c_{\mathrm{R}}(x,t)}{\partial x^2} \tag{10.51}$$

and for the product P

$$\frac{\partial c_{\mathrm{P}}(x,t)}{\partial t} = D_{\mathrm{P}}\frac{\partial^2 c_{\mathrm{P}}(x,t)}{\partial x^2} + k_{\mathrm{f}}c_{\mathrm{O}}(x,t) - k_{\mathrm{b}}c_{\mathrm{P}}(x,t) \tag{10.52}$$

The boundary conditions for an oxidation are such that we consider that only the species R is present in solution at the start of the scan

$$c_{\mathrm{R}}(x,0) = c_{\mathrm{R}} \quad \text{and} \quad c_{\mathrm{O}}(x,0) = c_{\mathrm{O}} \approx 0 \quad \text{and} \quad c_{\mathrm{P}}(x,0) = c_{\mathrm{P}} \approx 0 \tag{10.53}$$

We still consider that the volume of the reaction is large enough so that the bulk boundary conditions are

$$\lim_{x\to\infty} c_{\mathrm{R}}(x,t) = c_{\mathrm{R}} \quad \text{and} \quad \lim_{x\to\infty} c_{\mathrm{O}}(x,t) = c_{\mathrm{O}} \approx 0 \quad \text{and} \quad \lim_{x\to\infty} c_{\mathrm{P}}(x,t) = c_{\mathrm{P}} \approx 0 \tag{10.54}$$

We still make the hypothesis that the initial electrode potential is negative enough with respect to the standard redox potential such that the interfacial concentration of oxidised species is negligible at equilibrium at the initial electrode potential.

The equality of the fluxes of O and R at the interface remains valid, as does the Nernst law for the interfacial concentrations.

The best way of treating these problems of coupled reactions is to change the variables so that we return to something resembling Fick equations. Here, we define a virtual concentration c_{tot} as

$$c_{\mathrm{tot}}(x,t) = c_{\mathrm{O}}(x,t) + c_{\mathrm{P}}(x,t) = c_{\mathrm{O}}(x,t)[1+K] \tag{10.55}$$

where K is the chemical equilibrium constant. The hypothesis pertaining to equation (10.55) is that we have a post-equilibrium to the redox reaction where k_{f} and k_{b} are high.

In this way, the sum of the differential equations of O and P reduces to

$$\frac{\partial c_{\mathrm{tot}}(x,t)}{\partial t} = D\frac{\partial^2 c_{\mathrm{tot}}(x,t)}{\partial x^2} \tag{10.56}$$

if we make the hypothesis that the diffusion coefficients of O and P are both equal to D. Since the boundary conditions for the interfacial flux of P is

$$D\left(\frac{\partial c_{\mathrm{P}}(x,t)}{\partial x}\right)_{x=0} = 0 \tag{10.57}$$

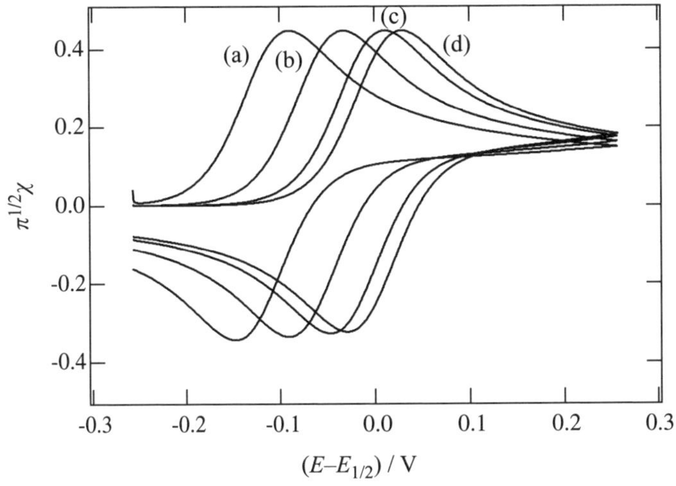

Fig. 10.8 Influence of the post-equilibrium constant on the cyclic voltammogram. (a) $K = 100$, (b) $K = 10$, (c) $K = 1$ and (d) $K = 0$.

the condition for the equality of the fluxes is

$$D\left(\frac{\partial c_{\text{tot}}(x,t)}{\partial x}\right)_{x=0} + D_{\text{R}}\left(\frac{\partial c_{\text{R}}(x,t)}{\partial x}\right)_{x=0} = 0 \tag{10.58}$$

By doing so, we are back to the case described in §10.1 and we can therefore calculate the interfacial concentrations as a function of the convoluted current

$$c_{\text{R}}(0,t) = c_{\text{R}} - \hat{I}(t)/\sqrt{\pi D_{\text{R}}} \tag{10.59}$$

and

$$c_{\text{tot}}(0,t) = \hat{I}(t)/\sqrt{\pi D} \tag{10.60}$$

The convoluted current is obtained by substituting equations (10.59) and (10.60) in the Nernst equation

$$\theta S(t) = \frac{c_{\text{R}} - \hat{I}(t)/\sqrt{\pi D_{\text{R}}}}{\hat{I}(t)/(1+K)\sqrt{\pi D_{\text{O}}}} \tag{10.61}$$

We can express the convoluted current as a function of the potential as before

$$\hat{I}(t) = \frac{(1+K)\sqrt{\pi D_{\text{R}}}\, c_{\text{R}}}{1+K+\xi\,\theta S(t)} \tag{10.62}$$

ξ is a dimensionless number defined by:

$$\xi = \sqrt{\frac{D_{\text{R}}}{D_{\text{O}}}} \tag{10.63}$$

The current is then obtained by numerical integration as before, and the results are shown in Figure 10.8. A displacement of about $(RT/nF)\ln(1+K)$ between the voltammograms can be seen.

If the chemical step is a pseudo-first order complexation reaction of the oxidised species O with m ligands L

$$\mathrm{O} + m\mathrm{L} \underset{k_\mathrm{b}}{\overset{k_\mathrm{f}}{\rightleftarrows}} \mathrm{OL}_m$$

cyclic voltammetry can be used to measure the complexation constant $K_\mathrm{c} = c_{\mathrm{OL}_m}/c_\mathrm{O}c_\mathrm{L}^m$

10.3.2 EC$_\mathrm{i}$ reactions

Another type of complex reactions includes post-redox non-chemically reversible chemical reactions and whose reaction kinetics may be slow

$$\mathrm{R} \rightleftarrows \mathrm{O} + ne^- \longrightarrow \mathrm{P}$$

We shall called these reactions EC$_\mathrm{i}$ reactions (**E**lectrochemically reversible reaction at the electrode followed by a **C**hemically **i**rreversible reaction).

A classic example is the oxidation of para-aminophenol in quinone imine followed by its hydrolysis to form benzoquinone. Para-aminophenol is a redox molecule often used in the conception of amperometric biosensors.

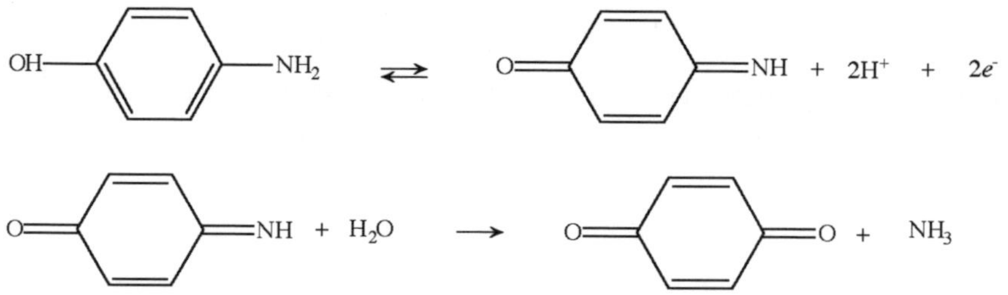

Here, a change of variables is not necessary. We shall start directly from equations (10.50)-(10.52) taking $k_\mathrm{b} = 0$.

The Laplace transform of the concentration of R is still

$$\bar{c}_\mathrm{R}(x,s) = \frac{c_\mathrm{R}}{s} + A(s)\,e^{-\sqrt{\frac{s}{D_\mathrm{R}}}x} \tag{10.64}$$

whilst the Laplace transform of equation (10.50) is

$$s\bar{c}_\mathrm{O}(x,s) - c_\mathrm{O}(x,0) \approx s\bar{c}_\mathrm{O}(x,s) = D_\mathrm{O}\frac{\partial^2\bar{c}_\mathrm{O}(x,s)}{\partial x^2} - k_\mathrm{f}\bar{c}_\mathrm{O} \tag{10.65}$$

so the Laplace transform of the concentration of O is

$$\bar{c}_\mathrm{O}(x,s) = B(s)\,e^{-\sqrt{\frac{s+k_\mathrm{f}}{D_\mathrm{O}}}x} \tag{10.66}$$

and that of the current is

$$\bar{I}(s) = nFA\,D_R\left[\frac{\partial \bar{c}_R(x,s)}{\partial x}\right]_{x=0} = nFA\,D_R\left[-A(s)\sqrt{\frac{s}{D_R}}\right] \tag{10.67}$$

and

$$\bar{I}(s) = -nFA\,D_O\left[\frac{\partial \bar{c}_O(x,s)}{\partial x}\right]_{x=0} = nFA\,D_O\left[B(s)\sqrt{\frac{s+k_f}{D_O}}\right] \tag{10.68}$$

The constant $A(s)$ can then be determined to be equal to

$$A(s) = -\frac{\bar{I}(s)}{nFA\,D_R}\sqrt{\frac{D_R}{s}} \tag{10.69}$$

The Laplace transform of the interfacial concentration of R is therefore

$$\bar{c}_R(0,s) = \frac{c_R}{s} - \frac{\bar{I}(s)}{nFA\,D_R}\sqrt{\frac{D_R}{s}} \tag{10.70}$$

and the constant $B(s)$ is given by:

$$B(s) = \frac{\bar{I}(s)}{nFA\,D_O}\sqrt{\frac{D_O}{s+k_f}} \tag{10.71}$$

The transform of the interfacial concentration of O is then simply :

$$\bar{c}_O(0,s) = \frac{\bar{I}(s)}{nFA\,D_O}\sqrt{\frac{D_O}{s+k_f}} \tag{10.72}$$

The interfacial concentrations are then

$$c_R(0,t) = c_R - \frac{1}{\sqrt{\pi\,D_R}}\int_0^t f(\tau)\,(t-\tau)^{-1/2}\,\mathrm{d}\tau \tag{10.73}$$

and knowing that $L^{-1}\{(s+a)^{-1/2}\} = e^{-at}/\sqrt{\pi t}$, we have

$$c_O(0,t) = \frac{1}{\sqrt{\pi\,D_O}}\int_0^t f(\tau)\,e^{-k_f(t-\tau)}(t-\tau)^{-1/2}\,\mathrm{d}\tau \tag{10.74}$$

By substituting in the Nernst equation, we get

$$\theta S(t) = \frac{c_R(0,t)}{c_O(0,t)} = \frac{c_R - \dfrac{1}{\sqrt{\pi\,D_R}}\displaystyle\int_0^t f(\tau)\,(t-\tau)^{-1/2}\,\mathrm{d}\tau}{\dfrac{1}{\sqrt{\pi\,D_O}}\displaystyle\int_0^t f(\tau)\,e^{-k_f(t-\tau)}(t-\tau)^{-1/2}\,\mathrm{d}\tau} \tag{10.75}$$

In order to integrate, we can expand this expression to give

$$\xi\theta S(t)\int_0^t f(\tau)\,e^{-k_f(t-\tau)}(t-\tau)^{-1/2}\,\mathrm{d}\tau = \left[c_R\sqrt{\pi\,D_R} - \int_0^t f(\tau)\,(t-\tau)^{-1/2}\,\mathrm{d}\tau\right] \tag{10.76}$$

which is again

$$\xi\theta S(\sigma t)\int_0^{\sigma t}\frac{e^{-\frac{k_\mathrm{f}}{\sigma}(\sigma t-z)}\chi(z)}{\sqrt{\sigma t-z}}\,\mathrm{d}z \;=\; \left[1-\int_0^{\sigma t}\frac{\chi(z)}{\sqrt{\sigma t-z}}\,\mathrm{d}z\right] \qquad (10.77)$$

To integrate the term on the left, we put $x=\sqrt{\lambda(\sigma t-z)}$.

$$\int_0^{\sigma t}\frac{e^{-\frac{k_\mathrm{f}}{\sigma}(\sigma t-z)}\chi(z)}{\sqrt{\sigma t-z}}\,\mathrm{d}z \;=\; 2\sqrt{\lambda^{-1}}\int_0^{\sqrt{\lambda\sigma t}}e^{-x^2}\chi(x)\,\mathrm{d}x \qquad (10.78)$$

λ is often called the kinetic parameter ($\lambda = k_\mathrm{f}/\sigma$). Thus, equation (10.77) becomes

$$2\xi\theta S(\sigma t)\sqrt{\lambda^{-1}}\int_0^{\sqrt{\lambda\sigma t}}e^{-x^2}\chi(x)\,\mathrm{d}x \;+\; 2\sqrt{\lambda^{-1}}\int_0^{\sqrt{\lambda\sigma t}}\chi(x)\,\mathrm{d}x = 1 \qquad (10.79)$$

knowing that

$$\mathrm{erf}(x)=\frac{2}{\sqrt{\pi}}\int_0^x e^{-y^2}\,\mathrm{d}y \qquad (10.80)$$

The first integral of equation (10.79) can be integrated by parts

$$\int_0^{\sqrt{\lambda\sigma t}}e^{-x^2}\chi(x)\,\mathrm{d}x \;=\; \frac{\sqrt{\pi}}{2}\left[\mathrm{erf}(x)\chi(x)\right]_0^{\sqrt{\lambda\sigma t}}-\frac{\sqrt{\pi}}{2}\int_0^{\sqrt{\lambda\sigma t}}\mathrm{erf}(x)\mathrm{d}\chi(x) \qquad (10.81)$$

and in the same way, the second integral is

$$\int_0^{\sqrt{\lambda\sigma t}}\chi(x)\,\mathrm{d}x \;=\; \left[x\chi(x)\right]_0^{\sqrt{\lambda\sigma t}}-\int_0^{\sqrt{\lambda\sigma t}}x\,\mathrm{d}\chi(x) \qquad (10.82)$$

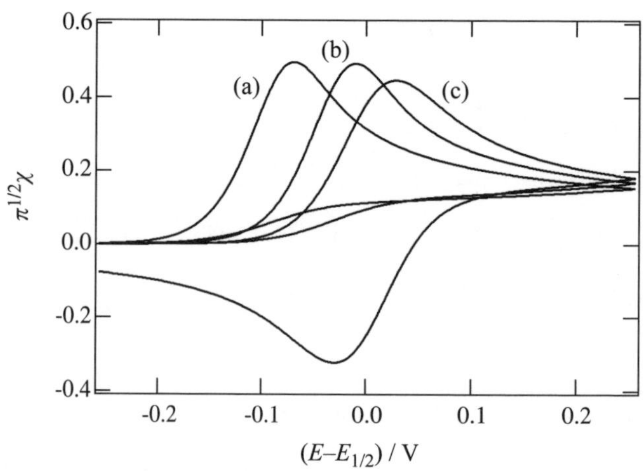

Fig. 10.9 Influence of the chemical reaction rate constant on the cyclic voltammogram. (a) λ=1000, (b) λ=10, (c) λ=0.1.

By substitution, we then have

$$\left[\left(\sqrt{\pi}\theta S(\sigma t)\sqrt{\lambda^{-1}}\operatorname{erf}\sqrt{\lambda\sigma t}+2\sqrt{\sigma t}\right)\chi\left(\sqrt{\lambda\sigma t}\right)\right] - \int_0^{\sqrt{\lambda\sigma t}}\left(\sqrt{\pi}\theta S(\sigma t)\sqrt{\lambda^{-1}}\operatorname{erf}(x)+2\sqrt{\lambda^{-1}}x\right)\mathrm{d}\chi(x) = 1 \tag{10.83}$$

When the kinetic parameter λ is small, we have

$$\operatorname{erf}(\sqrt{\lambda\sigma t}) = \frac{2}{\sqrt{\pi}}\sqrt{\lambda\sigma t} \tag{10.84}$$

and so we fall back on equation (10.25).

Equation (10.83) can be numerically integrated as before.

We see here that cyclic voltammetry is an excellent tool to study the kinetics of chemical reactions in the bulk. Using ultrafast cyclic voltammetry with ultra-microelectrodes, it is possible to access the nanosecond time domain.

10.3.3 $\mathbf{EC_{cat}}$ reactions

The last category of post-redox reactions that we shall look at is that of the catalytic reactions or EC_{cat} reactions (**E**lectrochemically reversible reactions at the electrode followed by a **C**hemical reaction **cat**alycally regenerating the original reduced species)

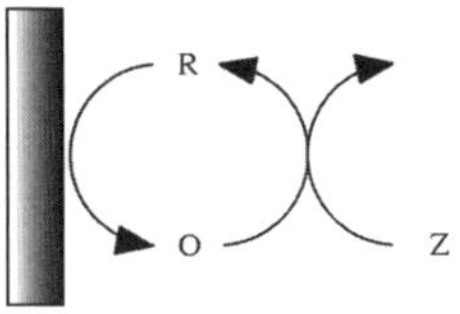

In order to simplify, consider that Z is always in excess, and then the differential equations are reduced to the following

$$\frac{\partial c_{\mathrm{O}}(x,t)}{\partial t} = D_{\mathrm{O}}\frac{\partial^2 c_{\mathrm{O}}(x,t)}{\partial x^2} - k_{\mathrm{f}}c_{\mathrm{O}}(x,t)c_{\mathrm{Z}}(x,t) \tag{10.85}$$

and

$$\frac{\partial c_{\mathrm{R}}(x,t)}{\partial t} = D_{\mathrm{R}}\frac{\partial^2 c_{\mathrm{R}}(x,t)}{\partial x^2} + k_{\mathrm{f}}c_{\mathrm{O}}(x,t)c_{\mathrm{Z}}(x,t) \tag{10.86}$$

with the initial boundary conditions for an oxidation given by

$$c_{\mathrm{R}}(x,0) = c_{\mathrm{R}} \quad \text{and} \quad c_{\mathrm{O}}(x,0) = c_{\mathrm{O}} \approx 0 \quad \text{and} \quad c_{\mathrm{Z}}(x,t) = c_{\mathrm{Z}} \tag{10.87}$$

with as before the hypothesis that the initial potential is far enough from the standard redox potential so that the initial concentration of the oxidised species can be neglected. Considering a large volume of the solution, we also have the following bulk boundary conditions

$$\lim_{x\to\infty} c_{\mathrm{R}}(x,t) = c_{\mathrm{R}} \quad \text{and} \quad \lim_{x\to\infty} c_{\mathrm{O}}(x,t) = c_{\mathrm{O}} \approx 0 \tag{10.88}$$

The Laplace transforms of reactions (10.85) and (10.86) are then

$$s\bar{c}_{\mathrm{O}}(x,s) - c_{\mathrm{O}}(x,0) \;\approx\; s\bar{c}_{\mathrm{O}}(x,s) \;=\; D_{\mathrm{O}}\,\frac{\partial^2 \bar{c}_{\mathrm{O}}(x,s)}{\partial x^2} - k_{\mathrm{f}} c_{\mathrm{Z}} \bar{c}_{\mathrm{O}}(x,s) \tag{10.89}$$

$$s\bar{c}_{\mathrm{R}}(x,s) - c_{\mathrm{R}}(x,0) \;=\; D_{\mathrm{R}}\,\frac{\partial^2 \bar{c}_{\mathrm{R}}(x,s)}{\partial x^2} + k_{\mathrm{f}} c_{\mathrm{Z}} \bar{c}_{\mathrm{O}}(x,s) \tag{10.90}$$

First, we have to solve equation (10.89) and bring the concentration of $\bar{c}_{\mathrm{O}}(x, s)$ into equation (10.90). The Laplace transform of the concentration of O is, as before (see equation (10.66))

$$\bar{c}_{\mathrm{O}}(x,s) \;=\; B(s)\, e^{-\sqrt{\frac{s+k_{\mathrm{f}}c_{\mathrm{Z}}}{D_{\mathrm{O}}}}\,x} \tag{10.91}$$

The Laplace transform of the current expressed as a function of the diffusion flux of O is

$$\bar{I}(s) \;=\; -nFA\,D_{\mathrm{O}}\left[\frac{\partial \bar{c}_{\mathrm{O}}(x,s)}{\partial x}\right]_{x=0} \;=\; nFA\,D_{\mathrm{O}}\left[B(s)\sqrt{\frac{s+k_{\mathrm{f}}c_{\mathrm{Z}}}{D_{\mathrm{O}}}}\right] \tag{10.92}$$

which allows us to determine the constant $B(s)$

$$B(s) \;=\; \frac{\bar{I}(s)}{nFA\,D_{\mathrm{O}}}\sqrt{\frac{D_{\mathrm{O}}}{s+k_{\mathrm{f}}c_{\mathrm{Z}}}} \tag{10.93}$$

Thus, the Laplace transform of the concentration of the oxidised species is

$$\bar{c}_{\mathrm{O}}(x,s) \;=\; \frac{\bar{I}(s)}{nFA\,D_{\mathrm{O}}}\sqrt{\frac{D_{\mathrm{O}}}{s+k_{\mathrm{f}}c_{\mathrm{Z}}}}\,\exp^{-\sqrt{\frac{s+k_{\mathrm{f}}c_{\mathrm{Z}}}{D_{\mathrm{O}}}}\,x} \tag{10.94}$$

Equation (10.90) then becomes

$$s\bar{c}_{\mathrm{R}}(x,s) - c_{\mathrm{R}}(x,0) \;=\; D_{\mathrm{R}}\,\frac{\partial^2 \bar{c}_{\mathrm{R}}(x,s)}{\partial x^2} + \frac{k_{\mathrm{f}}c_{\mathrm{Z}}\bar{I}(s)}{nFA\,D_{\mathrm{O}}}\sqrt{\frac{D_{\mathrm{O}}}{s+k_{\mathrm{f}}c_{\mathrm{Z}}}}\,\exp^{-\sqrt{\frac{s+k_{\mathrm{f}}c_{\mathrm{Z}}}{D_{\mathrm{O}}}}\,x} \tag{10.95}$$

which is a differential equation of the type:

$$\frac{\partial^2 y(x)}{\partial x^2} + a\cdot\exp^{-bx} - c\cdot y(x) = d \tag{10.96}$$

The solution of this equation is the sum of a particular solution and the solution of the homogeneous equation. To solve the homogeneous equation, we can use, for example, the Laplace transformation method. The transform of the homogeneous equation is:

$$s^2\bar{y}(s)-sy(0)-y'(0)+\frac{a}{s+b}-c\bar{y}(s)=0 \tag{10.97}$$

which, by factorising, yields

$$\bar{y}(s)=\left(sy(0)+y'(0)-\frac{a}{s+b}\right)/\left(s^2-c\right)=\frac{u(s)}{s+\sqrt{c}}+\frac{v(s)}{s-\sqrt{c}}+\frac{w(s)}{s+b} \tag{10.98}$$

Taking the inverse transform, we then get

$$y(x) \quad = \quad u(s)\exp^{-x\sqrt{s}}+v(s)\exp^{x\sqrt{s}}+w(s)\exp^{-bx} \tag{10.99}$$

Coming back to equation (10.95), the solution of the homogeneous equation is then

$$\bar{c}_R(x,s) \quad = \quad u(s)\exp^{-x\sqrt{\frac{s}{D_R}}}+v(s)\exp^{x\sqrt{\frac{s}{D_R}}}+w(s)\exp^{-\sqrt{\frac{s+k_f c_Z}{D_O}}\,x} \tag{10.100}$$

Applying the boundary conditions (10.88) shows that $v(s)=0$ since the concentration cannot tend to infinity.

The particular solution is also given by equation (10.88) and so we have

$$\bar{c}_R(x,s) \quad = \quad \frac{c_R}{s}+u(s)\exp^{-x\sqrt{\frac{s}{D_R}}}+w(s)\exp^{-\sqrt{\frac{s+k_f c_Z}{D_O}}\,x} \tag{10.101}$$

To determine the two other integration constants, we can use the definition of the current:

$$\bar{I}(s) \quad = \quad nFA\,D_R\left[\frac{\partial \bar{c}_R(x,s)}{\partial x}\right]_{x=0} \quad = \quad -nFA\,D_R\left[u(s)\sqrt{\frac{s}{D_R}}+w(s)\sqrt{\frac{s+k_f c_Z}{D_O}}\right] \tag{10.102}$$

We can also use the transform of the mass conservation law

$$\int_0^\infty\left[\bar{c}_R(x,s)-\frac{c_R}{s}\right]\mathrm{d}x+\int_0^\infty \bar{c}_O(x,s)\,\mathrm{d}x \quad = \quad 0 \tag{10.103}$$

which is, by replacing

$$\left[u(s)\sqrt{\frac{D_R}{s}}+w(s)\sqrt{\frac{D_O}{s+k_f c_Z}}\right]+\frac{\bar{I}(s)}{nFAD_O}\left(\frac{D_O}{s+k_f c_Z}\right) \quad = \quad 0 \tag{10.104}$$

In the simple case $D_O=D_R$, we find that $u(s)=0$ and that

$$w(s) \quad = \quad -\frac{\bar{I}(s)}{nFA\,D}\sqrt{\frac{D}{s+k_f c_Z}} \tag{10.105}$$

Finally, the Laplace transform of the concentration of the reduced species is

$$\bar{c}_R(x,s) \quad = \quad \frac{c_R}{s}-\frac{\bar{I}(s)}{nFA\,D}\sqrt{\frac{D}{s+k_f c_Z}}\exp^{-\sqrt{\frac{s+k_f c_Z}{D}}\,x} \tag{10.106}$$

The Laplace transform of the interfacial concentration in O is then:

$$\bar{c}_{\mathrm{O}}(0,s) = \frac{\bar{I}(s)}{nFA\,D}\sqrt{\frac{D}{s+k_{\mathrm{f}}c_{\mathrm{Z}}}} \tag{10.107}$$

and that of R

$$\bar{c}_{\mathrm{R}}(x,s) = \frac{c_{\mathrm{R}}}{\mathrm{s}} - \frac{\bar{I}(s)}{nFA\,D}\sqrt{\frac{D}{s+k_{\mathrm{f}}c_{\mathrm{Z}}}} \tag{10.108}$$

As before $L^{-1}\{(s+a)^{-1/2}\} = e^{-at}/\sqrt{\pi t}$, and the interfacial concentration of O is

$$c_{\mathrm{O}}(0,t) = \frac{1}{\sqrt{\pi D}}\int_0^t f(\tau)\,e^{-k_{\mathrm{f}}c_{\mathrm{Z}}(t-\tau)}(t-\tau)^{-1/2}\,\mathrm{d}\tau \tag{10.109}$$

and that of R

$$c_{\mathrm{R}}(0,t) = c_{\mathrm{R}} - \frac{k_{\mathrm{f}}c_{\mathrm{Z}}}{\sqrt{\pi D}}\int_0^t f(\tau)\exp^{-k_{\mathrm{f}}c_{\mathrm{Z}}(t-\tau)}(t-\tau)^{-1/2}\,\mathrm{d}\tau \tag{10.110}$$

Substituting in the Nernst equation, we get

$$\theta S(t) = \frac{c_{\mathrm{R}}(0,t)}{c_{\mathrm{O}}(0,t)} = \frac{c_{\mathrm{R}} - \dfrac{1}{\sqrt{\pi D}}\displaystyle\int_0^t f(\tau)\,e^{-k_{\mathrm{f}}c_{\mathrm{Z}}(t-\tau)}(t-\tau)^{-1/2}\,\mathrm{d}\tau}{\dfrac{1}{\sqrt{\pi D}}\displaystyle\int_0^t f(\tau)\,e^{-k_{\mathrm{f}}c_{\mathrm{Z}}(t-\tau)}(t-\tau)^{-1/2}\,\mathrm{d}\tau} \tag{10.111}$$

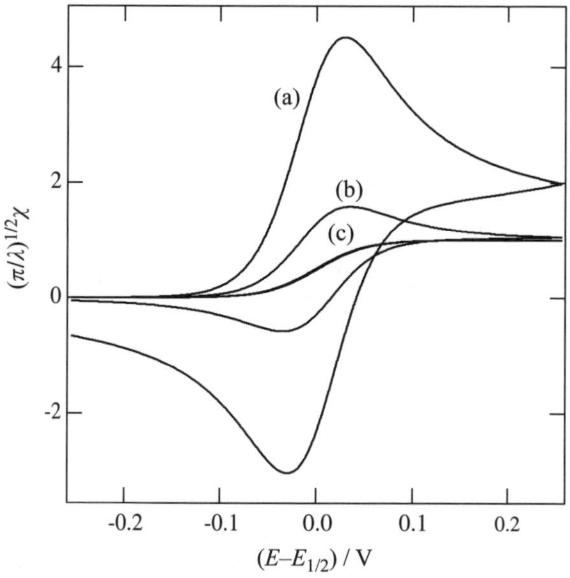

Fig. 10.10 Influence of the kinetic parameter λ on the cyclic voltammetry of catalytic reactions, the representation being normalised by the limiting current. (a) $\lambda = 0.01$, (b) $\lambda = 0.1$, (c) $\lambda = 10$.

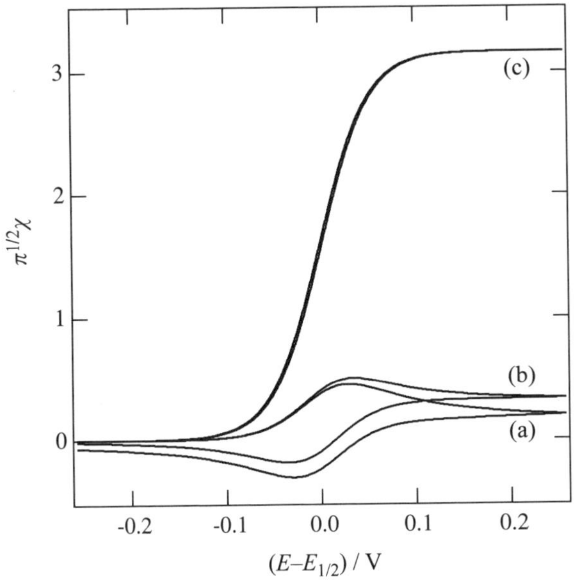

Fig. 10.11 Influence of the parameter λ on the cyclic voltammograms for EC_{cat} reactions. (a) $\lambda = 0.01$, (b) $\lambda = 0.1$, (c) $\lambda = 10$.

This equation can be integrated numerically by parts as before, by making the same change in the variables. Here, we define the kinetic parameter λ

$$\lambda = \sqrt{\frac{k_f c_Z}{\sigma}} = \sqrt{\frac{k_f c_Z RT}{nF v}} \tag{10.112}$$

We can see in Figures 10.10 and 10.11 that for high values of λ the current tends to a limiting value.

The limiting current is simply given by

$$I_\infty = nFAc_R\sqrt{D\,k_f c_Z} \tag{10.113}$$

The wave obtained for large catalytic recyclings is simply given by

$$I = \frac{nFAc_R\sqrt{D\,k_f c_Z}}{1+\exp^{nF(E-E_{1/2})/RT}} \tag{10.114}$$

For efficient catalytic recycling, the current potential response is similar to that obtained in steady state amperometry (see Figure 7.13). The major difference here is that the limiting plateau current is controlled by the kinetics of the catalytic reaction.

10.4 ELECTRON TRANSFER AT LIQUID | LIQUID INTERFACES

Electron transfer reactions at liquid | liquid interfaces, such as those shown in Figure 2.16, are interesting from a mass transfer point of view. In effect, we need to consider the arrival of two reactants at the surface and the departure of two products away from it.

For a reaction of the type

$$O_1^{w} + R_2^{o} \rightleftarrows R_1^{w} + O_2^{o}$$

where the O_1/R_1 redox couple is in the aqueous phase and the O_2/R_2 redox couple is in the organic phase, there are then four differential equations to solve. Two for the aqueous phase

$$\frac{\partial c_{O_1}(x,t)}{\partial t} = D_{O_1}\frac{\partial^2 c_{O_1}(x,t)}{\partial x^2} \tag{10.115}$$

$$\frac{\partial c_{R_1}(x,t)}{\partial t} = D_{R_1}\frac{\partial^2 c_{R_1}(x,t)}{\partial x^2} \tag{10.116}$$

and two for the organic pahse

$$\frac{\partial c_{O_2}(x,t)}{\partial t} = D_{O_2}\frac{\partial^2 c_{O_2}(x,t)}{\partial x^2} \tag{10.117}$$

$$\frac{\partial c_{R_2}(x,t)}{\partial t} = D_{R_2}\frac{\partial^2 c_{R_2}(x,t)}{\partial x^2} \tag{10.118}$$

The initial boundary conditions are

$$c_{O_1}(x,0) = c_{O_1} \quad \text{and} \quad c_{R_2}(x,0) = c_{R_2} \tag{10.119}$$

$$c_{R_1}(x,0) = c_{R_1} \quad \text{and} \quad c_{O_2}(x,0) = c_{O_2} \tag{10.120}$$

and the bulk boundary conditions are

$$\lim_{x\to\infty} c_{O_1}(x,t) = c_{O_1} \quad \text{and} \quad \lim_{x\to-\infty} c_{R_2}(x,t) = c_{R_2} \tag{10.121}$$

$$\lim_{x\to\infty} c_{R_1}(x,t) = c_{R_1} \quad \text{and} \quad \lim_{x\to-\infty} c_{O_2}(x,t) = c_{O_2} \tag{10.122}$$

The equalities of the diffusion fluxes at the interface are

$$D_{O_1}\left(\frac{\partial c_{O_1}(x,t)}{\partial x}\right)_{x=0} + D_{R_2}\left(\frac{\partial c_{R_2}(x,t)}{\partial x}\right)_{x=0} = 0 \tag{10.123}$$

$$D_{R_1}\left(\frac{\partial c_{R_1}(x,t)}{\partial x}\right)_{x=0} + D_{O_2}\left(\frac{\partial c_{O_2}(x,t)}{\partial x}\right)_{x=0} = 0 \tag{10.124}$$

As in §10.1, going via the Laplace transformations of the Fick equations and the introduction of the notion of convoluted current allows us to obtain the Laplace transforms of the concentrations of the reactants and products

$$\bar{c}_{O_1}(0,s) = \frac{c_{O_1}}{s} - \frac{\bar{I}(s)}{nFA\,D_{O_1}}\sqrt{\frac{D_{O_1}}{s}} \tag{10.125}$$

$$\bar{c}_{R_2}(0,s) = \frac{c_{R_2}}{s} - \frac{\bar{I}(s)}{nFA\,D_{R_2}}\sqrt{\frac{D_{R_2}}{s}} \tag{10.126}$$

$$\bar{c}_{R_1}(0,s) = \frac{c_{R_1}}{s} + \frac{\bar{I}(s)}{nFA\,D_{R_1}}\sqrt{\frac{D_{R_1}}{s}} \tag{10.127}$$

$$\bar{c}_{O_2}(0,s) = \frac{c_{O_2}}{s} + \frac{\bar{I}(s)}{nFA\,D_{O_2}}\sqrt{\frac{D_{O_2}}{s}} \tag{10.128}$$

The Nernst equation (2.67) allows us to establish a ratio between the surface concentrations

$$\Delta_o^w\phi = \Delta_o^w\phi^{\ominus\prime} + \frac{RT}{nF}\ln\left(\frac{c_{R_1}^w(0,t)c_{O_2}^o(0,t)}{c_{O_1}^w(0,t)c_{R_2}^o(0,t)}\right) \tag{10.129}$$

Again using the notations of equation (10.8), we write

$$\frac{c_{R_1}(0,t)c_{O_2}(0,t)}{c_{O_1}(0,t)c_{R_2}(0,t)} = \theta\,e^{-\sigma t} = \theta S(t) = \frac{\left[c_{R_1} + \dfrac{\hat{I}}{\sqrt{\pi D_{R_1}}}\right]\left[c_{O_2} + \dfrac{\hat{I}}{\sqrt{\pi D_{O_2}}}\right]}{\left[c_{O_1} - \dfrac{\hat{I}}{\sqrt{\pi D_{O_1}}}\right]\left[c_{R_2} - \dfrac{\hat{I}}{\sqrt{\pi D_{R_2}}}\right]} \tag{10.130}$$

By substitution, we find that the convoluted current is the root of the quadratic equation

$$a\hat{I}(t)^2 + b\hat{I}(t) + c = 0 \tag{10.131}$$

with, taking all the diffusion coefficients as equal,

$$a = \frac{\theta S(t) - 1}{\pi D} \tag{10.132}$$

$$b = \frac{\theta S(t)}{\sqrt{\pi D}}\left[c_{O_1} + c_{R_2}\right] + c_{R_1} + c_{O_2} \tag{10.133}$$

$$c = c_{O_1}c_{R_2}\theta S(t) - c_{R_1}c_{O_2} \tag{10.134}$$

By defining the following parameters

$$c_1 = c_{O_1} + c_{R_1} \qquad c_2 = c_{O_2} + c_{R_2} \qquad \kappa = c_2 / c_1 \tag{10.135}$$

$$\alpha \;=\; c_{\mathrm{R}_1}/c_1 \qquad \beta \;=\; c_{\mathrm{R}_2}/c_2 \tag{10.136}$$

the coefficients *a*, *b* and *c* reduce to

$$a \;=\; (\theta S(t)-1)/\pi D \tag{10.137}$$

$$b \;=\; \frac{c_1}{\sqrt{\pi D}}\big[[(1-\alpha)+\beta\kappa]\theta S(t)+\alpha+(1-\beta)\kappa\big] \tag{10.138}$$

$$c \;=\; \kappa c_1^2\big[(1-\alpha)\beta\theta S(t)-\alpha(1-\beta)\big] \tag{10.139}$$

So the convoluted current is now

$$\hat{I}(t) \;=\; \frac{-b+\sqrt{b^2-4ac}}{2a} \;=\; \frac{c_1}{\sqrt{\pi D}}f(\alpha,\beta,\kappa) \tag{10.140}$$

As before, we can do a numerical integration avoiding the discontinuity at $a=0$ and $\Delta_\mathrm{o}^\mathrm{w}\phi = \Delta_\mathrm{o}^\mathrm{w}\phi^{\ominus\prime}$. Thus, taking up the definitions in §10.1, we have

$$\int_0^{\sigma t}\frac{g(z)}{c_1\sqrt{\pi D\sigma}}\frac{\mathrm{d}z}{\sqrt{\sigma t-z}} \;=\; \int_0^{\sigma t}\frac{\chi(z)\,\mathrm{d}z}{\sqrt{\sigma t-z}} \;=\; f(\alpha,\beta,\kappa,\theta S(t)) \tag{10.141}$$

For the case $\alpha = 0.5$, i.e. when there is an equimolar mixture of oxidised and reduced species in a phase, e.g. the aqueous phase, for small values of κ, the mass transfer is limited by the diffusion of the species in an organic phase, and the situation reduces to that of redox reactions on solid electrodes. When κ increases, diffusion in the aqueous phase plays a role as demonstrated by the data in Figure 10.12.

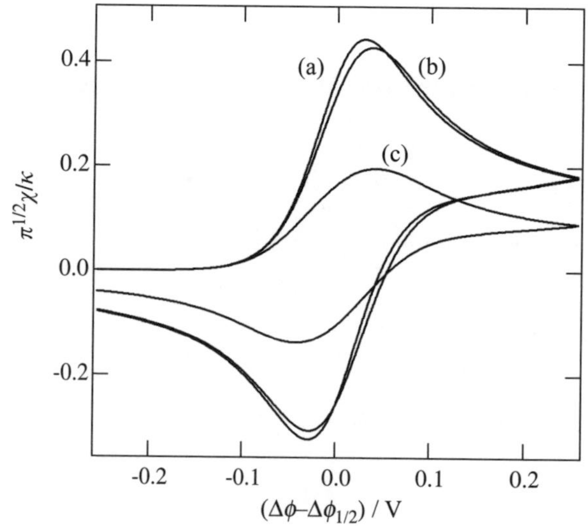

Fig. 10.12 Cyclic voltammogram for electron transfer reactions at liquid | liquid interfaces. The curves are normalised by the ratio κ of the concentrations. $\alpha = 0.5$, (a) $\kappa = 0.001$, (b) $\kappa = 0.1$ and (c) $\kappa = 1$.

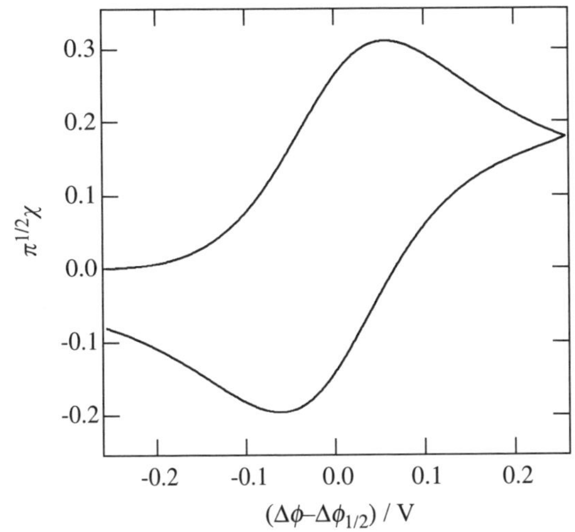

Fig. 10.13 Cyclic voltammogram of electron transfer reactions at liquid | liquid interfaces. $\alpha = 1$, $\beta = 0$ and $\kappa = 1$.

Another interesting case is when $\alpha \approx 1$, $\beta \approx 0$ and $\kappa = 1$. Here, there is a separation of about 120 mV between the peaks as if each reactant diffusing towards the interface carried a charge of 1/2. The forward peak current is such that $\sqrt{\pi}\chi = 0.4463 \cdot \sqrt{2}$.

Apart from these two extreme cases, there are no particular criteria for studying electron transfer reaction at liquid | liquid interfaces, and a numerical integration is necessary to analyse each experimental case.

10.5 ASSISTED ION TRANSFER AT LIQUID | LIQUID INTERFACES

Ion transfer reactions assisted by a ligand, either hydrophilic or lipophilic, represent an interesting class of electrochemical reactions where cyclic voltammetry provides precious information on the type of reaction.

It is possible to distinguish 3 types of ion transfer reactions from the aqueous phase to the organic phase assisted by the presence of a ligand or ionophore:

- Aqueous Complexation reactions followed by Transfer (ACT)
- Transfer reactions followed by Organic phase Complexation reactions (TOC),
- Transfer by Interfacial Complexation, Transfer by Interfacial Dissociation (TIC/TID)

In the simple case of the transfer of an ion M^+ by a ligand L, partially soluble in the two phases, we can consider three equilibria: the two equilibria of the complexation reactions in the adjacent phases and the equilibrium for the distribution of the ligand between the two phases

$$K_a^w = \frac{c_{M^+L}^w}{c_{M^+}^w c_L^w} \qquad K_a^o = \frac{c_{M^+L}^o}{c_{M^+}^o c_L^o} \qquad K_D = \frac{c_L^o}{c_L^w} \tag{10.142}$$

From a mass transfer point of view, we need to consider the diffusion of the ion, of the ligand and of the complex in the two phases. Therefore, we have three equations per phase.

$$\frac{\partial c_{M^+}}{\partial t} = D_{M^+} \frac{\partial^2 c_{M^+}}{\partial x^2} - k_f c_{M^+} c_L + k_b c_{ML} \tag{10.143}$$

$$\frac{\partial c_L}{\partial t} = D_L \frac{\partial^2 c_L}{\partial x^2} - k_f c_{M^+} c_L + k_b c_{ML} \tag{10.144}$$

$$\frac{\partial c_{ML}}{\partial t} = D_{ML} \frac{\partial^2 c_{ML}}{\partial x^2} + k_f c_{M^+} c_L - k_b c_{ML} \tag{10.145}$$

with the following initial conditions

$$c_{M^+}^w(x,0) = c_{M^+}^w \qquad c_L^w(x,0) = c_L^w \qquad c_{ML^+}^w(x,0) = c_{ML^+}^w \tag{10.146}$$

$$c_{M^+}^o(x,0) = c_{M^+}^o \qquad c_L^o(x,0) = c_L^o \qquad c_{ML^+}^o(x,0) = c_{ML^+}^o \tag{10.147}$$

and the following bulk boundary conditions for the aqueous phase

$$\lim_{x\to\infty} c_{M^+}^w(x,t) = c_{M^+}^w \quad \lim_{x\to\infty} c_L^w(x,t) = c_L^w \quad \lim_{x\to\infty} c_{ML^+}^w(x,t) = c_{ML^+}^w \tag{10.148}$$

and for the organic phase

$$\lim_{x\to-\infty} c_{M^+}^o(x,t) = c_{M^+}^o \quad \lim_{x\to-\infty} c_L^o(x,t) = c_L^o \quad \lim_{x\to-\infty} c_{ML^+}^o(x,t) = c_{ML^+}^o \tag{10.149}$$

If we make the hypothesis that the kinetics of the complexation and decomplexation are rapid, it is convenient as in §10.3.1, to consider virtual total concentrations of the metal and the ligand.

$$c_{Mtot} = c_{M^+} + c_{ML} \qquad c_{Ltot} = c_L + c_{ML} \tag{10.150}$$

Thus, by considering that all the diffusion coefficients in a phase are equal, equations (10.143)-(10.145) reduce to

$$\frac{\partial c_{Mtot}}{\partial t} = D \frac{\partial^2 c_{Mtot}}{\partial x^2} \tag{10.151}$$

$$\frac{\partial c_{Ltot}}{\partial t} = D \frac{\partial^2 c_{Ltot}}{\partial x^2} \tag{10.152}$$

The conditions for the interfacial concentrations are given by the Nernst equation for the transfer of the M^+ ion, i.e.

$$\frac{c_{M^+}^o}{c_{M^+}^w} = \exp^{F(\Delta_o^w\phi - \Delta_o^w\phi_{M^+}^{\ominus\prime})/RT} = \theta_M S(t) \tag{10.153}$$

and by the Nernst equation for the complex

$$\frac{c_{\mathrm{ML^+}}^{\mathrm{o}}}{c_{\mathrm{ML^+}}^{\mathrm{w}}} = \exp^{F(\Delta_{\mathrm{o}}^{\mathrm{w}}\phi - \Delta_{\mathrm{o}}^{\mathrm{w}}\phi_{\mathrm{ML^+}}^{\ominus\prime})/RT} = \theta_{\mathrm{ML}}S(t) \tag{10.154}$$

The standard transfer potential of the complex is calculated from that of the transfer of the ion and the different equilibrium constants. Thus, we have

$$\theta_{\mathrm{ML}} = \theta_{\mathrm{M}}\left(\frac{K_{\mathrm{a}}^{\mathrm{o}}c_{\mathrm{L}}^{\mathrm{o}}}{K_{\mathrm{a}}^{\mathrm{w}}c_{\mathrm{L}}^{\mathrm{w}}}\right)_{x=0} = \theta_{\mathrm{M}}\frac{K_{\mathrm{a}}^{\mathrm{o}}K_D}{K_{\mathrm{a}}^{\mathrm{w}}} \tag{10.155}$$

With the total concentrations defined above, the equations for the equality of the fluxes are

$$-D^{\mathrm{w}}\frac{\partial c_{\mathrm{Mtot}}^{\mathrm{w}}}{\partial x} = D^{\mathrm{o}}\frac{\partial c_{\mathrm{Mtot}}^{\mathrm{o}}}{\partial x} = f(t) \tag{10.156}$$

and

$$-D^{\mathrm{w}}\frac{\partial c_{\mathrm{Ltot}}^{\mathrm{w}}}{\partial x} = D^{\mathrm{o}}\frac{\partial c_{\mathrm{Ltot}}^{\mathrm{o}}}{\partial x} = g(t) \tag{10.157}$$

$f(t)$ is the total flux of ions and $g(t)$ the total flux of the ligand.

As before, we can solve the Laplace transforms of equations (10.156)-(10.157) to obtain

$$c_{\mathrm{Mtot}}^{\mathrm{w}}(0,t) = c_{\mathrm{Mtot}}^{\mathrm{w}} - \frac{1}{\sqrt{\pi D^{\mathrm{w}}}}\int_0^t \frac{f(t)}{\sqrt{t-\tau}}\mathrm{d}\tau \tag{10.158}$$

$$c_{\mathrm{Mtot}}^{\mathrm{o}}(0,t) = c_{\mathrm{Mtot}}^{\mathrm{o}} - \frac{1}{\sqrt{\pi D^{\mathrm{o}}}}\int_0^t \frac{f(t)}{\sqrt{t-\tau}}\mathrm{d}\tau \tag{10.159}$$

and

$$c_{\mathrm{Ltot}}^{\mathrm{w}}(0,t) = c_{\mathrm{Ltot}}^{\mathrm{w}} - \frac{1}{\sqrt{\pi D^{\mathrm{w}}}}\int_0^t \frac{g(t)}{\sqrt{t-\tau}}\mathrm{d}\tau \tag{10.160}$$

$$c_{\mathrm{Ltot}}^{\mathrm{o}}(0,t) = c_{\mathrm{Ltot}}^{\mathrm{o}} - \frac{1}{\sqrt{\pi D^{\mathrm{o}}}}\int_0^t \frac{g(t)}{\sqrt{t-\tau}}\mathrm{d}\tau \tag{10.161}$$

This results in a system of two equations

$$c_{\mathrm{Mtot}}^{\mathrm{w}}(0,t) + \xi c_{\mathrm{Mtot}}^{\mathrm{o}}(0,t) = c_{\mathrm{Mtot}}^{\mathrm{w}} + \xi c_{\mathrm{Mtot}}^{\mathrm{o}} = c_{\mathrm{MT}} \tag{10.162}$$

$$c_{\mathrm{Ltot}}^{\mathrm{w}}(0,t) + \xi c_{\mathrm{Ltot}}^{\mathrm{o}}(0,t) = c_{\mathrm{Ltot}}^{\mathrm{w}} + \xi c_{\mathrm{Ltot}}^{\mathrm{o}} = c_{\mathrm{LT}} \tag{10.163}$$

which is by substitution

$$c_{\mathrm{M}}^{\mathrm{w}}(0,t)\left[\left(1+\xi\theta_{\mathrm{M}}S(t)\right) + c_{\mathrm{L}}^{\mathrm{w}}(0,t)K_{\mathrm{a}}^{\mathrm{w}}\left(1+\xi\theta_{\mathrm{ML}}S(t)\right)\right] = c_{\mathrm{MT}} \tag{10.164}$$

$$c_{\mathrm{L}}^{\mathrm{w}}(0,t)\left[\left(1+\xi K_D\right) + c_{\mathrm{M}}^{\mathrm{w}}(0,t)K_{\mathrm{a}}^{\mathrm{w}}\left(1+\xi\theta_{\mathrm{ML}}S(t)\right)\right] = c_{\mathrm{LT}} \tag{10.165}$$

We have therefore a quadratic equation which allows us to calculate $c_{\mathrm{M}}^{\mathrm{w}}(0,t)$ and $c_{\mathrm{L}}^{\mathrm{w}}(0,t)$. From this, we then deduce the total interfacial concentration

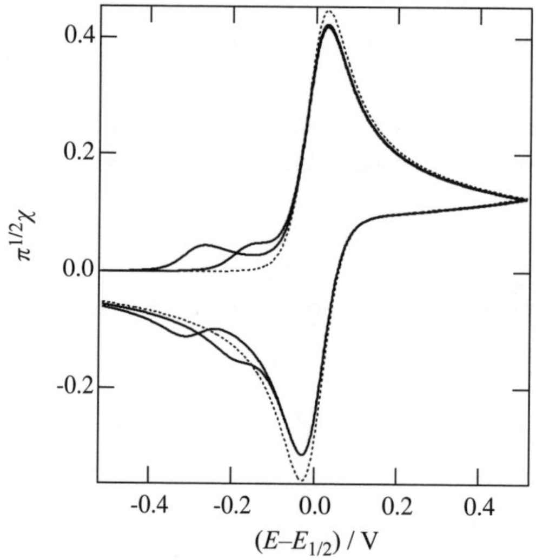

Fig. 10.14 Cyclic voltammograms for ion transfer assisted by a ligand. Without the ligand is the dotted line, from left to right $K_{ao} = 10^5$ and $K_{ao} = 10^3$ ($K_D = 1000$, $K_{aw} = 0.001$).

$$c_{\text{Mtot}}^{\text{w}}(0,t) = c_{\text{M}}^{\text{w}}(0,t)\left[1 + K_{\text{a}}^{\text{w}} c_{\text{L}}^{\text{w}}(0,t)\right] \tag{10.166}$$

By replacement in equation (10.158), it is possible to calculate the total charge flux by numerical integration. The example in Figure 10.14 shows clearly the assisted transfer of the ion before the transfer of the ion itself. The higher the complexation constant in the organic phase, the more the transfer is facilitated and thus happens at more negative potentials.

10.6 SURFACE REACTIONS

10.6.1 Thin layer cell

We saw in chapters 8 and 9 the influence of a finite diffusion layer on the impedance. For cyclic voltammetry, the effect of a diffusion layer with a limited thickness has also important consequences.

As in §10.1.1, consider a system comprising a reduced species with the following initial conditions

$$c_{\text{R}}(x,0) = c_{\text{R}} \quad \text{and} \quad c_{\text{O}}(x,0) = c_{\text{O}} \approx 0 \tag{10.167}$$

Again, let us make the hypothesis that the redox reaction is reversible, and that the interfacial concentrations obey the Nernst equation

$$\frac{c_{\text{R}}(0,t)}{c_{\text{O}}(0,t)} = \theta\, e^{-\sigma t} = \theta S(t) \tag{10.168}$$

with

$$\theta = \exp\left[-\frac{nF}{RT}(E_i - E^{\ominus\prime})\right] = \frac{c_R(x,0)}{c_O(x,0)} = \frac{c_R}{c_O} \tag{10.169}$$

and

$$\sigma = \frac{nF}{RT}\nu \tag{10.170}$$

For restricted volumes of solution, we can consider an equation for the conservation of mass such that the concentrations are considered to be homogeneous throughout the volume

$$c_{tot} = c_R(t) + c_O(t) \tag{10.171}$$

By substitution, we get

$$c_R(t) = c_{tot}\frac{\exp\left[-\frac{nF}{RT}(E - E^{\ominus\prime})\right]}{1+\exp\left[-\frac{nF}{RT}(E - E^{\ominus\prime})\right]} \tag{10.172}$$

The current can be defined from the amount of charge passed which is

$$I = -nFV\left(\frac{dc_R(t)}{dt}\right) \tag{10.173}$$

By differentiating equation (10.173), we get

$$I = \frac{n^2F^2\,\nu\,V\,c_{tot}}{RT}\frac{\exp\left[-\frac{nF}{RT}(E - E^{\ominus\prime})\right]}{\left[1+\exp\left[-\frac{nF}{RT}(E - E^{\ominus\prime})\right]\right]^2} \tag{10.174}$$

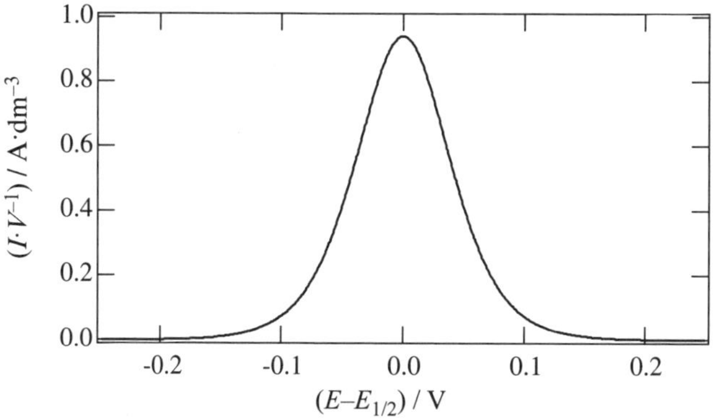

Fig. 10.15 Linear sweep voltammogram for an oxidation in a thin layer cell. $c_R = 1$ mM, $\upsilon = 1$ mV·s^{-1}.

One of the characteristics of cyclic voltammetry in a thin layer cell is that the peak current is directly proportional to the scan rate. Another major characteristic is the symmetry of the peak as shown in Figure 10.15.

The current on the return sweep is the opposite of that on the forward cycle. The forward and return peaks are therefore both centred on the formal potential.

10.6.2 Cyclic voltammetry for adsorbed species

For adsorbed species, equation 10.174 can be expressed as a function of the total interfacial concentration. By the same demonstration, we can show that

$$I = \frac{n^2F^2\, v\, A\, \Gamma_{\text{tot}}}{RT} \frac{\exp\left[-\frac{nF}{RT}(E-E^{\ominus\prime})\right]}{\left[1+\exp\left[-\frac{nF}{RT}(E-E^{\ominus\prime})\right]\right]^2} \qquad (10.175)$$

10.7 HEMI-SPHERICAL DIFFUSION

When carrying out cyclic voltammetry on hemispherical electrodes (or even on microdisc electrodes) we should obtain at low scan rates a voltammogram resembling the steady state current-potential response as obtained in §7.4.2. To demonstrate this, we shall consider the differential Fick equations in spherical coordinates

$$\frac{\partial c_{\text{O}}(r,t)}{\partial t} = D_{\text{O}}\left[\frac{\partial^2 c_{\text{O}}(r,t)}{\partial r^2} + \frac{2}{r}\frac{\partial c_{\text{O}}(r,t)}{\partial r}\right] \qquad (10.176)$$

and

$$\frac{\partial c_{\text{R}}(r,t)}{\partial t} = D_{\text{R}}\left[\frac{\partial^2 c_{\text{R}}(r,t)}{\partial r^2} + \frac{2}{r}\frac{\partial c_{\text{R}}(r,t)}{\partial r}\right] \qquad (10.177)$$

with the initial boundary conditions

$$c_{\text{R}}(r,0) = c_{\text{R}} \quad \text{and} \quad c_{\text{O}}(r,0) = c_{\text{O}} \approx 0 \qquad (10.178)$$

and the following bulk conditions

$$\lim_{r\to\infty} c_{\text{R}}(r,t) = c_{\text{R}} \quad \text{and} \quad \lim_{r\to\infty} c_{\text{O}}(r,t) = c_{\text{O}} \approx 0 \qquad (10.179)$$

The equality of diffusion fluxes at the interface simply reads

$$D_{\text{O}}\left(\frac{\partial c_{\text{O}}(r,t)}{\partial r}\right)_{r=r_{\text{e}}} + D_{\text{R}}\left(\frac{\partial c_{\text{R}}(r,t)}{\partial r}\right)_{r=r_{\text{e}}} = 0 \qquad (10.180)$$

As before, to use the Nernst equation as boundary conditions, it is always preferable to define the following dimensionless parameters:

$$\frac{c_{\text{R}}(r_{\text{e}},t)}{c_{\text{O}}(r_{\text{e}},t)} = \theta\, e^{-\sigma t} = \theta S(t) \qquad (10.181)$$

with

$$\theta = \exp\left[-\frac{nF}{RT}(E_i - E^{\ominus\prime})\right] = \frac{c_R(r_e,0)}{c_O(r_e,0)} = \frac{c_R}{c_O} \tag{10.182}$$

and again

$$\sigma = \frac{nF}{RT}v \tag{10.183}$$

To solve the Fick equations in spherical coordinates, there are several methods involving changes of variables. The simplest method is the one used in §8.1.2, which consists of putting

$$u_O(r,t) = r\left[c_O(r,t) - c_O\right] \tag{10.184}$$

and

$$u_R(r,t) = r\left[c_R - c_R(r,t)\right] \tag{10.185}$$

Thus, the Fick differential equations become

$$\frac{\partial u_O(r,t)}{\partial t} = D_O \frac{\partial^2 c_O(u,t)}{\partial r^2} \tag{10.186}$$

and

$$\frac{\partial u_R(r,t)}{\partial t} = D_R \frac{\partial^2 u_R(r,t)}{\partial r^2} \tag{10.187}$$

The Laplace transforms of these functions are

$$\bar{u}_O(r,s) = \alpha(s)\, e^{(r_e-r)\sqrt{\frac{s}{D_O}}} \tag{10.188}$$

and

$$\bar{u}_R(r,s) = \beta(s)\, e^{(r_e-r)\sqrt{\frac{s}{D_R}}} \tag{10.189}$$

To find the constants $\alpha(s)$ and $\beta(s)$, as in linear diffusion we can use the definition of the current

$$\bar{I}(s) = nFA\,D_R\left[\frac{\partial \bar{c}_R(r,s)}{\partial r}\right]_{r=r_e} = -nFA\,D_O\left[\frac{\partial \bar{c}_R(r,s)}{\partial r}\right]_{r=r_e} \tag{10.190}$$

From the definition of u, we have

$$\frac{\partial \bar{c}_O(r,s)}{\partial r} = \frac{\bar{u}_O(r,s)}{r^2} - \frac{1}{r}\left(\frac{\partial \bar{u}_O(r,s)}{\partial r}\right) = \frac{\bar{u}_O(r,s)}{r^2} + \frac{1}{r}\alpha(s)\sqrt{\frac{s}{D_O}}\, e^{(r_e-r)\sqrt{\frac{s}{D_O}}} \tag{10.191}$$

and

$$\frac{\partial \bar{c}_R(r,s)}{\partial r} = \frac{\bar{u}_R(r,s)}{r^2} - \frac{1}{r}\left(\frac{\partial \bar{u}_R(r,s)}{\partial r}\right) = \frac{\bar{u}_R(r,s)}{r^2} + \frac{1}{r}\beta(s)\sqrt{\frac{s}{D_R}}\, e^{(r_e-r)\sqrt{\frac{s}{D_R}}} \tag{10.192}$$

From this, we deduce that

$$\alpha(s) = \frac{\bar{I}(s)}{nFAD_O}\left[\frac{r_e}{r_e^{-1}+\sqrt{\dfrac{s}{D_O}}}\right] \tag{10.193}$$

and

$$\beta(s) = \frac{\bar{I}(s)}{nFAD_R}\left[\frac{r_e}{r_e^{-1}+\sqrt{\dfrac{s}{D_R}}}\right] \tag{10.194}$$

If we make the hypothesis that $D_R = D_O$, then $\alpha(s) = \beta(s)$. In this particular case, given that $\alpha(s)$ and $\beta(s)$ are the Laplace transforms of u_O and u_R with $r = r_e$, the functions u_O and u_R are equal at the surface of the electrode. Thus, expanding and introducing the Nernst equation, we have

$$c_R - c_R(r_e,t) = c_O(r_e,t) - c_O = \theta\left[c_O - c_O(r_e,t)S(t)\right] \tag{10.195}$$

which is

$$c_O(r_e,t) = c_O\left[\frac{1+\theta}{1+\theta S(t)}\right] \tag{10.196}$$

So the u functions at the electrode surface are

$$u_O(r_e,t) = r_e c_O\,\theta\left[\frac{1-S(t)}{1+\theta S(t)}\right] = u_R(r_e,t) = r_e c_R\left[\frac{1-S(t)}{1+\theta S(t)}\right] \tag{10.197}$$

The current is then given by

$$I = nFAD\left(\frac{\partial c_R}{\partial r}\right)_{r=r_e} = nFAD\left[\frac{u_R(r_e,t)}{(r_e)^2} - \frac{1}{r_e}\left(\frac{\partial u_R}{\partial r}\right)_{r=r_e}\right] \tag{10.198}$$

The first term corresponds to the spherical contribution and the second to the current that we would have with a planar electrode

$$I = \frac{nFADc_R}{r_e}\left[\frac{1-S(t)}{1+\theta S(t)}\right] - \frac{nFAD}{r_e}\left(\frac{\partial u_R}{\partial r}\right)_{r=r_e} = I_s + I_p \tag{10.199}$$

Notice that the spherical contribution converges to the limiting value of the diffusion current given by equation (7.41)

$$I_s = 2\pi nFDc_R r_e\left[\frac{1-S(t)}{1+\theta S(t)}\right] = I_d\left[\frac{1-S(t)}{1+\theta S(t)}\right] = I_d\phi \tag{10.200}$$

Let's demonstrate that the second term in equation (10.199) really corresponds to the current of a cyclic voltammogram on a planar electrode. Effectively, the Laplace transform of the u function is

$$\bar{u}_{\mathrm{O}}(r,s) = \bar{u}_{\mathrm{O}}(r_{\mathrm{e}},s)\, e^{(r_{\mathrm{e}}-r)\sqrt{\frac{s}{D_{\mathrm{O}}}}} = r_{\mathrm{e}} c_{\mathrm{R}} \bar{\phi}\, e^{(r_{\mathrm{e}}-r)\sqrt{\frac{s}{D_{\mathrm{O}}}}} \tag{10.201}$$

and the Laplace transform of the spherical contribution is

$$\bar{I}_{\mathrm{p}} = -\frac{nFAD}{r_{\mathrm{e}}}\left[\left(\frac{\partial \bar{u}(r,s)}{\partial r}\right)_{r=r_{\mathrm{e}}}\right] \tag{10.202}$$

Differentiating equation (10.201) we get

$$\bar{I}_{\mathrm{p}} = nFAD\bar{\phi}\sqrt{\frac{s}{D}} \tag{10.203}$$

We then have an expression which is independent of the size of the electrode and which is identical to equation (10.14). The current I_{p} can be calculated as in §10.1 to get the function χ.

The graphs in Figure 10.16 show the influence of the different terms on the voltammetry. Of course, the linear contribution depends on the scan rate, and we can draw these voltammograms with the aid of a proportionality factor ρ and the dimensionless current χ.

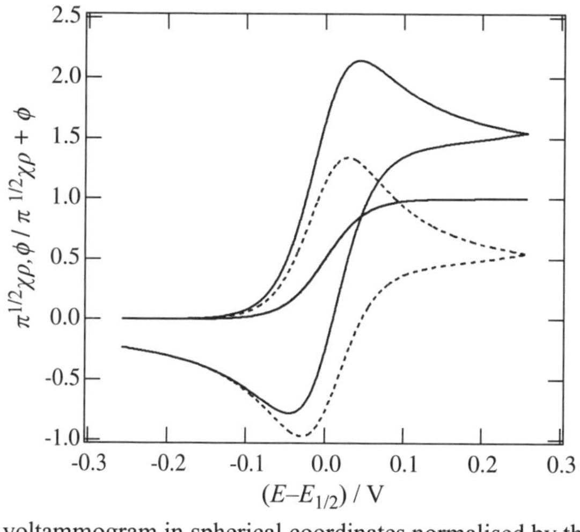

Fig. 10.16 Cyclic voltammogram in spherical coordinates normalised by the limiting diffusion current. The dotted line is the planar contribution, the solid line the spherical contribution and the total current ($\rho = 3$).

10.8 VOLTABSORPTOMETRY

This is an optical technique that can be used during cyclic voltammetry, which consists of measuring the absorbance or fluorescence linked to the presence of

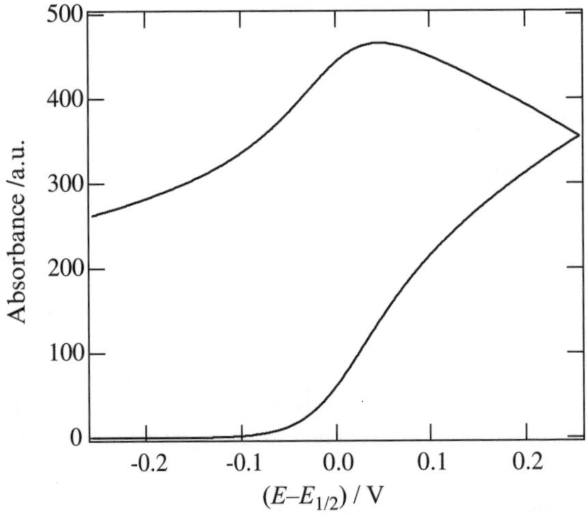

Fig. 10.17 Voltabsorptogram for the data in Figure 10.1.

absorbent or fluorescent ions in the diffusion layer. The basic principle is shown in Figure 9.38, since the setup is the same as for reflectance with a modulated potential.

As in §9.4.2, a first approximation of the absorbance is given by the Lambert-Beer law, which, by integrating over the whole optical path gives

$$A \quad = \quad 2\varepsilon \int_0^\infty \frac{c(x,t)}{\cos\theta}\,\mathrm{d}x \tag{10.204}$$

where ε is the molar absorption coefficient, and θ the angle of reflection. The quantity of absorbing species present in the diffusion layer is equal to the charge passed as shown in Figure 10.17

$$\int_0^\infty \frac{c(x,t)}{\cos\theta}\,\mathrm{d}x \quad = \quad \frac{1}{nFA}\int_0^t I(t)\,\mathrm{d}t \tag{10.205}$$

Thus, the absorbance is directly proportional to the charge passed.

The maximum of the curve on the return sweep corresponds to the isosbestic point of zero current.

10.9 SEMI-INTEGRATION

From an electrical point of view, we saw in chapter 9 that an electrochemical system often behaves as a linear system. Thus, for different excitation functions, we have different responses.

In fact, it is interesting to note that the electrochemical responses that we have studied are all linked by semi-integration as shown in Figure 10.18.

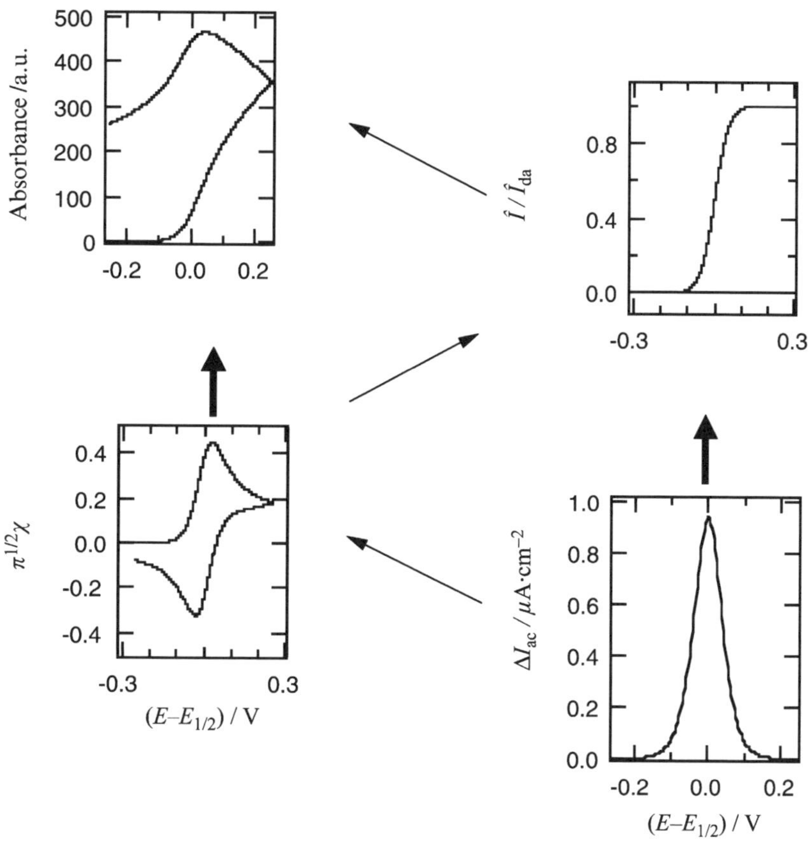

Fig. 10.18 The response of a reversible electrochemical system. The oblique arrows correspond to semi-integrations, the vertical arrows to classic integrations.

From a practical point of view, it is always easier to integrate numerically rather than to differentiate, and therefore it is preferable if the experimental method can give us directly either the bell curve of AC voltammetry described in §9.4.1, or a cyclic voltammogram; all the other curves being then obtainable by semi-integration or integration.

ANNEX A

VECTOR ANALYSIS

1 COORDINATE SYSTEMS

In a three-dimensional space, a vector defined by a point M can be represented in Cartesian coordinates by:

$$\mathbf{OM} \;=\; x\,\hat{\boldsymbol{i}} + y\,\hat{\boldsymbol{j}} + z\,\hat{\boldsymbol{k}}$$

In cylindrical coordinates, we have:

$$\mathbf{OM} \;=\; r\,\hat{\boldsymbol{u}}_{\mathrm{r}} + z\,\hat{\boldsymbol{k}}$$

as illustrated in Figure A1,

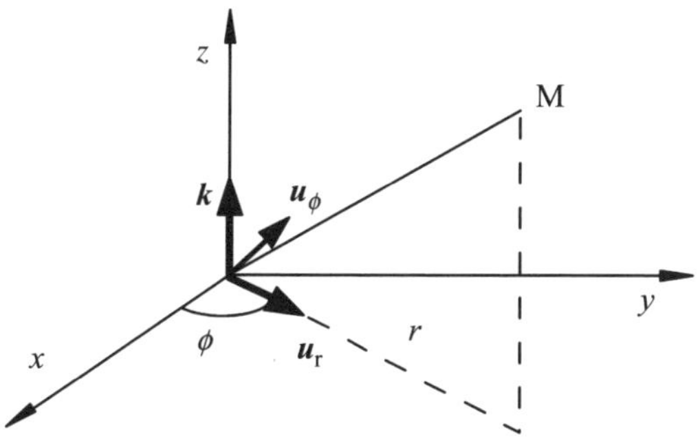

Fig. A.1 Representation in cylindrical coordinates.

and in spherical coordinates:

$$\mathbf{OM} \;=\; r\,\hat{\boldsymbol{u}}_{\mathrm{r}}$$

as illustrated in Figure A2. The angle ϕ from the x-axis is called the azimuthal angle and θ the polar angle from the z-axis. The vector $\hat{\boldsymbol{k}}$ orients the angle vector ϕ and the vector $\hat{\boldsymbol{u}}_\phi$, perpendicular to $\hat{\boldsymbol{k}}$ and $\hat{\boldsymbol{u}}_{\mathrm{r}}$, orients the angle ϕ. The vector $\hat{\boldsymbol{u}}_\theta$ is defined by the cross product $\hat{\boldsymbol{u}}_\theta = \hat{\boldsymbol{u}}_\phi \times \hat{\boldsymbol{u}}_{\mathrm{r}}$.

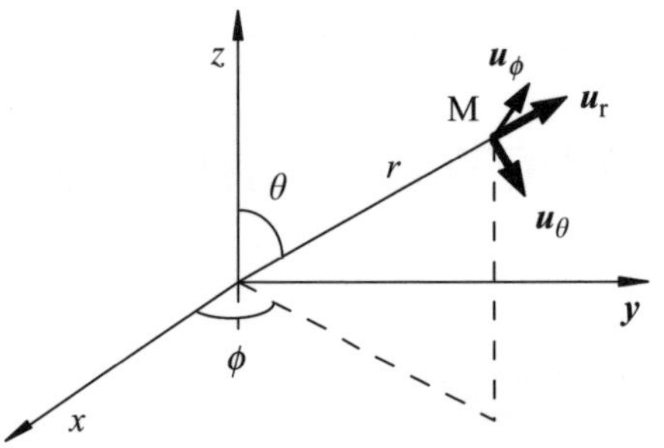

Fig. A.2 Representation in spherical coordinates.

The line element **dM** shown in Figure A3 is then

$$\mathbf{dM} = \mathbf{d}x\,\hat{\boldsymbol{i}} + \mathbf{d}y\,\hat{\boldsymbol{j}} + \mathbf{d}z\,\hat{\boldsymbol{k}}$$

$$\mathbf{dM} = \mathbf{d}r\,\hat{\boldsymbol{u}}_{\mathrm{r}} + r\mathbf{d}\phi\,\hat{\boldsymbol{u}}_{\phi} + \mathbf{d}z\,\hat{\boldsymbol{k}}$$

$$\mathbf{dM} = \mathbf{d}r\,\hat{\boldsymbol{u}}_{\mathrm{r}} + r\mathbf{d}\theta\,\hat{\boldsymbol{u}}_{\theta} + r\sin\theta\mathbf{d}\phi\,\hat{\boldsymbol{u}}_{\phi}$$

2 CIRCULATION OF THE FIELD VECTOR

Let $\boldsymbol{a}$ the field vector at point M and **dM** the line element along the curve C. By definition, the following differential form is called the circulation of the the vector $\boldsymbol{a}$.

$$\delta C = \boldsymbol{a} \cdot \mathbf{dM}$$

Expressed in Cartesian, cylindrical and spherical coordinates, we have

$$\delta C = a_x dx + a_y dy + a_z dz$$

$$\delta C = a_r dr + a_\phi r d\phi + a_z dz$$

$$\delta C = a_r dr + a_\theta r d\theta + a_\phi r \sin\theta d\phi$$

The following path integral is called the circulation of $\boldsymbol{a}$ along the path C, also called the flow of $\boldsymbol{a}$ along C, and the integral is called the flow integral.

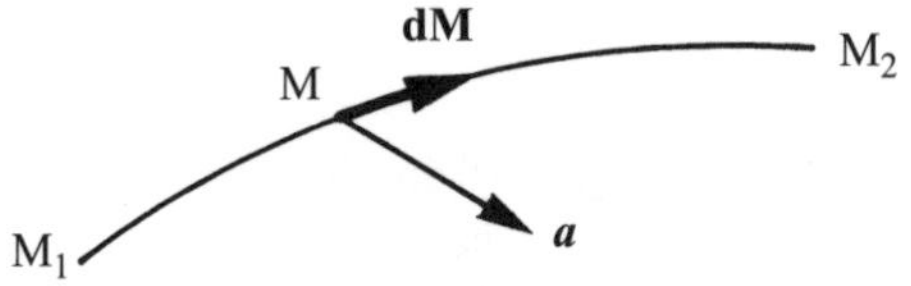

Fig. A.3 Circulation of a vector along a path.

$$C = \int_{(c)} \boldsymbol{a}(\mathrm{M}) \cdot \mathbf{dM}$$

It can be shown that if we know the parametric equation of the curve (C) as a function of t, this integral reduces to

$$C = \int_{t_1}^{t_2} \boldsymbol{a}(t) \cdot \frac{\mathbf{dM}}{\mathrm{d}t}\,\mathrm{d}t$$

3 THE VECTOR GRADIENT

The vector gradient of the scalar function V(M) is a vector such that

$$\mathrm{d}V = \mathrm{grad}V(\mathrm{M}) \cdot \mathbf{dM}$$

Given that $\mathrm{d}V$ is a total differential (also called the exact differential), we obtain the following equalities:

$$\mathrm{d}V = \frac{\partial V}{\partial x}\mathrm{d}x + \frac{\partial V}{\partial y}\mathrm{d}y + \frac{\partial V}{\partial z}\mathrm{d}z$$

$$\mathrm{d}V = \frac{\partial V}{\partial r}\mathrm{d}r + \frac{\partial V}{\partial \phi}\mathrm{d}\phi + \frac{\partial V}{\partial z}\mathrm{d}z$$

$$\mathrm{d}V = \frac{\partial V}{\partial r}\mathrm{d}r + \frac{\partial V}{\partial \theta}\mathrm{d}\theta + \frac{\partial V}{\partial \phi}\mathrm{d}\phi$$

As a result, we can derive the expression for the vector gradient in Cartesian coordinates

$$\nabla V = \mathbf{grad}V = \frac{\partial V}{\partial x}\hat{\boldsymbol{i}} + \frac{\partial V}{\partial y}\hat{\boldsymbol{j}} + \frac{\partial V}{\partial z}\hat{\boldsymbol{k}}$$

Likewise, in cylindrical coordinates, we have:

$$\nabla V = \mathbf{grad}V = \frac{\partial V}{\partial r}\hat{\boldsymbol{u}}_{\mathrm{r}} + \frac{1}{r}\frac{\partial V}{\partial \phi}\hat{\boldsymbol{u}}_{\phi} + \frac{\partial V}{\partial z}\hat{\boldsymbol{k}}$$

And finally, in spherical coordinates

$$\nabla V = \mathbf{grad}V = \frac{\partial V}{\partial r}\hat{\boldsymbol{u}}_{\mathrm{r}} + \frac{\partial V}{r\partial \theta}\hat{\boldsymbol{u}}_{\theta} + \frac{\partial V}{r\sin\theta\partial \phi}\hat{\boldsymbol{u}}_{\phi}$$

4 FLUX OF THE FIELD VECTOR

Let's consider a surface element dS in a field vector. The flux of the field vector passing through this surface element is defined by :

$$\mathrm{d}\Phi = \boldsymbol{a}(\mathrm{M}) \cdot \hat{\boldsymbol{n}} \mathrm{d}S$$

The flux leaving a surface S is obtained by integrating over the entire surface.

$$\Phi = \iint_S \boldsymbol{a}(\mathrm{M}) \cdot \hat{\boldsymbol{n}} \mathrm{d}S$$

The sign of the flux depends on the choice of the unitary vector. In the case of a close surface, the flux is positive.

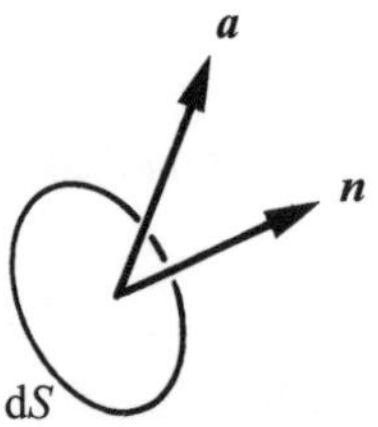

Fig. A.4 Vector flux through a surface element.

5 THE GREEN-OSTROGRADSKI THEOREM

The Green-Ostrogradski theorem states that, for any closed surface S from which a vector field is leaving, there is a scalar function of M called the ***divergence*** of the vector ***a*** such that:

$$\iint_S \boldsymbol{a}(\mathrm{P}) \cdot \hat{\boldsymbol{n}}(\mathrm{P})\, \mathrm{d}S = \iiint_V \mathrm{div}\boldsymbol{a}(\mathrm{M})\, \mathrm{d}\tau$$

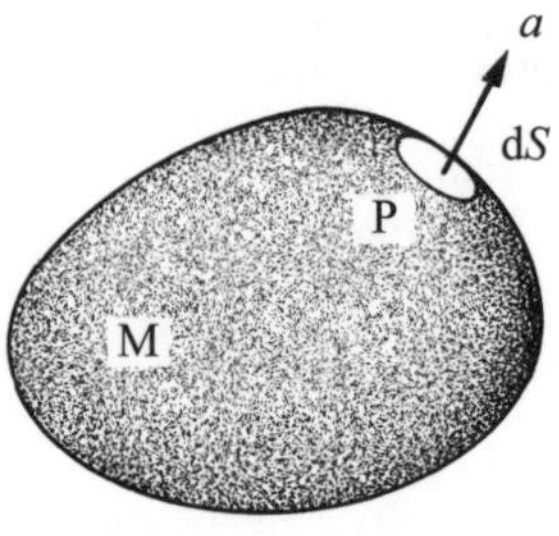

Fig. A.5 Flux of a vector a through a closed surface.

In Cartesian coordinates, the volume element $\mathrm{d}\tau$ is defined as the cube $\mathrm{d}x\mathrm{d}y\mathrm{d}z$. The flux leaving through the planes of abscissa x and $x+\mathrm{d}x$, and through the surface $\mathrm{d}y\mathrm{d}z$ is given by:

$$\mathrm{d}\Phi_x = \left[a_x(x+\mathrm{d}x) - a_x\right]\mathrm{d}y\mathrm{d}z = \frac{\partial a_x}{\partial x}\mathrm{d}\tau$$

By using the same reasoning for the other planes, we have:

$$\mathrm{d}\Phi = \left(\frac{\partial a_x}{\partial x} + \frac{\partial a_y}{\partial y} + \frac{\partial a_z}{\partial z}\right)\mathrm{d}\tau$$

According to the Green-Ostrogradski theorem, we obtain

$$\nabla \cdot \boldsymbol{a} = \operatorname{div} \boldsymbol{a} = \frac{\partial a_x}{\partial x} + \frac{\partial a_y}{\partial y} + \frac{\partial a_z}{\partial z}$$

In the same way, it can be shown that in cylindrical and spherical coordinates:

$$\nabla \cdot \boldsymbol{a} = \operatorname{div} \boldsymbol{a} = \frac{a_r}{r} + \frac{\partial a_r}{\partial r} + \frac{1}{r}\frac{\partial a_\phi}{\partial \phi} + \frac{\partial a_z}{\partial z}$$

and

$$\nabla \cdot \boldsymbol{a} = \operatorname{div} \boldsymbol{a} = \frac{2a_r}{r} + \frac{\partial a_r}{\partial r} + \frac{a_\theta}{r\tan\theta} + \frac{1}{r}\frac{\partial a_\theta}{\partial \theta} + \frac{1}{r\sin\theta}\frac{\partial a_\phi}{\partial \phi}$$

We refer to a scalar function defined as the divergence of the gradient as the ***Laplacian*** of the scalar function V(M)

$$\Delta V = \nabla^2 V = \operatorname{div} \mathbf{grad}\, V = \frac{\partial^2 V}{\partial x^2} + \frac{\partial^2 V}{\partial y^2} + \frac{\partial^2 V}{\partial z^2}$$

In the same way, we obtain for cylindrical and spherical coordinates

$$\Delta V = \nabla^2 V = \operatorname{div} \mathbf{grad}\, V = \frac{1}{r}\frac{\partial V}{\partial r} + \frac{\partial^2 V}{\partial r^2} + \frac{1}{2r}\frac{\partial^2 V}{\partial \phi^2} + \frac{\partial^2 V}{\partial z^2}$$

and

$$\Delta V = \nabla^2 V = \operatorname{div} \mathbf{grad}\, V$$

$$= \frac{1}{r^2}\frac{\partial}{\partial r}\left(r^2\frac{\partial V}{\partial r}\right) + \frac{1}{r^2\sin\theta}\frac{\partial}{\partial \theta}\left(\sin\theta\frac{\partial V}{\partial \theta}\right) + \frac{1}{r^2\sin^2\theta}\frac{\partial^2 V}{\partial \phi^2}$$

ANNEX B

WORK FUNCTIONS AND STANDARD REDOX POTENTIALS

Table B. 1 Work function of metals.

Metal	Φ/eV	Metal	Φ/eV
Ag	4.30	Nd	3.1
Al	4.19	Ni	4.73
Au	5.32	Os	4.83
Ba	2.35	Pb	4.18
Be	5.08	Pd	5.00
Bi	4.36	Po	4.6
Ca	2.71	Pr	2.7
Cd	4.12	Pt	5.40
Ce	2.80	Rb	2.20
Co	4.70	Re	4.95
Cr	4.40	Ru	4.80
Cs	1.90	Sb	4.56
Cu	4.70	Sc	3.5
Fe	4.65	Sm	2.95
Ga	4.25	Sn	4.35
Hf	3.65	Sr	2.76
Hg	4.50	Ta	4.22
K	2.30	Te	4.70
In	4.08	Th	3.71
La	3.40	Ti	4.10
Li	3.10	Tl	4.02
Mg	3.66	U	3.50
Mn	3.90	V	4.44
Mo	4.30	W	4.55
Na	2.70	Zn	4.30
Nb	4.20	Zr	4.00

Data taken from *Standard Potentials in Aqueous Solutions*, edited by Bard, Parsons and Jordan, Marcel Dekker, 1985, New York, USA.

Table B.2 Standard redox potentials in acid solutions.

Redox couple	$E^{\ominus}$/V
$(3/2)N_2 + H^+ + e^- \rightarrow HN_3$	−3.10
$Li^+ + e^- \rightarrow Li$	−3.045
$K^+ + e^- \rightarrow K$	−2.925
$Rb^+ + e^- \rightarrow Rb$	−2.925
$Cs^+ + e^- \rightarrow Cs$	−2.923
$Ba^{2+} + 2e^- \rightarrow Ba$	−2.92
$Ra^{2+} + 2e^- \rightarrow Ra$	−2.916
$Sr^{2+} + 2e^- \rightarrow Sr$	−2.89
$Ca^{2+} + 2e^- \rightarrow Ca$	−2.84
$Na^+ + e^- \rightarrow Na$	−2.714
$La^{3+} + 3e^- \rightarrow La$	−2.37
$Ce^{3+} + 3e^- \rightarrow Ce$	−2.34
$Mg^{2+} + 2e^- \rightarrow Mg$	−2.356
$Lu^{3+} + 3e^- \rightarrow Lu$	−2.30
$1/2H_2 + e^- \rightarrow H^-$	−2.25
$AlF_6^{3-} + 3e^- \rightarrow Al + 6F^-$	−2.067
$Be^{2+} + 2e^- \rightarrow Be$	−1.97
$Th^{4+} + 4e^- \rightarrow Th$	−1.83
$Np^{3+} + 3e^- \rightarrow Np$	−1.79
$Zr^{4+} + 4e^- \rightarrow Zr$	−1.70
$Al^{3+} + 3e^- \rightarrow Al$	−1.67
$U^{3+} + 3e^- \rightarrow U$	−1.66
$Ti^{2+} + 2e^- \rightarrow Ti$	−1.63
$Hf^{4+} + 4e^- \rightarrow Hf$	−1.56
$SiF_6^{2-} + 4e^- \rightarrow Si + 6F^-$	−1.2
$TiF_6^{2-} + 4e^- \rightarrow Ti + 6F^-$	−1.191
$Mn^{2+} + 2e^- \rightarrow Mn$	−1.18
$V^{2+} + 2e^- \rightarrow V$	−1.13
$Nb^{3+} + 3e^- \rightarrow Nb$	−1.1
$H_3BO_3 + 3H^+ + 3e^- \rightarrow B + 3H_2O$	−0.890
$SiO_2(vit) + 4H^+ + 4e^- \rightarrow Si + 2H_2O$	−0.888
$TiO^{2+} + 2H^+ + 2e^- \rightarrow Ti + H_2O$	−0.882
$Ta_2O_5 + 10H^+ + 10e^- \rightarrow 2Ta + 5H_2O$	−0.81
$Zn^{2+} + 2e^- \rightarrow Zn$	−0.7626
$Te + 2H^+ + 2e^- \rightarrow H_2Te$	−0.740
$Nb_2O_5 + 10H^+ + 10e^- \rightarrow 2Nb + 5H_2O$	−0.65

Table B.2 *(continued)*

Redox couple	$E^{\ominus}$/V
$Ga^{3+} + 3e^- \rightarrow Ga$	–0.529
$U^{4+} + e^- \rightarrow U^{3+}$	–0.52
$H_3PO_2 + H^+ + e^- \rightarrow P(w) + 2H_2O$	–0.508
$H_3PO_3 + 2H^+ + 2e^- \rightarrow H_3PO_2 + H_2O$	–0.499
$Fe^{2+} + 2e^- \rightarrow Fe$	–0.44
$Cr^{3+} + e^- \rightarrow Cr^{2+}$	–0.424
$Cd^{2+} + 2e^- \rightarrow Cd$	–0.4025
$Ti^{3+} + e^- \rightarrow Ti^{2+}$	–0.37
$PbI_2 + 2e^- \rightarrow Pb + 2I^-$	–0.365
$PbSO_4 + 2e^- \rightarrow Pb + SO_4^{2-}$	–0.3505
$Eu^{3+} + e^- \rightarrow Eu^{2+}$	–0.35
$In^{3+} + 3e^- \rightarrow In$	–0.3382
$Tl^+ + e^- \rightarrow Tl$	–0.3363
$PbBr_2 + 2e^- \rightarrow Pb + 2Br^-$	–0.280
$Co^{2+} + 2e^- \rightarrow Co$	–0.277
$H_3PO_4 + 2H^+ + 2e^- \rightarrow H_3PO_3 + H_2O$	–0.276
$PbCl_2 + 2e^- \rightarrow Pb + 2Cl^-$	–0.268
$Ni^{2+} + 2e^- \rightarrow Ni$	–0.257
$V^{3+} + e^- \rightarrow V^{2+}$	–0.255
$2SO_4^{2-} + 4H^+ + 4e^- \rightarrow S_2O_6^{2-} + 2H_2O$	–0.253
$SnF_6^{2-} + 4e^- \rightarrow Sn + 6F^-$	–0.25
$N_2 + 5H^+ + 4e^- \rightarrow N_2H_5^+$	–0.23
$As + 3H^+ + 3e^- \rightarrow AsH_3$	–0.225
$Mo^{3+} + 3e^- \rightarrow Mo$	–0.2
$CuI + e^- \rightarrow Cu + I^-$	–0.182
$CO_2 + 2H^+ + 2e \rightarrow HCOOH(aq)$	–0.16
$AgI + e^- \rightarrow Ag + I^-$	–0.1522
$Si + 4H^+ + 4e^- \rightarrow SiH_4$	–0.143
$Sn^{2+} + 2e^- \rightarrow Sn$	–0.136
$Pb^{2+} + 2e^- \rightarrow Pb$	–0.1251
$P(w) + 3H^+ + 3e^- \rightarrow PH_3$	–0.063
$O_2 + H^+ + e^- \rightarrow HO_2$	–0.046
$Hg_2I_2 + 2e^- \rightarrow 2Hg + 2I^-$	–0.0405
$Se + 2H^+ + 2e^- \rightarrow H_2Se$	–0.028
$2H^+ + 2e^- \rightarrow H_2$	0.000
$CuBr + e^- \rightarrow Cu + Br$	0.033

Table B.2 *(continued)*

Redox couple	$E^{\ominus}$/V
$HCOOH(aq) + 2H^+ + 2e^- \rightarrow HCHO(aq) + H_2O$	0.056
$AgBr + e^- \rightarrow Ag + Br^-$	0.0711
$TiO^{2+} + 2H^+ + e^- \rightarrow Ti^{3+} + H_2O$	0.100
$CuCl + e^- \rightarrow Cu + Cl^-$	0.121
$C + 4H^+ + 4e^- \rightarrow CH$	0.132
$Hg_2Br_2 + 2e^- \rightarrow 2Hg + 2Br^-$	0.13920
$S + 2H^+ + 2e^- \rightarrow H_2S$	0.144
$Np^{4+} + e^- \rightarrow Np^{3+}$	0.15
$Sn^{4+} + 2e^- \rightarrow Sn^{2+}$	0.15
$Sb_4O_6 + 12H^+ + 12e^- \rightarrow 4Sb + 6H_2O$	0.1504
$SO_4^{2-} + 2H^+ + 2e^- \rightarrow H_2SO_3 + H_2O$	0.158
$Cu^{2+} + e^- \rightarrow Cu^+$	0.159
$UO_2^{2+} + e^- \rightarrow UO_2^+$	0.16
$2H_2SO_3^- + 3H^+ + 2e^- \rightarrow HS_2O_4^- + 2H_2O$	0,173
$AgCl + e^- \rightarrow Ag + Cl^-$	0.2223
$HCHO(aq) + 2H^+ + 2e^- \rightarrow CH_3OH(aq)$	0.232
$(CH_3)_2SO_2 + 2H^+ + 2e^- \rightarrow (CH_3)_2SO + 2H_2O$	0.238
$UO_2^{2+} + 4H^+ + 2e^- \rightarrow U^{4+} + 2H_2O$	0.27
$HCNO + H^+ + e^- \rightarrow 1/2C_2N_2 + H_2O$	0.330
$Cu^{2+} + 2e^- \rightarrow Cu$	0.340
$AgIO_3 + e^- \rightarrow Ag + IO_3^-$	0.354
$Fe(CN)_6^{3-} + e^- \rightarrow Fe(CN)_6^{4-}$	0.3610
$C_2N_2 + 2H^+ + 2e^- \rightarrow 2HCN(aq)$	0.373
$UO_2^+ + 4H^+ + e^- \rightarrow U^{4+} + 2H_2O$	0.38
$H_2N_2O_2 + 6H^+ + 4e^- \rightarrow 2NH_3OH^+$	0.387
$2H_2SO_3 + 2H^+ + 4e^- \rightarrow S_2O_3^{2-} + 3H_2O$	0.400
$PdBr_4^{2-} + 2e^- \rightarrow Pd + 4Br^-$	0.49
$H_2SO_3 + 4H^+ + 4e^- \rightarrow S + 3H_2O$	0.500
$2H_2SO_3 + 4H^+ + 6e^- \rightarrow S_4O_6^{2-} + 6H_2O$	0.507
$Cu^+ + e^- \rightarrow Cu$	0.520
$I_2 + 2e^- \rightarrow 2I^-$	0.5355
$I_3^- + 2e^- \rightarrow 3I^-$	0.536
$AgBrO_3 + e^- \rightarrow Ag + BrO_3^-$	0.546
$Cu^{2+} + Cl^- + e^- \rightarrow CuCl$	0.559
$MnO_4^- + e^- \rightarrow MnO_4^{2-}$	0.56
$S_2O_6^{2-} + 4H^+ + 2e^- \rightarrow 2H_2SO_3$	0.569

Table B.2 *(continued)*

Redox couple	$E^{\ominus}$/V
$CH_3OH(aq) + 2H^+ + 2e^- \rightarrow CH_4 + H_2O$	0.59
$Au(SCN)_4^- + 3e^- \rightarrow Au + 4SCN^-$	0.636
$PdCl_4^{2-} + 2e^- \rightarrow Pd + 4Cl^-$	0.64
$Cu^{2+} + Br^- + e^- \rightarrow CuBr$	0.654
$Ag_2SO_4 + 2e^- \rightarrow 2Ag + SO_4^{2-}$	0.654
$O_2 + 2H^+ + 2e^- \rightarrow H_2O_2$	0.695
$HN_3 + 11H^+ + 8e^- \rightarrow 3NH_4^+$	0.695
$PtBr_4^{2-} + 2e^- \rightarrow Pt + 4Br^-$	0.698
$2NO + 2H^+ + 2e^- \rightarrow H_2N_2O_2$	0.71
$PtCl_4^{2-} + 2e^- \rightarrow Pt + 4Cl^-$	0.758
$Rh^{3+} + 3e^- \rightarrow Rh$	0.76
$(SCN)_2 + 2e^- \rightarrow 2SCN^-$	0.77
$Fe^{3+} + e^- \rightarrow Fe^{2+}$	0.771
$Hg_2^{2+} + 2e^- \rightarrow 2Hg$	0.7960
$Ag^+ + e^- \rightarrow Ag$	0.7991
$2NO_3^- + 4H^+ + 2e^- \rightarrow N_2O_4 + 2H_2O$	0.803
$AuBr_4^- + 3e^- \rightarrow Au + 4Br^-$	0.854
$2HNO_2 + 4H^+ + 4e^- \rightarrow H_2N_2O_2$	0.86
$IrCl_6^{3-} + 3e^- \rightarrow Ir + 6Cl^-$	0.86
$Cu^{2+} + I^- + e^- \rightarrow CuI$	0.861
$IrCl_6^{2-} + e^- \rightarrow IrCl_6^{3-}$	0.867
$2Hg^{2+} + 2e^- \rightarrow Hg$	0.9110
$Pd^{2+} + 2e^- \rightarrow Pd$	0.915
$NO_3^- + 3H^+ + 2e^- \rightarrow HNO_2 + H_2O$	0.94
$NO_3^- + 4H^+ + 3e^- \rightarrow NO + 2H_2O$	0.957
$AuBr_2^- + e^- \rightarrow Au + 2Br^-$	0.960
$PtO + 2H^+ + 2e^- \rightarrow Pt + H_2O$	0.980
$HNO_2 + H^+ + e^- \rightarrow NO + H_2O$	0.996
$AuCl_4^- + 3e^- \rightarrow Au + 4Cl^-$	1.002
$Pu^{4+} + e^- \rightarrow Pu^{3+}$	1.01
$PuO_2^{2+} + e^- \rightarrow PuO_2^+$	1.02
$PuO_2^{2+} + 4H^+ + 2e^- \rightarrow Pu^{4+} + 2H_2O$	1.03
$N_2O_4 + 4H^+ + 4e^- \rightarrow NO + 2H_2O$	1.039
$PuO_2^+ + 4H^+ + e^- \rightarrow Pu^{4+} + 2H_2O$	1.04
$Br_2(l) + 2e^- \rightarrow 2Br^-$	1.065
$N_2O_4 + 2H^+ + 2e^- \rightarrow 2HNO_3$	1.07

Table B.2 *(continued)*

Redox couple	$E^{\ominus}$/V
$Cu^{2+} + 2CN^- + e^- \rightarrow Cu(CN)_2^-$	1.12
$H_2O_2 + H^+ + e^- \rightarrow OH + H_2O$	1.14
$ClO_3^- + 3H^+ + 2e^- \rightarrow HClO_2 + H_2O$	1.181
$ClO_2 + H^+ + e^- \rightarrow HClO_2$	1.188
$S_2Cl_2 + 2e^- \rightarrow 2S + 2Cl^-$	1.19
$IO_3^- + 6H^+ + 5e^- \rightarrow 1/2I_2 + 3H_2O$	1.195
$ClO_4^- + 2H^+ + 2e^- \rightarrow ClO_3^- + H_2O$	1.201
$O_2 + 4H^+ + 4e^- \rightarrow 2H_2O$	1.229
$MnO_2 + 4H^+ + 2e^- \rightarrow Mn^{2+} + 2H_2O$	1.23
$NpO_2^{2+} + e^- \rightarrow NpO_2^+$	1.24
$N_2H_5^+ + 3H^+ + 2e^- \rightarrow 2NH_4^+$	1.275
$PdCl_6^{2-} + 2e^- \rightarrow PdCl_4^{2-} + 2Cl^-$	1.288
$2HNO_2 + 4H^+ + 4e^- \rightarrow N_2O + 3H_2O$	1.297
$NH_3OH^+ + 2H^+ + 2e^- \rightarrow NH_4^+ + H_2O$	1.35
$Cl_2 + 2e^- \rightarrow 2Cl^-$	1.3583
$Cr_2O_7^{2-} + 14H^+ + 6e^- \rightarrow 2Cr^{3+} + 7H_2O$	1.36
$2NH_3OH^+ + H^+ + 2e^- \rightarrow N_2H_5^+ + 2H_2O$	1.41
$HO_2 + H^+ + e^- \rightarrow H_2O_2$	1.44
$PbO_2(\alpha) + 4H^+ + 2e^- \rightarrow Pb^{2+} + 2H_2O$	1.468
$BrO_3^- + 6H^+ + 5e^- \rightarrow 1/2Br_2 + 3H_2O$	1.478
$Mn^{3+} + e^- \rightarrow Mn^{2+}$	1.5
$MnO_4^- + 8H^+ + 5e^- \rightarrow Mn^{2+} + 4H_2O$	1.51
$Au^{3+} + 3e^- \rightarrow Au$	1.52
$NiO_2 + 4H^+ + 2e^- \rightarrow Ni^{2+} + 2H_2O$	1.593
$H_5IO_6 + H^+ + 2e^- \rightarrow IO_3^- + 3H_2O$	1.603
$HBrO + H^+ + e^- \rightarrow 1/2Br_2 + H_2O$	1.604
$HClO + H^+ + e^- \rightarrow 1/2Cl_2 + H_2O$	1.630
$HClO_2 + 2H^+ + 2e^- \rightarrow HClO + H_2O$	1.674
$PbO_2(\alpha) + SO_4^{2-} + 4H^+ + 2e^- \rightarrow PbSO_4 + 2H_2O$	1.698
$MnO_4^- + 4H^+ + 3e^- \rightarrow MnO_2 + 2H_2O$	1.70
$Ce^{4+} + e^- \rightarrow Ce^{3+}$	1.72
$H_2O_2 + 2H^+ + 2e^- \rightarrow 2H_2O$	1.763
$Au^+ + e^- \rightarrow Au$	1.83
$Co^{3+} + e^- \rightarrow Co^{3+}$	1.92
$HN_3 + 3H^+ + 2e^- \rightarrow NH_4^+ + N_2$	1.96
$S_2O_8^{2-} + 2e^- \rightarrow 2SO_4^{2-}$	1.96

Table B.2 *(continued)*

Redox couple	$E^{\ominus}$ /V
$Ag^{2+} + e^- \rightarrow Ag^+$	1.980
$O_3 + 2H^+ + 2e^- \rightarrow O_2 + H_2O$	2.075
$F_2O + 2H^+ + 4e^- \rightarrow 2F^- + H_2O$	2.153
$OH + H^+ + e^- \rightarrow H_2O$	2.38
$O(g) + 2H^+ + 2e^- \rightarrow H_2O$	2.430
$F_2 + 2e^- \rightarrow 2F^-$	2.87
$F_2 + 2H^+ + 2e^- \rightarrow 2HF(aq)$	3.053

Data taken from *Standard Potentials in Aqueous Solutions*, edited by Bard, Parsons and Jordan, Marcel Dekker, 1985, New York, USA.

SYMBOLS

a	activity	no units
A	area	m^2
c	concentration	$M = mol \cdot l^{-1}$ (molar)
C	capacity	$F = C \cdot V^{-1}$(Farad)
d	density	$kg \cdot m^{-3}$ or $kg \cdot l^{-1}$
D	diffusion coefficient	$m^2 \cdot s^{-1}$ or $cm^2 \cdot s^{-1}$
$\mathbf{D}$	electric displacement vector	$C \cdot m^{-2}$
e	elementary charge	$= 1.602177 \cdot 10^{-19}$ C
E	energy	J
E	electrode potential	V
$E^{\ominus}_{ox/red}$	standard redox potential	V
$E^{\ominus\prime}_{ox/red}$	formal redox potential	V
$[E^{\ominus}_{ox/red}]_{SHE}$	standard redox potential with respect to the standard hydrogen electrode	V
$[E^{\ominus}_{ox/red}]_{Ag\|AgCl\|KCl_{sat}}$	standard redox potential with respect to the silver-silver chloride electrode in a saturated KCl solution	V
E_{abs}	absolute electrode potential	V
E_{eq}	equilibrium electrode potential	V
E_D	Donnan potential	V
E_{SHE}	electrode potential on the SHE scale	V
$\mathbf{E}$	electric field vector	$V \cdot m^{-1}$
f	fugacity	Pa
$f(\,)$	distribution function	
F	Faraday's constant	$= 96485\ C \cdot mol^{-1}$
F_V	volumic flow rate	$m^3 \cdot s^{-1}$
$g(r)$	radial distribution function	
G	Gibbs energy	J
$\widetilde{G}$	electrochemical Gibbs energy	J
$\bar{G}_i$	partial molar Gibbs energy for the species i	J
G	conductance	Ω^{-1}
G_m	molar Gibbs energy	$J \cdot mol^{-1}$
$\Delta G^{\ominus}$	standard Gibbs energy of a reaction	$J \cdot mol^{-1}$
$\Delta G^{\ominus}_{act}$	standard Gibbs energy of activation	$J \cdot mol^{-1}$
$\Delta G^{\ominus\, \alpha \to \beta}_{tr,i}$	standard Gibbs energy of transfer for the species i from α to β	$J \cdot mol^{-1}$

h	Planck's constant	$= 6.626\ 10^{-34}\ J \cdot s$
H	enthalpy	J
H_m	molar enthalpy	$J \cdot mol^{-1}$
ΔH_f	formation enthalpy	$J \cdot mol^{-1}$
ΔH_{hyd}	hydration enthalpy	$J \cdot mol^{-1}$
ΔH_i	ionisation enthalpy	$J \cdot mol^{-1}$
ΔH_{sol}	solvation enthalpy	$J \cdot mol^{-1}$
ΔH_{sub}	sublimation enthalpy	$J \cdot mol^{-1}$
ΔH_R	reticulation or lattice formation enthalpy	$J \cdot mol^{-1}$
I	electric current	A
$\bar{I}$	Laplace transform of the electric current	
$\hat{I}$	convoluted current	$M \cdot m \cdot s^{-1/2}$
I_o	exchange current	A
I_a	anodic current	A
I_c	cathodic current	A
I_d	limiting diffusion current	A
j	current density	$A \cdot m^{-2}$
$\boldsymbol{J}_i$	flux vector for the species *i*	$mol \cdot m^{-2} \cdot s^{-1}$
k	Boltzmann's constant	$= 1.38066 \cdot 10^{-23}\ J \cdot K^{-1}$
$k^{\ominus}$	standard rate constant	$m \cdot s^{-1}$
k_a	anodic rate constant	$m \cdot s^{-1}$
k_c	cathodic rate constant	$m \cdot s^{-1}$
K_A	association constant	no units
K_D	distribution constant	no units
K_i	Henry's constant for the species *i*	Pa
K_S	solubility product	no units
K_{ij}^{pot}	selected coefficient in potentiometry	
m	molality	$mol \cdot kg^{-1}$
M	molar mass	$kg \cdot mol^{-1}$
N_A	Avogadro's constant	$= 6.02214 \cdot 1023\ mol^{-1}$
p	pressure	Pa
P	distribution coefficient or partition coefficient	no units
$P^{\ominus}$	standard distribution coefficient	no units
$p^{\ominus}$	standard pressure	= 1 bar = 100 kPa
p^*	saturation vapour pressure	Pa
$\boldsymbol{p}$	dipole moment vector	$C \cdot m$
P	vecteur polarisation dielectrique	$C \cdot m^{-2}$
q	charge	C
R	gas constant	$= 8.34151 J \cdot K^{-1} \cdot mol^{-1}$
R	resistance	Ω
R_{ct}	charge transfer resistance	Ω
S	entropy	J·K–1
S_m	molar entropy	$J \cdot mol^{-1} \cdot K^{-1}$
t	transport number	no units

T	temperature	K
U	internal energy	J
u_i	electric (or electrophoretic) mobility	$m^2 \cdot V^{-1} \cdot s^{-1}$
$\tilde{u}_i$	electrochemical mobility	$m^2 \cdot J^{-1} \cdot s^{-1}$
$\boldsymbol{v}$	velocity vector	$m \cdot s^{-1}$
V	volume	m^3
V	electrostatic potential	V
V_m	molar volume	$m^3 \cdot mol^{-1}$
x_i	molar fraction for the species i	no units
Y	admittance	Ω^{-1}
Z	impedance	Ω
Z_W	Warburg impedance	Ω

α	real chemical potential	$J \cdot mol^{-1}$
χ	electric susceptibility	no units
χ	surface potential	V
δ	diffusion layer thickness	m
ε	molar absorption coefficient	$m^2 \cdot mol^{-1}$
ε	permittivity	$J^{-1} \cdot C^2 \cdot m^{-1}$
ε_0	vacuum permittivity	$= 8.85419 \cdot 10^{-12}$ $J^{-1} \cdot C^2 \cdot m^{-1}$
ε_r	relative permittivity	no units
γ	interfacial tension	$N \cdot m^{-1}$
γ	activity coefficient (in the molarity scale unless specified otherwise)	
Γ	surface concentration	$mol \cdot m^{-2}$
$\Gamma()$	surface excess concentration	$mol \cdot m^{-2}$
η	viscosity	$kg \cdot m^{-1} \cdot s^{-1}$
κ	reciprocal Debye's length	m^{-1}
λ	Solvent re-organisation Gibbs energy	$kJ \cdot mol^{-1}$
λ_i	molar ionic conductivity	$S \cdot m^2 \cdot mol^{-1}$
λ^o	limiting molar ionic conductivity	$S \cdot m^2 \cdot mol^{-1}$
Λ_m	molar conductivity	$S \cdot m^2 \cdot mol^{-1}$
Λ_m^o	limiting molar conductivity	$S \cdot m^2 \cdot mol^{-1}$
μ	chemical potential	$J \cdot mol^{-1}$
$\mu^{\ominus}$	standard chemical potential (for the gas phase in chapter 1, in the molarity scale otherwise)	$J \cdot mol^{-1}$
$\tilde{\mu}$	electrochemical potential	$J \cdot mol^{-1}$
ν_i	stœchiometric coefficient for the species i	no units
ν	kinematic viscosity	$m^2 \cdot s^{-1}$
ν	scan rate	$V \cdot s^{-1}$
φ	fugacity coefficient	no units
ϕ	Inner potential or Galvani potential	V

$\Delta_{\mathrm{o}}^{\mathrm{w}}\phi_i^{\ominus}$	standard transfer potential for the species *i* from the phase w to the phase o	V
Φ	work function	J
ρ	volumic charge density	$\mathrm{C \cdot m^{-3}}$
σ	surface charge density	$\mathrm{C \cdot m^{-2}}$
σ	conductivity	$\mathrm{S \cdot m^{-1}}$
τ_{L}	longitudinal relaxation time	s
τ_{D}	Debye relaxation time	s
ψ	outer potential	V
ω	angular velocity	$\mathrm{rad \cdot s^{-1}}$
ζ	zeta (or electrokinetic) potential	V
ζ	friction coefficient	$\mathrm{kg \cdot s^{-1}}$

INDEX

absolute redox potential, 48
AC voltammetry, 368, 409
accumulation layer, 213-215
admittance, 341
angular frequency, 340
anode, 39, 50-53
anodic current, 266
anodic stripping voltammetry, 331
apparent standard redox potential, 42, 265
assisted ion transfer at liquid | liquid interfaces, 398-399
Avogadro constant, 17, 177

band electrode, 285-286
Barker's square wave voltammetry, 325
Bode diagram, 343, 346-351, 356-357, 359, 363, 368
Butler-Volmer equation, 273-275, 294, 353, 383

calomel electrode, 43-44, 63-65, 120
cathode, 39, 50-53, 229
cathodic current, 266
cathodic stripping voltammetry, 332
charge transfer coefficient, 270, 274-275
chronoamperometry, 301
concentration impedances, 352-354, 356, 360-361, 366
conduction band, 24, 209
convoluted current, 377, 381-382, 387, 396-398
convolution theorem, 377
Cottrell equation, 304, 313
Coulomb's theorem, 10-11
cyclic voltammetry, 217, 326, 339, 375

depletion layer, 211, 214-215
differential capacity, 189
differential pulse polarography, 315, 318, 325, 329
differential pulse voltammetry, 318, 323, 331-332
diffuse layer, 197-198, 200
diffusion layer, 275-278
distribution coefficient, 58, 256, 258
distribution diagrams, 60
distribution potential, 58-60
Donnan dialysis, 262
Donnan exclusion principle, 79, 257-258
Donnan potential, 79-80

EC_{cat} reactions, 391, 395
EC_i reactions, 388
EC_r reactions, 385
electric displacement vector, 15, 154
electric susceptibility, 12, 16
electro-capillary curve, 189
electrochemical potential, 20-23, 25-27
electrode potential, 36
electrokinetic Phenomena, 221
electrokinetic potential, 223
electron hole, 209, 213
electron transfer reactions at liquid | liquid interfaces, 63, 396, 398-399
Electro-osmosis, 221-222, 228, 234-237, 244
electro-osmotic mobility, 228-231
electrophoresis, 221-222, 225, 228-231, 233-235, 243-245, 249, 256
exchange current, 272-275, 356

Fermi-Dirac distribution, 25, 208
Fermi level, 24-27, 30-31, 48-49, 209, 213-215
Fermi level in solution, 30
Fermi level of the electron in solution, 31, 49
Fick's second equation, 302

flow cell or wall jet cell, 337
formal redox potential, 42, 45, 265, 268, 270, 272, 277, 279, 308, 310, 312, 325, 331, 382
fugacity, 4, 36-37, 43-44, 73
fugacity coefficient, 4

galvanic cell, 36, 44
Gauss equation, 197
Gauss theorem, 9
gaussian distribution, 233
Gibbs adsorption equation, 180-182, 185-188, 191-194
Gouy–Chapman theory, 195, 205, 207

half-wave potential, 48, 279-280, 282, 285, 288-289, 313, 326, 380
Helmholtz capacity, 200-203
hemi-spherical diffusion, 283-284, 404
Henry's law, 5

impedance, 339-368
inner potential, 19-22, 26, 28, 35, 37, 64, 80
interfacial tension, 177, 180, 183, 188-190
inversion layer, 215
ion exchange chromatography, 256
ion exchange membranes, 78-79, 260-262
ion pair chromatography, 259
ion selective electrodes, 70-73
irreversible reaction, 267
isoelectric focusing, 234, 252-254
isoelectric point, 252-253
isotachophoresis, 234, 242

Kelvin probe, 28
kinematic viscosity, 281

Laplace transformation, 232, 302, 320, 392
Leclanché cell, 50-51
limiting anodic convoluted current, 381
limiting diffusion current, 277-280, 282, 284, 286, 288, 407
line tension, 179-180
Lippmann's equation, 189, 194
liquid | liquid micro-interfaces, 287
liquid junction potential, 65, 67
Luggin capillary, 296-298

membrane-covered electrode, 286, 360
micellar electrokinetic capillary chromatography, 233
microdiscs, 282-284
microelectrodes, 239, 266, 276, 282-283, 285, 287, 290, 298
microhemispheres, 282-283
molar absorption coefficient, 371, 408
moving boundary, 239-241

Nernst layer, 275, 277
Nikolsky Equation, 76-77
normal pulse polarography, 314-315, 329
Nyquist diagram, 342-343, 345-346, 348-350, 356-359, 363, 367

ohmic drop, 285, 292-293, 296-299, 328, 339, 359-360, 380, 384
OTTLE, 45-46
outer potential, 11-12, 19, 21, 49
overpotential, 272-274, 288, 295, 298

p|n junction, 208-213
PAGE Electrophoresis, 244
Parsons-Zobel, 202
partition coefficient, 58
permittivity, 9, 12-17, 94-96
pH electrode, 70, 73-74
pKa of an acid in the organic phase, 60
Poisson–Boltzmann equation, 195
polarisable interface, 184
polarisation vector, 12, 15, 102
polarisation window, 185, 191
polarography, 311-315
potential of zero charge, 190, 203, 205, 208
potential-modulated reflectance, 371
potential-pH diagram, 54
potentiometric titration, 67
potentiostat, 265-266
Pourbaix diagram, 54, 56

quasi-reversible reaction, 267

Randles-Ershler circuit, 358-359
Randles-Sevcik equation, 379
Raoult's Law, 4-5
real chemical potential, 23, 26, 49, 95
recessed microdiscs, 284
reciprocal Debye length, 112, 148, 150, 198, 224
redox buffer effect, 68, 70
redox Fermi level, 30
reference electrode, 35-37, 44-46
relative permittivity, 12-17, 94-96
reptation theory, 246, 249
reversible reaction, 267
rotating disc electrode, 280, 282

selectivity coefficient, 76, 78, 258
semi-integrated current, 381
silver|silver chloride electrode, 41-45
space charged region, 198
spectroscopy, 17, 45, 371
square wave voltammetry, 319, 323, 325, 329-331
staircase voltammetry, 323, 326, 328-329
standard cell potential, 41
standard distribution coefficient, 58
standard electrochemical potential, 21-22, 269
standard Gibbs transfer energy, 57
standard hydrogen electrode, 35-36
standard molality, 7
standard pressure, 3-4, 36-37, 43-44, 73
standard redox potential, 38-46
standard transfer potential, 57, 401
Stern layer, 203, 215, 224
streaming electrode, 205
streaming potential, 222, 226, 228
superposition principle, 319, 323-324, 326, 329, 340
surface concentration, 182-183
surface excess charge, 189-190, 198, 206
surface excess concentration, 182-183, 187-189, 193
surface Gibbs energy, 177-180
surface Gibbs energy density, 178-180
surface potential, 12, 16-20, 26, 28-29, 48, 92-93, 204
surface reconstruction, 219
surface relaxation effect, 219
surface tension, 94, 178, 180

Tafel plots, 274
thin layer cell, 334-335, 365, 367-368, 402-404
thin layer voltammetry, 333

valence band, 24, 209, 215
variance, 186, 188, 191-192, 233, 254
viscosity, 135-136, 172, 222-223
Volta potential difference, 11-12, 27-29, 49
voltabsorptometry, 407
Warburg impedance, 356-357, 359-360, 370
work function, 26-29, 203-204, 417

zeta potential, 223-225, 229-230, 233
zone electrophoresis, 231, 243-244